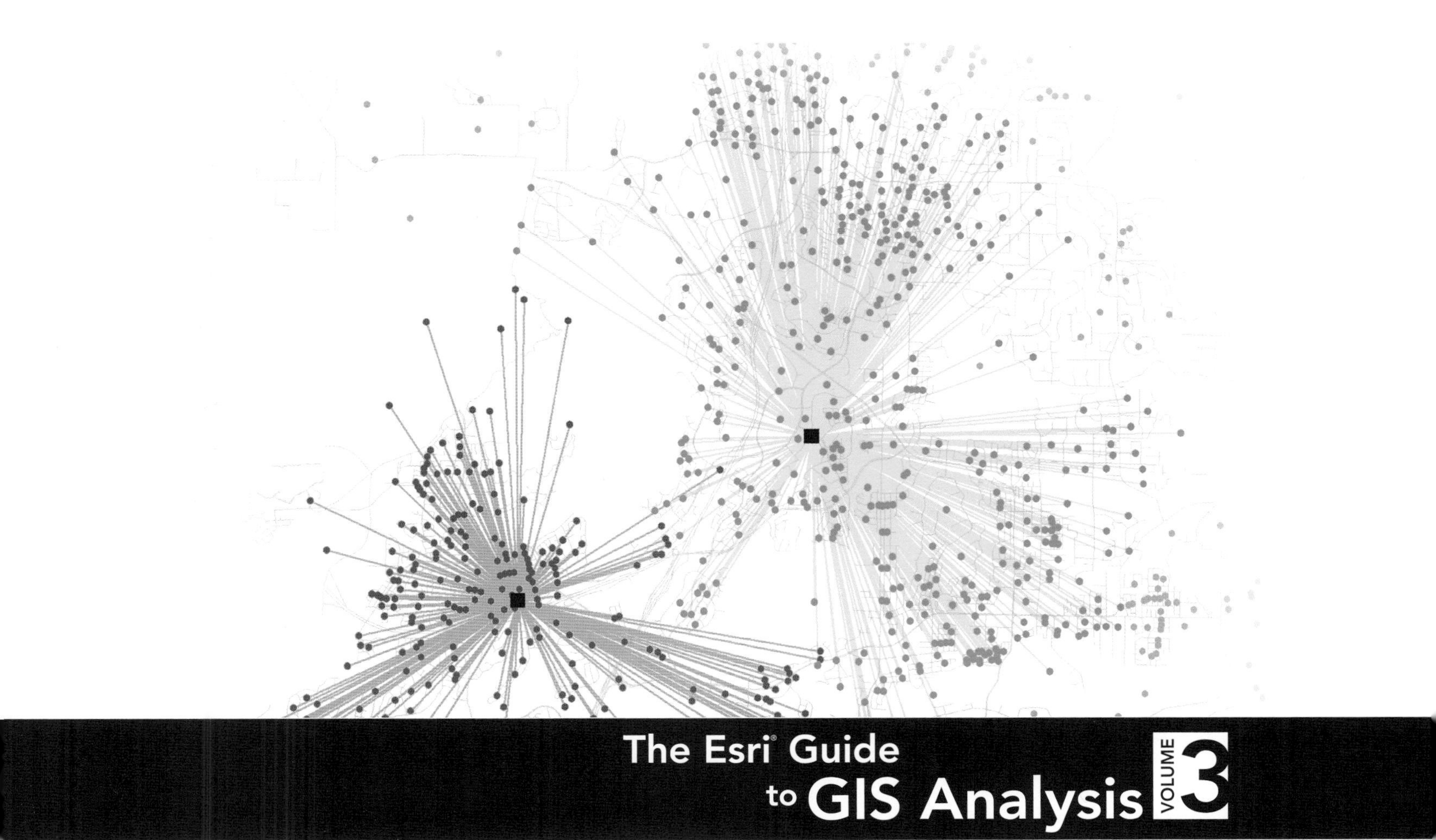

The Esri® Guide to GIS Analysis VOLUME 3

Modeling Suitability, Movement, and Interaction

Andy Mitchell

Esri Press
REDLANDS | CALIFORNIA

Esri Press, 380 New York Street, Redlands, California 92373-8100

Printed in the United States of America
18 17 16 15 14   2 3 4 5 6 7 8 9 10

Ask for Esri Press titles at your local bookstore or order by calling 800-447-9778, or shop online at esri.com/esripress. Outside the United States, contact your local Esri distributor or shop online at eurospanbookstore.com/esri.

Esri Press titles are distributed to the trade by the following:

In North America:
Ingram Publisher Services
Toll-free telephone: 800-648-3104
Toll-free fax: 800-838-1149
E-mail: customerservice@ingrampublisherservices.com

In the United Kingdom, Europe, Middle East and Africa, Asia, and Australia:
Eurospan Group
3 Henrietta Street
London WC2E 8LU
United Kingdom
Telephone: 44(0) 1767 604972
Fax: 44(0) 1767 601640
E-mail: eurospan@turpin-distribution.com

Contents

Foreword

As GIS has grown more powerful, it has also grown more confusing; a contemporary technology such as Esri's ArcGIS presents a bewildering array of functions, data models, and options. Achieving the promise of GIS by making sense of all of this challenges our abilities as users, and even more as instructors of GIS courses, especially with respect to the analysis and modeling functions.

This third book in Andy Mitchell's series follows the pattern of its siblings in presenting a very complex topic in a very simple and intuitive way. Unlike approaches that focus on navigating the user interface or on detailing functionality, the series begins with the basic questions GIS users need to answer and on the concepts that underlie those questions. I find this approach very attractive, as it strikes to the heart of what GIS is all about: using a powerful technology to address inherently simple questions about the geographic world. The user is encouraged to think first about the problem to be addressed, to identify the relevant issues, and to lay out an approach, all before any effort is made to interact with the technology. In this respect the series lies close to what I would call critical spatial thinking, the mindset of an informed and enquiring user. It emphasizes the essential collaboration that must exist between user and technology, since GIS cannot of itself solve problems; on the other hand, it can provide extremely valuable and powerful assistance to a user who has a well-defined problem to solve.

This third volume covers problems of spatial interaction, site selection, routing, and scheduling. Many of these problems involve informed decisionmaking, and fall under the heading of "geodesign," the use of geographic technologies to find the best solutions to spatial problems, or what Carl Steinitz of Harvard University has termed "geography by design." This is a more complex domain than the previous two books, but one with great practical relevance. As usual, the author approaches the topic with clarity, using abundant illustrations, and maintaining a very high quality of presentation. The book is a worthy addition to the series. In the preface I wrote to the second book, I said that it "should be required reading for everyone who ventures into the world of spatial analysis with GIS." Now there's more required reading, but as with the previous books, the illustrations and text make the experience unusually pleasurable.

Michael F. Goodchild
Center for Spatial Studies and Department of Geography
University of California, Santa Barbara

Acknowledgments

Many people contributed their knowledge to this book. The GIS professionals who have presented maps and papers at the annual Esri User Conference over the years provided the real-world context for many of the concepts and methods presented in the book. Others, mainly from academia, contributed through their publications, which are listed at the end of each chapter.

Dr. Michael Goodchild of the University of California at Santa Barbara (who also wrote the foreword), Dr. Thomas Balstrøm of the University of Aalborg at Copenhagen, Dr. Gary Raines of the U.S. Geological Survey (retired), Doug Walker of Placeways, and Dr. David Maidment of the University of Texas at Austin reviewed all or portions of the manuscript and provided valuable comments.

Steve Kopp, of Esri, was the primary technical advisor for the book. A number of other people at Esri offered their expertise, both through their review of portions of the manuscript and in informal discussions. These include Linda Beale, Clint Brown, Garry Burgess, Ryan DeBruyn, Dean Djokic, Drew Flater, Witold Fraczek, Charlie Frye, Steve Grise, Alan Hatakeyama, Kevin Johnston, Deelesh Mandloi, Bill Miller, Scott Murray, Nawajish Noman, Lauren Rosenshein, and Jay Sandhu.

A number of organizations provided the data used to create the map examples in the book. They are listed in the "Data credits" section. Several people at Esri, including Canserina Kurnia, Earl Sarow, and Lori Armstrong, provided help in obtaining datasets.

Tim Ormsby edited the book, and Joyce Frye was the copyeditor. Catherine Ortiz was the project manager. Thanks to everyone who helped at Esri Press.

Once again, thanks to Jack Dangermond and Clint Brown, who recognized the value of publishing a guide to GIS analysis, and provided the support for writing it. And, finally, special thanks to Dr. Dana Tomlin and Dr. Carl Steinitz for their teaching and guidance.

Introducing GIS modeling 1

The first book in *The Esri Guide to GIS Analysis* series showed you how to get answers to straightforward spatial queries—mapping the most and least, finding what's nearby, and so on. The second book showed you how to use statistics to find patterns and relationships in your geographic data, beyond simply gleaning them visually from a map. Together these volumes cover a broad swath of commonly used GIS analysis methods.

Beyond spatial query and pattern analysis, you can use GIS to answer questions that are more subjective and dependent on criteria and factors that you define:

- Where is the best location?
- Which is the most efficient path?
- Where will the flow go?
- Which place will people travel to?

In these analyses, you define what constitutes "best" or "least cost," and you specify the factors that influence flow or the way in which a facility can best serve the surrounding population. These analyses are used to support decisions: Where should the new housing development be built? Which is the best path for the pipeline? Which areas are at risk from storm runoff? Where should we build the new branch library? While the GIS can't make the decision for you, based on the criteria and parameters you specify, it can provide valuable insight. And by changing the criteria and parameters, you can explore alternate scenarios and the impacts of different plans before you implement them. Most of these analyses predate GIS (or even the use of computers), but GIS has made the process faster and more flexible, and made it easier to visualize and comprehend the results.

Historically, the different types of analysis have been the purview of different disciplines—finding the best location has traditionally been done by land planners; finding the least-cost path has commonly been used in the world of shipping and logistics; modeling flow has been done by hydrologists as well as by public works departments (for water and sewer networks); and modeling where people will travel (and shop) has been done in the retail and commercial sector.

The goal of this book is to gather the methods used to perform these analyses in one place and present them to a wider audience of GIS users. If you've used GIS primarily for mapping or simple spatial queries and pattern analysis, you'll find many new methods to support your decisions. If you're familiar with one or a few of the methods included here, you'll find best practices as well as other methods, or even other types of analysis, you may not have previously considered for your GIS applications.

The analyses in the book broadly fall into two categories—evaluating locations and analyzing movement.

Evaluating locations involves looking at various characteristics of a set of locations to determine how suitable they are for a particular use. The characteristics are represented by layers of information. These include characteristics of the locations themselves (such as the slope steepness or vegetation), as well as what's surrounding or nearby (such as the distance of the locations from highways or streams). Evaluating locations is used for finding the best site to build something or the best use for each location in a study area. (These kinds of uses are collectively called "suitability analysis.") It's also used to predict areas that might be susceptible to risk, such as from landslides, fires, or insect infestations, based on the conditions at a site. The same type of analysis is used to identify where something (such as a particular bird species) will likely be found, by combining the preferred conditions of the species (certain type of vegetation, distance to water, presence of food sources, and so on).

Analyzing movement covers a broad range of methods. What they all have in common is that the analysis identifies the least travel cost between two or more locations. Cost is defined in terms of money, time, or distance. The movement can be by people or animals, in which case some decision process is involved, or by material that simply follows the path of least resistance, such as rainfall runoff after a storm or motor oil dumped into a drain through a stormwater system.

The book explores methods to do the following:

- Identify existing locations—such as lots or a section of roadway—that are suitable for an intended use. *"Show me all the lots that are suitable for a shipping distribution center."*
- Find locations anywhere in your study area that are suitable for a particular use. *"Show me all the areas that are suitable mountain lion habitat."*
- Rate locations as more or less suitable. *"Show me the most and least suitable locations for a new housing development."*
- Find the best path or corridor between two locations traveling overland. *"Show me the path for a new transmission line that costs the least to construct while minimizing environmental impacts."*

- Find the shortest, quickest, or least expensive path between a series of stops, optionally picking up and dropping off people or goods. *"Show me the driving route that requires the least travel time while allowing me to visit all my customers within their available time windows."*
- Find out where water will flow and accumulate over the land surface. *"Show me the total area upstream from the stream gauge."*
- Trace the flow of water or other material through pipes from an origin location. *"Show me which manholes are in the path of motor oil dumped into a storm drain."*
- Find the optimal location for a facility that serves the surrounding population such that overall travel time is minimized, the most people are served, or everyone is within a specified distance of a facility. *"Show me the location for the new library that ensures no one in the community has to travel more than ten minutes to reach a library."*
- Predict which facility people are likely to use, given the attractiveness of the facility, how far away it is, and the locations of competitors. *"Show me the likelihood that people will visit a regional park, given the size of the park and its amenities."*

ANALYSIS	METHODS	CHAPTER
Evaluate suitable existing locations	*Logical & spatial selection*	*Chapter 2, "Finding suitable locations"*
Identify suitable locations anywhere in the study area	*Vector overlay* *Raster overlay*	*Chapter 2, "Finding suitable locations"*
Rate suitable locations	*Weighted overlay* *Fuzzy overlay*	*Chapter 3, "Rating suitable locations"*
Find the best overland path	*Path-distance analysis*	*Chapter 4, "Modeling paths"*
Find the most efficient route over fixed infrastructure	*Least-cost path* *Vehicle Routing Problem*	*Chapter 4, "Modeling paths"*
Find out where water will flow over the land surface	*Flow direction analysis*	*Chapter 5, "Modeling flow"*
Trace where fluid will flow through pipes	*Geometric network trace*	*Chapter 5, "Modeling flow"*
Find the optimum location for a facility that serves the surrounding population	*Location-Allocation*	*Chapter 6, "Modeling interaction"*
Predict visitation to a facility	*Spatial interaction model*	*Chapter 6, "Modeling interaction"*

Most of the methods for evaluating locations and analyzing movement involve a process in which you run a series of functions, or tools, in sequence to get the end result. For each tool, you specify parameters and the input datasets. (In many cases, the output dataset from one step becomes the input dataset for the next step.) Together, the data, tools, and parameters—along with the sequence of steps—comprise a GIS model.

The book compares methods and explains the theory behind them so you can better interpret the results of your analysis. It also describes how a particular method is implemented in the GIS. Understanding the implementation will help you decide which method is best suited for the task at hand. In this book the implementation described is the one used by Esri's ArcGIS software. Comparable tools can be found in other GIS software. The book assumes familiarity with basic GIS concepts, as well as the GIS analysis concepts presented in the earlier volumes.

The methods presented here are some of the most commonly used and can be implemented using tools contained within typical GIS software.

Of course, additional methods exist for evaluating locations and analyzing movement. Typical of such methods are dispersion models that predict the spread of a phenomenon, such as a wildfire over the land, a pollutant through groundwater, or a toxic gas through the air. Other methods, such as simulation models, predict behavior. Examples include predicting the home range of mountain lions given the behavior of the lions, the presence of other mountain lions, prey, and preferred habitat; or models that predict urban growth over several decades given various economic scenarios, population growth, and people's preferences for places to live, work, and shop.

In general, these types of models are implemented outside a GIS, often using mathematical or statistical software. In some cases, GIS is used to process data for input to a model, the model is run outside the GIS, and the results of the model are imported back to the GIS for display as maps and charts. In other cases, the linkage with GIS is more integrated: the model sends data to the GIS for spatial processing using a particular tool or function—or perhaps even a complete model built inside the GIS—and the results are returned to the external model for the next step. Since these methods are primarily implemented outside of a GIS, they are beyond the scope of this book.

The process for creating a GIS model is much like that for other GIS analyses, with a few additional considerations. Each method presented in the book essentially follows the process described below.

DEFINE THE GOAL OF THE ANALYSIS

The goal of the analysis can be stated as a problem to be solved or as a question to be answered. The more specific the question, the more likely it is you will get meaningful results.

- What is the best location for my company's new shipping distribution center?
- I need to find the areas that are good mule deer summer habitat.
- I need to find the best path for a new highway connecting two towns.
- Which of the three candidate locations for the new library will best serve the city's residents?
- Given a choice of several regional parks, which ones are hikers most likely to use?

DEFINE THE CRITERIA

When you evaluate locations or analyze movement, you first define the criteria for what makes a place suitable or what constitutes the least cost. You also decide which criteria are more important and which less, and weight the criteria accordingly.

What makes for a good location to build something? Usually, it is a combination of factors—for a shipping distribution center it might be a site that is currently vacant, at least five acres in size, within a mile of a highway, and not in a flood zone.

What makes a location suitable habitat? Again, it is often a combination of factors. For summer mule deer habitat, the factors might include elevation, vegetation type, and distance from water.

What defines the best path—is it the shortest one, the quickest one, the one that costs the least money to build, or the one that has the least impact on the environment?

If you're meeting demand for services, as with a library, do you want to serve the most people, minimize the total travel time to the facility from all locations, or ensure all the residents of the community are equally close to a facility?

If you're predicting travel to facilities such as regional parks, you define the factors that make one facility more attractive than another (and hence more likely to draw people from farther away). Hikers might be attracted to parks with the greatest total length of trails, those having a lake or stream, and those having a wide variety of habitats.

Defining the criteria and their weights is the core of creating your model. There are many sources of information for determining criteria and weights.

You might have firsthand knowledge or experience of the phenomenon you're modeling, based on your research, observation, or the expertise you've developed over years of working in your particular field. Even then, you'll likely want to get some verification from others knowledgeable in the field.

Published research, industry standards, or regulations may also guide you in defining criteria and weights. For example, research on mountain lions may indicate that the presence of vegetation cover, steep slopes, and proximity to water are more important in identifying suitable habitat than other factors, such as distance from roads or urban areas. Similarly, a city's building regulations might stipulate the permissible steepness of slopes for different types of development or how close construction can be to a stream or wetland.

If there is no published literature to help you define your criteria, you can attempt to develop a consensus among a group of experts on the issue. One such technique is called the Delphi process, in which experts independently submit their best knowledge of the phenomenon you're modeling and the associated criteria. The results are compiled and redistributed for the experts to assess and reevaluate. Through several such iterations (hopefully not too many), a consensus may emerge.

Another approach is to convene a workshop in which experts discuss the criteria and try to reach a consensus on the spot. One technique for use in a workshop (or even by an individual attempting to define criteria) is pairwise comparison. In this method, all potential criteria are identified, and each criterion is compared to each of the other criteria in turn. The first criterion in the pair is compared to the second on a scale ranging from, for example, "much more important" to "equally important." Eventually, the most important criteria rise to the top. (If one particular criterion is listed as "much more important" when compared to each of the other criteria, it is clearly *the* most important.) See "References and further reading" at the end of this chapter for sources that discuss these and other methods for defining criteria.

Some criteria are difficult to quantify. This is especially true of social values, for example when trying to determine the value to a community in protecting a scenic view (although you can undertake a survey to try to determine this). To the extent you estimate the values associated with such criteria, the results of your model will also be an estimate.

When assigning criteria and weights, subjectivity and value judgments are often involved. For example, in selecting a site for a new housing subdivision, there are economic and social criteria to consider: protection of wildlife habitats, increased traffic, higher demand on schools and other public services. By emphasizing certain criteria over others, you can create alternative scenarios—one that takes greater account of economic factors, for example, and another that takes greater account of environmental factors.

COLLECT THE DATA

The criteria you define will determine the data you need for your model. This being a GIS model, the data will consist of map layers.

CRITERIA	MAP LAYER
Currently vacant	*Parcels*
At least five acres in size	*Parcels*
Within one mile of a highway	*Highways*
Outside a flood zone	*Flood zones*

Criteria for finding the best location for a shipping distribution center, along with the associated data layer for each criterion.

Data is available from a variety of sources—within your organization, from other organizations in your region, from state and national agencies, or from commercial suppliers. Much of this data is available via the Internet.

In some cases, a map layer can be used as is in the model. In other cases you'll need to modify or enhance the layer before it can be used. You may need to add or calculate additional attributes. You may need to edit or process the data to make it usable for the model. You may also need to derive new map layers. For example, if your model calls for a layer of slope steepness (and one doesn't exist), you'd create one from an elevation layer.

GIS data takes the form of either raster or vector data. Generally, the methods for evaluating locations can use either type, although, depending on the specific method, one or the other might be more appropriate. For analyzing movement, the methods are geared specifically to either raster or vector data.

Geographic data has unique characteristics which can impact the results of your model, and hence your decisions. These include spatial bias in the data, issues associated with geographic scale, and locational accuracy (whether things in the data are where they actually are on the ground).

Using GIS data in the modeling process is discussed later in this chapter.

RUN THE MODEL

The goal of your analysis, and the type of data you're using, will determine which analysis method you'll use. In some cases more than one method may be appropriate—the choice depends on personal preference or the type of output you want to obtain. You can also use alternate methods to verify the results of your analysis.

In most cases, each step in the model has a corresponding tool or function—you specify the parameters for each tool and then run the tools in sequence. Depending on the method and the GIS software you're using, you might run the tools interactively via the software's interface. Often, you can automate this process by specifying all the data inputs, tools, and parameters, and the order in which to run the tools, and then run them all at one time.

In ArcGIS, automating the process is usually done by creating a model document using ModelBuilder. ModelBuilder allows you to create a flow diagram with all the model components, and then run the model. Building a model document has a number of advantages—you can see a visual representation of the model that makes the steps easy to follow; it is relatively easy to modify the model, if necessary, and rerun it; in the process of building the model, you are documenting the analysis; and a model document is easy to share with others.

In this book, we show simplified versions of ModelBuilder flow diagrams, with map graphics of the actual input and output data layers, to illustrate the concepts behind the methods. (The diagrams may not include all the required data processing steps and corresponding tools.) The tool names used are, in most cases, those used in ArcGIS.

The application examples themselves are also simplified versions of real-world models. Your models may include more criteria and input layers. However, the process will be the same.

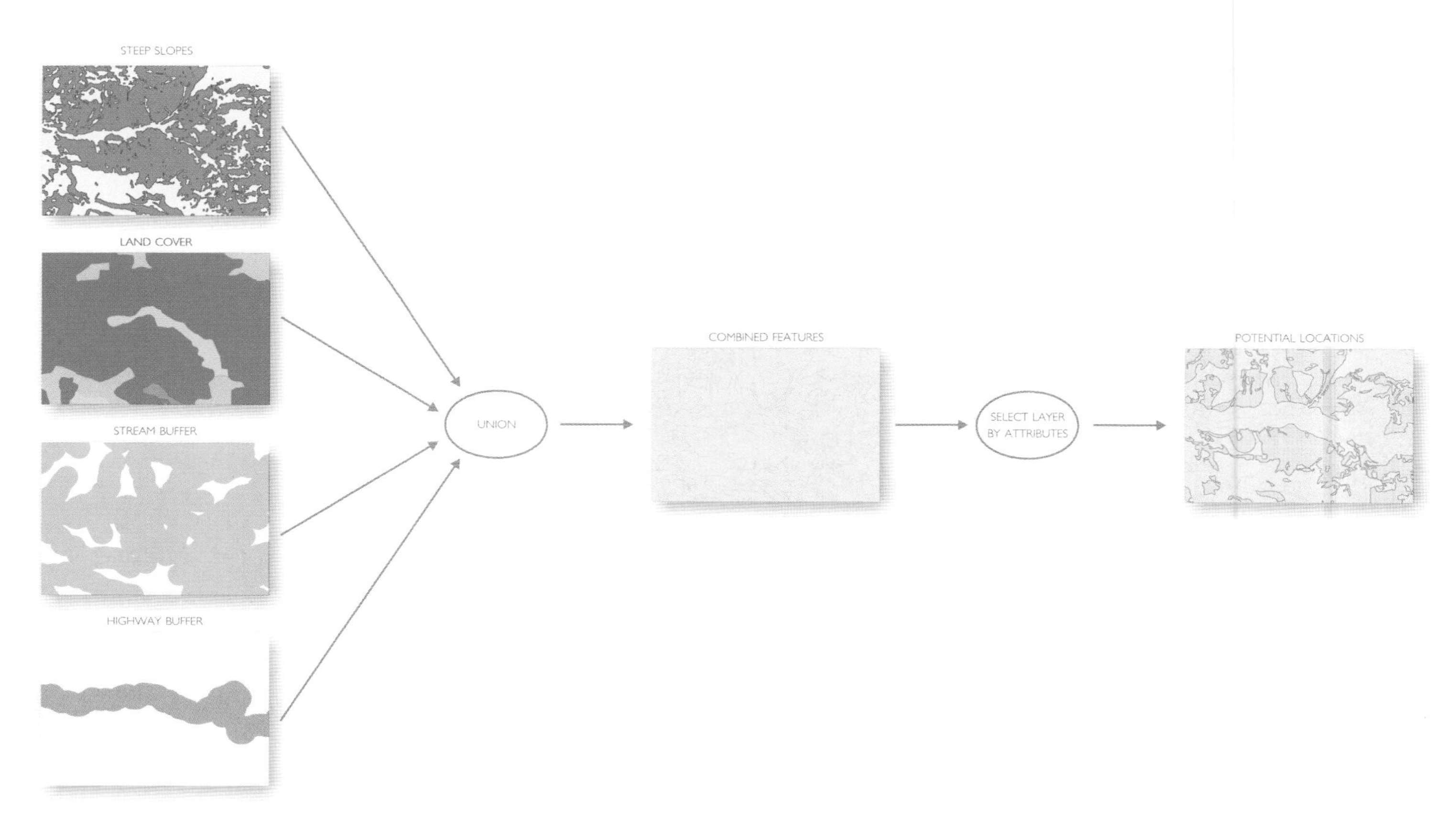

Simplified flow diagram for a typical model (in this case, for identifying likely mountain lion habitat). The input layers, analysis tools, and output layers are shown.

VERIFY THE RESULTS

Once you've obtained the results from the model, usually in the form of a map layer and associated tables, you need to determine if the results are valid. There are several ways to do this. One is to see if the results make sense given your knowledge of the study area and the problem you're solving. You'll also want to share the results with colleagues or experts in the field who were not involved in developing the model, to see if they think the results make sense.

Another way to verify the results is to check the results against your original criteria. You can do this by displaying the results layer on top of the layers representing your criteria (or by comparing them side-by-side). The results should reflect the criteria and weights you specified in the model.

You can also modify the model criteria slightly and run the model again (see below) to see how much the results vary. If the results vary little, this will give you confidence that the results are stable, and you'll have more confidence in any decisions based on the results. If the results vary depending on the criteria you specify, it means the model is more sensitive to some criteria than others. You'll want to look at these sensitive criteria closely and make sure you've weighted them correctly in the model.

Of course, the most effective way to verify the results of your model (if you have the time and resources) is to go into the field and see if they reflect reality. If you're modeling mountain lion habitat, you'd go to areas that the model indicates are suitable habitat and look for evidence of mountain lions. You'd also go to areas that the model indicates are not suitable habitat and look for evidence of lions. If you find evidence of lions in unsuitable habitat (or no evidence of lions in suitable habitat), you'll want to do additional research to find out which criteria are missing from your model.

MODIFY AND RERUN THE MODEL

If your analysis of the results shows errors in the model or if you find that some of your assumptions about the criteria were incorrect, you can modify the model inputs and parameters and rerun the model to produce a new result. If you've created a model document, this is fairly straightforward, as all the inputs and parameters are saved—you merely need to edit the document, make the necessary changes, and run the model again.

The goal of your analysis may require alternative scenarios. For example, if the goal is to find the optimal location for elementary schools in a new planned community, you can first run the model assuming you will build three schools, and then rerun the model to find out how much farther students will have to travel if you only build two new schools.

Alternative scenarios can also be based on different social and economic values. For example, when finding the best location for the new planned community, you might favor minimizing construction costs in one scenario and maximizing the amount of greenspace in another. By creating alternatives, you can see where the results of the scenarios are the same, so discussions and compromises can focus on the areas where the results differ.

DOCUMENT THE ANALYSIS

An important step in the process is to document the analysis and the model. You may need to revisit your work in the future, and good documentation will allow you to recall what you previously did. You may also want to share the model with others inside or outside your organization who will need clear documentation so they can replicate your analysis. Perhaps most importantly, peers, decision makers, and citizens will want to see and understand how the results were obtained.

A ModelBuilder document is a useful way to store and present your analysis process. Including graphics and labels in the diagram will help others visualize the criteria that went into the model as well as the inputs and outputs at each step of the process.

DISPLAY AND APPLY THE RESULTS

The results from the analysis can feed directly into a decision process via maps, tables, and charts. The results can also be used in further analyses or to provide additional understanding of the problem.

Similarly, the results of a model can be used as input to another model. For example, you might use a suitability model to find potential locations for a new school, and then use the resulting locations as input to a location-allocation model to find the school that will serve the most students.

Bear in mind that the result of your model is just another tool to help with decisions or with understanding the phenomenon you're modeling—it's not the definitive answer. There may be other political or economic factors involved in the decision process. New data or future research may change the criteria or weights for your model, making your results obsolete. But the results you obtain can help inform decisions in the present and advance understanding of the problem at hand.

The nature of GIS data has implications for evaluating locations and analyzing movement. These include the type of spatial and attribute data required by a particular method, whether there is spatial bias in the data, issues of geographic scale, and issues of spatial and attribute data quality. By understanding the data you're working with, you'll be able to avoid pitfalls in developing your analysis and interpreting the results.

SPATIAL DATA TYPES

Most of the analysis methods described in this book require a particular type of spatial data. The type of data you have available at least partially determines the method you use. (Conversely, if there's a particular method you want to use, you'll need to obtain the appropriate data.)

Vector data

Vector data uses coordinate pairs to define both the location and shape of geographic features. Point features are defined by a single pair. Such pairs also define the vertices of a line feature or the boundary that defines an enclosed area (a polygon).

Vector data is often used to represent features that are discrete in space—individual buildings, streets, or parks.

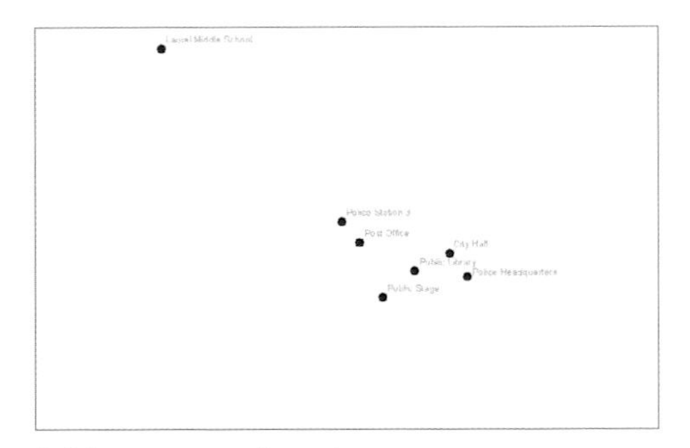

Buildings represented as points.

Streets represented as lines.

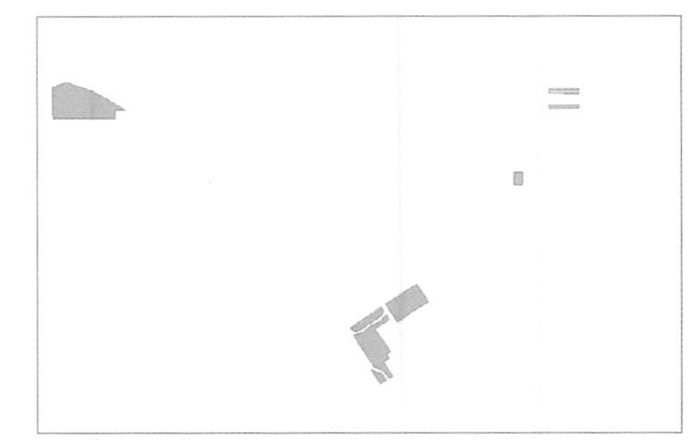

Parks represented as polygons.

It can also be used to represent spatially continuous phenomena such as elevation or land cover.

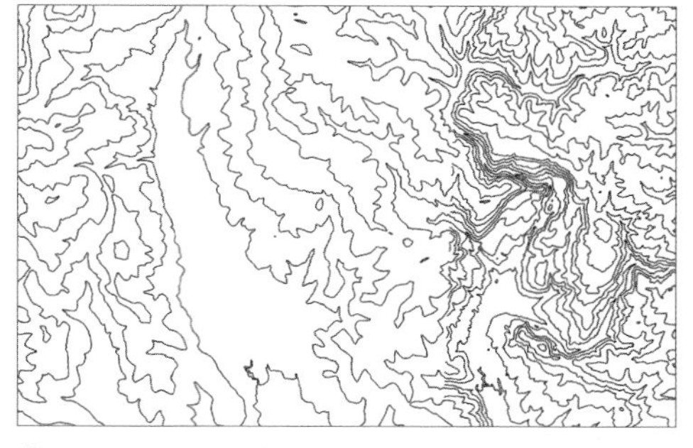

Elevation represented as contour lines.

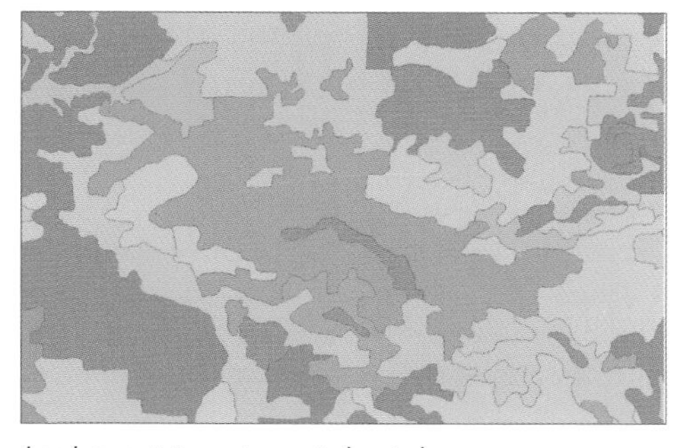

Land-cover types represented as polygons.

Features of the same type are stored as a layer. Each layer has an associated table that stores the attributes for the features.

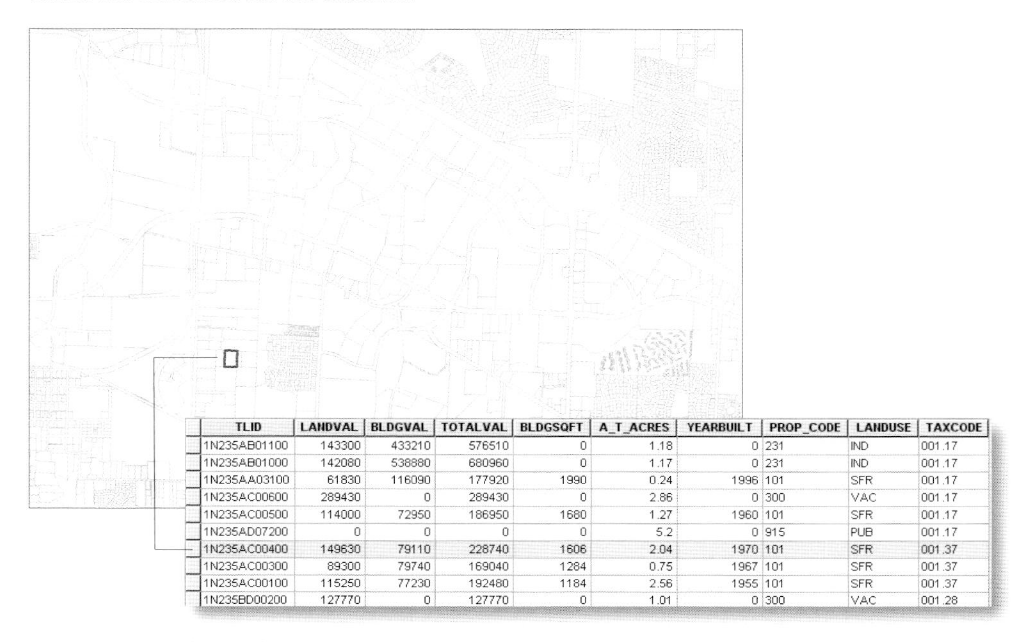

TLID	LANDVAL	BLDGVAL	TOTALVAL	BLDGSQFT	A_T_ACRES	YEARBUILT	PROP_CODE	LANDUSE	TAXCODE
1N235AB01100	143300	433210	576510	0	1.18	0	231	IND	001.17
1N235AB01000	142080	538880	680960	0	1.17	0	231	IND	001.17
1N235AA03100	61830	116090	177920	1990	0.24	1996	101	SFR	001.17
1N235AC00600	289430	0	289430	0	2.86	0	300	VAC	001.17
1N235AC00500	114000	72950	186950	1680	1.27	1960	101	SFR	001.17
1N235AD07200	0	0	0	0	5.2	0	915	PUB	001.17
1N235AC00400	149630	79110	228740	1606	2.04	1970	101	SFR	001.37
1N235AC00300	89300	79740	169040	1284	0.75	1967	101	SFR	001.37
1N235AC00100	115250	77230	192480	1184	2.56	1955	101	SFR	001.37
1N235BD00200	127770	0	127770	0	1.01	0	300	VAC	001.28

A layer of land parcels (polygons) with its attribute table. The attributes are displayed when you identify a feature. They can also be used to symbolize features, create reports, and do analysis.

Vector data can be used for some types of analysis that involve finding suitable locations (chapter 2, "Finding suitable locations").

Features represented as vector data are often used when displaying the results of your analysis, to add context to maps—such as with highways, rivers, county boundaries, and so on.

Networks

Networks are a specific type of vector data. They are comprised of a set of connected features in a system—usually one that some material or object flows through or over. These include utility networks—such as a system of water pipes and valves—and transportation networks—such as a system of streets, turns, and stops.

A network is primarily composed of lines, known as edges, and the junctions, or nodes, where edges connect. The GIS knows which edges connect at which junctions. That allows you to model the movement through or over the network. You create a network from existing features using tools in the GIS.

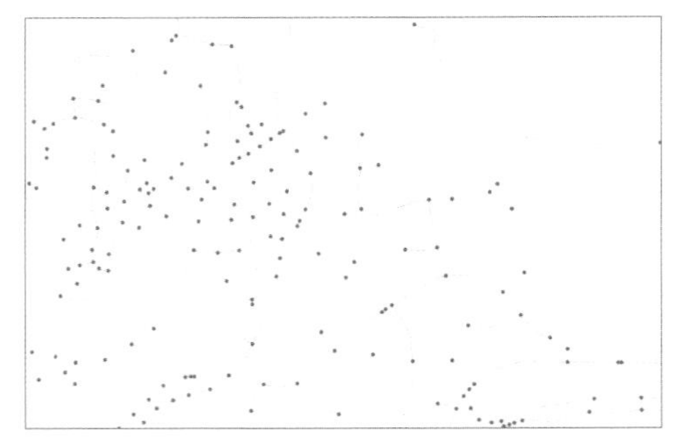

A street network showing edges (gray lines) and junctions (gray dots).

The edges and junctions in the network have associated attributes that can affect the movement through the network, such as the diameter of stormwater pipes or the time required to travel each street segment.

Transportation networks are used to model routes for visiting various locations or making deliveries (chapter 4, "Modeling paths") and for locating facilities (chapter 6, "Modeling interaction"). Utility networks are used to trace upstream or downstream flow (chapter 5, "Modeling flow").

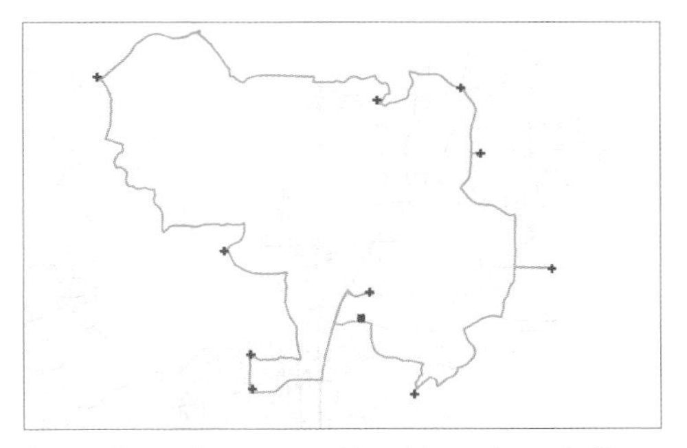

A route for a salesperson making visits to doctors' offices (blue crosses), starting from the sales office (blue square).

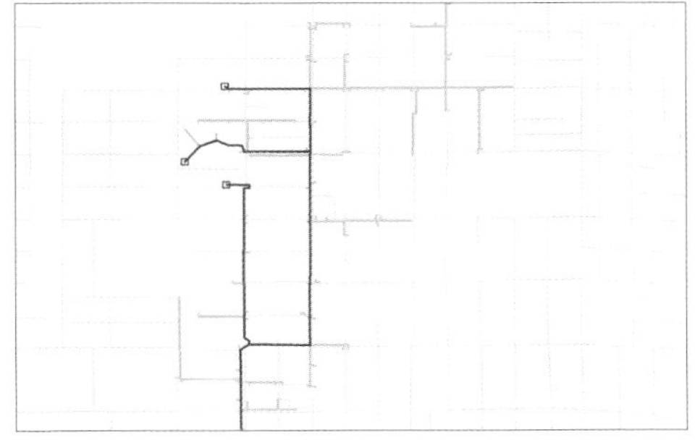

A downstream trace of motor oil through stormwater pipes (blue lines). The oil was dumped into three inlets (blue boxes); the red lines show the pipes the oil will flow through. Street centerlines are also shown in gray.

Raster data

Raster layers represent geographic space as a grid of cells, with each cell being assigned a value representing the attribute for that particular location. Each layer essentially represents a single characteristic.

Raster data is often used to represent geographic phenomena that are spatially continuous, such as elevation or soil moisture—at any location within your study area the elevation above sea level or the amount of moisture in the soil can be measured. In these two examples, the values are also continuous—for example, the elevation values for a particular study area may range from 230 meters to 600 meters. However, raster data is also used to represent categorical data that is spatially continuous, such as land-cover categories or soil types.

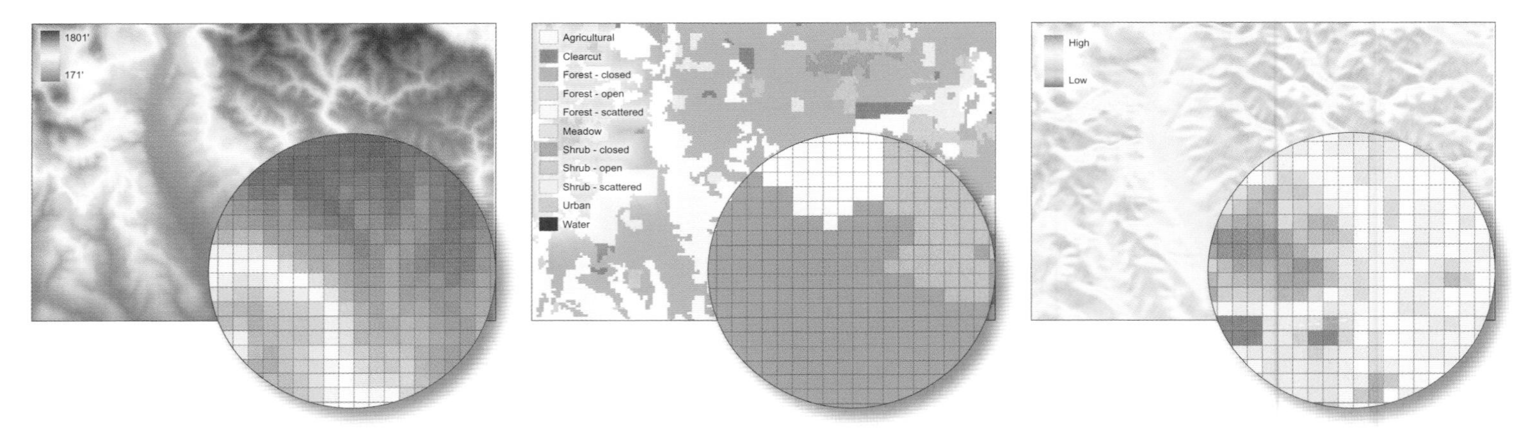

Each raster layer represents one characteristic—elevation, land cover, or solar radiation, in this example. The land surface is divided into a grid of cells, as shown by the close-ups, with each cell assigned an attribute value for that particular characteristic. All three of these layers represent spatially continuous phenomena.

Discrete features—such as rivers or protected areas—can also be represented as raster data, in which case a cell either contains a feature, or does not. Discrete features are typically represented with vector data, but raster data may be used if the analysis method requires it.

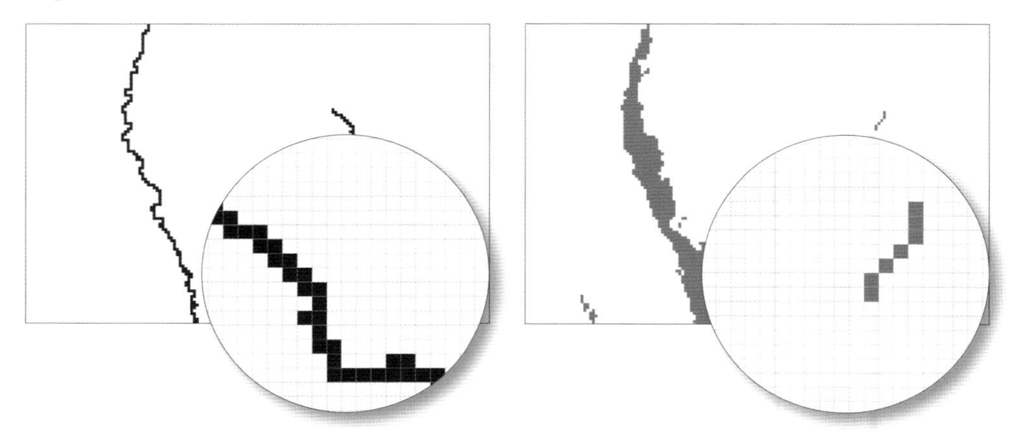

Discrete features such as rivers (left) and protected areas (right) can be represented as raster layers.

Rasters are often used for finding the best location for something (chapter 2, "Finding suitable locations" and chapter 3, "Rating suitable locations") as well as for creating overland paths and corridors (chapter 4, "Modeling paths"). Rasters are also used in hydrologic modeling to identify stream channels and drainage basin boundaries (chapter 5, "Modeling flow").

ATTRIBUTE DATA

Attribute data—the descriptive information about geographic features—is used to specify and filter criteria. (For vector data, attributes are stored in tables associated with a layer; for raster data, they are the values associated with the cells in the raster.) Attribute values are nominal, ordinal, interval, or ratio. The latter three types are measured on scales of high to low (or more to less). Some analysis methods require you to use a particular scale.

Nominal

Nominal values describe features by name or type and are often used for categorical attributes. Parcels can be displayed or selected by land-use code, for example.

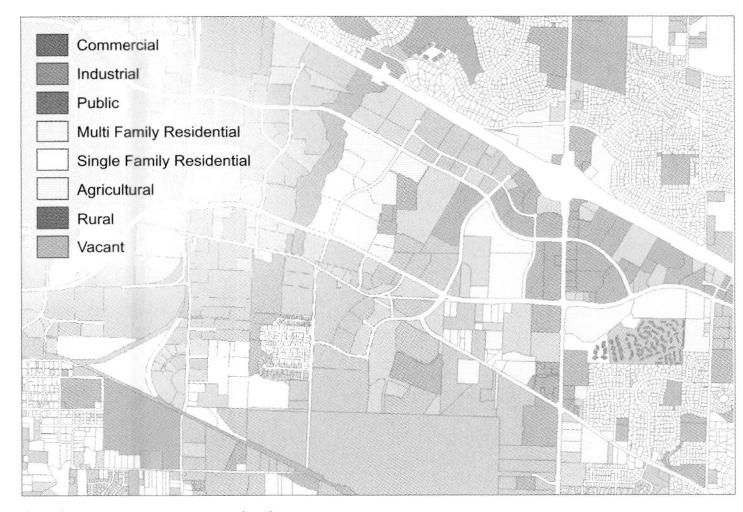

Land-use types are nominal values.

Ordinal

An ordinal scale describes attribute values that are ordered from high to low. You know where a value falls in the order, but you don't know how much higher or lower it is than another value. The difference between a value of 2 and a value of 3 may not be the same as the difference between 3 and 4. For example, soil types with a value of 3 for suitability for crops (on a scale of 1 for best and 8 for worst) are better than soils ranked 4 and worse than soils ranked 2, but you don't know how much better or worse.

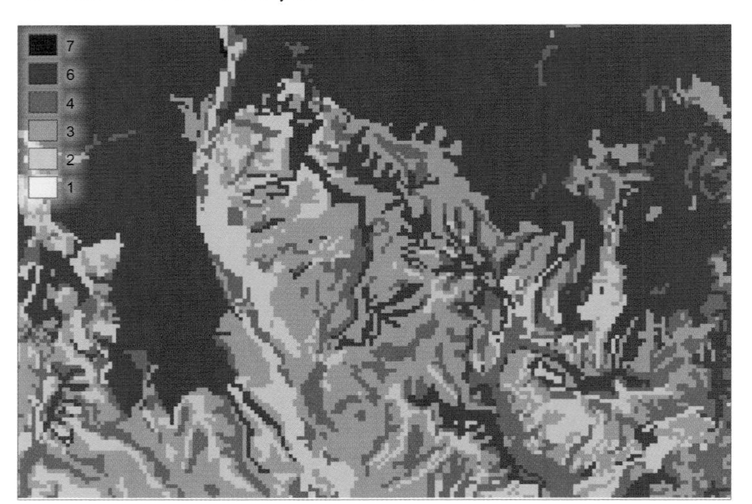

Layer of soil class, indicating suitability for development (with 7 being least suitable). Soil classes are ordinal data.

Interval

With an interval scale, the difference between each value on the scale is the same. So the measured difference between a value of 2 and a value of 3 *is* equal to the difference between 3 and 4. A typical interval scale is temperature in degrees Fahrenheit or Celsius (the difference in temperature between 19° and 20° is equal to the difference between 20° and 21°—one degree).

In an interval scale, the value of 0 is merely an arbitrary starting point—it does not represent the absence of the thing being measured (so 0° C is not the absence of temperature, merely a value to start counting from, both higher and lower). Because of this, you cannot determine how much more or less one value is than another, percentage-wise—you cannot state, for example, that a temperature of 40° is 25 percent warmer than a temperature of 30° (you only know that it is 10° warmer).

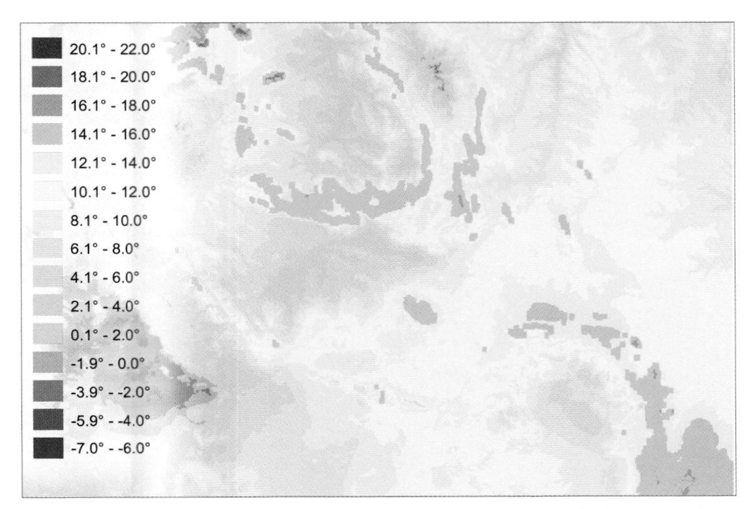

Layer of minimum annual temperature in degrees Celsius—an example of an interval scale.

Ratio

An interval scale that has an absolute and meaningful zero is known as a ratio scale. On this scale, zero means the absence of the thing being measured—a census block group with a population of zero has no people living in it, for example. Because of this, you can determine relative quantities—a block group having a population of 1,500 has twice as many people in it as a block group having a population of 750. Many measurements use a ratio scale (distance from streams or the monetary cost of clearing land). Using a ratio scale is important when measuring costs, such as when modeling a least-cost path (as you'll see in chapter 4, "Modeling paths").

Layer of census block groups color-coded by total population. Population is measured on a ratio scale.

ACCOUNTING FOR SPATIAL BIAS

Since geographic data represents real-world phenomena and processes, certain characteristics of geography can lead to spatial bias in the data. You'll need to be aware of any bias before conducting your analysis and take steps to counter it if the bias will affect the analysis results. Two common forms of bias are spatial autocorrelation and correlated data layers.

Spatial autocorrelation

With geographic data, it's often the case that things near each other are more alike than things far apart (think of house values in a city, crime rates in a county, vegetation in a national park, or temperature across a country). This phenomenon is known as spatial autocorrelation. Many analyses rely on sampled data. If some samples are close together and others spread out, and your data is spatially autocorrelated, the samples that are close together could overemphasize the values found in those areas, leading to incorrect results.

Spatial autocorrelation can be measured using a spatial statistics tool such as Moran's *I*. If spatial autocorrelation exists in a particular data layer, you should at least be aware that autocorrelated values in that layer may be overemphasized in your analysis.

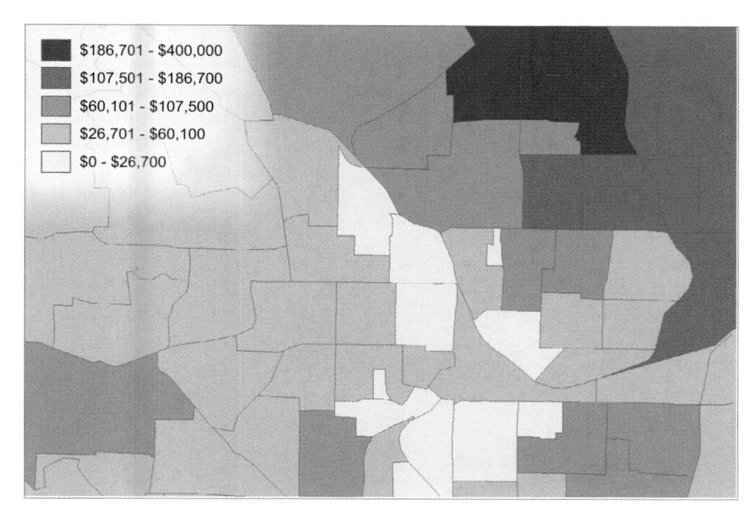

Map showing median property value in each census tract. Tracts having a high value occur near each other (upper right) as do tracts having a low value (center and bottom).

Correlated layers

Another form of spatial bias involves two or more data layers that may be correlated. If you include in your analysis several layers related to the same phenomenon, that phenomenon will be overrepresented. For example, if you're modeling the distribution of a plant species, including layers for aspect and solar radiation would overweight the influence of sun exposure, since these two layers are correlated. (In the northern hemisphere, south facing slopes receive a lot of sun). As a result, the influence of other layers in the model (such as soil type and rainfall) would be underweighted.

You can test to see if layers are correlated using a statistics tool such as Pearson's correlation coefficient or Spearman's rank correlation coefficient or by using a regression analysis such as Ordinary Least Squares. If two or more of your layers are strongly correlated, consider using only one of the layers in your analysis.

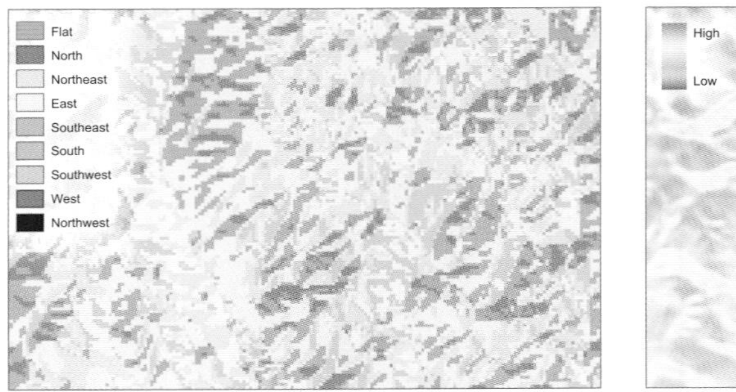

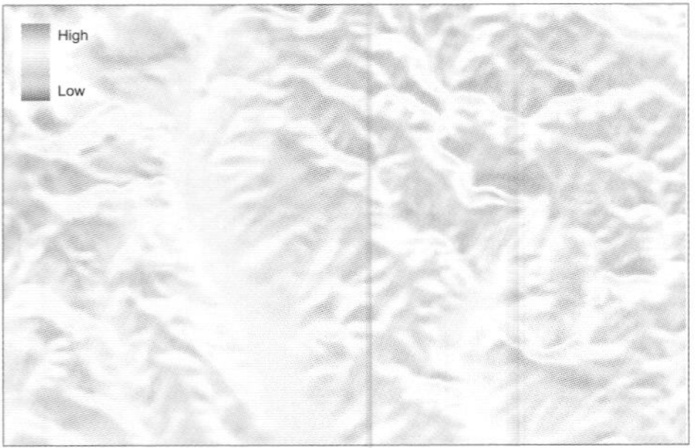

Aspect layer (left) and solar radiation layer (right). Including both in your model could overweight the influence of sun exposure, since the layers are correlated.

GEOGRAPHIC EXTENT AND RESOLUTION

The geographic extent and resolution of your analysis is determined by the phenomenon you're modeling. If the goal of your model is to identify suitable mountain lion habitat, for example, you'd likely use a broad extent and a coarse data resolution. The extent would include data that covers hundreds of square kilometers since that is the range of mountain lions. The resolution would be based on the minimum area of a patch of habitat (a patch less than, say, twenty acres may not be viable, even if connected to other patches via a corridor). You'll want to use data at a comparable resolution—a minimum polygon or cell area of twenty acres. Using a smaller than appropriate patch size adds unnecessary complexity and processing time to your model.

Conversely, if you're modeling suitable habitat for beavers, you'd use a narrow extent of perhaps a single watershed (since beavers have a limited range) and a fine resolution—a minimum polygon or cell area of a few hundred square meters. A larger minimum area might result in potential habitat not being identified, since a small patch of habitat would be subsumed within the surrounding unsuitable habitat.

The extent and resolution will also impact your choice of datasets to use. For example, if one criterion is that suitable areas are closer to streams, you need to decide *which* streams. The distance layer will be quite different if you use streams created from 1:250,000 scale data versus 1:100,000 scale data. Which dataset you use depends on the extent of your study area, the purpose of your analysis, and the resolution you require.

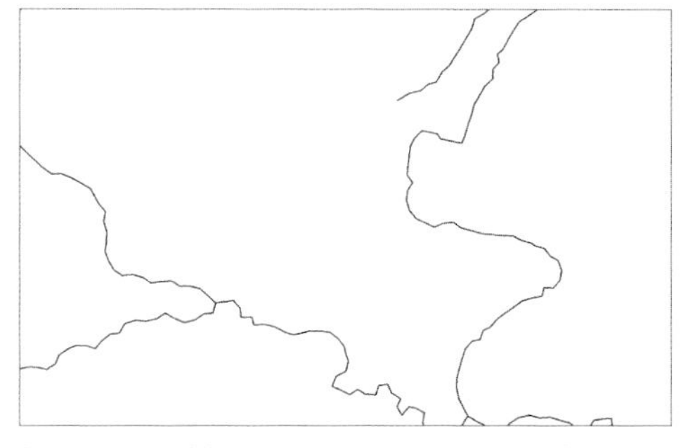

Rivers captured from a printed map at a scale of 1:250,000—the locations of only the largest rivers are included.

Distance to rivers using data from the 1:250,000 scale map.

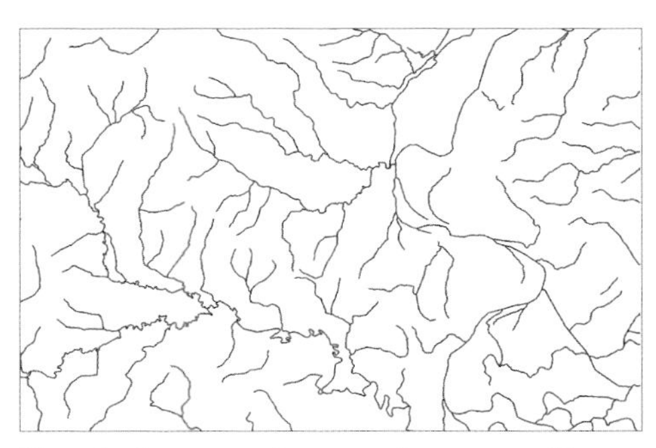

Rivers for the same study area as above, captured from a printed map at a scale of 1:100,000—smaller streams, as well as larger rivers, are included.

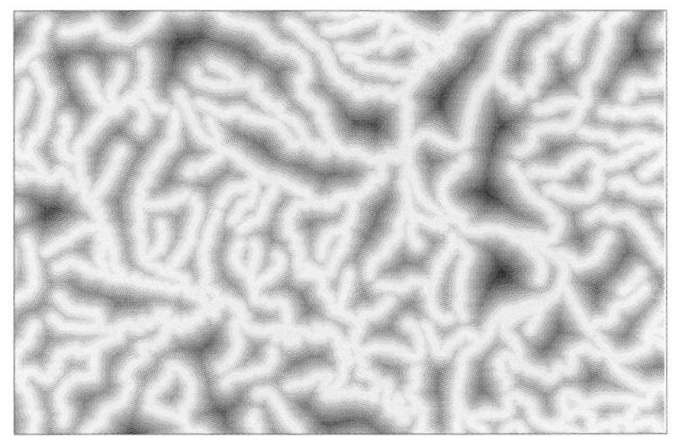

Distance to rivers using data from the 1:100,000 scale map.

Related to geographic extent is the location and configuration of your study area boundary. In some cases, how you delimit the study area can impact the results of your analysis. This is especially true of methods that involve distance calculations. For example, if one criterion is that locations should be close to streams, and your study area is limited to a particular watershed, a location might appear to be far from streams but, in fact, be close to a stream which is just outside the watershed boundary (and study area).

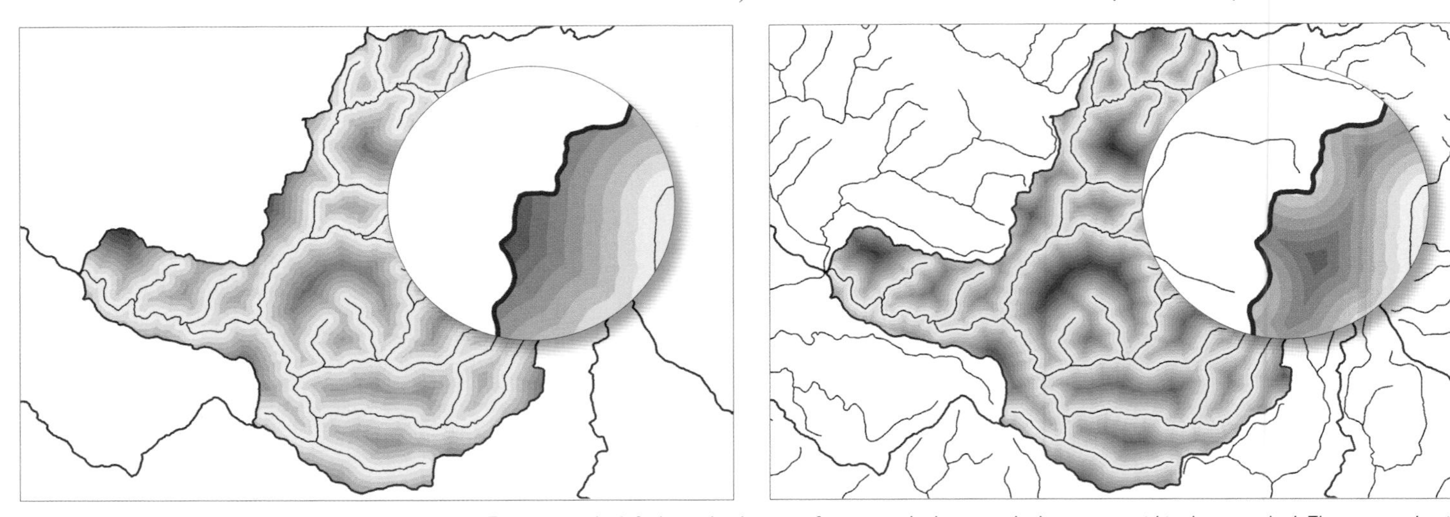

Distance to rivers (darker blue indicates farther distance). The map on the left shows the distances for a watershed, using only the streams within the watershed. The map on the right shows the distances within the watershed, including streams outside the watershed in the calculation. The close-up views show that some areas at the edge of the watershed are, in fact, close to streams when the streams outside the watershed are included.

In a spatial interaction model, the study area boundary also directly impacts your results—destinations (including competitors) outside the boundary will be excluded from the analysis, but, in reality, they may draw visits from origin locations within the study area. In a location-allocation model to find the best location for a new facility, if your study area is a city boundary, people outside the study area will be excluded even though they might in fact use the planned facility.

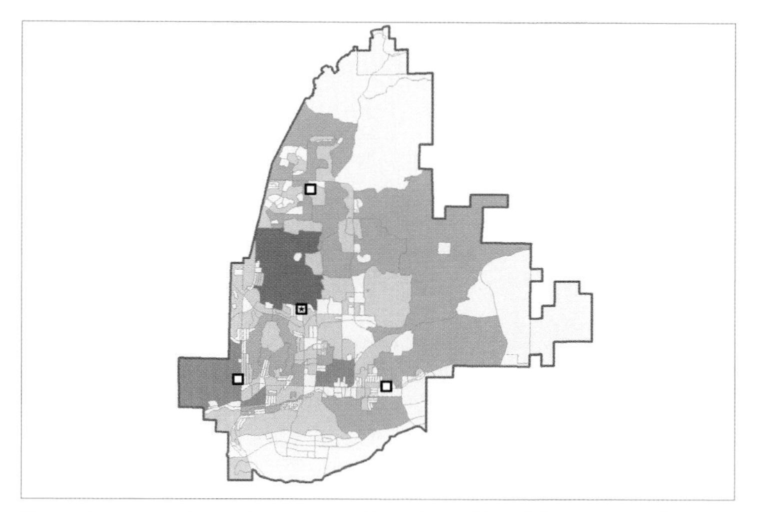

This analysis shows the best location for a library (box with star) from four possible locations. The best location provides access to the most people while minimizing travel distance. (Census block groups are color-coded by population, with darker blue indicating higher population.) Only census blocks within the city limit were considered.

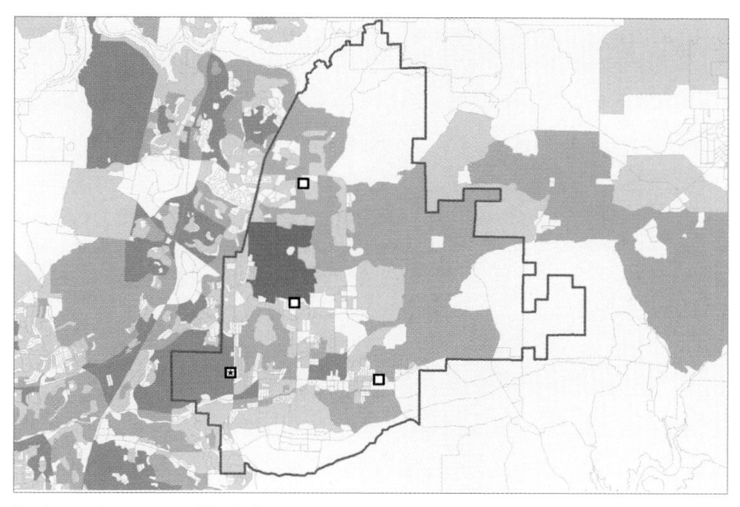

In this analysis, census blocks beyond the city limit were included in the study area (assuming people outside the city might visit the library). The best location shifts to the one in the lower left.

Another issue related to geographic extent is the way in which distances are calculated. The issue arises from the fact that GIS data (which represents geography on the surface of the curved earth) is projected to a flat plane for display as a map, thereby distorting distances. The map projection you use will affect the distance calculations and hence your results. If your study area is limited to a city, county, or state, the curvature of the earth is minimal, so distance distortion is generally not an issue. At the continental or global scale, however, the earth's curvature is a factor in calculating distance. If you're working at this scale and distance is one of the criteria in your model (or if distance calculations are inherent in the analysis method, such as with spatial interaction models), you'll want to use a map projection that minimizes distance distortion.

DATA QUALITY

GIS data has both a spatial component and an attribute component. Features have to have both the correct location and the correct attributes to ensure the results of your model are valid.

For network-based models, such as routing delivery trucks over streets or modeling flow through a water system, an error in the network will invalidate the results. Examples of errors might be a missing street or one that is incorrectly tagged as one-way, or a valve in a water system that is tagged as open when it's actually closed.

In suitability models, several data layers are combined, in a series of steps, to create a result. Small errors in each of several layers can be compounded when the layers are combined. In addition, the errors are carried through subsequent steps—the results of each step are based on the previous steps, so before long your results will differ significantly from the results you'd obtain using better data.

For example, in a suitability analysis, a missing stream feature will result in an incorrect stream distance layer. The error will be compounded if the distance layer is combined with a land-cover layer on which the wrong land-cover code has been assigned to one area. Your final results will show some suitable areas as unsuitable, and vice versa.

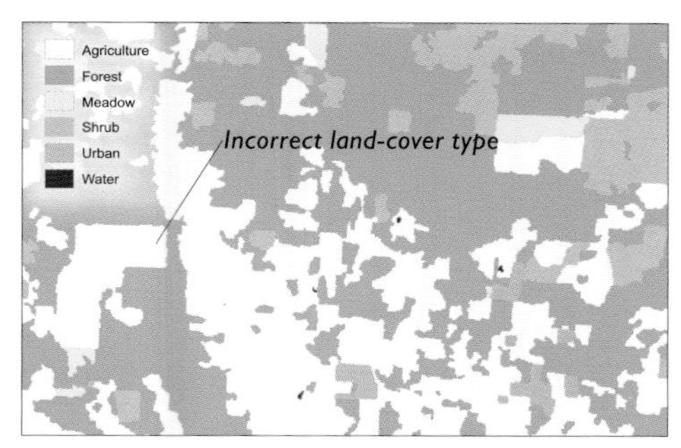

The goal of this analysis is to identify locations that are suitable for a new housing development. Land-cover types that are easier to build on, such as agriculture, are more suitable. But in the land-cover data layer, one large area has been misidentified as agricultural when it is not.

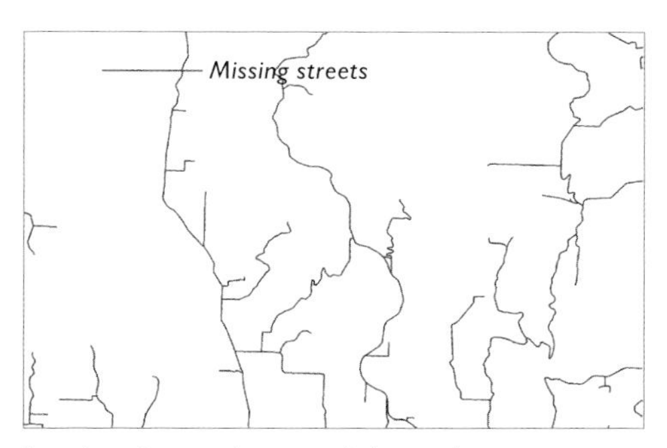

Locations that are closer to existing roads are more suitable. But in the streets data layer, some streets are missing in the northwest corner of the study area.

The results of the analysis using the bad data show suitable locations in dark green and unsuitable locations in dark red. The area with the missing streets is shown as unsuitable, while the area to the south of that—where the land cover was misidentified—includes some apparently suitable locations.

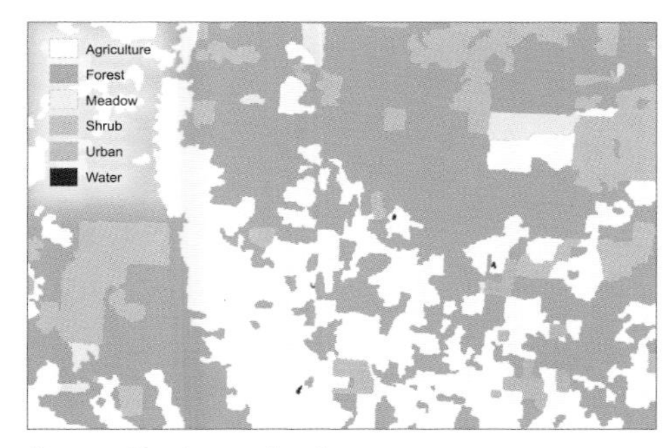

Corrected land-cover data layer.

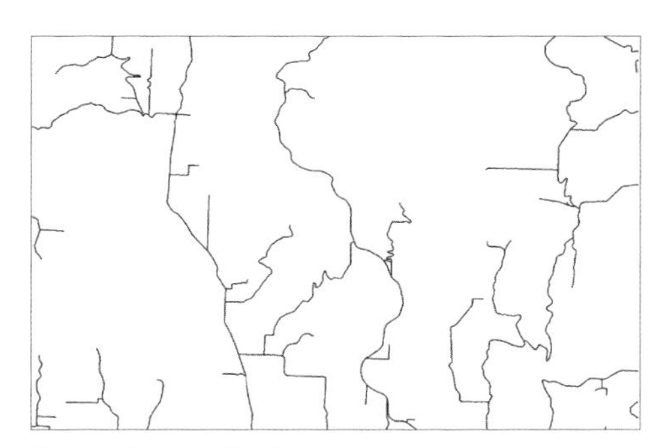

Corrected streets data layer.

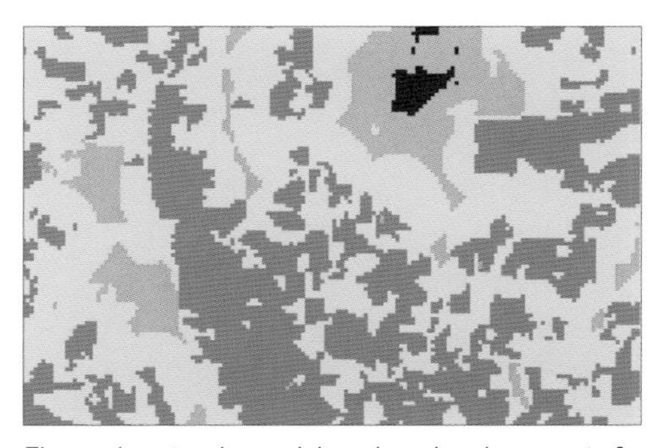

The results using the good data show that there are, in fact, some suitable locations in the northwest corner of the study area, and that the area to the south is less suitable than it appeared when the bad data was used.

You'll want to ensure you're using the most accurate and current data possible. If the data is from within your own organization, you can rely on the data quality controls that are in place. However, it's increasingly common to gather data from a wide variety of sources, including the Internet. You'll want to research the quality of the data and check it against alternate sources, if possible, to ensure it meets the requirements of your analysis. At the very least, by assessing the quality of the data you'll know what level of confidence you can have in the results of your model.

references and further reading

Allen, David. 2011. *Getting to Know ArcGIS ModelBuilder*. Redlands, CA: Esri Press. The book presents tutorials that show you how to automate GIS analysis tasks (as well as data processing tasks) using ArcGIS ModelBuilder.

Carr, Margaret H. and Paul D. Zwick. 2007. *Smart Land-Use Analysis: The LUCIS Model.* Redlands, CA: Esri Press. The book presents an approach to land-use suitability analysis. Includes a discussion of the Delphi process, pairwise comparison, and several other techniques for defining criteria.

Mitchell, Andy. 2005. *The Esri Guide to GIS Analysis, Volume 2: Spatial Measurements and Statistics*. Redlands, CA: Esri Press. Includes discussions of spatial autocorrelation and other issues related to spatial bias in GIS data.

Zeiler, Michael. 2010. *Modeling Our World: The Esri Guide to Geodatabase Design*. Redlands, CA: Esri Press. This authoritative book on GIS data describes in detail the various types of spatial and attribute data used in GIS mapping and analysis.

Finding suitable locations

Use a suitability model to find locations for a particular use—for example, finding the best location for a regional mall or deciding where to locate a new planned community. Your model includes layers representing the characteristics of locations within the study area. You then identify the locations that have the characteristics you're seeking—that is, your criteria for what constitutes a "suitable" location for the use. "Suitability" is a broadly defined term. Because they identify locations having a particular set of characteristics, these types of models can also be used to find areas susceptible to an event such as a wildfire, insect infestation, or landslide (in a sense, they identify locations with "suitable" conditions for such an event). Similarly, they are used to identify areas that are considered desirable. For example, a business would use a suitability model to identify neighborhoods that may have many potential customers and in which they should advertise. A wildlife biologist studying mountain lions would use a suitability model to find areas that are potential mountain lion habitat.

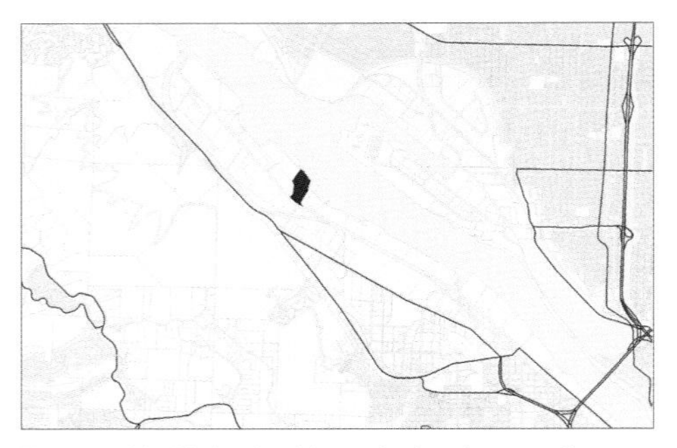

The parcel highlighted in blue is the best location for a regional mall.

Tan areas represent potential mountain lion habitat.

One method for modeling suitability identifies locations that meet all your criteria (and eliminates locations that fail to meet one or more criteria). This type of suitability model divides locations into two groups, or sets: those that are suitable and those that aren't. The model uses Boolean logic to determine whether a location is part of the set of suitable locations. It evaluates whether each location meets each of your criteria, on a yes/no basis. The answer must be "yes" for all criteria for a location to be included in the set of suitable locations.

Boolean suitability models are useful if your criteria have distinct thresholds (for example, "no development within 200 feet of a stream") and if all criteria must be met for a location to be considered suitable. This is often the case when the criteria are defined by regulations, such as when finding a location for a new school. Boolean models are relatively simple to create, use, and understand. They are the subject of this chapter.

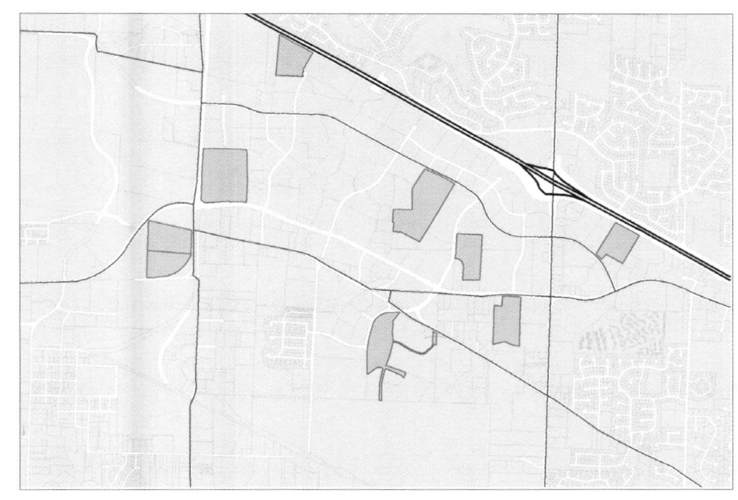

Areas in blue indicate best locations for a shipping distribution center, based on selection criteria. This model was created using Boolean selection.

A couple of other methods for modeling suitability allow you to rate locations on a scale, from more suitable to less suitable. These methods are discussed in Chapter 3, "Rating suitable locations."

designing a boolean suitability model

To design a Boolean suitability model, you state the problem you're trying to solve, state the specific criteria that define a suitable location, and choose the appropriate tools.

DEFINE THE PROBLEM

The first step is to define—as specifically as possible—the problem you're trying to solve. This will help you define the criteria and guide the development of your model.

Your problem statement might look like these examples:

"I need to find the parcels in this city that are suitable for building an elementary school."

"I need to find locations in this study area that are suitable mountain lion habitat."

Finding suitable locations is sometimes an initial step in solving a larger problem. For example, if your goal is to protect mountain lion habitat, once you've identified patches of suitable habitat you'd want to link them by creating corridors (modeling paths and corridors is discussed in Chapter 4, "Modeling paths").

DEFINE THE CRITERIA

The criteria—how you define what is "suitable"—form the heart of the model. You'll want to list the specific characteristics that make a location suitable. For example, when looking for a location to build a small shopping center, the criteria might include:

- Currently vacant
- Between two and five acres in size
- Within a half-mile of a highway
- Inside the city limits

Each criterion has a corresponding source data layer.

CRITERIA	SOURCE LAYER
Currently vacant	*Parcels*
Between two and five acres	*Parcels*
Within a half-mile of highway	*Highways*
Inside city limits	*City boundary*

Generally, criteria are based on a characteristic of the location itself or on the spatial relationship of the location to other features.

Criteria that are based on a characteristic of the location rely on attribute values. For example, the first two criteria require that the parcels layer has attributes for land use and size (in acres). Your criteria specify the values that constitute suitable parcels (vacant and between two and five acres, respectively). The values can be categories, can be within a range of numeric values, or can be a threshold value.

For category values, the criteria specify which categories are considered suitable for the use you're modeling. For example, when looking for a location to build a new shopping center, you would look for parcels that have a land-use category of "vacant."

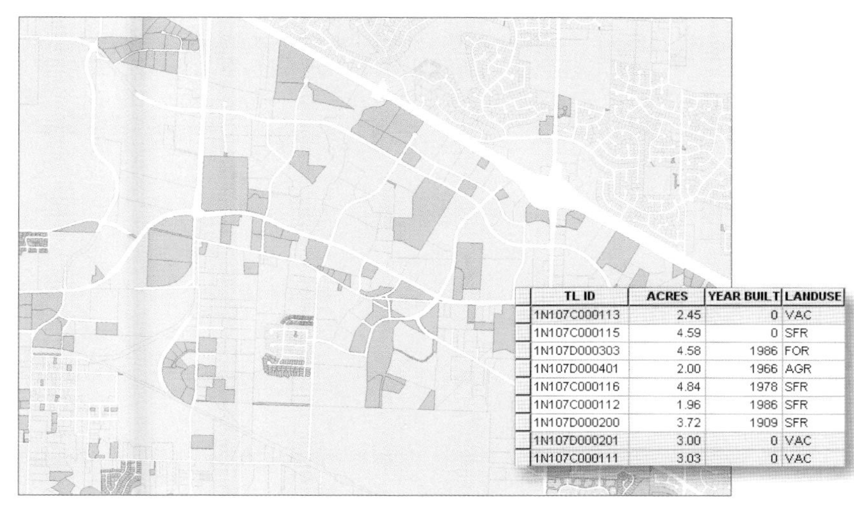

TL ID	ACRES	YEAR BUILT	LANDUSE
1N107C000113	2.45	0	VAC
1N107C000115	4.59	0	SFR
1N107D000303	4.58	1986	FOR
1N107D000401	2.00	1966	AGR
1N107C000116	4.84	1978	SFR
1N107C000112	1.96	1986	SFR
1N107D000200	3.72	1909	SFR
1N107D000201	3.00	0	VAC
1N107C000111	3.03	0	VAC

Parcels shown in blue are vacant. "Vacant" is one category of land use associated with parcels.

For continuous values, criteria can be a threshold value or can be within a range of values. For example, a location for a new subdivision must be on slopes of less than seven degrees (a threshold); a shopping center must be on a parcel of between two and five acres in size (a range).

TL ID	ACRES	YEAR BUILT	LANDUSE
1N2100000607	3.06	1926	AGR
1N2100000602	15.52	0	AGR
1N2110001300	45.27	1950	AGR
1N2110001190	50.62	0	AGR
1N2110001191	3.58	0	AGR
1N2110001400	17.82	1930	AGR
1N2110001500	19.96	0	AGR
1N2110001701	6.17	1945	AGR
1N2110001603	5.00	1885	AGR

Parcels between two and five acres in size—a value range criterion.

Criteria based on spatial relationships include how close a location is to other features and whether a location is inside or outside an area. In the shopping center example, the criteria require that the location be within a half-mile of a highway and inside the city limit.

Parcels less than a half-mile from a highway (shown in red).

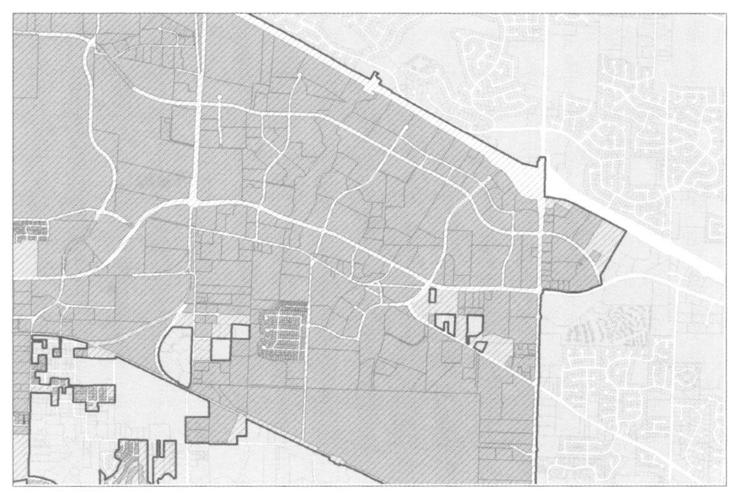

Parcels within the city limit (brown border and hatching).

When listing the criteria for your model, consider possible alternative models you may want to create to ensure you collect all the necessary source layers, with the required attributes, up front. For example, you may need to create models for a set of alternative solutions, such as potential locations for a subdivision given different priorities—a low-cost solution, and one that favors protection of resources such as streams, wetlands, and agricultural land. The first requires source layers for slope, roads, and land cover, while the second requires land cover, streams, and wetlands. During the design stage you'd identify all five layers as required for your models.

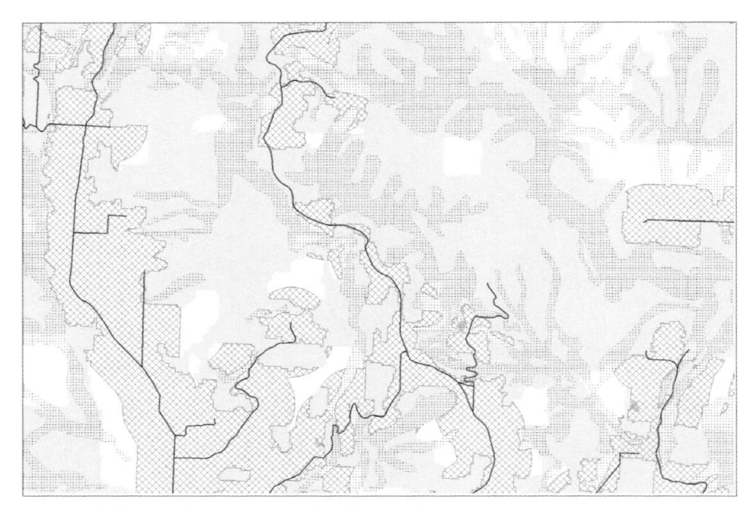

Areas suitable for a housing subdivision (hatching) using a low development cost model. These areas avoid forest (green) and steep slopes (brown stippling) and are near existing roads.

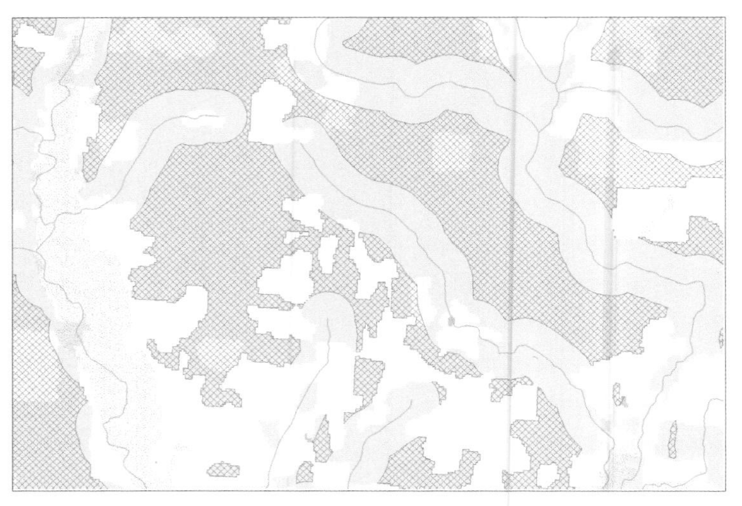

Areas suitable for a housing subdivision (hatching) using a model that protects agricultural land (pale yellow), wetlands (blue stippling), and streams.

CHOOSE THE APPROPRIATE TOOLS

Once you've specified the criteria you'll use in your model, you'll choose the appropriate tools.

One set of tools allows you to do logical and spatial selections—corresponding to the various criteria—to filter out unsuitable locations, leaving the suitable locations remaining. This method is good for evaluating existing features, such as parcels, census tracts, cities, or roads to determine if they are suitable. For example, to find a location to build a new store, you'd start with a layer of parcels, select the vacant ones, select the remaining ones zoned commercial, select from these the ones of at least an acre in size, and finally select the ones within a quarter-mile of a major highway. With each selection the number of parcels decreases, resulting in the set of parcels that meet all the criteria.

Another set of tools lets you overlay map layers representing the individual criteria to combine the criteria values. In the process, new features are created, having all the associated criteria. To identify suitable locations, you select those features that meet all your criteria. For example, to create a mountain lion habitat model, you would overlay layers of slope, land cover, distance to streams, and distance from highways. You'd then select the areas having attribute values that meet your criteria to identify those areas that are suitable mountain lion habitat. Some of the tools are used with vector data, while others are used with raster data.

Selected parcels show potential store locations. This model uses sequential logical and spatial selections on a set of discrete features.

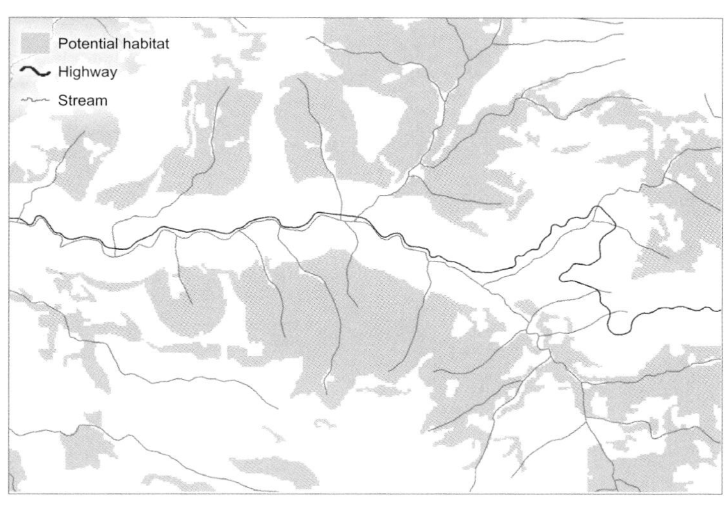

Tan areas indicate suitable mountain lion habitat. This model relies on layer overlay and selection of contiguous areas meeting the selection criteria.

The next two sections present examples of using logical and spatial selection tools to find suitable features and examples of using overlay and logical selection tools to identify suitable locations.

Logical and spatial selection is useful if you have a set of features representing potential locations and you want to find the ones that meet your suitability criteria. The locations could be specific sites—such as parcels—on which to build or locate something, or they could be larger administrative areas you need to evaluate, such as census tracts, ZIP Codes, cities, or counties. For example, you may have a set of parcels and you want to find the ones that are suitable for building a shipping distribution center; or you may have a group of towns and you want to find the best one for building a regional medical center; or you may want to identify sections of highway that would be good locations for a roadside restaurant. In all these cases, the features (parcels, towns, or highways) are contained in an existing source layer. Your model applies a set of filters (representing your criteria) to exclude the features that don't meet the criteria. A feature is filtered out if it fails to meet any one of the criteria. The remaining features are the ones that meet all your criteria and are suitable for the use you're modeling.

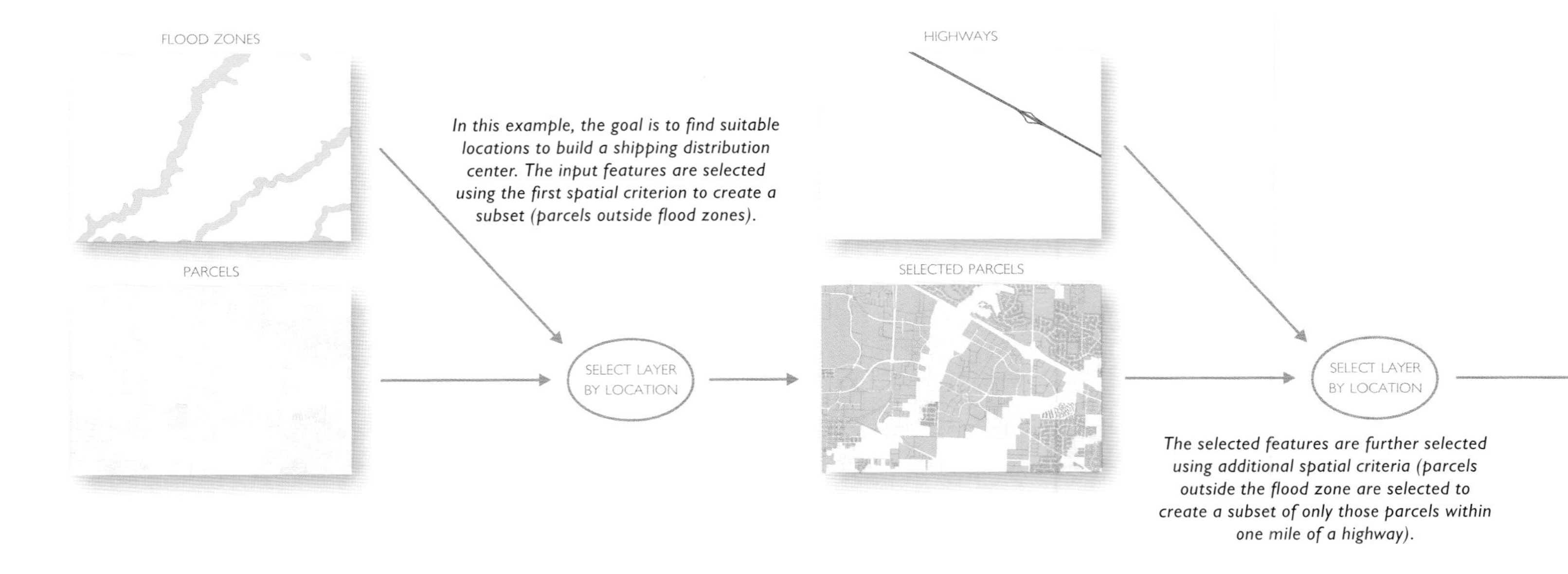

In this example, the goal is to find suitable locations to build a shipping distribution center. The input features are selected using the first spatial criterion to create a subset (parcels outside flood zones).

The selected features are further selected using additional spatial criteria (parcels outside the flood zone are selected to create a subset of only those parcels within one mile of a highway).

The process for evaluating locations is:

1. Collect the source layers
2. Perform the selections
3. Evaluate the results
4. Display and apply the results

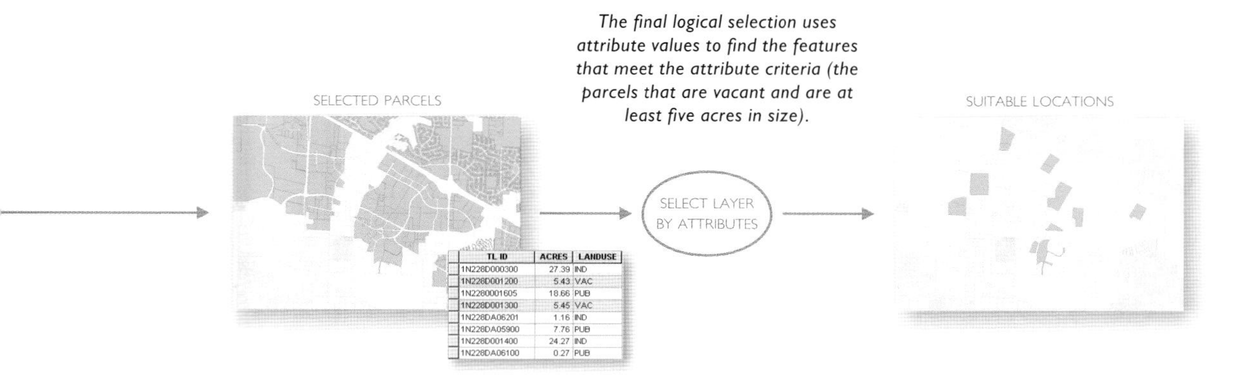

TL ID	ACRES	LANDUSE
1N228D000300	27.39	IND
1N228D001200	5.43	VAC
1N2280001605	18.66	PUB
1N228D001300	5.45	VAC
1N228DA06201	1.16	IND
1N228DA05900	7.76	PUB
1N228D001400	24.27	IND
1N228DA06100	0.27	PUB

The final logical selection uses attribute values to find the features that meet the attribute criteria (the parcels that are vacant and are at least five acres in size).

COLLECT THE SOURCE LAYERS

In the simplest form of the model, all of your criteria would be represented by attributes associated with a single source layer (such as the size and current use of parcels). In this case, you'd do the logical selections using only this layer—no spatial selections would be necessary.

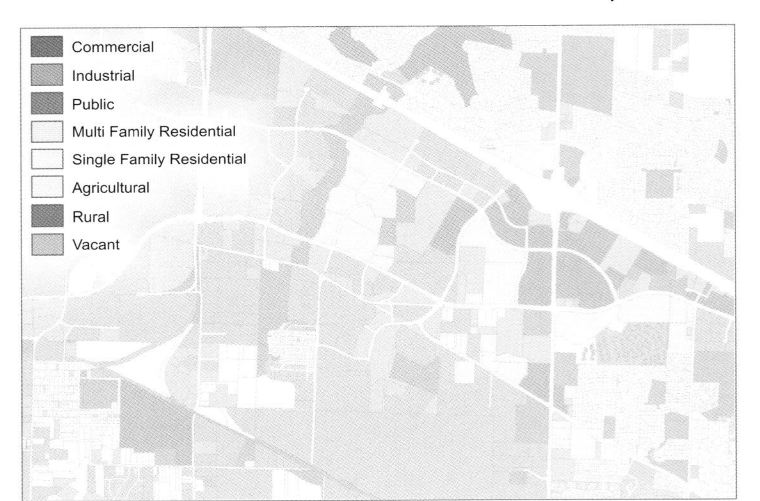

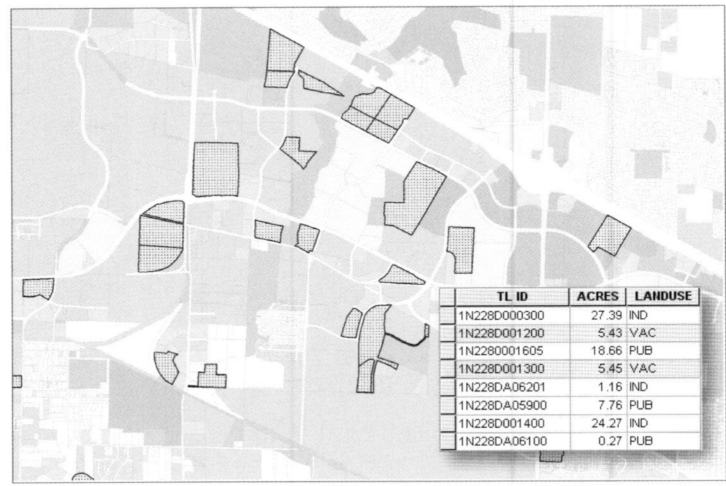

The left map shows parcels color-coded by land-use category. The map on the right shows the subset of parcels that are at least five acres in size and currently vacant (stippled parcels). The selection requires only one layer (parcels) with its associated attribute table.

In many cases, though, the model will include spatial as well as logical criteria. You'll want to state each criterion, then list the corresponding source layer you'll use to do the selection, along with the specific selection statement. For example, when evaluating parcels for a shipping distribution center, your list might look like that in the following table.

CRITERIA	SOURCE LAYER	SELECTION STATEMENT	SELECTION TYPE
Site must be outside a flood zone	*Flood zones*	*Select parcels that are outside the flood zone*	*Spatial*
Site must be within a mile of a highway	*Highways*	*Select parcels less than 5,280 feet from a highway*	*Spatial*
Site must be currently vacant	*Parcels*	*Select parcels with land use of "Vacant"*	*Logical*
Site must be at least five acres	*Parcels*	*Select parcels with area greater than or equal to five acres*	*Logical*

The list will help you determine which source layers you need to collect (in this example, parcels, flood zones, and highways), which attributes the features will need to have (land use and area for parcels), and what type of selections you'll perform.

Parcels layer.

Flood zones layer.

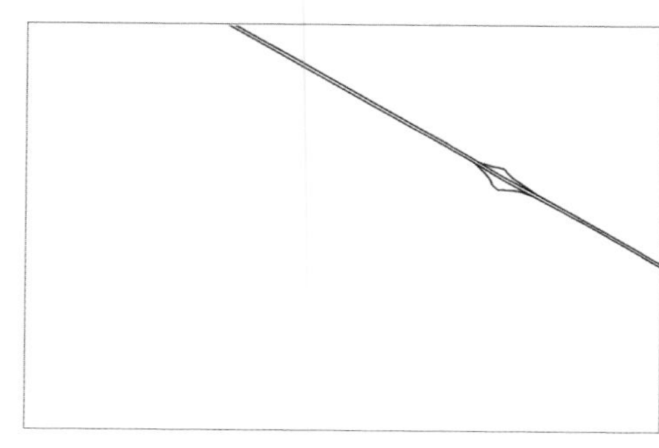

Highways layer.

If the source layer does not have the necessary attributes you may need to edit the layer to add them (if the attribute data is available and has been verified to be accurate), or join a table to the layer if the attributes are in a separate table. For example, you'd join a table containing land-use codes to the parcel layer to assign a land-use code to each parcel.

TL ID	ACRES	VALUE
1N1W08B 1500	37.71	$1,069,280.00
1N1W08A 700	19.11	$1,055,040.00
1N1W08B 1400	19.08	$1,790,200.00
1N1W08C 300	86.59	$362,560.00
1N1W08C 100	25.63	$694,320.00
1N1W08D 400	36.98	$183,320.00
1N1W08C 300	86.59	$362,560.00

TL ID	LANDUSE
1N1W08B 1500	SFR
1N1W08A 700	SFR
1N1W08B 1400	SFR
1N1W08C 300	SFR
1N1W08C 100	SFR
1N1W08D 400	VAC
1N1W08C 300	SFR

Land-use codes are joined to the layer attributes table, using a unique identifier as a key.

TL ID	ACRES	VALUE	LANDUSE
1N1W08B 1500	37.71	$1,069,280.00	SFR
1N1W08A 700	19.11	$1,055,040.00	SFR
1N1W08B 1400	19.08	$1,790,200.00	SFR
1N1W08C 300	86.59	$362,560.00	SFR
1N1W08C 100	25.63	$694,320.00	SFR
1N1W08D 400	36.98	$183,320.00	VAC
1N1W08C 300	86.59	$362,560.00	SFR

After the join, the table includes the land-use codes along with the other required attributes.

In addition to joining tables, you might need to calculate new attribute values in order to perform the selection. For example, if one of your criteria is that parcels must be less than $150,000 per acre, you'd add a new field to the attribute table and calculate value per acre.

PARCEL ID	LANDUSE	VALUE	ACRES	VALUE PER ACRE
1N223DD05100	VAC	$1,743,820.00	13.46	$129,555.72
1N225BA02300	VAC	$1,219,000.00	9.82	$124,134.42
1N225BD00200	VAC	$913,170.00	5.42	$168,481.55
1N225CA00100	VAC	$5,281,090.00	28.85	$183,053.38
1N225DC01600	VAC	$2,380,000.00	13.16	$180,851.07
1N225CD00300	VAC	$1,054,400.00	6.05	$174,280.99
1N226AA00100	VAC	$746,050.00	5.10	$146,284.32

Value per acre is calculated by dividing the VALUE field by the ACRES field.

PERFORM THE SELECTIONS

To find the features that meet the criteria, you'll perform a series of selections. Many models include both spatial and logical selections. Spatial selections select features using their location in relation to other features. Logical selections select features having particular attribute values that correspond to site characteristics.

Spatial selections

A common spatial selection is finding features that are inside or outside a particular area or zone, such as parcels outside the flood zone, or watersheds containing areas of coniferous forest. Another typical selection is finding features that are close to, or far from, other features (such as parcels near highways or census tracts far from a hospital).

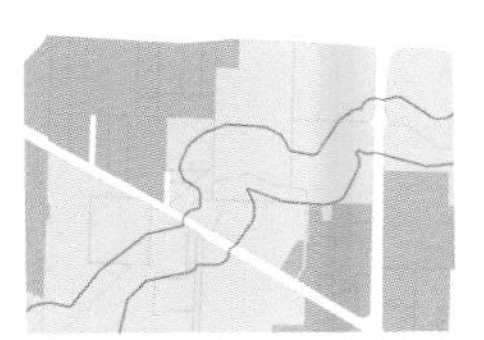
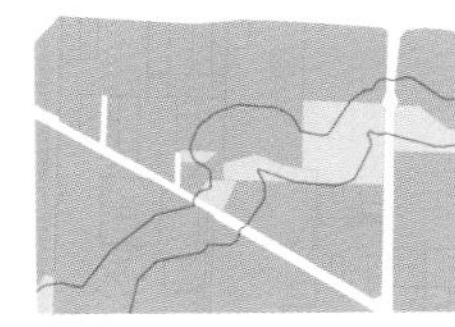

In the left diagram, selected parcels (dark green) are completely outside the flood zone—no portion of a selected parcel falls within the zone. In the right diagram, selected parcels overlap the flood zone, but most of each parcel is outside the zone. For the unselected parcels (light green), most of each parcel is inside the zone.

Selecting features inside—or outside—an area

The key to creating this type of selection is to define what "inside" is. A feature can be completely inside a zone, mostly inside, or simply partially overlap the zone. Which definition you use, and how you structure the selection statement, depends on your criterion. For example, if you're selecting a site for a school, you may want to select only parcels completely outside the flood zone; if you're selecting a site for a parking lot, finding a parcel mostly, but not completely, outside the flood zone would suffice (the lot can easily be closed if a flood threatens). If you're selecting watersheds for research on, say, a bird species that requires only a small patch of forest for nesting you'd select those watersheds that contain any coniferous forest. If your research is on a mammal species that is forest-dwelling and has a large home range, you'd select watersheds that are mostly or completely covered by coniferous forest.

You may need to complete several selections to ultimately select the features you want for a particular criterion. For example, to select watersheds containing any shrub land, you'd first select the shrub polygons on the land-cover layer (using a logical selection) and then select the watersheds that overlap the selected land-cover polygons (a spatial selection).

Shrub polygons have been selected (dark brown) on a layer of land cover.

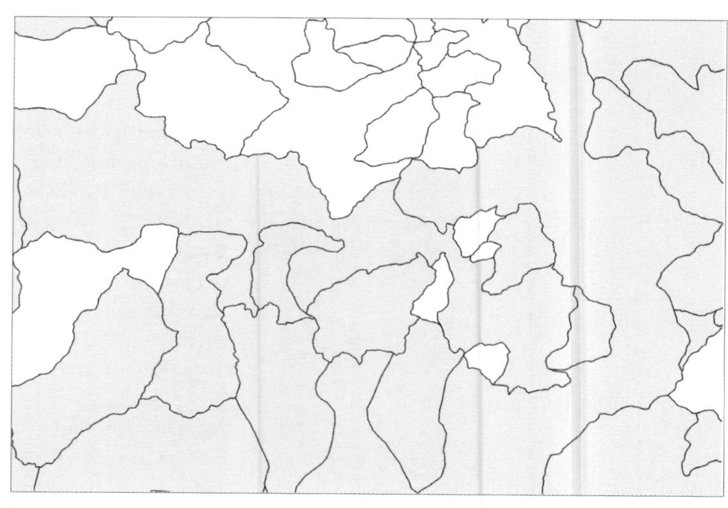

The selected shrub polygons are then used to select watersheds that contain shrub (blue polygons).

Selecting features close to or far from other features

Your criteria might require that a suitable feature be either adjacent to or within a certain distance of other features (or conversely, not adjacent or farther than a certain distance).

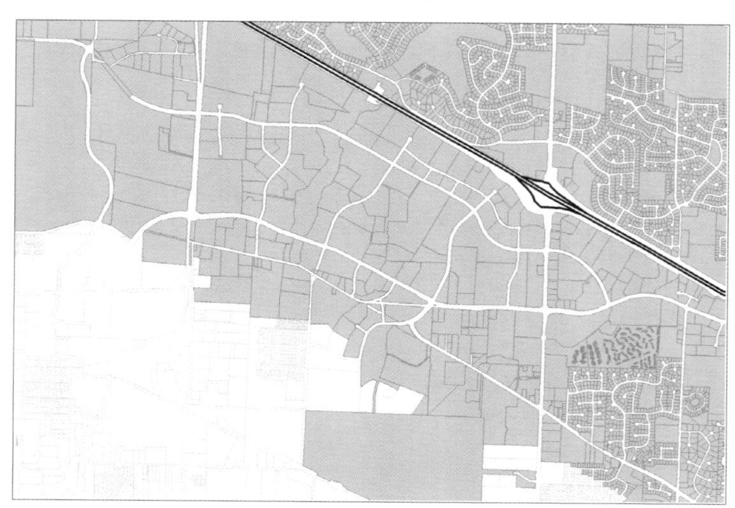

Parcels within one mile of a highway. Features are selected if any portion is within the specified distance.

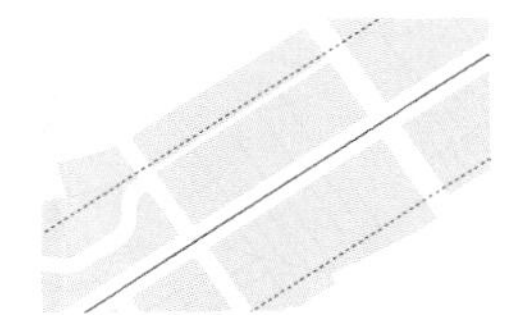

Parcels that fall partially within the distance (dotted line) of the road centerline are selected.

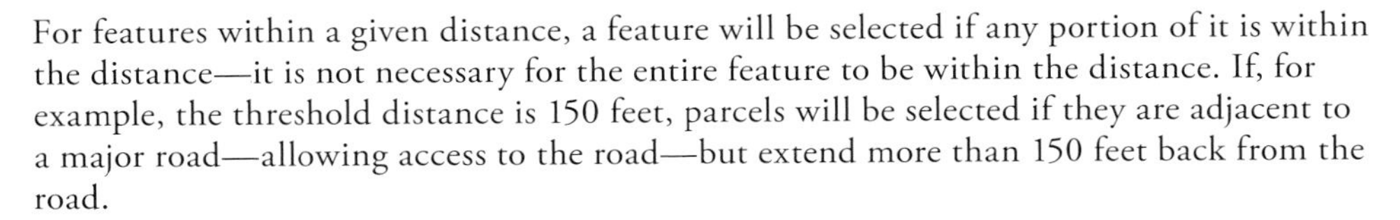

For features within a given distance, a feature will be selected if any portion of it is within the distance—it is not necessary for the entire feature to be within the distance. If, for example, the threshold distance is 150 feet, parcels will be selected if they are adjacent to a major road—allowing access to the road—but extend more than 150 feet back from the road.

To select features that are beyond a certain distance, you will need to first select the features that are within the distance and then switch the selected set. For example, when selecting a site for a convenience store, you might want to find parcels that are within a quarter mile of a school, but not closer than 500 feet. To do this, you'd select the parcels within 500 feet of a school, switch the selected set, and then from these parcels select the ones within a quarter mile.

The initial selection—parcels within 500 feet of a school.

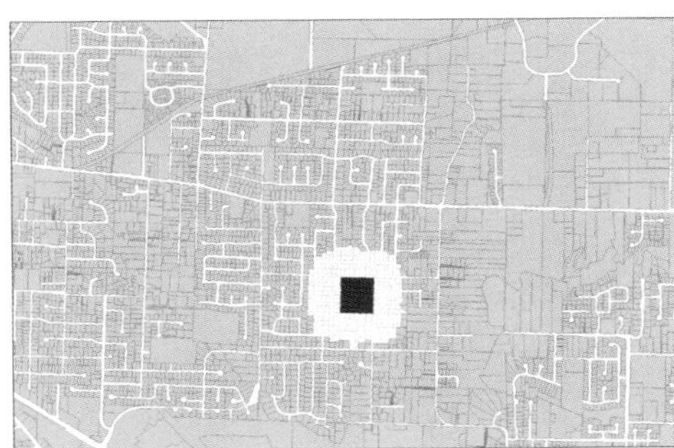

The switched set—parcels that are more than 500 feet from a school.

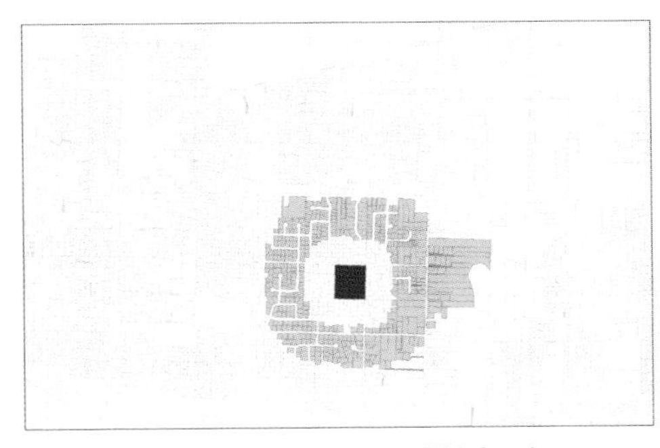

The final selection—parcels at least 500 feet from a school but within a quarter mile.

Logical selections

Logical selections are performed using the attributes of the features. The attribute values, which are stored in the layer's attribute table, represent the characteristics of the feature. Your criteria statement translates directly into a logical selection in the GIS. Logical selections use Boolean operators, such as "equal to," "greater than," "less than," and so on. Use the "and" operator to string together the individual statements corresponding to each criterion. For example, if the site must be at least five acres and vacant, the logical selection statement would look like this (depending on the names of the fields and the land-use codes in the attribute table):

```
ACRES >= 5 AND LANDUSE = 'VAC'
```

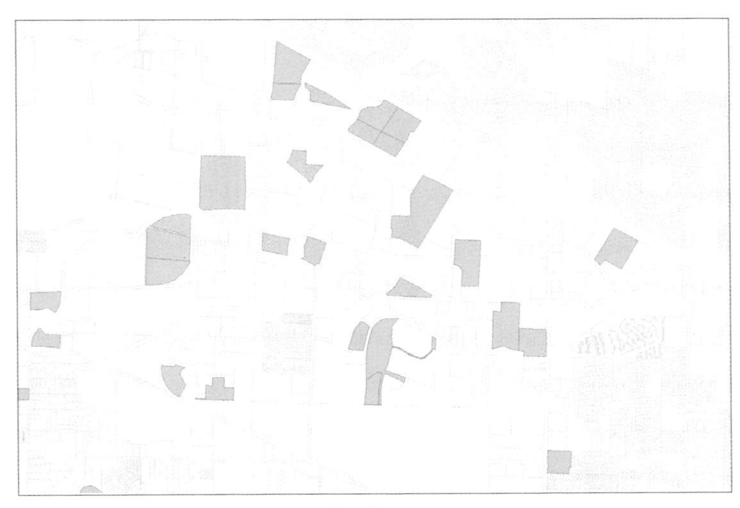

Parcels that are vacant and are larger than five acres.

The "and" operator is also used to select a value range for a particular attribute. If you want to select parcels between 5,000 and 10,000 square feet, the statement would be

AREA > 5000 AND AREA < 10000

If you need to select several categories for a single attribute, use the "or" operator. For example, if the required land use for a site is forest, shrub, or meadow, the statement would be

LANDUSE = 'FOREST' OR LANDUSE = 'SHRUB' OR LANDUSE = 'MEADOW'

In most cases, you'll create a single selection statement to select the features that meet all your criteria at one time. Use parentheses to organize the sections of the statement. For example, entering

(AREA > 5000 AND AREA < 10000) AND (LANDUSE = 'FOREST' OR LANDUSE = 'SHRUB' OR LANDUSE = 'MEADOW')

would select sites that are between 5,000 and 10,000 square feet and are either forest, shrub, or meadow.

DEFINITION	OPERATOR
Equal to	=
Not equal to	<>
Greater than	>
Greater than or equal to	>=
Less than	<
Less than or equal to	<=

The selection order

In general, the order in which you perform the spatial and logical selections will not affect the results. There are a couple common ways of structuring the order of the selections.

One approach makes the model run faster. To do this, order the selections so that the greatest number of features is filtered out in the first one or two selections. Subsequent selections will involve a smaller set of features and will complete faster than they would if they were run on the full set of features.

In general, logical selections are faster than spatial selections (since they merely involve searching a table), so you might do these first, especially if you are selecting from a large set of features. For example, you'd start by selecting those parcels that are currently vacant and at least five acres in size, then perform the subsequent spatial selections.

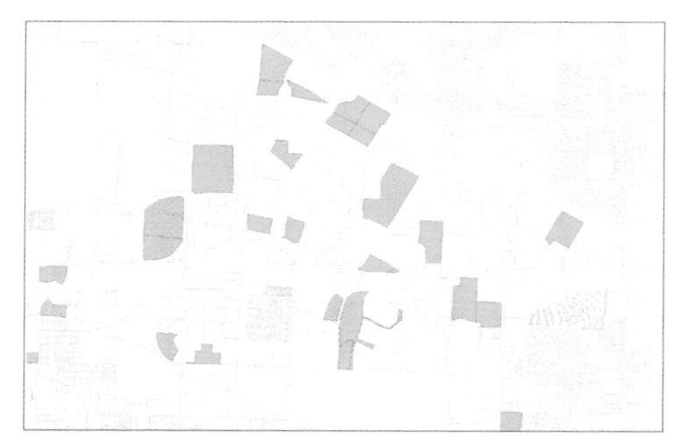

Parcels that are vacant and more than five acres in size...

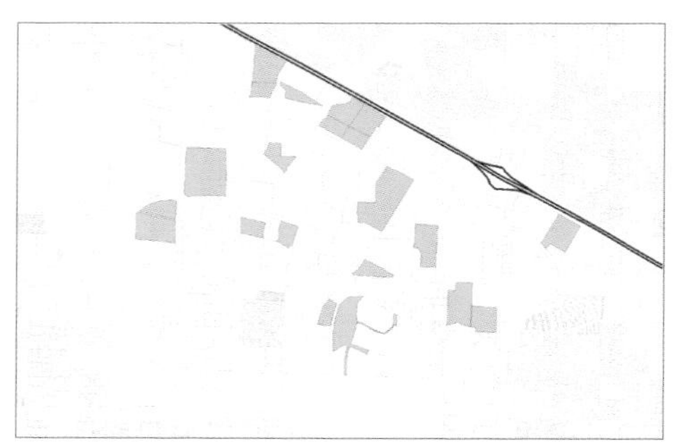

...and within one mile of a highway...

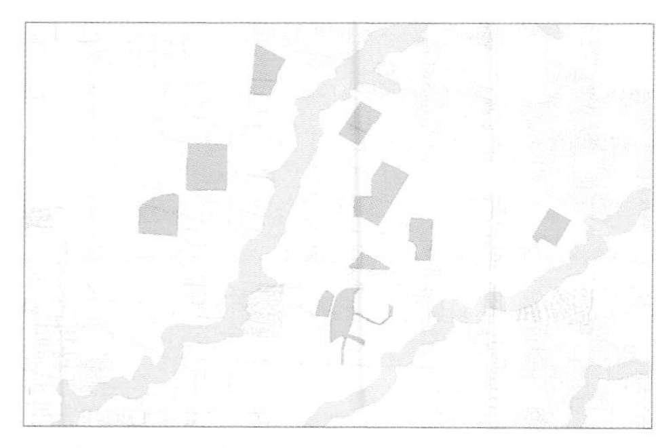

...and outside the flood zones.

Another common approach allows you to see the effect of each spatial selection as it is performed, which can provide insight to your criteria. For example, if one of your criteria is that parcels are within a mile of a highway and, after doing the selection, you can see that many parcels meet this criterion, you might decide to decrease the distance to a half-mile to find the parcels that are even more preferable.

For this approach, do all the spatial selections first, then create a single attribute selection statement for the final selection.

Parcels that are outside the flood zones...

...and within one mile of a highway...

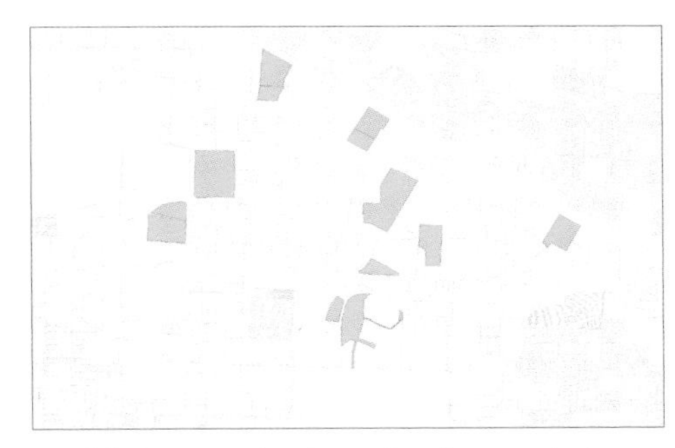

...and that are vacant and more than five acres in size.

Regardless of the approach you use, the selected set of features created by each step is temporary—once you've made the next selection, there is a new selected set consisting of a subset of the previously selected features. If you're doing the selections interactively, it's a good idea to save each intermediate set of selected features as a layer so you can document what you've done and so you can walk back through the model (if, for example, you end up with no sites that meet your criteria, you can modify the criteria and rerun the model). Alternatively, if you're building a model document (as described in Chapter 1), then the selection steps are saved, even though the intermediate sets of selected features are not. You may still want to save each intermediate set of selected features as a layer (by including the save function in your model document at each appropriate step). Reviewing the intermediate selected sets may help you evaluate the results of your model or improve it.

EVALUATE THE RESULTS

While the selection process is fairly straightforward, you'll want to check your results to make sure your selections were valid. To check your model, display the attributes for the selected sites to make sure they meet each of your selection criteria, along with the features used for the spatial selections. This is made easier of you've created your model as a ModelBuilder model document (as described in Chapter 1, "Introducing GIS modeling").

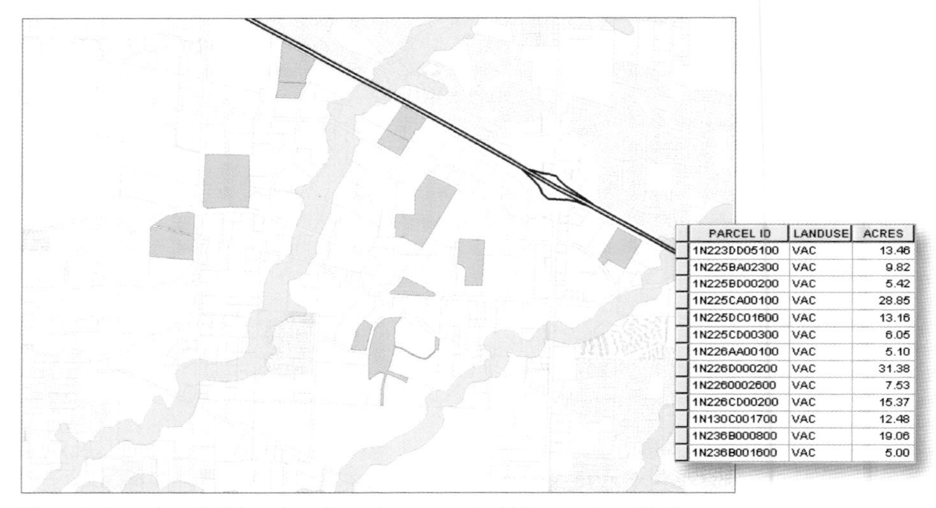

PARCEL ID	LANDUSE	ACRES
1N223DD05100	VAC	13.46
1N225BA02300	VAC	9.82
1N225BD00200	VAC	5.42
1N225CA00100	VAC	28.85
1N225DC01600	VAC	13.16
1N225CD00300	VAC	6.05
1N226AA00100	VAC	5.10
1N226D000200	VAC	31.38
1N2260002600	VAC	7.53
1N226CD00200	VAC	15.37
1N130C001700	VAC	12.48
1N236B000800	VAC	19.06
1N236B001600	VAC	5.00

The map shows that all of the selected parcels are near the highway and outside the flood zones, while the table shows that they are, in fact, larger than five acres and vacant.

In Boolean suitability models there may be potential sites that meet most, but not all, the criteria and are filtered out. For example, a parcel that is otherwise suitable might be just beyond the maximum distance from a highway.

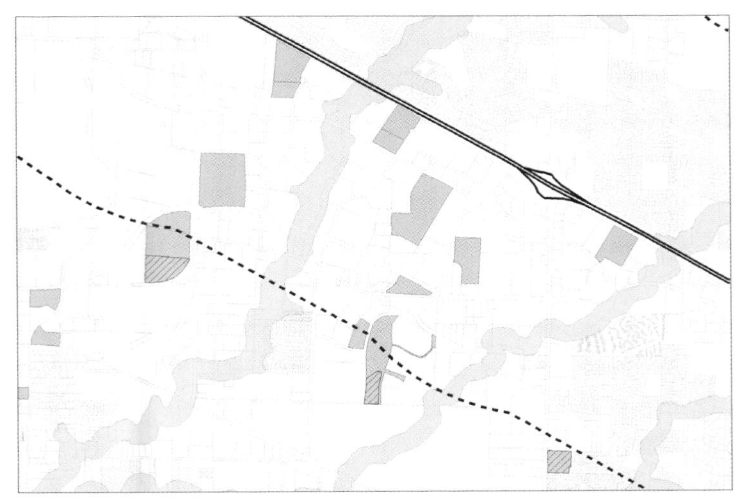

The three parcels with hatching are suitable locations except for being slightly farther than a mile from the highway, as indicated by the dashed red line.

If you've saved each selected set that resulted from a selection, or if you've created a model document, it's easy to modify your criteria to find alternative locations (for example, by slightly increasing the maximum distance a parcel can be from a highway). Alternatively, you can use one of the suitability index methods described in chapter 3, "Rating suitable locations" to rate locations based on how closely they meet the criteria.

DISPLAY AND APPLY THE RESULTS

Once you've identified the suitable features, you'll want to display and highlight them on a map. You may also want to create a report listing the suitable locations along with any other pertinent information. Reports can summarize attribute information for the features (for example, the average value per acre of parcels). Reports are created from the layer's attribute table. Labeling the features on the map allows you to quickly reference them in the report.

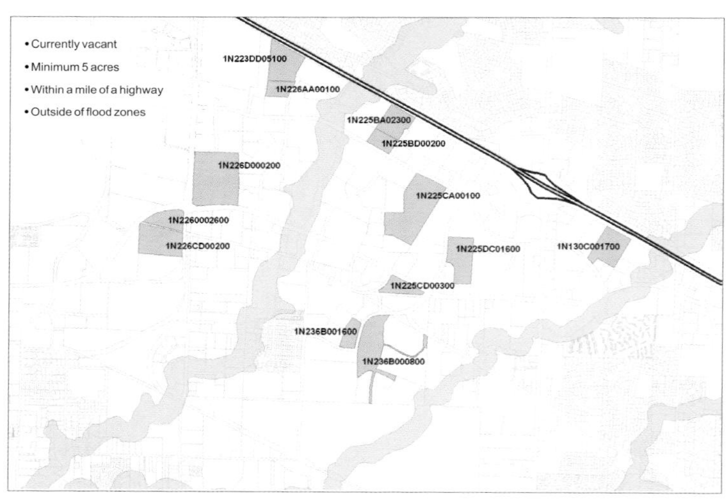

Distribution Center Potential Locations

PARCEL ID	LANDUSE	ACRES	VALUE	VALUE PER ACRE
1N236B001600	VAC	5.00	$646,870.00	$129,374.00
1N226AA00100	VAC	5.10	$746,050.00	$146,284.32
1N225BD00200	VAC	5.42	$913,170.00	$168,481.55
1N225CD00300	VAC	6.05	$1,054,400.00	$174,280.99
1N2260002600	VAC	7.53	$343,070.00	$45,560.42
1N225BA02300	VAC	9.82	$1,219,000.00	$124,134.42
1N130C001700	VAC	12.48	$4,371,720.00	$350,298.09
1N225DC01600	VAC	13.16	$2,380,000.00	$180,851.07
1N223DD05100	VAC	13.46	$1,743,820.00	$129,555.72
1N226CD00200	VAC	15.37	$669,940.00	$43,587.51
1N236B000800	VAC	19.06	$862,930.00	$45,274.40
1N225CA00100	VAC	28.85	$5,281,090.00	$183,053.38
1N226D000200	VAC	31.38	$3,662,160.00	$116,703.64
Average VALUE PER ACRE				$141,341.50

Map showing suitable parcels for a shipping distribution center, with each parcel labeled with its parcel ID. The highway and flood zones are also shown. The text describes the criteria for a suitable location. The accompanying report lists each parcel along with related information. Including the average value per acre in the report lets you quickly see which parcels are above the average value and which are below.

Using graphs to display information about the suitable locations can give you additional insight. For example, creating a histogram of parcels by size allows you to get a better sense of the number of available large and small parcels.

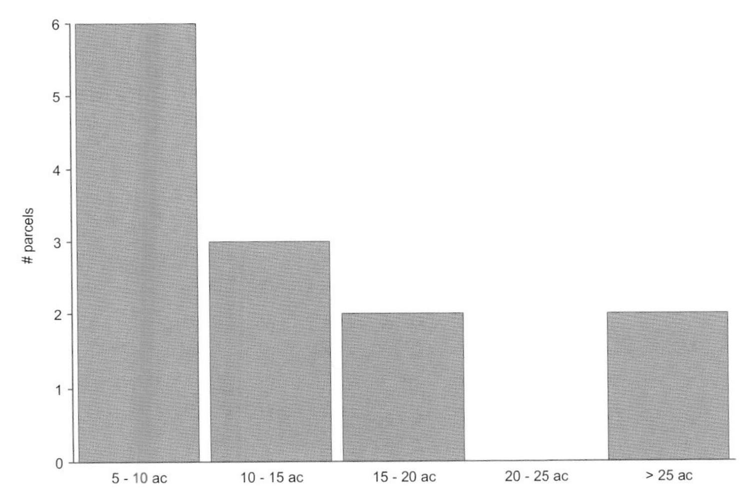

This histogram shows the distribution of parcels based on acreage. Most of the parcels (nine) are less than fifteen acres, while only four are more than fifteen acres (and none are between twenty and twenty-five acres). You have fewer locations to choose from if you decide you need a large parcel for your distribution center.

Displaying the suitable locations using attribute values helps you see spatial patterns. For example, you might display parcels color coded by value per acre to see if parcels tend to be more expensive in one area versus another. Similarly, displaying the suitable locations with related information can reveal opportunities and constraints. You might, for example, display parcels that are suitable for a shipping distribution center on top of a zoning map to see which parcels have the desired zoning (industrial) or which might be in residential areas, making them less attractive for a distribution center.

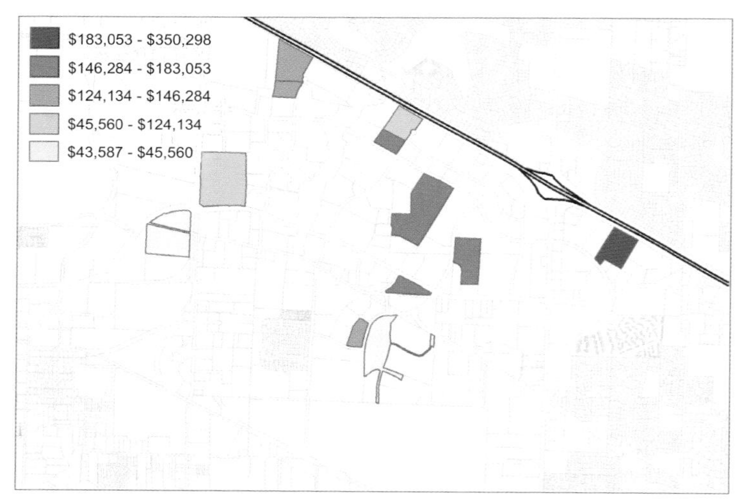

Suitable parcels color-coded by relative value (dollars per acre). Generally, the parcels to the west (which is toward the center of the metropolitan area) and closer to the highway are more expensive on a per-unit basis. You can use the map to explore the trade-offs between accessibility and cost.

Suitable parcels displayed with zoning map. Most of the parcels are in an industrial zone; two are in a commercial zone and one is in a residential zone and would require a rezoning application. Several parcels are adjacent to residential areas, which could result in noise or traffic issues.

Use overlay to find suitable locations if your criteria are in multiple map layers. For example, you may want to find areas that are good habitat for mountain lions. You first define the characteristics that make good habitat and identify corresponding source data layers representing each characteristic. By overlaying the layers, you can query and identify the locations that meet all your criteria. The model is based on Boolean logic, as with the model described in the previous section.

You build your model using either vector input layers or raster input layers, depending on the data you have. While the concepts are essentially the same for both data types, the implementation of the overlay process is slightly different, as you'll see in this section.

The process for identifying potential locations is:

1. Collect the source layers
2. Create derived layers
3. Overlay the layers and identify the suitable locations
4. Evaluate the results
5. Display and apply the results

COLLECT THE SOURCE LAYERS

You first compile the layers representing your criteria. Often the layers include spatially continuous data such as land-cover type, slope, elevation or average annual rainfall. Discrete (or noncontinuous) features, including areas (such as urban area boundaries) and linear features (such as roads or streams), can also be used.

If you're modeling mountain lion habitat, your criteria, and corresponding source data layers, might include the ones in this table.

CRITERIA	SOURCE LAYER
Steep slopes (greater than 19 degrees)	*Elevation*
Covered by forest	*Land cover*
Within 2500 feet of a stream	*Streams*
More than 2500 feet from highways	*Highways*

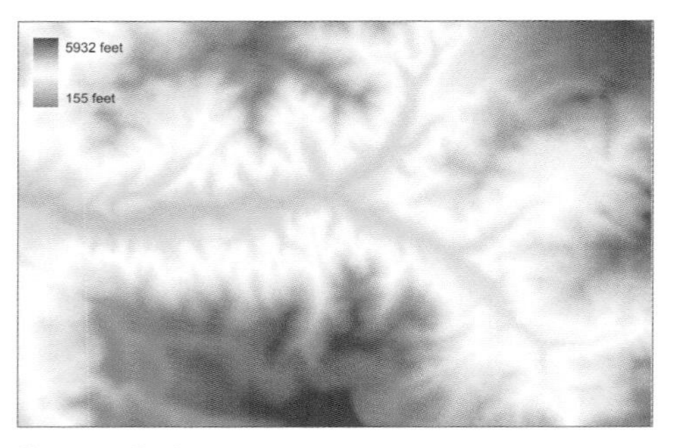

Elevation (feet).

Land cover.

Streams.

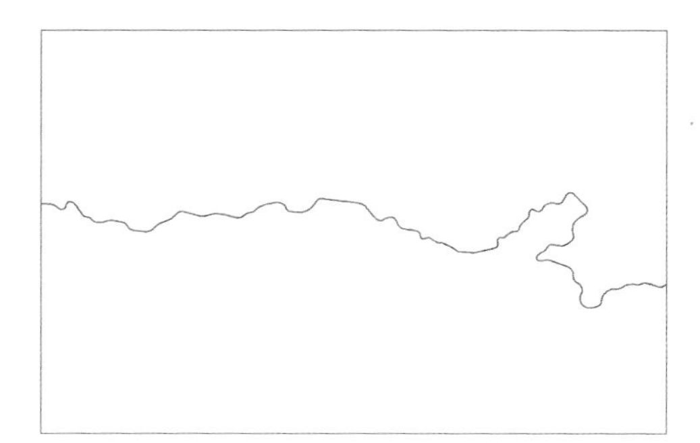

Highway.

Some of the source layers will be used as they are in the model. Other layers will need additional processing or will be used to create derived layers that will become input to the model. In this example, land cover is used as is, streams and highways are buffered to derive layers of distance from streams and distance from highways, and the slope steepness layer is derived from the elevation layer.

CREATE DERIVED LAYERS

In some cases, you'll need to derive the layer corresponding to a particular site characteristic from a source data layer. This is particularly true if your criteria include a threshold value, such as distance to or from geographic features. For example, if one criterion is that suitable locations are within 2,500 feet of streams, you'd use the streams layer to create a distance layer. (A derived distance layer is required since you will be overlaying it with other layers to identify suitable locations—this is unlike when selecting existing features, described earlier in this chapter, where an on-the-fly spatial selection was used to select parcels within a given distance of a highway.)

When using the vector overlay method you'd buffer stream features to create a layer of areas within 2,500 feet of a stream. When using the raster overlay method, you'd create a distance surface (alternatively, you can create a stream buffer polygon and then convert it to a raster layer).

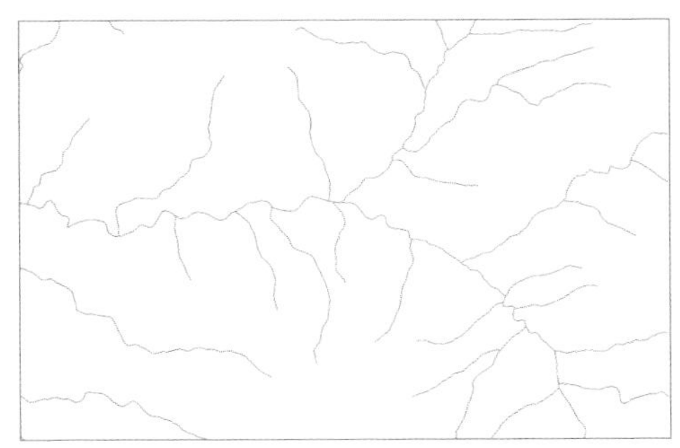

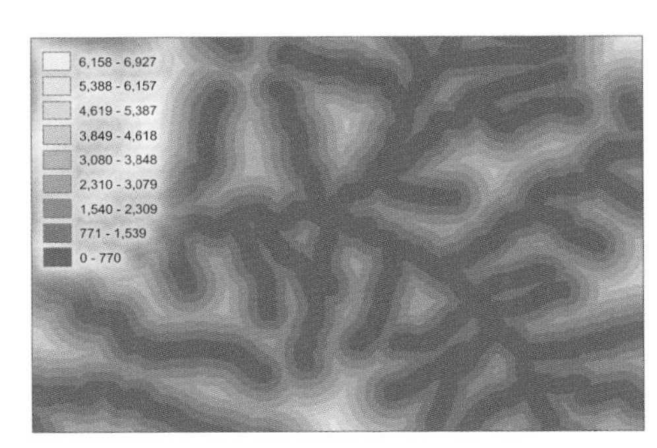

In the center is the streams layer. On the left, 2,500-foot buffer polygons have been created from the streams layer for use in a vector overlay model (areas inside the buffer are suitable). On the right, a raster surface showing distance from streams has been created for use in a raster overlay model (cells with a value of 2,500 or less are suitable).

Similarly, if suitable locations should be more than 2,500 feet from a highway, you'd use the highway layer to create a distance layer. This time, areas farther from the features (highways) are suitable. When using the vector overlay method you'd buffer highway features to identify areas more than 2,500 feet from a highway. When you combine the criteria layers, areas on the output layer that are outside the buffer are identified as such. When using the raster overlay method, you'd again create a distance surface.

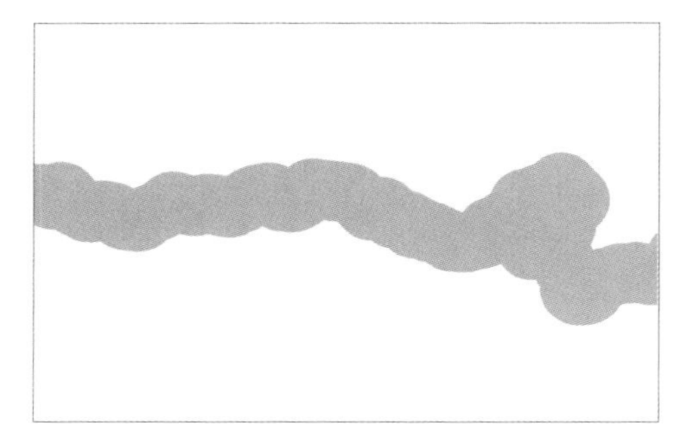

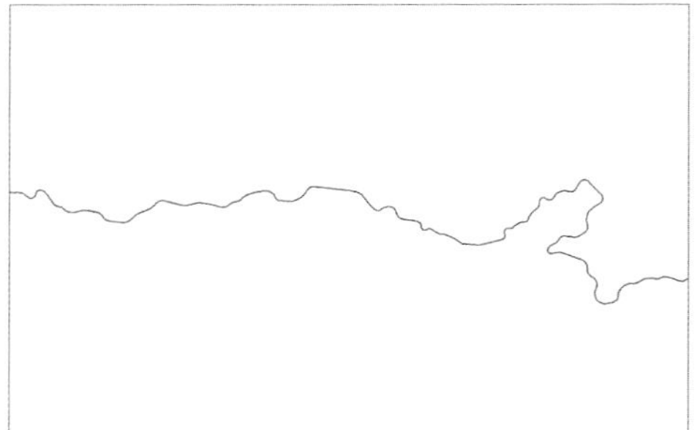

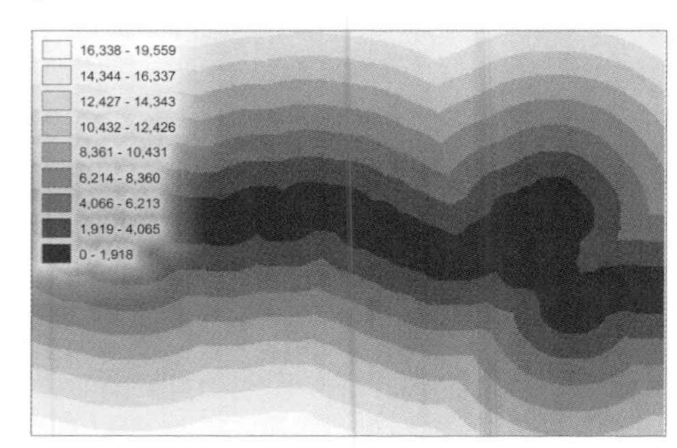

In the center is the highway layer. On the left, a 2,500-foot buffer polygon has been created from the highways layer for use in a vector overlay model (areas outside the buffer are suitable). On the right, a raster distance surface has been created for use in a raster overlay model (cells with a value greater than 2,500 are suitable).

In addition to distance layers, you may need to derive other layers from a source data layer for use in your model. You can derive a slope steepness layer from an elevation layer, for example, using tools in the GIS.

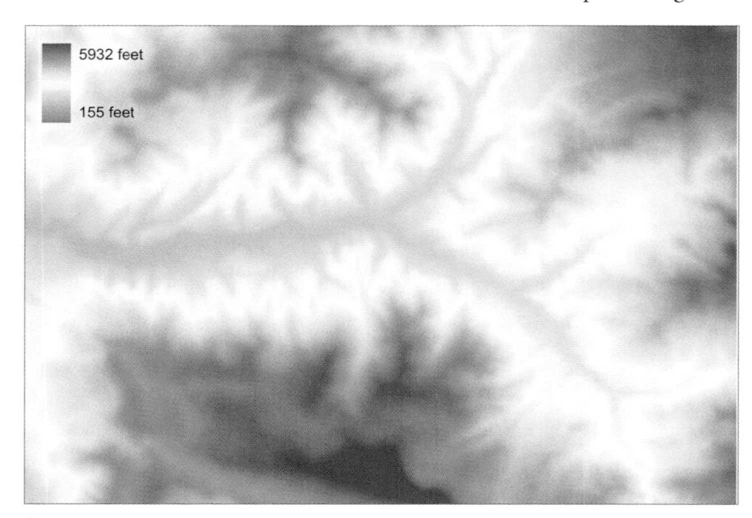

Elevation.

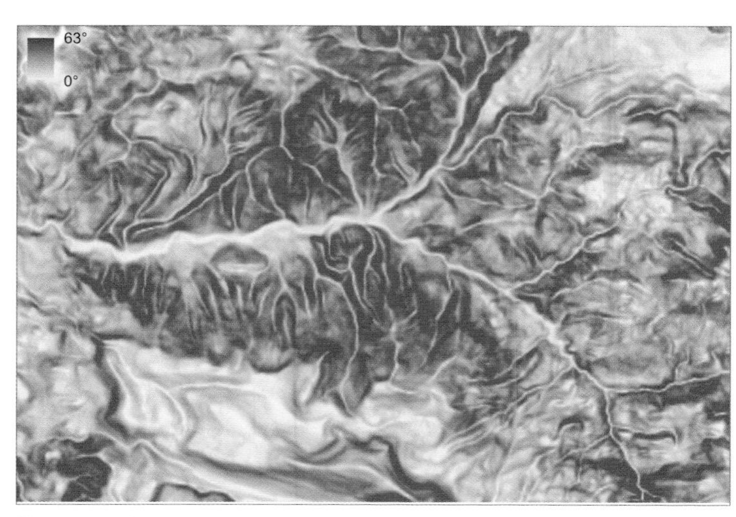

This layer of slope steepness was derived from an elevation surface (left) using tools in ArcGIS.

If you're using the vector overlay method, any of your source layers that are in raster data format will need to be converted to polygons. For continuous values, such as elevation or slope, the values will need to be classified into ranges before conversion. For example, if slopes greater than 18° are suitable, you'd reclassify a slope raster having values ranging from 0° to 63° into two classes: 0°–18° (unsuitable slopes) and 19°–63° (suitable slopes). You'd then convert the reclassified raster to a polygon layer.

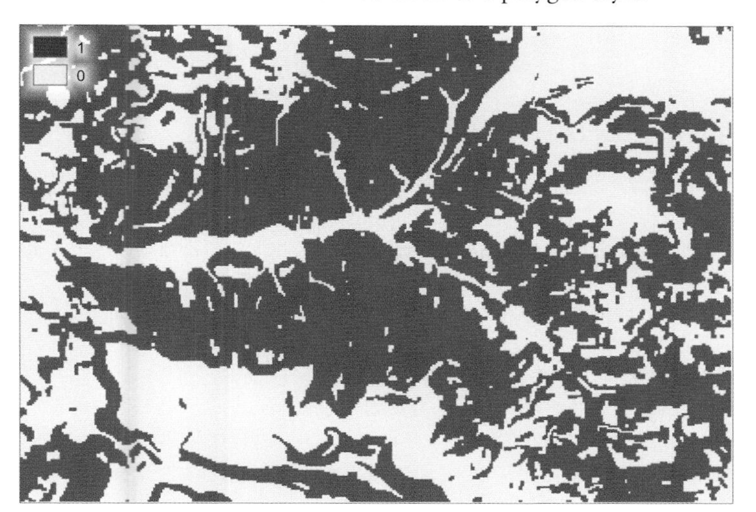

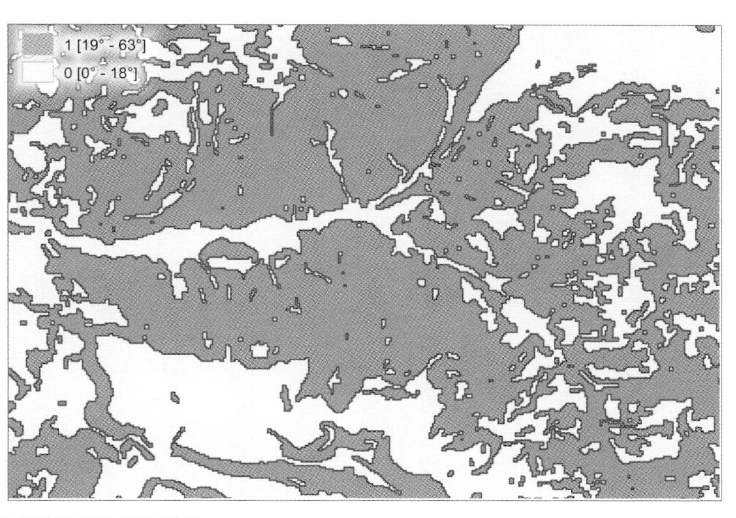

The map of polygons representing steep slopes (right) was created by reclassifying the slope raster into two categories (left) and then converting the reclassified raster to polygon features. When you reclassify the raster, the cells are assigned codes (in this example, 0 for slopes of 0°–18° and 1 for slopes of 19°–63°). The same codes are assigned to the corresponding polygons.

Conversely, if you're using the raster overlay method, any vector layers will need to be converted to raster layers. For example, land-cover polygons would need to be converted to a land-cover raster.

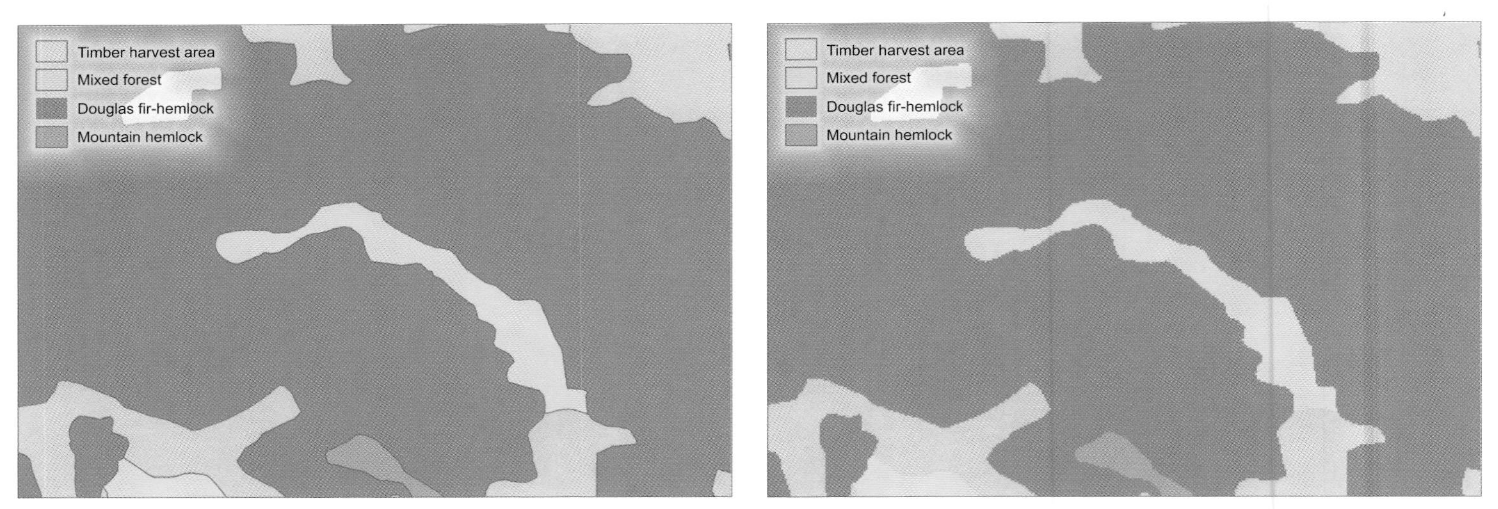

Land-cover category polygons (left) were converted to a raster of land-cover categories (right) for use in a raster overlay model.

OVERLAY THE LAYERS AND IDENTIFY SUITABLE LOCATIONS

To find the suitable locations, you overlay the layers in the GIS. When using the vector overlay method, you combine the source layers to create a new layer containing all of the site characteristics. With all of the characteristics on a single layer you can then query the layer's attribute table to identify the locations that have the characteristics meeting your criteria. When using the raster overlay method, you query the cells on the source layers to identify the ones that meet your criteria (the coincident cell on each layer represents the same location on the earth's surface). You can query the individual layers or combine them into a single layer, which you then query. A new raster layer is created that shows which cells meet all the criteria and which don't.

As stated earlier, the method you use depends mainly on your data. If most of your source layers are vector layers, use the vector overlay method. If most are raster layers, use the raster overlay method. However, each method has some advantages. The vector overlay method allows you to maintain boundaries (such as landownership) in the combined layer, if necessary. The raster method usually requires less data processing.

Using the vector overlay method

When you perform a vector overlay operation, new polygon features are created where polygons overlap between source layers. The site characteristics of all the source layers are combined in the output layer attribute table, with one record for each resulting polygon feature. You then select the attribute values that match your criteria. The selected polygons are the locations that are suitable.

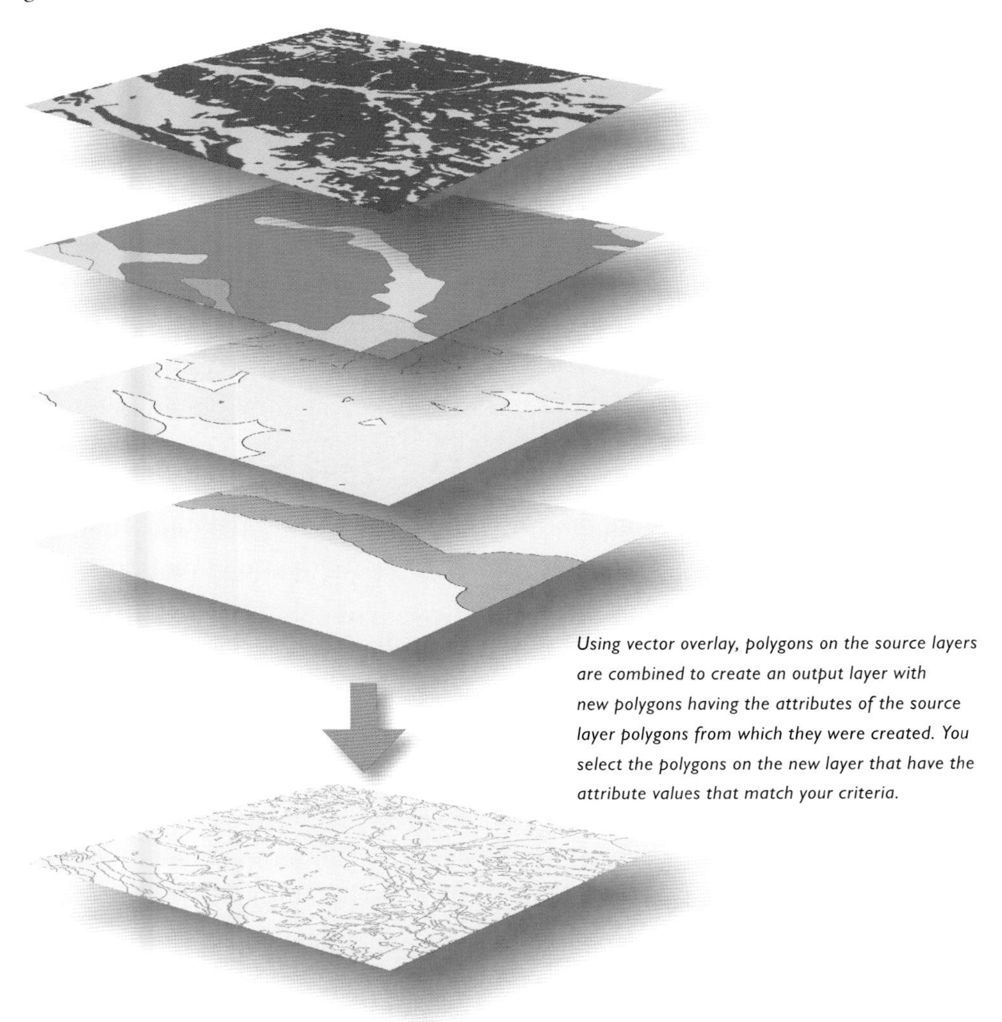

Using vector overlay, polygons on the source layers are combined to create an output layer with new polygons having the attributes of the source layer polygons from which they were created. You select the polygons on the new layer that have the attribute values that match your criteria.

The diagram below shows the process for finding suitable locations using vector overlay.

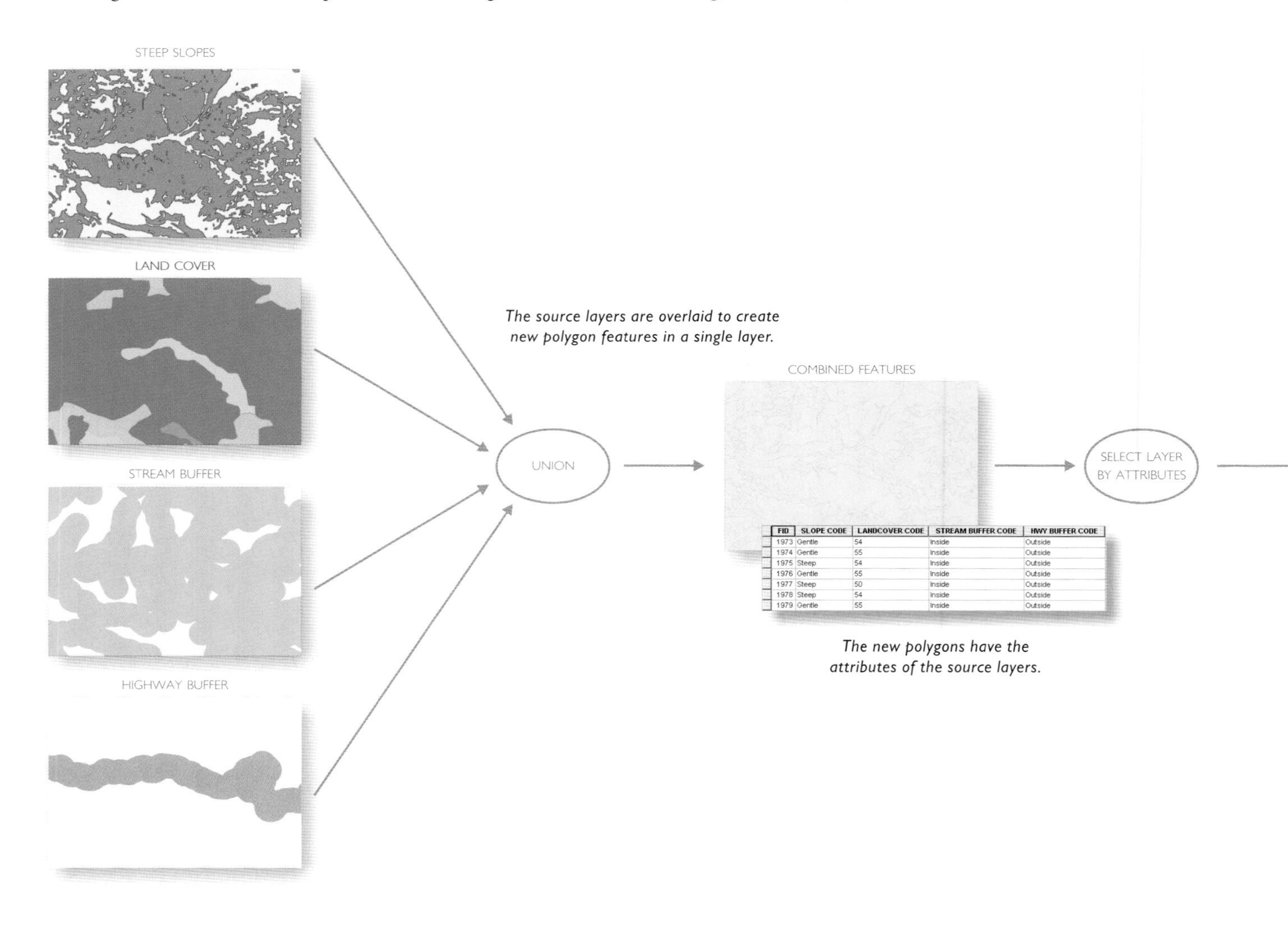

FID	SLOPE CODE	LANDCOVER CODE	STREAM BUFFER CODE	HWY BUFFER CODE
1973	Gentle	54	Inside	Outside
1974	Gentle	55	Inside	Outside
1975	Steep	54	Inside	Outside
1976	Gentle	55	Inside	Outside
1977	Steep	50	Inside	Outside
1978	Steep	54	Inside	Outside
1979	Gentle	55	Inside	Outside

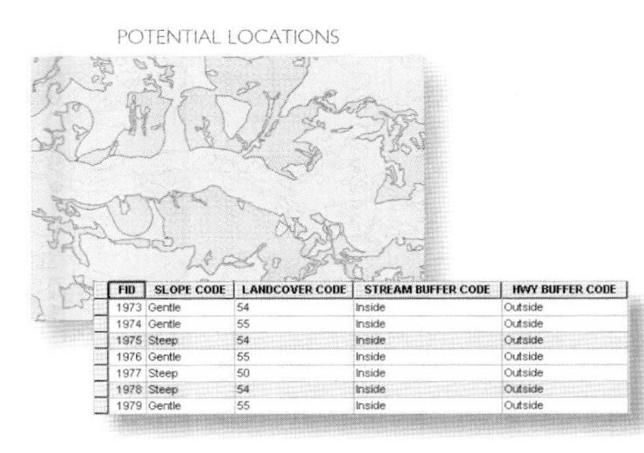

FID	SLOPE CODE	LANDCOVER CODE	STREAM BUFFER CODE	HWY BUFFER CODE
1973	Gentle	54	Inside	Outside
1974	Gentle	55	Inside	Outside
1975	Steep	54	Inside	Outside
1976	Gentle	55	Inside	Outside
1977	Steep	50	Inside	Outside
1978	Steep	54	Inside	Outside
1979	Gentle	55	Inside	Outside

The areas meeting the criteria are identified by selecting the features having the required attribute values.

The first step is to combine the source layers, using an overlay tool such as Intersect or Union. Either tool will create the attribute table with the criteria values you need—the difference is in how the geographic extents of the input layers are handled.

The Intersect tool creates an output layer with an extent equal to the area coincident on all the source layers. Use Intersect if the layers have different geographic extents (it also works if they all have the same extent).

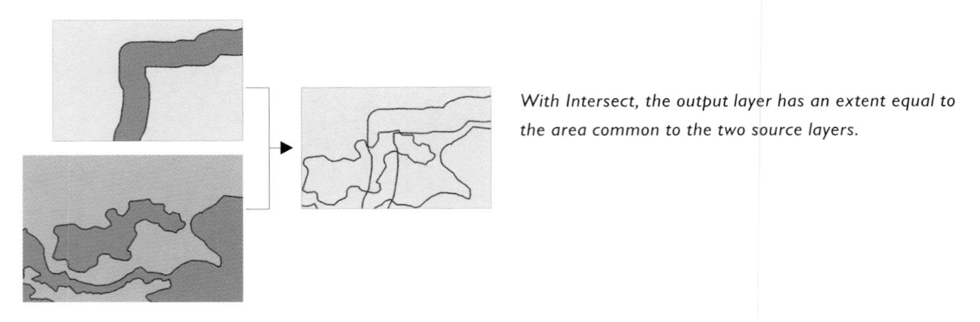

With Intersect, the output layer has an extent equal to the area common to the two source layers.

The Union tool creates an output layer that includes the full extent of all the source layers. Use Union if all the source layers have the same extent or if you plan to clip out your study area from the output layer (using a layer consisting of the study area boundary).

The diagram opposite shows how vector overlay creates the new features and assigns attributes.

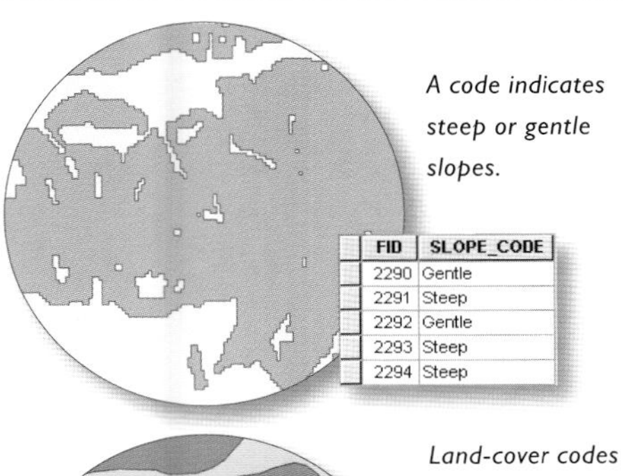

A code indicates steep or gentle slopes.

FID	SLOPE_CODE
2290	Gentle
2291	Steep
2292	Gentle
2293	Steep
2294	Steep

Land-cover codes indicate vegetation type (55 is Douglas fir - hemlock, for example).

FID	LANDCOVER CODE
5	5
6	50
7	51
8	54
9	55

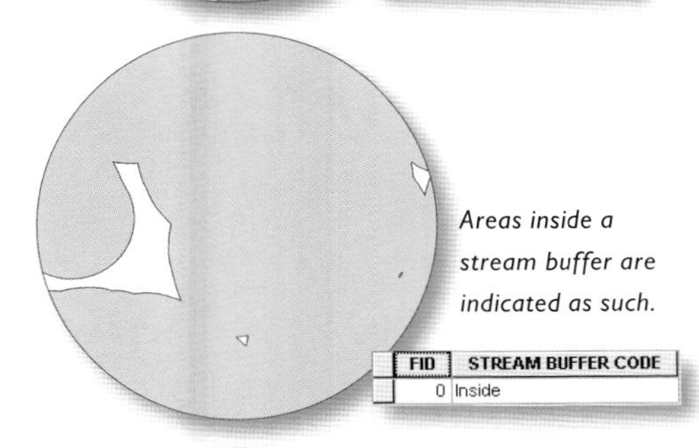

Areas inside a stream buffer are indicated as such.

FID	STREAM BUFFER CODE
0	Inside

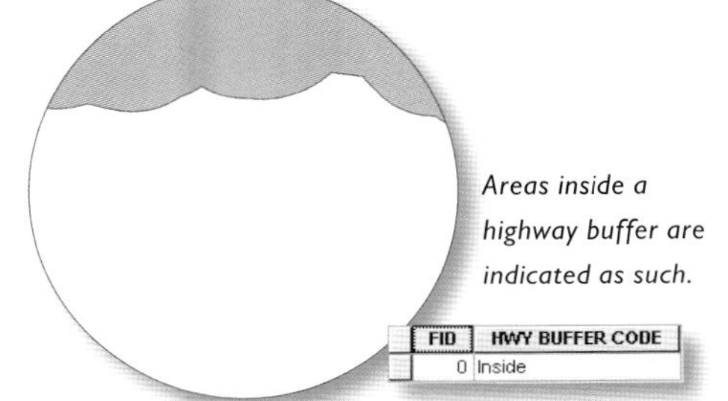

Areas inside a highway buffer are indicated as such.

FID	HWY BUFFER CODE
0	Inside

The polygon boundaries for the source layers are overlaid on top of each other in a single layer. In the diagram, the outlines of the slope polygons are in brown, the land-cover polygons in green, the stream buffer polygons in blue, and the highway buffer polygons in red.

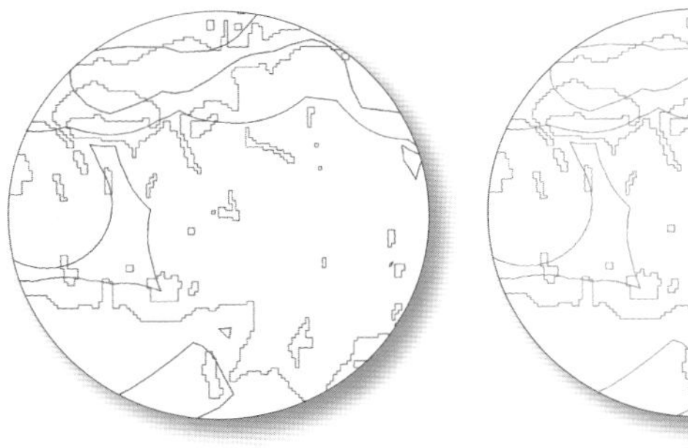

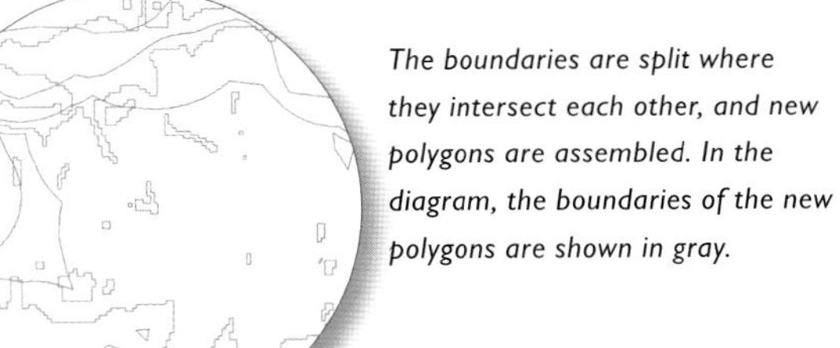

The boundaries are split where they intersect each other, and new polygons are assembled. In the diagram, the boundaries of the new polygons are shown in gray.

The new polygons have the attributes of the polygons they were constructed from. The polygons are shown here color-coded by unique combinations of site characteristics. The purple polygons, for example, are on steep slopes, have land cover of Douglas fir - hemlock, are outside the stream buffer, and outside the highway buffer.

FID	SLOPE CODE	LANDCOVER CODE	STREAM BUFFER CODE	HWY BUFFER CODE
1255	Gentle	55	Outside	Outside
1256	Steep	4	Outside	Outside
1257	Steep	54	Outside	Outside
1258	Steep	55	Outside	Outside
1259	Steep	65	Outside	Outside

As a result of the overlay operation, you'll likely end up with very small polygons—called slivers—where the borders of polygons from different source layers are almost coincident. To avoid selecting and including these small areas in your results, you'll want to resolve any slivers at this point. Usually this is done by selecting polygons below a minimum size and merging them with neighboring polygons. The minimum size depends on your application and the use you're modeling. For example, if a minimum area that would comprise viable mountain lion habitat is twenty acres (that is, an isolated patch of suitable habitat less than 20 acres is unlikely to be inhabited by any lions), you'd merge any polygons smaller than this.

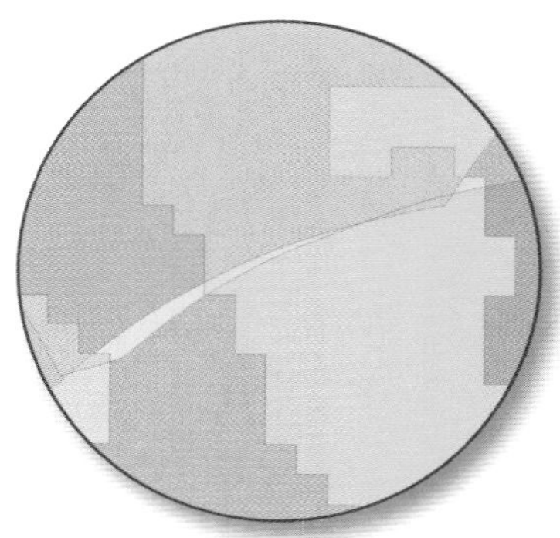

Slivers (small polygons) are created where polygon boundaries from different source layers are slightly offset.

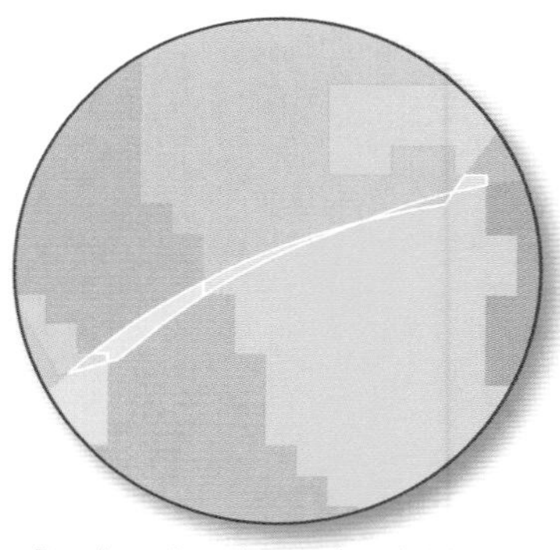

Five slivers have been selected (white borders).

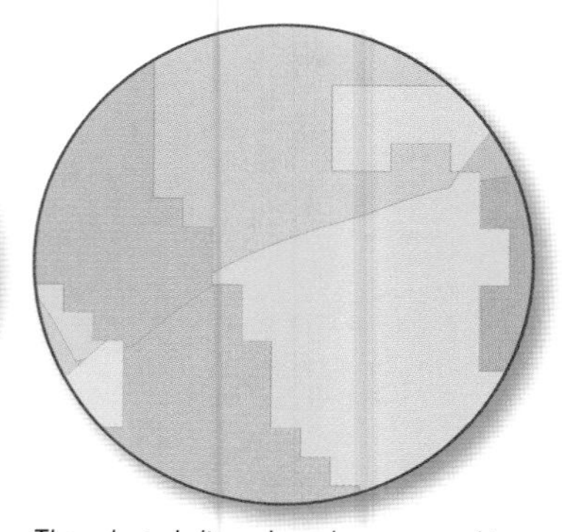

The selected slivers have been merged into adjacent larger polygons.

Once you've created the output layer, you select the polygons that meet your criteria. To do this, you create a logical selection (as discussed in the previous section, "Finding suitable locations using selection.") For the mountain lion habitat model, for example, the selection statement would look like this:

SLOPE CODE = 1 AND (LANDCOVER CODE = 54 OR LANDCOVER CODE = 55 OR LANDCOVER CODE = 65) AND STREAM BUFFER CODE = INSIDE AND HWY BUFFER CODE = OUTSIDE

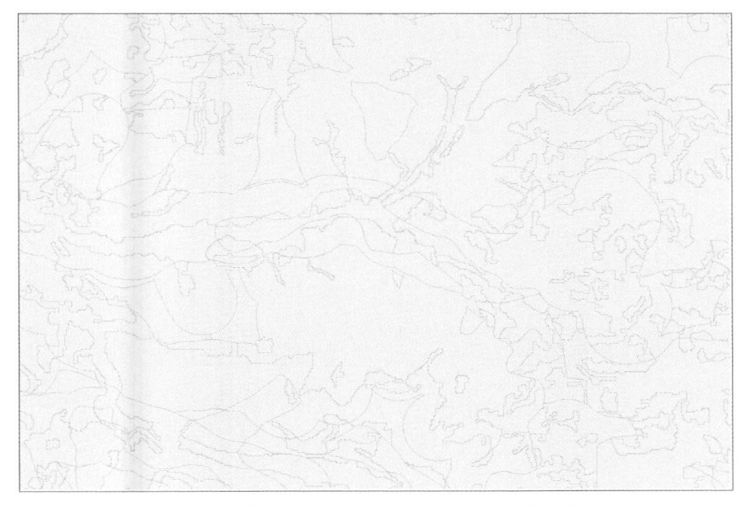

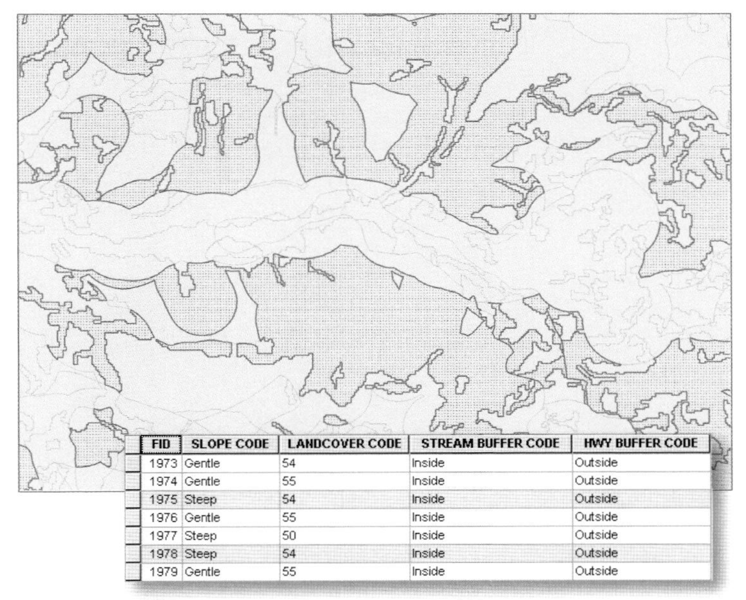

FID	SLOPE CODE	LANDCOVER CODE	STREAM BUFFER CODE	HWY BUFFER CODE
1973	Gentle	54	Inside	Outside
1974	Gentle	55	Inside	Outside
1975	Steep	54	Inside	Outside
1976	Gentle	55	Inside	Outside
1977	Steep	50	Inside	Outside
1978	Steep	54	Inside	Outside
1979	Gentle	55	Inside	Outside

The map above shows the polygons resulting from the overlay of the four source layers. On the right, selected polygons are highlighted in the map and in the associated attribute table.

Using the raster overlay method

The raster overlay method results in a single layer of suitable and unsuitable cells—known as a suitability layer—rather than a vector layer with many polygons and an attribute table containing the attribute values for each polygon.

There are a couple of ways to create a raster suitability layer. One approach is to query the individual source layers to create the suitability layer.

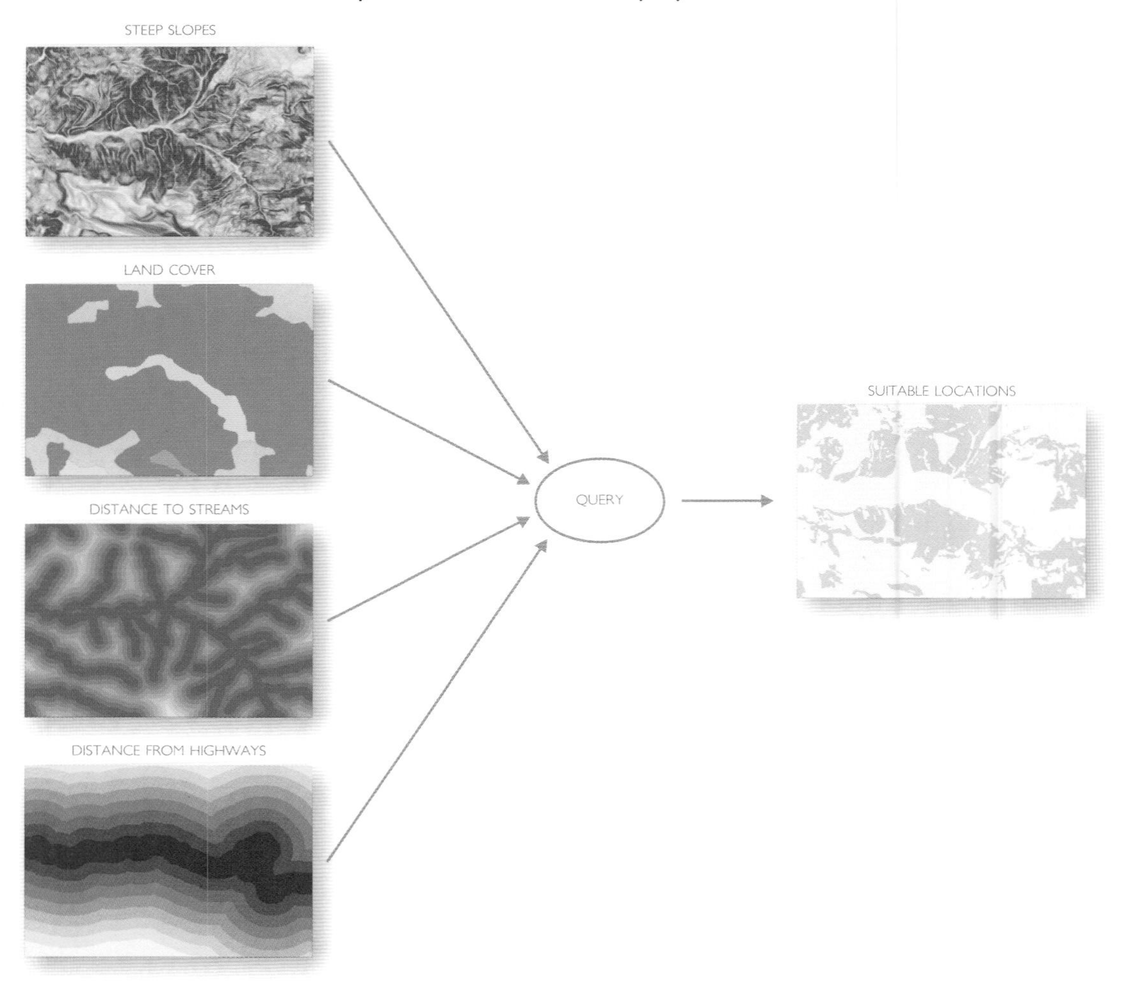

You can do this by creating a single query statement or by performing a series of sequential queries. The query essentially filters a source layer into cells that meet the criteria and those that don't. So, for a mountain lion habitat model, your queries may look like this:

SLOPE > 18

LANDCOVER CODE = 54 OR LANDCOVER CODE = 55 OR LANDCOVER CODE = 65

STREAM DIST < 2500

HWY DIST > 2500

If you do sequential queries, the layer created by the first query is used in the second query, and so on.

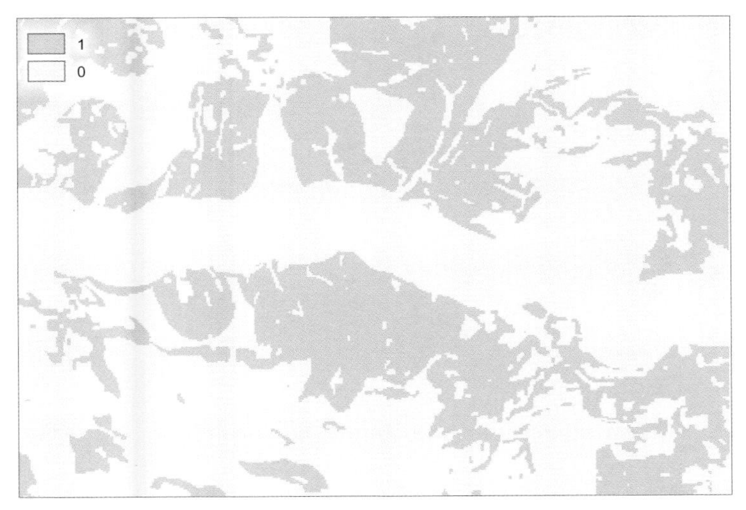

The source layers have been queried to create the suitability layer with two values—1 for cells meeting all the criteria (suitable habitat) and 0 for all other cells.

The advantage of this method is that your model can be flexible—if you want to modify your criteria by using a different threshold value for slope or different land-cover categories, you simply modify the query statements.

Another approach is to assign new values to each source layer to create a new layer with two values—cells that meet the criteria are assigned one value and cells that don't are assigned another. (Because they consist of only two values, indicating suitable and unsuitable cells, these are termed binary suitability layers.) You then combine the binary suitability layers into a single layer and identify the cells that were defined as suitable on all the binary layers (and thus are suitable locations).

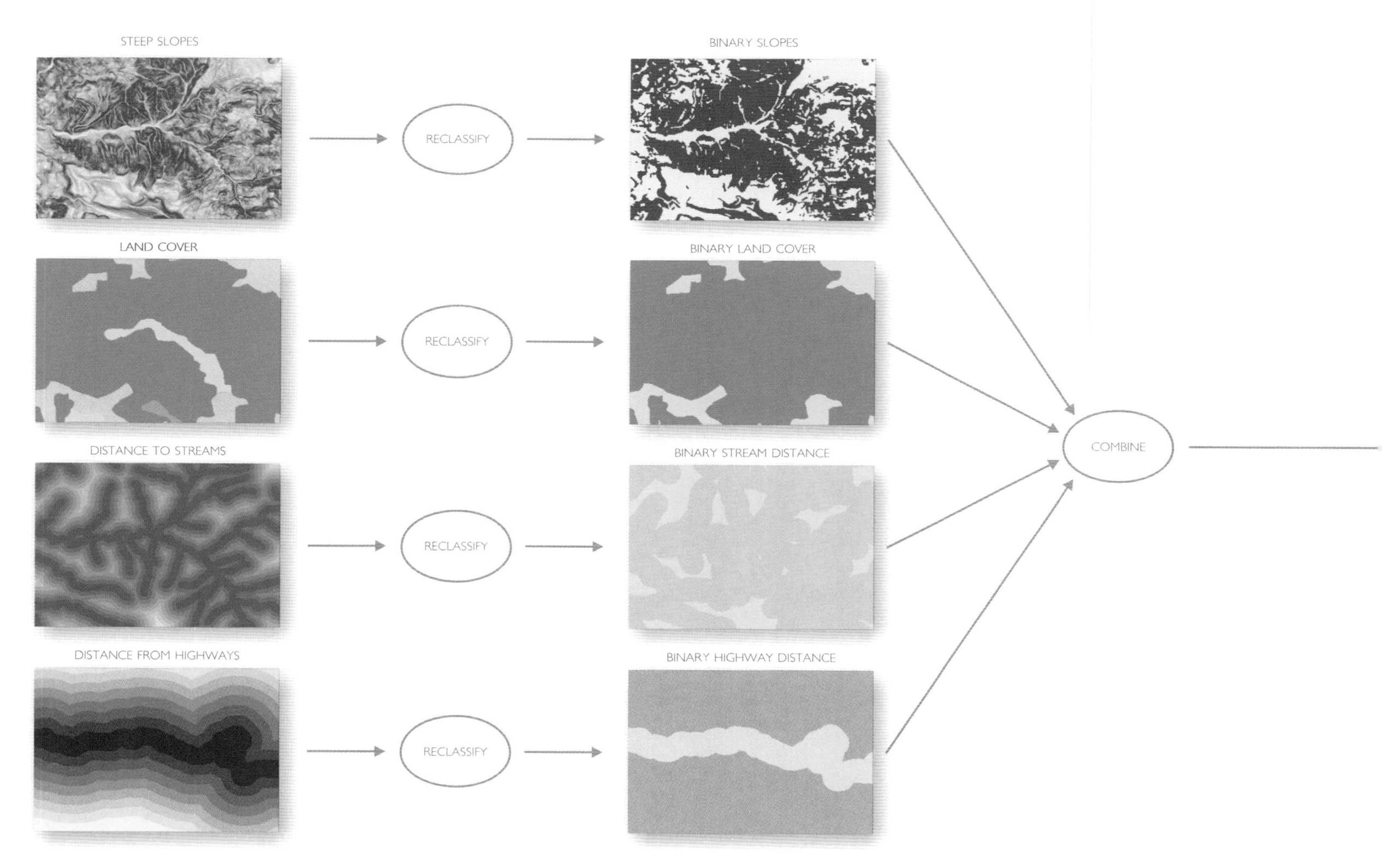

An advantage of this method is that you can determine which criteria each cell (or location) met and which it didn't. A disadvantage is that if you want to modify your model by changing the criteria values that are considered suitable, you'd have to re-create the binary suitability layers using the new values and recombine them.

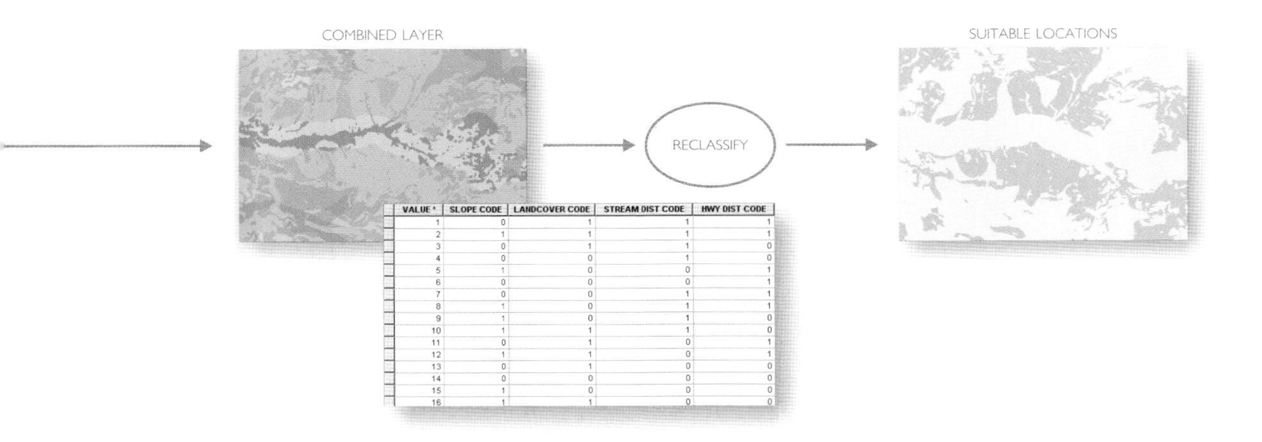

VALUE *	SLOPE CODE	LANDCOVER CODE	STREAM DIST CODE	HWY DIST CODE
1	0	1	1	1
2	1	1	1	1
3	0	1	1	0
4	0	0	1	0
5	1	0	0	1
6	0	0	0	1
7	0	0	1	1
8	1	0	1	1
9	1	0	1	0
10	1	1	1	0
11	0	1	0	1
12	1	1	0	1
13	0	1	0	0
14	0	0	0	0
15	1	0	0	0
16	1	1	0	0

When creating the binary suitability layers, values of 1 and 0 are often used by convention to identify cells that meet the criteria and those that don't. For example, if slopes greater than 18° are suitable, you'd assign cells with values of 19° or greater a value of 1. Cells with a value of 18° or less would be assigned a value of 0.

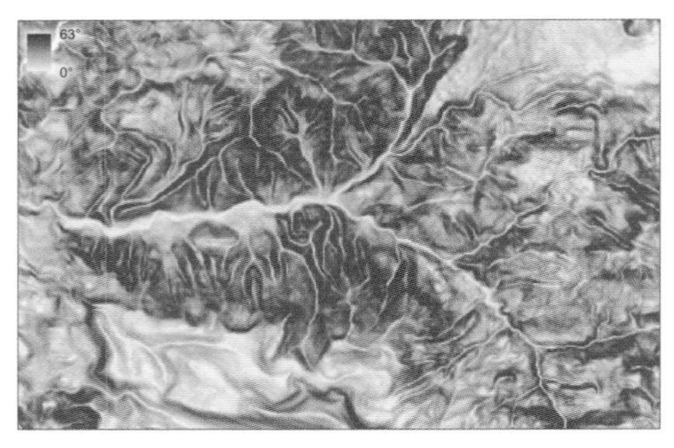

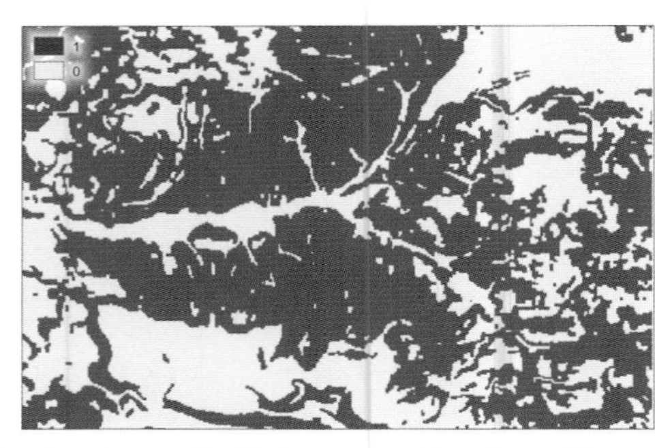

Slope as a continuous raster (left). On the right, the slope values have been reclassified to create a binary layer with values of 1 (slopes greater than 18°) and 0 (slopes of 18° or less). Cells with a value of 1 are suitable.

For a source layer of distance from streams, with areas within 2,500 feet of a stream being suitable, you'd create a binary stream distance layer by assigning a value of 1 to cells with a value less than or equal to 2,500, and 0 to all cells with a value greater than 2,500.

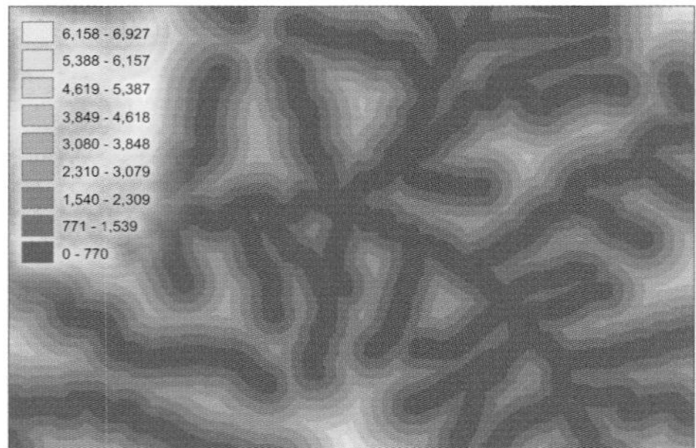

A layer of stream features was used to create a raster of distance from streams (center). The distance raster was then reclassified into areas less than 2,500 feet from streams (1) and greater than 2,500 feet (0).

If areas more than 2,500 feet from a highway are suitable, you'd create a binary highway distance layer by assigning a value of 1 to cells with a value greater than or equal to 2,500, and 0 to all cells with a value less than 2,500.

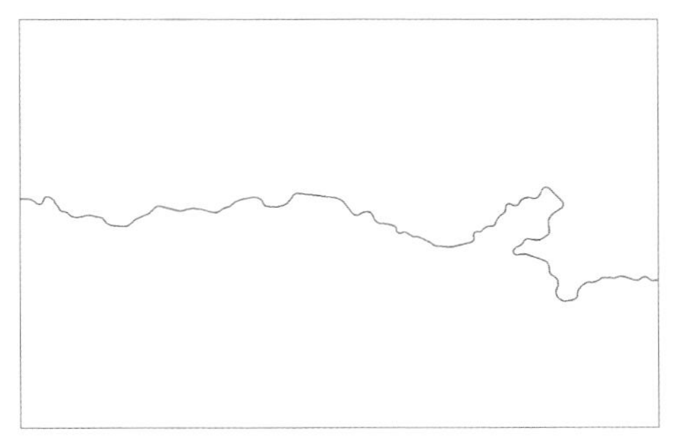

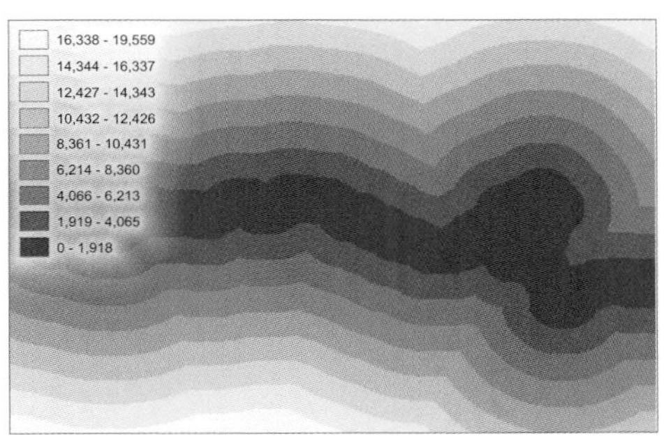

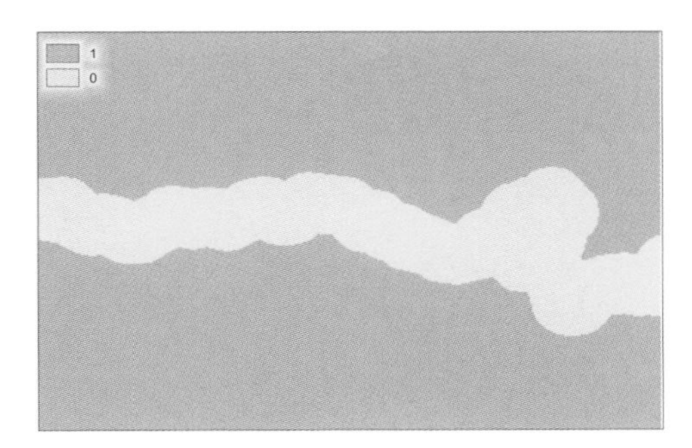

The highway layer was used to create a distance raster (center). The distance raster was then reclassified into areas more than 2,500 feet from the highway (1) and less than 2,500 feet (0).

As mentioned earlier, an alternative to creating a distance raster is to buffer the linear features (say streams or highways) using the required distance and then convert the buffer polygon to a raster layer. You'd then create the binary suitability layer with values of 1 and 0. The advantage of creating a distance raster, though, is that you can easily change your criteria by creating a new binary suitability layer using a different threshold distance. You avoid having to go through the process of creating a new buffer at the new distance, converting it to raster data format, and then creating the new binary suitability layer.

For categorical data, such as land cover, suitable categories are assigned a value of 1, and all other categories get a value of 0. For example, if the suitable land-cover values are mixed forest, Douglas fir - hemlock, and mountain hemlock, you'd create a binary suitability layer with a value of 1 for cells having those land-cover values and a value of 0 for all other cells.

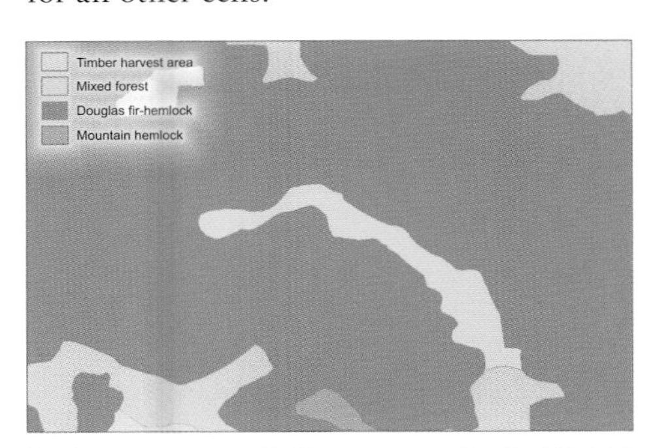

Land-cover categories (left) have been reclassified (right) into suitable habitat (1) and unsuitable (0).

Once you've created the binary suitability layers, you combine them. A common way of doing this is to create an output layer where each unique combination of values is assigned a specific value by the GIS—any cell with a particular combination is assigned that value. The attribute table for the output layer shows the combinations of values from the binary suitability layers and the assigned output value.

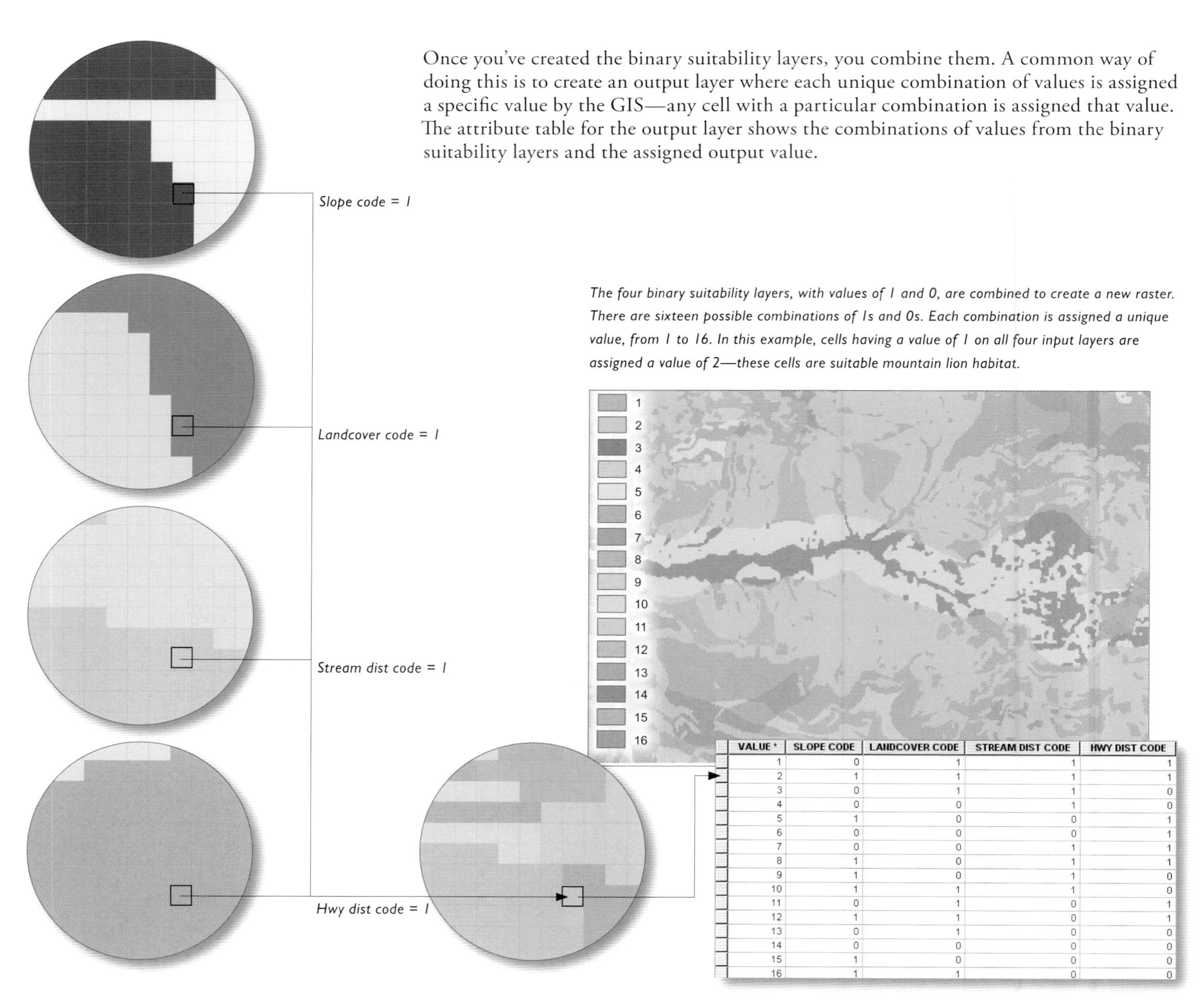

The four binary suitability layers, with values of 1 and 0, are combined to create a new raster. There are sixteen possible combinations of 1s and 0s. Each combination is assigned a unique value, from 1 to 16. In this example, cells having a value of 1 on all four input layers are assigned a value of 2—these cells are suitable mountain lion habitat.

VALUE *	SLOPE CODE	LANDCOVER CODE	STREAM DIST CODE	HWY DIST CODE
1	0	1	1	1
2	1	1	1	1
3	0	1	1	0
4	0	0	1	0
5	1	0	0	1
6	0	0	0	1
7	0	0	1	1
8	1	0	1	1
9	1	0	1	0
10	1	1	1	0
11	0	1	0	1
12	1	1	0	1
13	0	1	0	0
14	0	0	0	0
15	1	0	0	0
16	1	1	0	0

To identify the cells that are suitable—that is, meet all the criteria—you select the cells that have a value of 1 for all input criteria layers, using a logical selection statement:

SLOPE CODE = 1 AND LANDCOVER CODE = 1 AND STREAM DIST CODE = 1 AND HWY DIST CODE = 1

This will show you the value assigned to suitable cells. (In this example there are only four input layers, so there are sixteen possible combinations of 1s and 0s and sixteen values in the combination layer. It is relatively easy to look through the attribute table and see which value is assigned to the combination of all 1s. But with six input layers, for example, there are sixty-four possible combinations—using the logical selection would be quicker than looking through the table to find the value.)

You then create the overall suitability layer by assigning new values to the combination layer to show suitable and unsuitable cells (again, values of 1 and 0 are commonly used). Cells having a value corresponding to the combination of all 1s (a value of 2, in this example) are assigned 1, and all other cells are assigned 0.

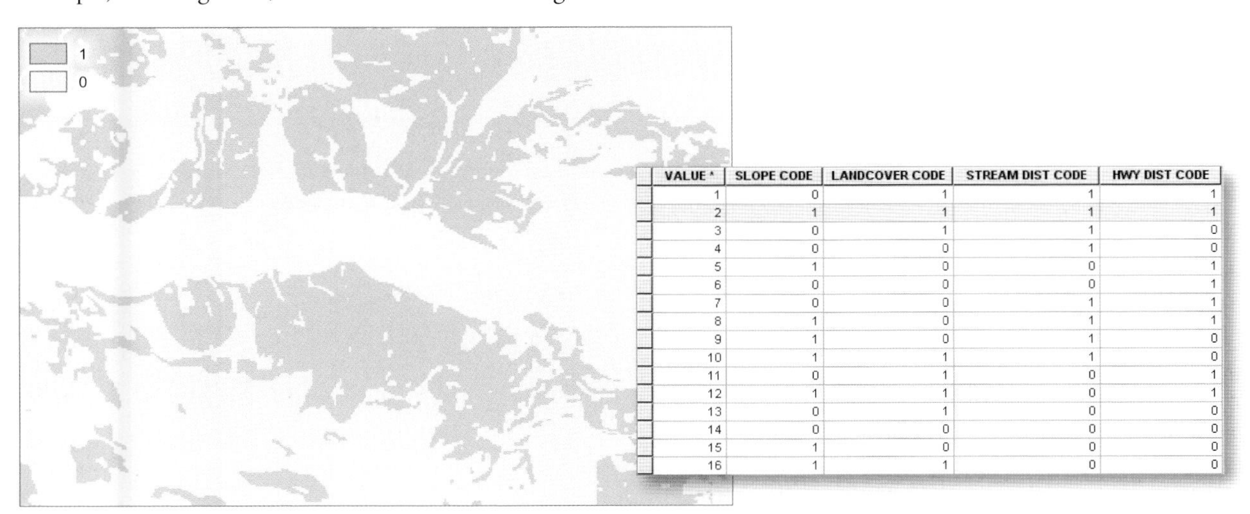

VALUE *	SLOPE CODE	LANDCOVER CODE	STREAM DIST CODE	HWY DIST CODE
1	0	1	1	1
2	1	1	1	1
3	0	1	1	0
4	0	0	1	0
5	1	0	0	1
6	0	0	0	1
7	0	0	1	1
8	1	0	1	1
9	1	0	1	0
10	1	1	1	0
11	0	1	0	1
12	1	1	0	1
13	0	1	0	0
14	0	0	0	0
15	1	0	0	0
16	1	1	0	0

The output combination layer has been reclassified to create the overall suitability layer with two values—1 for cells with a value of 2 on the combination layer (suitable habitat) and 0 for all other cells.

Since you know which criteria each cell met and which it didn't, you can easily relax your criteria or explore various combinations of criteria. Additional research might show, for example, that distance from highways is not a critical factor. You'd select and display the cells having values of 1 or 0 for the distance from highways criterion and a value of 1 for the other criteria.

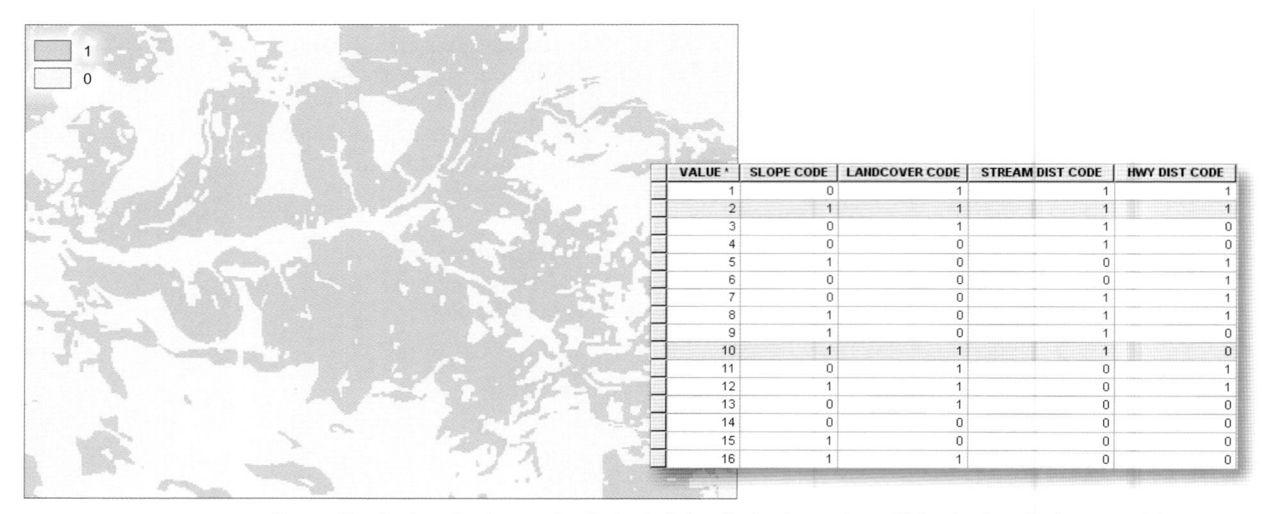

VALUE *	SLOPE CODE	LANDCOVER CODE	STREAM DIST CODE	HWY DIST CODE
1	0	1	1	1
2	1	1	1	1
3	0	1	1	0
4	0	0	1	0
5	1	0	0	1
6	0	0	0	1
7	0	0	1	1
8	1	0	1	1
9	1	0	1	0
10	1	1	1	0
11	0	1	0	1
12	1	1	0	1
13	0	1	0	0
14	0	0	0	0
15	1	0	0	0
16	1	1	0	0

The combination layer has been reclassified to include cells that have values of 1 for the slope, land cover, and distance from streams criteria layers. This corresponds to cells with values of 2 or 10 on the combination layer—these cells have been assigned a value of 1 (suitable).

EVALUATE THE RESULTS

Once you've identified the suitable locations, you'll want to verify the results of your model. If you used the vector overlay method, identify and list the attributes for at least some of the suitable polygons to make sure your selection was successful. The features should have attribute values that match your criteria.

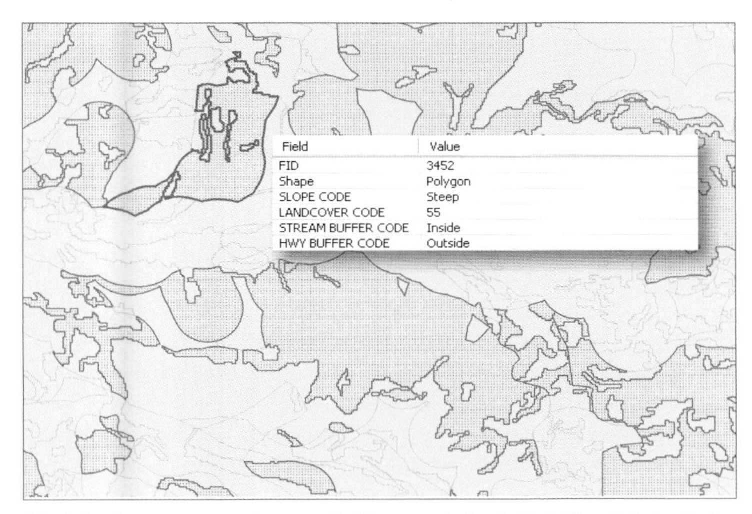

Stippled polygons are areas that are suitable mountain lion habitat. The attributes for the highlighted polygon (dark gray border) show that it does, in fact, meet all the criteria.

You can also display the selected polygons with the various source layers to ensure the selected areas do in fact reflect the attributes for that piece of land. It will be easier to do this if you save the selected polygons as a layer—the layer will contain only those polygons representing suitable areas. You'll want to look for suitable polygons that overlap unsuitable areas on the source layers. These may or may not be an issue for your analysis, but you will at least want to investigate them.

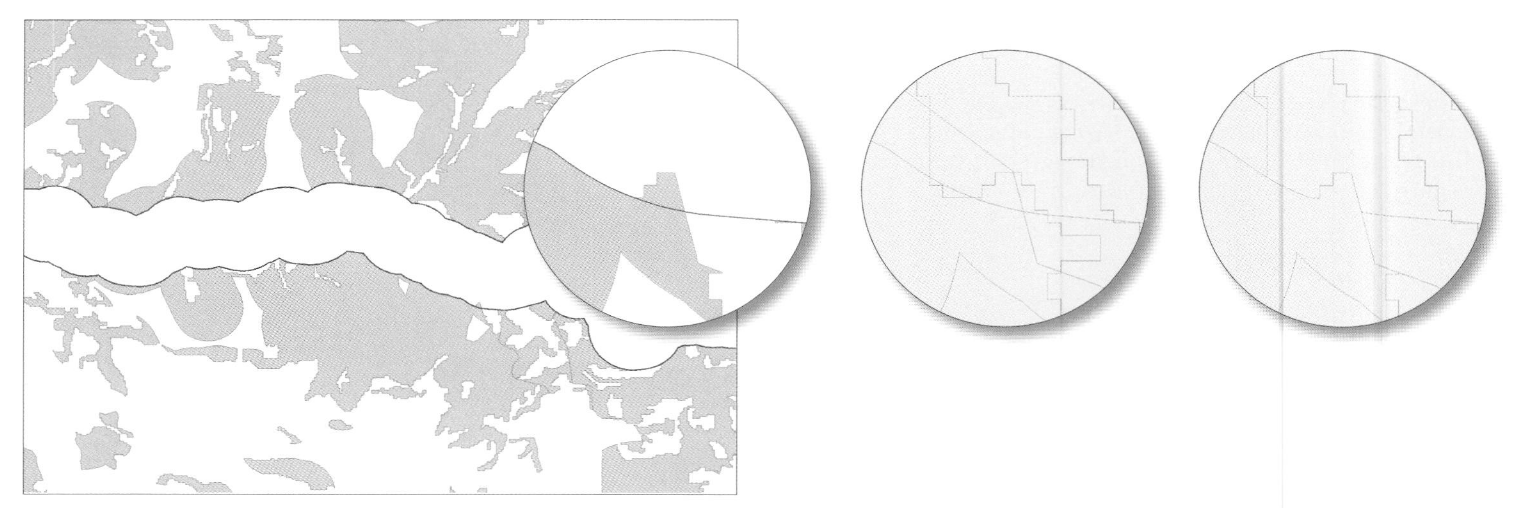

The map shows polygons representing suitable areas and the highway buffer. The boundaries of the buffer are displayed so you can see if it overlaps the polygons. All suitable areas should be outside the buffer. The close-up on the left shows that one polygon crosses into the highway buffer. This particular case turns out to be a legacy of merging polygons less than twenty acres in size into larger adjacent polygons to remove slivers. The middle close-up shows the polygon overlay layer before polygons were merged—the polygon that is in the highway buffer was originally a separate small polygon (about 2.5 acres) that did indeed fall inside the highway buffer. When it was merged with the adjacent larger polygon to the south (close-up on right), that area took on the characteristics of the adjacent polygon, including being characterized as being outside the highway buffer (even though it is in fact slightly inside the buffer). In this case you'd likely conclude that since the area is contiguous with the large suitable area outside the buffer, this is not an issue for your analysis.

If you used the raster overlay method, compare the suitable cells to the source or binary suitability layers to ensure you identified and reclassified the suitable areas correctly. There are several ways to do this, such as displaying the overall suitability layer on top of each binary suitability layer in turn using a transparency setting. Or you can convert the raster areas to polygons and display them with the binary suitability layers (as shown on the next page).

You'll also want to ensure that the results correspond with your knowledge of the study area and have the results reviewed by other experts. The model is an attempt to quantify what is often knowledge gained through experience in the field. You may need to modify the model, based on the initial results, to better capture this knowledge. You can do this by changing the threshold values of the criteria or by adding or eliminating source layers.

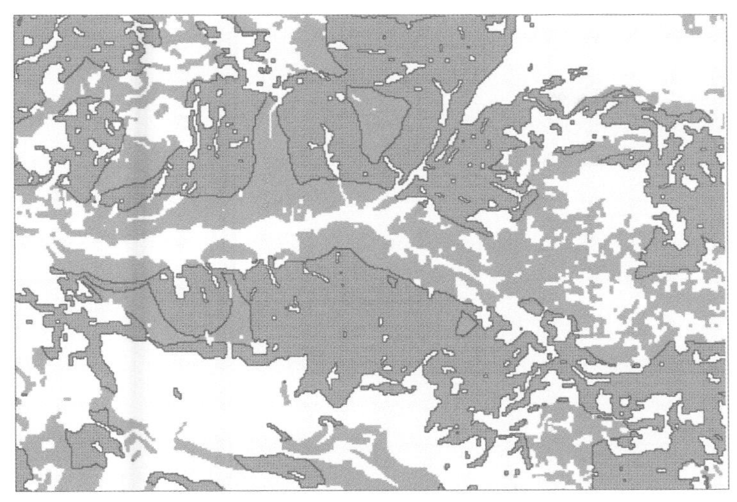

Stippled polygons are areas that are suitable mountain lion habitat. Suitable areas are on steep slopes (brown).

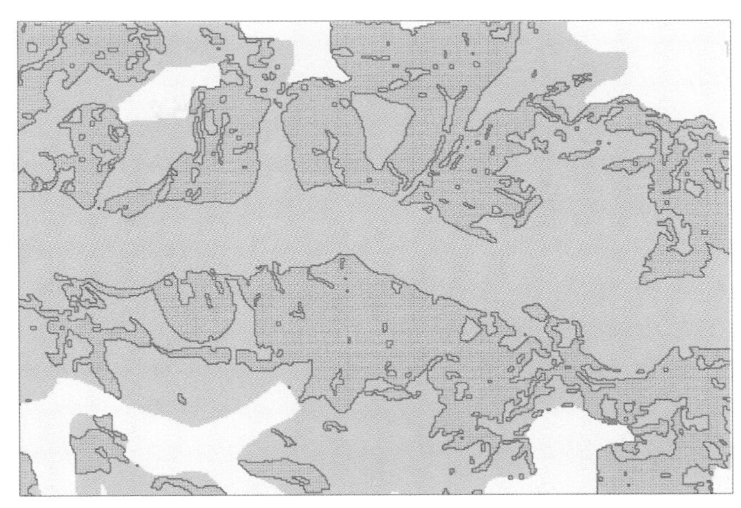

Suitable areas are forested (darker green).

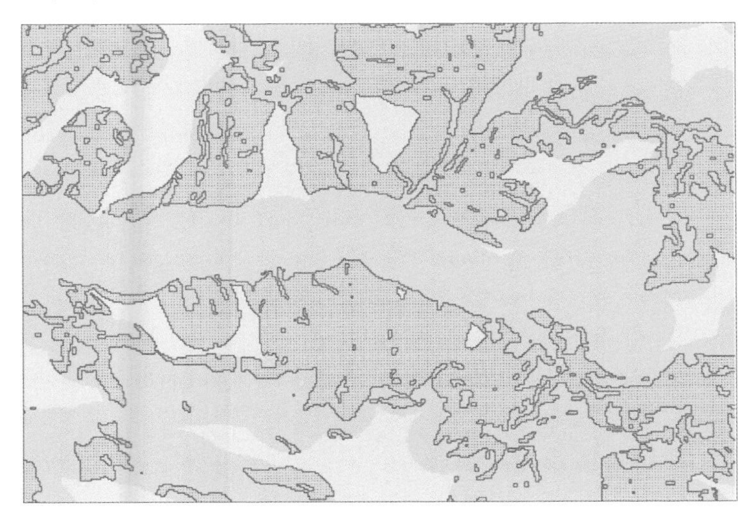

Suitable areas are within 2,500 feet of a stream (darker blue).

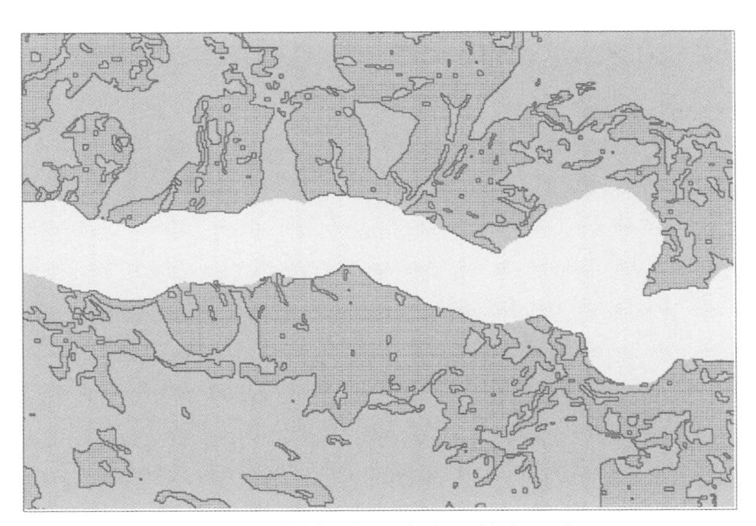

Suitable areas are more than 2,500 feet from a highway (darker red).

DISPLAY AND APPLY THE RESULTS

Once you've evaluated and verified the results of your model, you'll likely want to present the results to others—researchers, decision makers, or the general public.

It may be useful to generalize the overall suitability layer to simplify the final map and make the information easier to read and interpret. (Of course, you may lose some detail in the process.) As when eliminating slivers, you may want to remove other small areas that are not of consequence to your analysis. For a polygon suitability layer, you'd remove these polygons by selecting the target polygons and merging them with an adjacent polygon. For a raster suitability layer, you remove small areas using raster generalization tools or by converting the raster layer to a polygon layer and selecting and merging the target polygons.

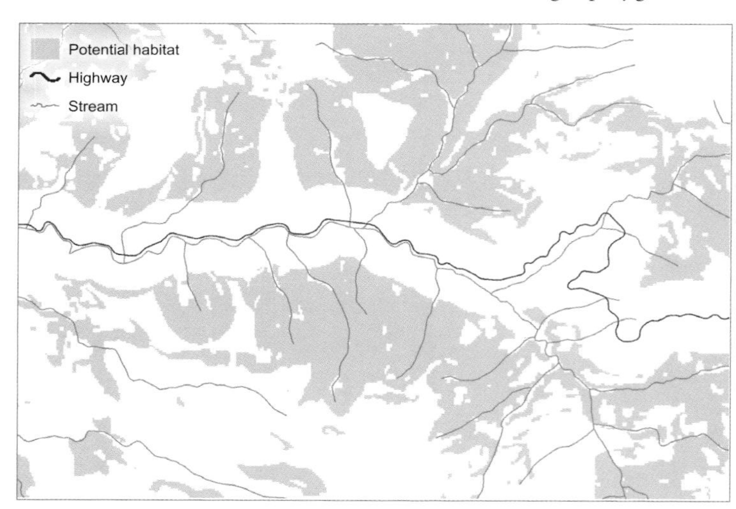

Original raster suitability layer (with highways and streams).

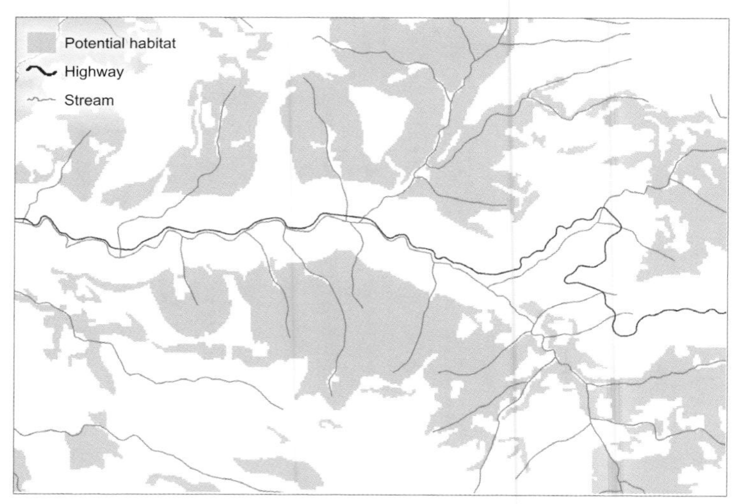

The raster suitability layer has been converted to polygons, and polygons less than twenty acres in size have been merged with surrounding larger polygons. The suitability layer is more continuous (with holes filled in) and fewer small stand-alone polygons, resulting in a map that is less cluttered and easier to read.

When using the vector overlay method, you may end up with adjacent polygons that are suitable. This can happen if your criteria include several categories for the same attribute, such as two or more suitable land-cover types. If it's important to know, for example, which land-cover type comprises each habitat area, you can map the adjacent sites color-coded by land cover.

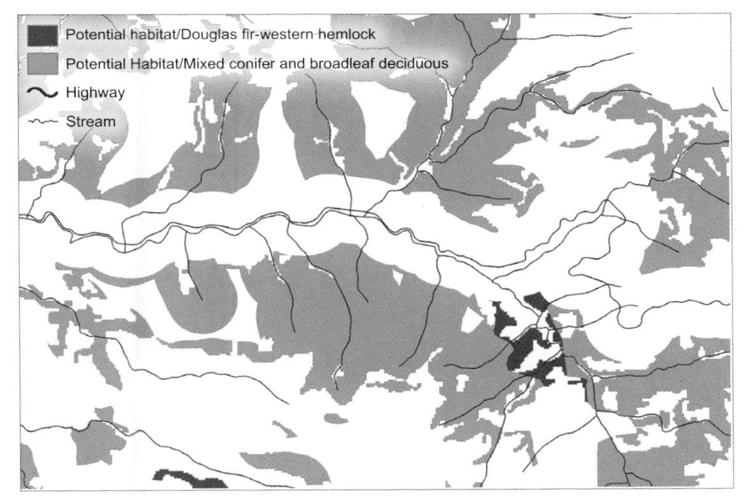

Potential mountain lion habitat color-coded by land-cover type—some of the light and dark brown areas are adjacent.

If distinguishing between these categories is not important, dissolving the boundaries between adjacent polygons to create a single contiguous area will simplify the final map. This will also let you easily obtain the areal extent of each contiguous suitable area.

To use the results of your model, you'll want to display the suitable areas with other layers so you, and others, can explore patterns and relationships, draw conclusions, or consider additional research. Which layers you present—and how you present them—depends on your audience. For a mountain lion habitat model, scientists and researchers will want to see the source layers and criteria values that went into the model. The public, on the other hand, may be more interested in seeing features—such as roads, rivers, towns, and so on—that help them locate the habitat areas.

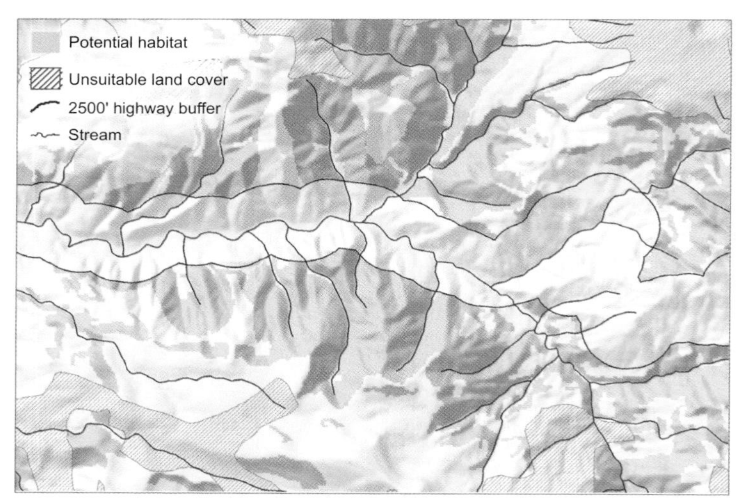

Map showing criteria for identifying potential mountain lion habitat. Researchers can see that the habitat areas are outside the highway buffer, avoid unsuitable land cover, and are near streams. The shaded relief shows that habitat areas occur on steep slopes.

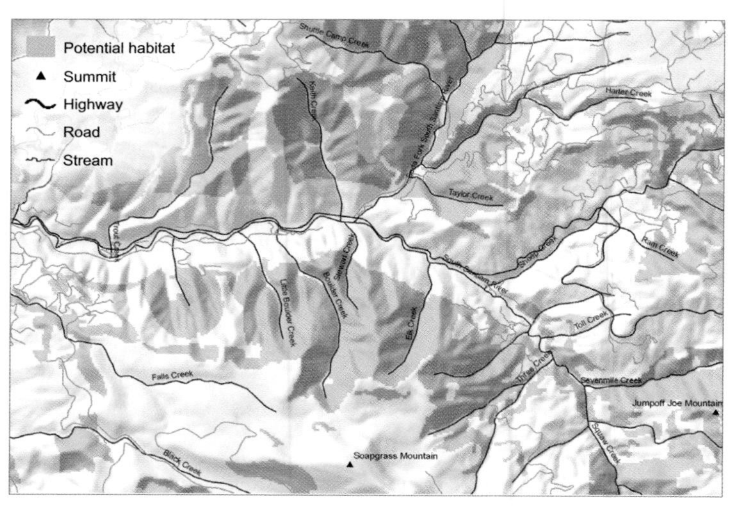

Map showing potential mountain lion habitat with landmarks such as mountain peaks, streams, highways, and roads.

Decision makers will want to see related information in conjunction with the suitability layer. For a mountain lion habitat model, this might be landownership or population density. In addition to creating maps, you can use the results of the model to generate summary information that will be useful for decision makers or researchers. For example, you might calculate statistics such as the total amount of suitable habitat in the study area or the amount of suitable habitat in each landownership category.

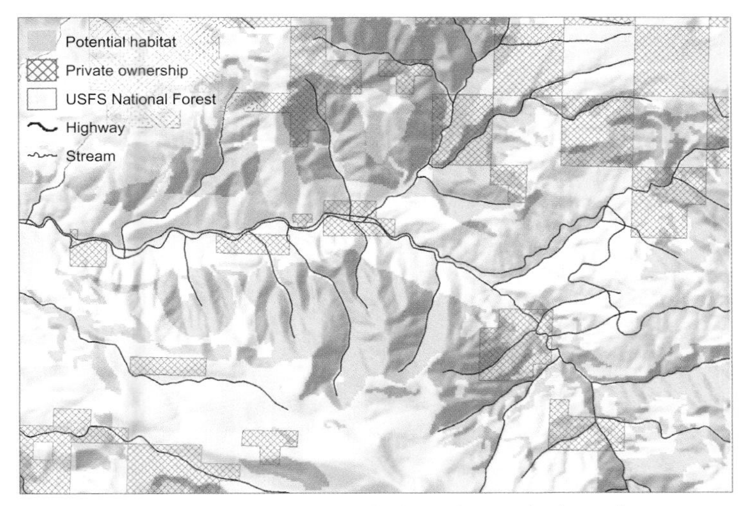

Potential mountain lion habitat with areas of public and private landownership.

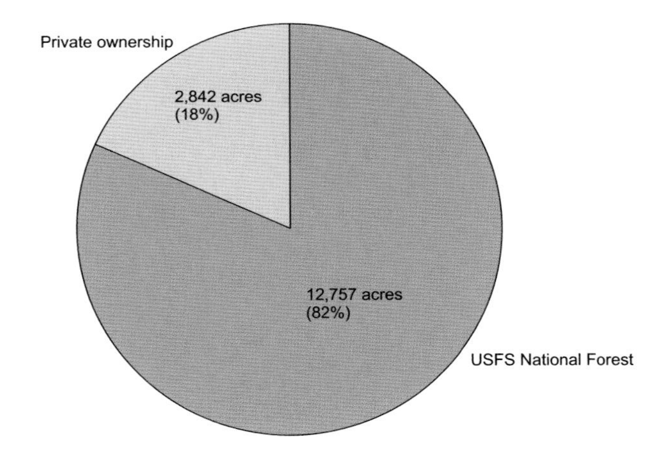

references and further reading

Mitchell, Andy. 1999. *The Esri Guide to GIS Analysis, Volume 1: Geographic patterns and relationships*. Redlands, CA: Esri Press. The basics of feature selection and polygon overlay, among other topics.

Tomlin, C. Dana. 1990. *Geographic Information Systems and Cartographic Modeling.* Englewood Cliffs, NJ: Prentice Hall. Classic work on raster analysis. Tomlin also describes the approach to modeling map information.

Rating suitable locations 3

You can use GIS to rate locations that are suitable for a particular use such as a housing development or mule deer habitat. These methods assign a relative value to each location to indicate how suitable it is for the particular use.

As with the Boolean suitability methods discussed in chapter 2, you create a model that combines various layers of information about a place. You then identify the locations within the study area that have the characteristics you're seeking. The Boolean methods show you which sites are suitable (those that meet all of your criteria) and which aren't. In contrast, rating locations based on their relative suitability gives you more flexibility in comparing locations as well as in considering alternatives and trade-offs when making decisions. As you'll see, however, the models used to rate locations require additional design and processing as compared to Boolean suitability models.

Rating locations is useful when you're looking for a site to place something. For example, when looking for a site for a new housing development, you might want to look at a range of locations—ones that are very suitable, somewhat suitable, and not suitable. A very suitable location might cost less to build on (if it is on flatter slopes, for example) and allow you to set higher home prices (if it is closer to stores and restaurants) but might have a high purchase price. Another slightly less suitable location might cost a little more to build on and fetch lower home prices but cost much less to purchase. A model that rates locations allows you to consider the trade-offs between the cost of land and the cost of construction.

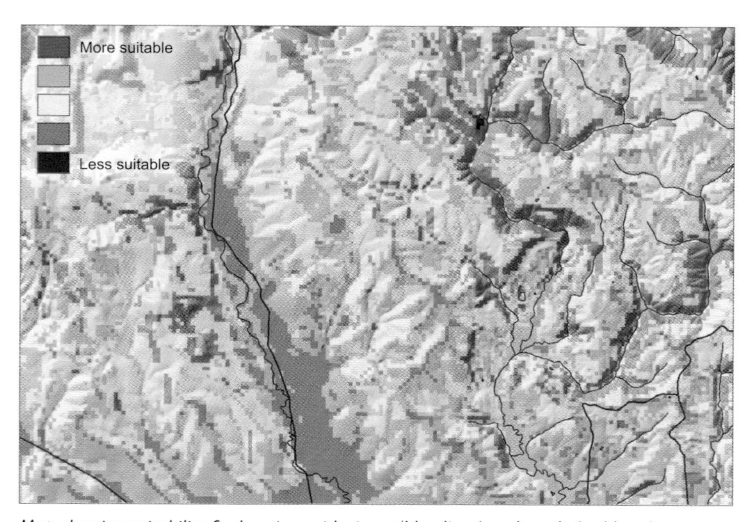

Map showing suitability for housing, with rivers (blue lines) and roads (red lines).

You'd also use a model that rates locations when trying to identify where something will be found—for example, you're studying mule deer and you want to find where they're likely to congregate in your study area. The deer will likely be in the most suitable areas but a few might also be found at any given time in slightly less suitable areas. If you used a Boolean suitability model you might exclude some areas that would be acceptable (if not prime) mule deer habitat. By using a model that rates locations you can not only identify areas where the deer are most likely to congregate, but also other areas where they will possibly be (and which may be more accessible).

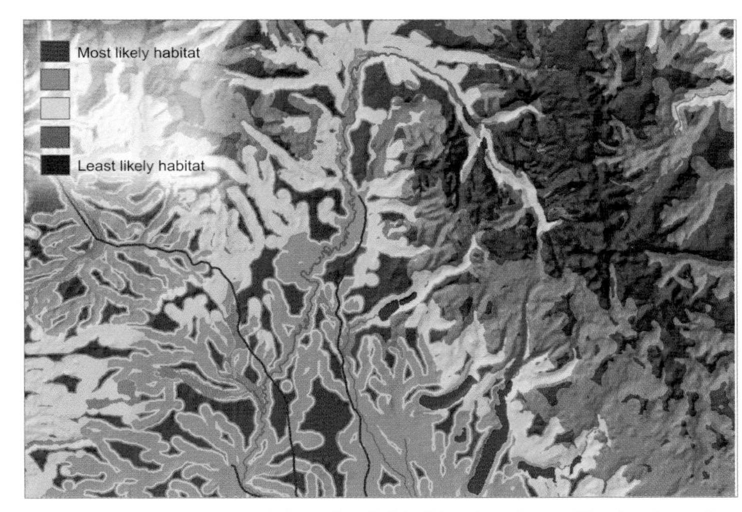

Map showing likely mule deer habitat. Roads (black lines) and rivers (blue lines) are also shown.

As with Boolean suitability models, suitability models that rate locations have a wide range of applications, from assessing locations at risk from an event such as a wildfire to finding desirable locations, such as finding the best cities in a region to host a convention. Indeed, these models can be used for any application that consists of finding and rating locations that meet a set of criteria that you define.

To design a model to rate suitable locations, you first define the problem you're trying to solve and determine what results you need from the model. You then consider the criteria that define what makes a location more or less suitable. Based on these decisions, you choose an appropriate modeling method.

DEFINE THE PROBLEM

As with any suitability model, you first state the problem you wish to solve or the question you wish to answer. For example,

"I want to find potential sites for a housing development in this county, and rate the sites based on how suitable they are."

"Where in this watershed is the most and least likely mule deer habitat?"

You'll also want to consider the results you want from your model. For example, you may want a rating of locations showing the best, second best, third best, and so on. This would be the case if you're looking for a site to build something, such as a housing development.

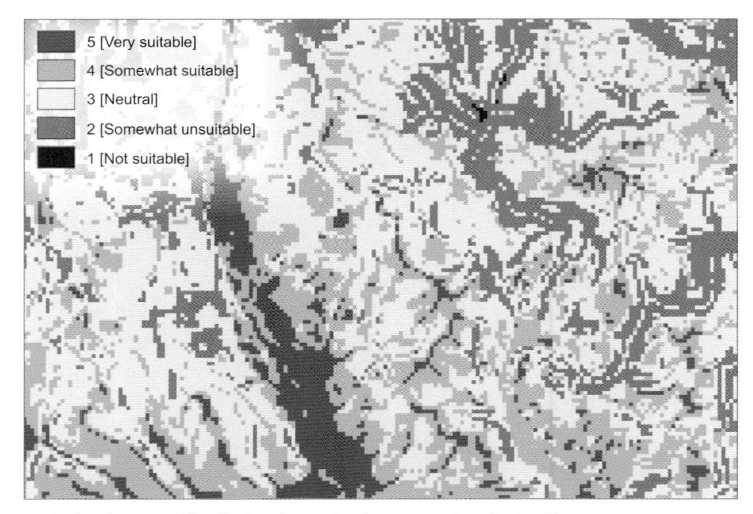

Map showing suitability for housing, using five categories of suitability.

Or, you may want to rate the locations on a continuous scale from most suitable to least suitable. This would be the case if you're trying to identify the most and least likely locations for something—such as mule deer habitat—that occurs across the landscape.

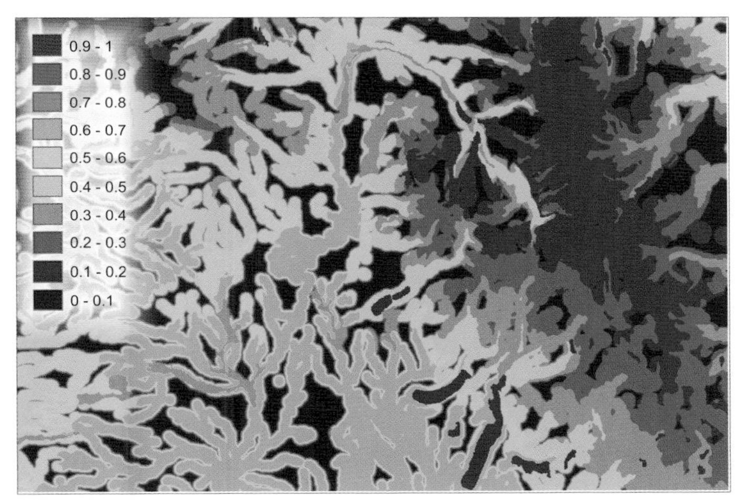

Map showing likely mule deer habitat on a continuous scale from 1 (very likely) to 0 (very unlikely).

DEFINE THE CRITERIA

Based on the problem definition, you list the specific criteria that make a location more or less suitable for the particular use. As described in chapter 2, "Finding suitable locations," criteria can pertain either to the characteristics of the site itself or to the proximity of the location to other features. Criteria can include physical characteristics of the location as well as economic, demographic, and cultural characteristics of the locations and their surroundings.

For example, the criteria for a housing development might include the following:

- Areas requiring less land clearing are more suitable
- Soils that support housing construction are more suitable
- Areas on flat slopes are more suitable
- Areas farther from streams are more suitable
- Areas near existing roads are more suitable

For a model of mule deer summer habitat, the criteria might include the following:

- Areas that provide cover are more suitable
- Areas near streams are more suitable
- Lower elevations are more suitable

Each criterion has a corresponding source data layer. When creating a model to rate suitable locations, you assign each value on a source layer a relative suitability value, based on your criteria. You need to consider how you will assign suitability values to source layer values. There are more possibilities than when using a Boolean suitability model because you're assigning relative suitability along a scale rather than just assigning an either/or value.

For source layers having categorical values, such as land-cover types, you'll assign each category a suitability value (the same value may be assigned to more than one category).

For source layers having continuous values, it may be that the criteria have well-defined value ranges (either by convention or regulation), each of which will be assigned a particular suitability value. For example, for a model of housing suitability, slopes of less than 7 percent may be considered very suitable, 7 percent to 15 percent somewhat suitable, 15 percent to 30 percent somewhat unsuitable, and slopes over 30 percent not at all suitable.

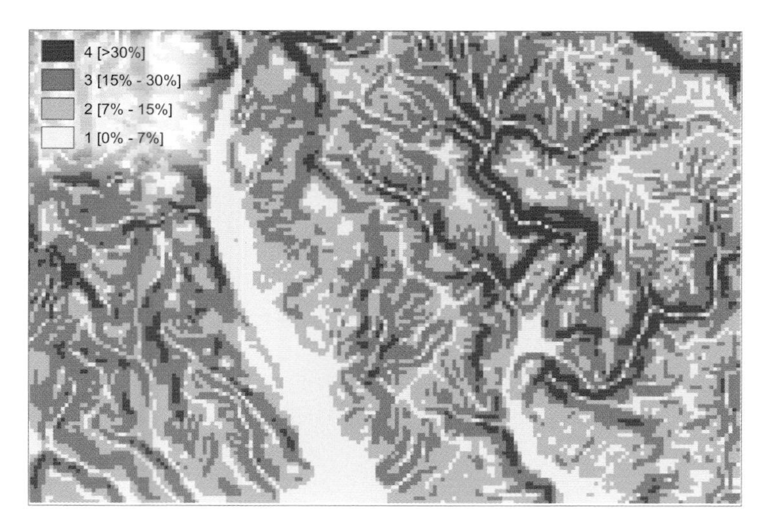

Slopes have been classified into four levels of suitability.

Alternatively, it may be that the criteria do not have defined value ranges. For example, for a mule deer habitat model, you may not be able to state that 0 to 2,300 meters elevation has a suitability value of 1, 2,300 meters to 3,100 meters a value of 2, and so on. You may only know that low elevations are more suitable than high elevations and that suitability decreases gradually as elevation increases. Thinking about the nature of the source layer values, and the relationship between these values and suitability, will help you choose the appropriate modeling method.

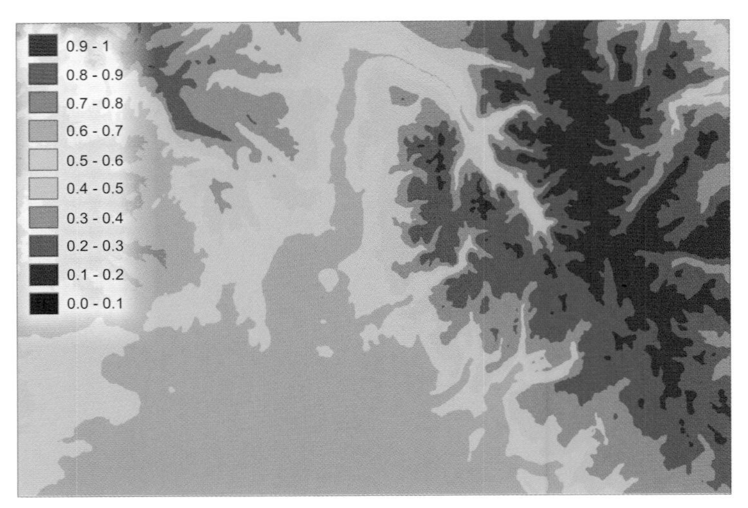

Elevation has been assigned suitability values on a continuous scale (from 1 to 0) with the lowest elevations (most suitable, in this example) assigned a value of 1 and the highest assigned a value of 0.

In addition to specifying the suitability of each source layer value, you can specify that certain criteria are more important than others, and thus carry more weight in the model. For example, in a model to rate locations for a housing development gentle slopes and proximity to roads may be more important than the land cover or soil type found at each location.

As described in chapter 1, there are a number of ways of determining the suitability values for the criteria as well as the relative importance of each criterion in the model, including your own experience, industry standards, and expert consensus.

CHOOSE A METHOD

Once you've considered the criteria you'll use in your model, you choose the appropriate modeling method. Two common methods for rating suitable locations are weighted overlay and fuzzy overlay.

With weighted overlay you assign a suitability value to each value (or range of values) in each source layer using a common scale that you specify, creating a new layer in the process (termed a suitability layer). The suitability layers are then overlaid and the suitability values summed. The result is a new layer with an overall suitability value for each location. This method allows you to assign more importance to some criteria than others, by assigning a weight to each suitability layer when the layers are overlaid—hence "weighted overlay." Weighted overlay is useful if you want to evaluate alternative scenarios by changing the relative importance of the various criteria. The method is also useful when most or all of your criteria have distinct break points between suitability values, as defined by regulations, industry standards, or accepted research.

Fuzzy overlay involves defining the relationship between source layer values and suitability on a continuous scale of 1 (very suitable) to 0 (not suitable). You specify a mathematical function to assign suitability values to source layer values. You then combine the layers to assign an overall suitability value to each location. This method is useful when the relationship between specific criteria and suitability is not well-defined. The method also gives you several options for combining layers using logical or mathematical operators (as discussed later in this chapter). Because of the flexibility it provides in assigning suitability values and in combining layers, fuzzy overlay is particularly good for creating a suitability model that attempts to capture the knowledge of experts in a particular field (if, for example, the experts are able to describe, but not necessarily quantify, the relationship between source layer values and suitability).

METHOD	WHAT IT'S GOOD FOR	HOW IT WORKS
Weighted overlay	*Rating suitable locations when the level of suitability corresponds to well-defined value ranges on source layers, and when you want to assign more importance to some criteria than others.*	*You specify suitability values for classes within each source layer using a scale of your choosing; you then assign weights to the resulting suitability layers and overlay the layers.*
Fuzzy overlay	*Rating suitable locations when criteria are hard to quantify and when you want flexibility in combining layers to capture expert opinion.*	*You specify a mathematical function (line or curve) that is used to assign suitability values to source layer values along a continuous scale of 0 to 1; you then combine the resulting layers using one or more logical or mathematical operators.*

These methods are discussed in more detail in the following sections. In fact, either method may be applicable for your analysis (and should give comparable results). The nature of your criteria may lead you to favor one method over the other.

Use weighted overlay to rate suitable locations if your criteria are defined by distinct categories and class ranges. The method is based on an approach to regional planning developed in the 1960s by landscape architect Ian McHarg. In McHarg's original method, acetate map sheets were used—each sheet represented a source layer, such as land cover, slope, soil type, and so on. Suitable areas on each sheet were left clear; unsuitable areas were shaded in dark tones (so, for example, gentle slopes that were suitable for construction of houses were clear; moderate slopes shaded in light gray, and slopes too steep to build on shaded dark gray). When the translucent sheets were registered (using pins) and overlaid on a light table, the lightest areas indicated the most suitable locations and the darkest the least suitable.

You implement McHarg's method in the GIS by mathematically overlaying the source layers corresponding to the criteria. You first assign each category or class in each layer a numeric value corresponding to how suitable it is for the proposed purpose. For example, in a model to identify suitable areas for a new housing development, using a scale of 1 (unsuitable) to 3 (very suitable), you'd assign a value of 1 to steep slopes, a value of 2 to moderate slopes, and a value of 3 to gentle slopes to create a slope suitability layer. You then combine the various suitability layers to assign each location an overall numeric value on the output layer indicating how suitable each is. One advantage to this mathematical overlay method is that you can assign more importance to some criteria than others by specifying a weight for each source layer.

You can use either polygon source layers or raster source layers with the weighted overlay method. When using polygon layers, the method is similar to the Boolean overlay method described in chapter 2, but involves assigning suitability values and weights to each polygon on each source layer. After the layers are combined you calculate the overall suitability value for each polygon on the output layer by performing calculations in the layer's attribute table.

The raster weighted overlay method is more commonly used. Raster data is well suited to mathematical overlay since the raster format uses cells that are coincident between layers, so the cell values can simply be summed to create the overall suitability layer. This makes the overlay computations more efficient and the results easier to analyze than when using the polygon overlay method. The raster method is discussed in this chapter.

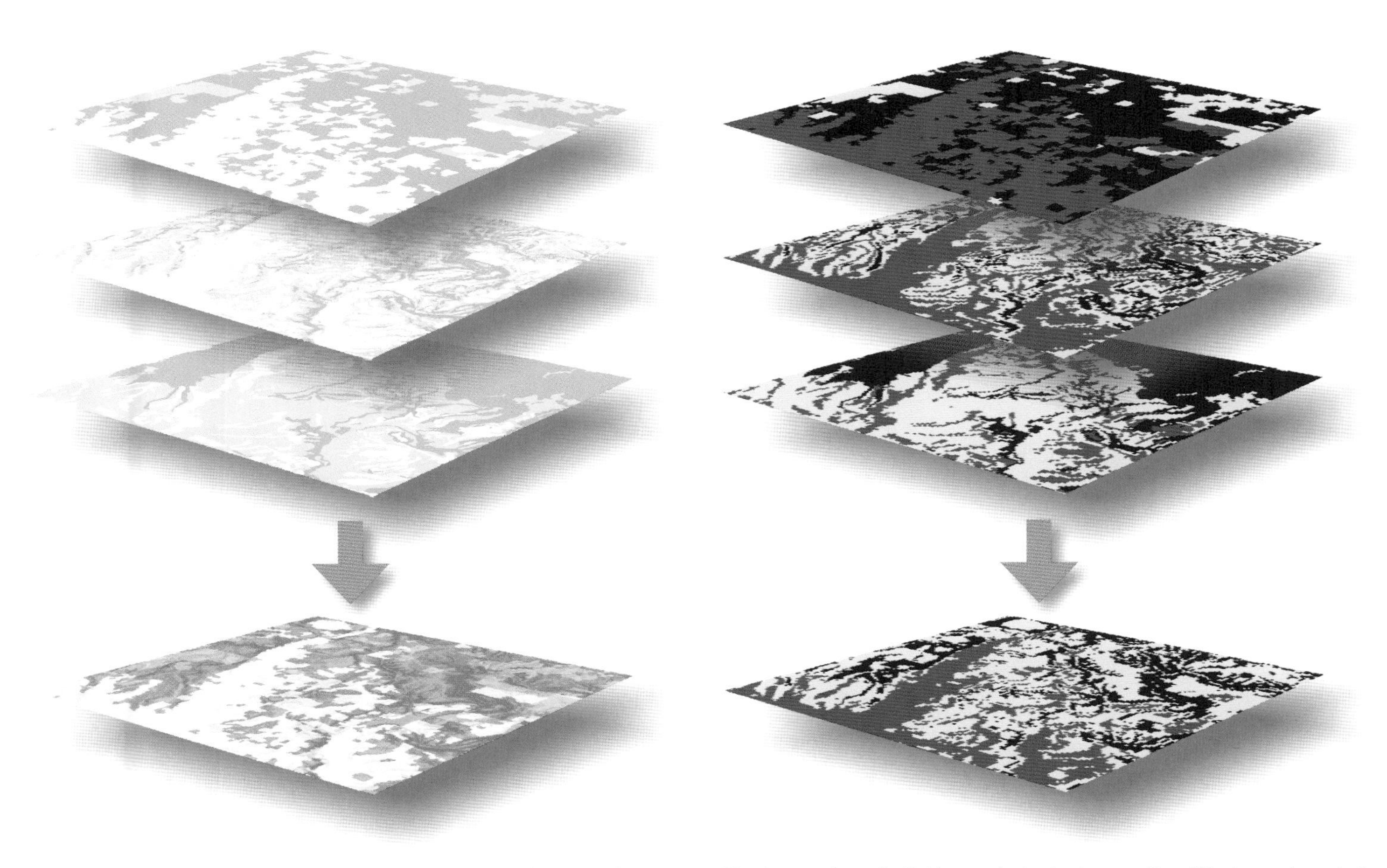

This diagram shows a conceptual—and simplified—version of the McHarg overlay method. Land cover (top), slopes, and soils are shaded based on how suitable they are for housing development. White indicates high suitability, light gray medium suitability, and dark gray low suitability. When the layers are graphically overlaid—that is, displayed on top of one another—the lightest areas indicate highest suitability and the darkest areas indicate lowest suitability.

This diagram shows the McHarg method as implemented in a GIS using mathematical overlay with rasters. Each cell is assigned a numeric value indicating its suitability—cells with high suitability are assigned a value of 3 (shown here as green), medium suitability a value of 2 (yellow), and low suitability a value of 1 (red). The layers are mathematically overlaid by summing the cell values on each layer and reassigning the summed values on a scale of 1 to 3. The resulting layer has values indicating which cells have overall high, medium, and low suitability. (In this example, only three levels of suitability are used, but five or nine levels are also commonly used and provide more variation in the resulting overall suitability layer.)

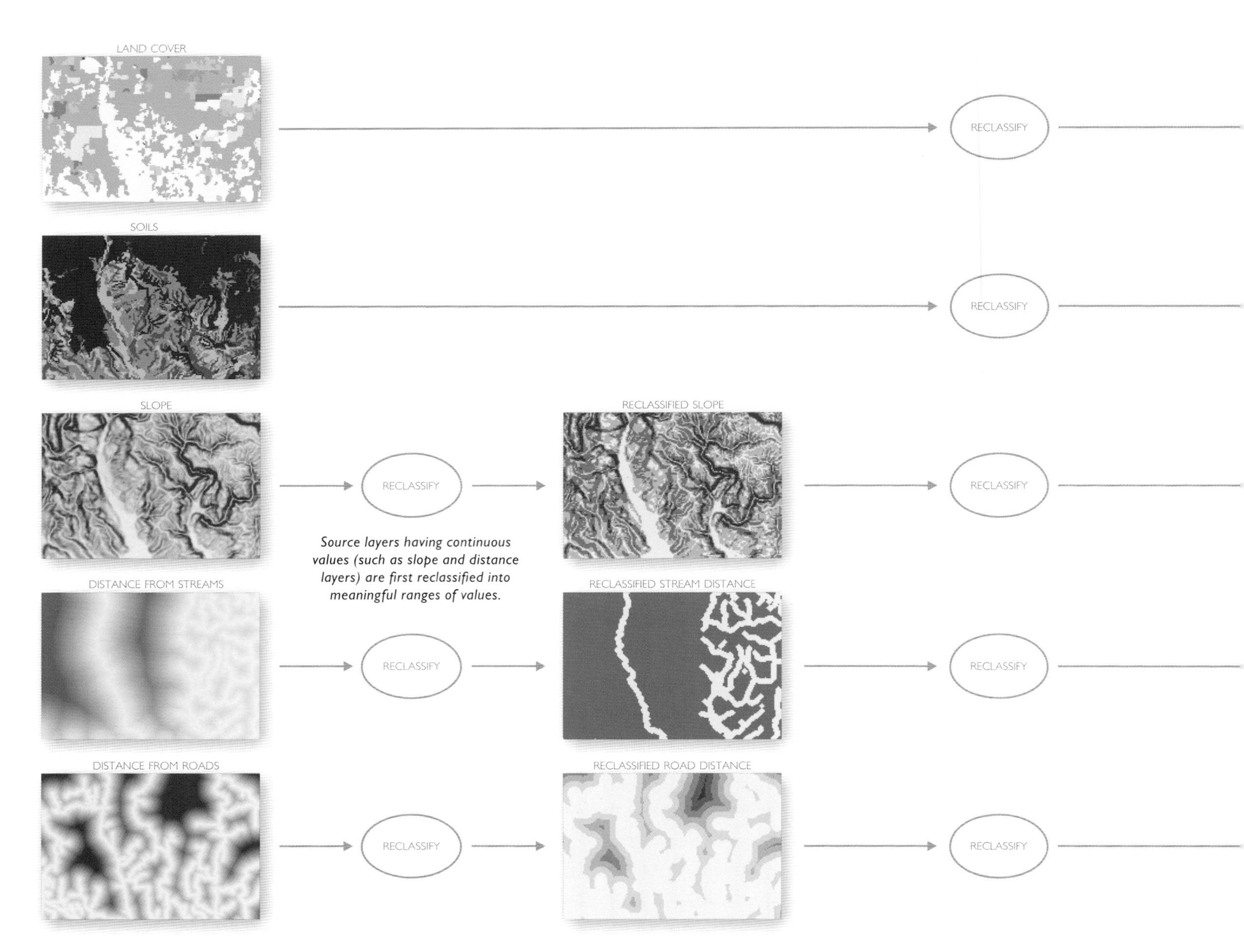

Source layers having continuous values (such as slope and distance layers) are first reclassified into meaningful ranges of values.

The process for using weighted overlay to rate locations is:

1. Collect the source layers
2. Create the suitability layers
3. Assign weights and combine the layers
4. Evaluate the results
5. Display and apply the results

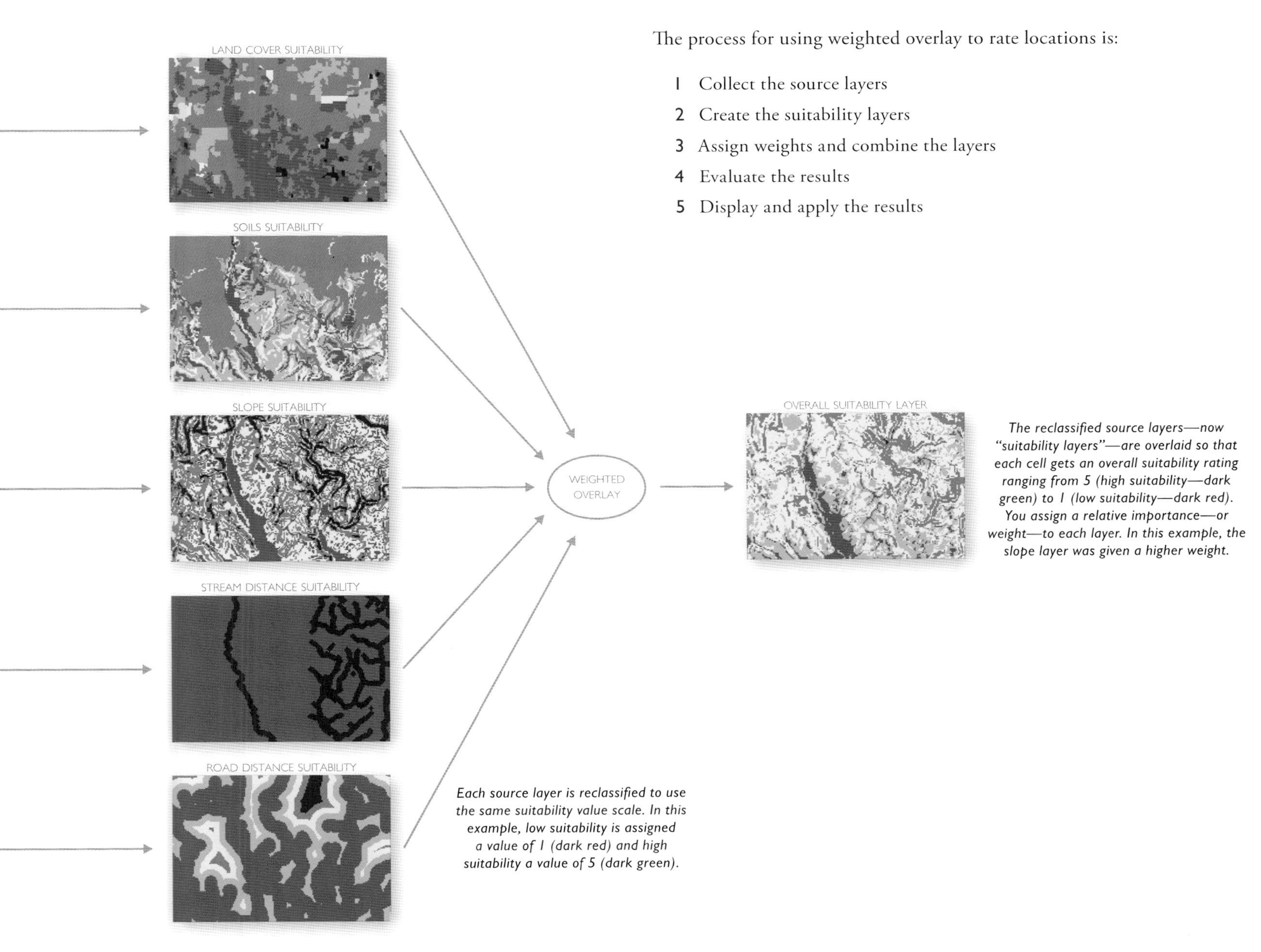

The reclassified source layers—now "suitability layers"—are overlaid so that each cell gets an overall suitability rating ranging from 5 (high suitability—dark green) to 1 (low suitability—dark red). You assign a relative importance—or weight—to each layer. In this example, the slope layer was given a higher weight.

Each source layer is reclassified to use the same suitability value scale. In this example, low suitability is assigned a value of 1 (dark red) and high suitability a value of 5 (dark green).

COLLECT THE SOURCE LAYERS

You first list the source layers representing the criteria for your model. For example, the criteria and corresponding source layers for a model to rate suitable locations for a housing development might include the ones in this table.

CRITERIA	SOURCE LAYER
Areas requiring less land clearing are more suitable	*Land cover*
Soils that support housing construction are more suitable	*Soils*
Areas on flat slopes are more suitable	*Slopes*
Areas farther from streams are more suitable	*Streams*
Areas near existing roads are more suitable	*Roads*

You then collect the layers and perform any required processing. For example, the weighted overlay model requires that the data be in raster format so the layers can be mathematically overlaid, so you'll need to convert any feature data to raster data.

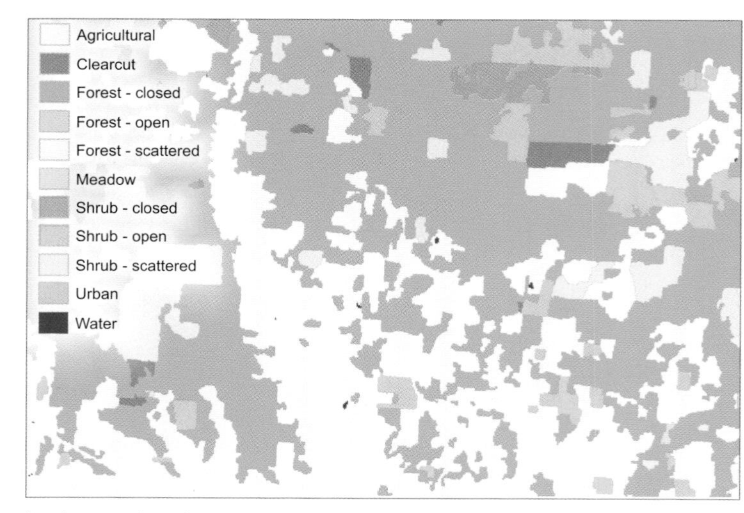

Land-cover polygon layer.

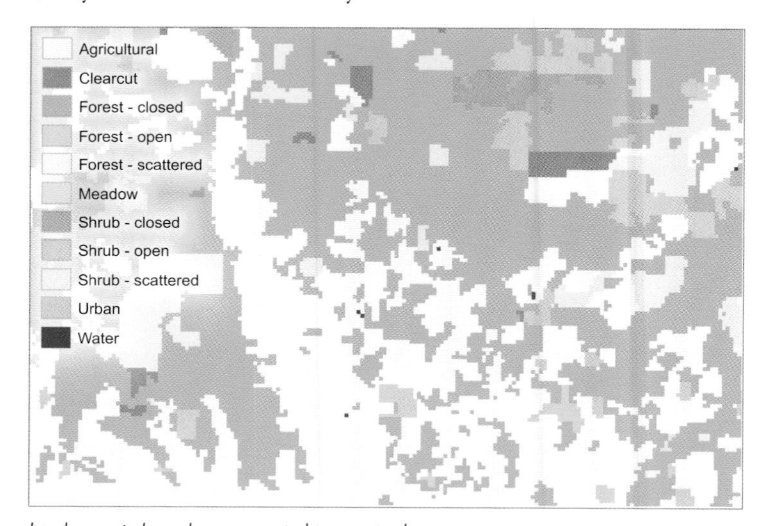

Land-cover polygon layer converted to a raster layer.

You may need to assign additional attributes to a layer. For example, a layer of soil types may require soil class codes that indicate their capability for development. If the codes don't exist in the soil layer's attribute table, you will need to add them. If the codes are in another table, you can join the table to the attribute table using a common item (as discussed in chapter 2, "Finding suitable locations").

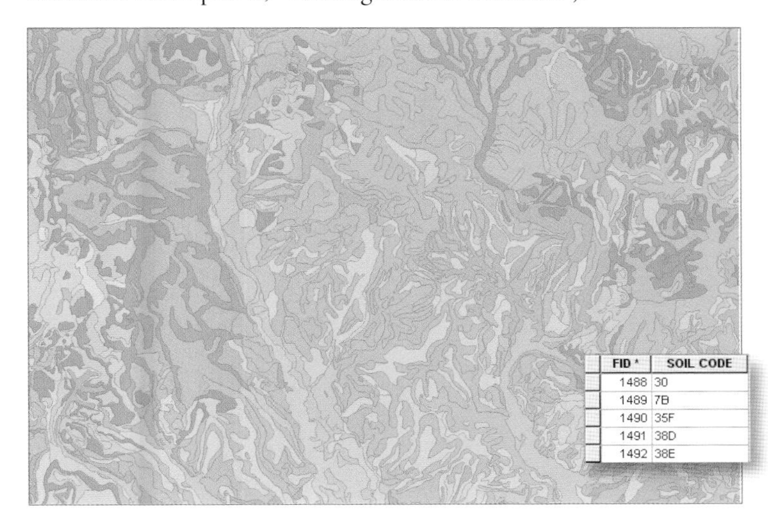

Soil type polygon layer.

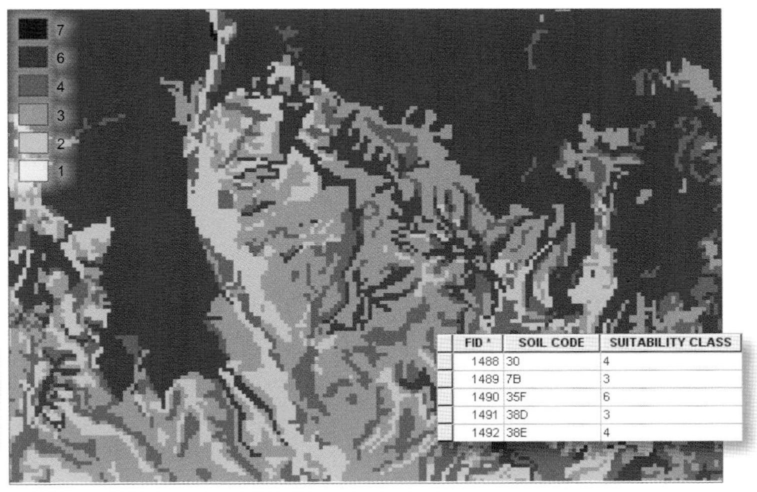

Raster layer of soil suitability class created from soil polygon layer. The layer is color-coded by suitability class, indicating suitability for development (with 1 being high suitability). The suitability class values were joined to the soil layer's attribute table.

You may also need to create derived datasets for some criteria. For example, if your model includes slope steepness, you can create the layer from an elevation source layer using tools in the GIS.

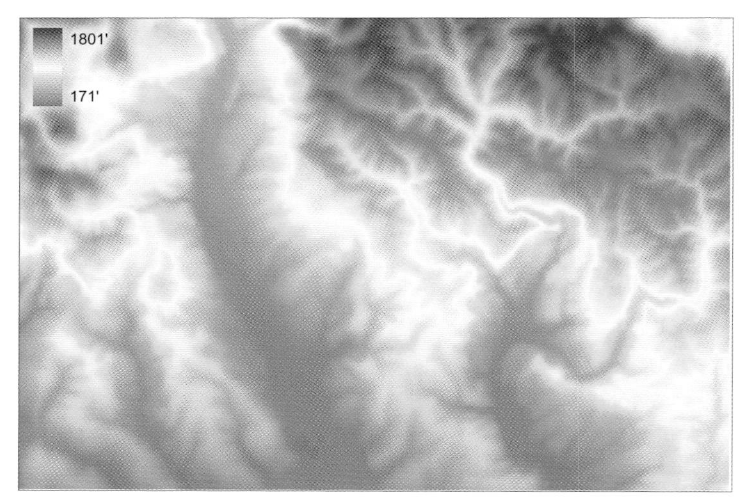

Elevation raster layer.

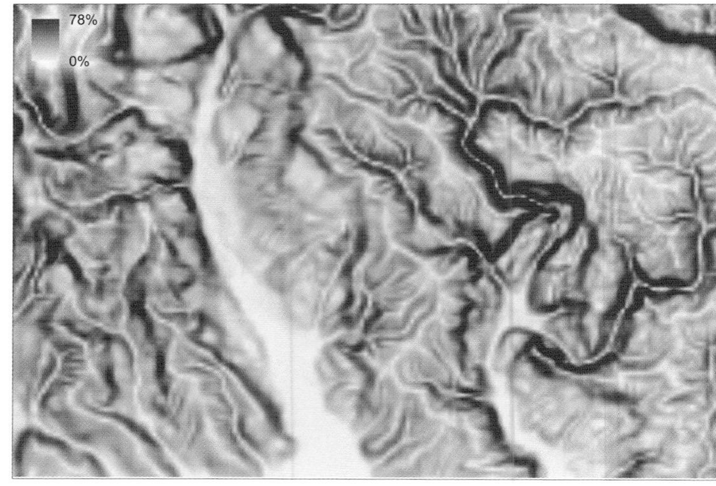

Slope raster layer derived from elevation layer.

Other terrain layers representing site characteristics such as aspect, solar radiation, or viewsheds (showing which areas can be seen from a location) can also be derived from an elevation layer.

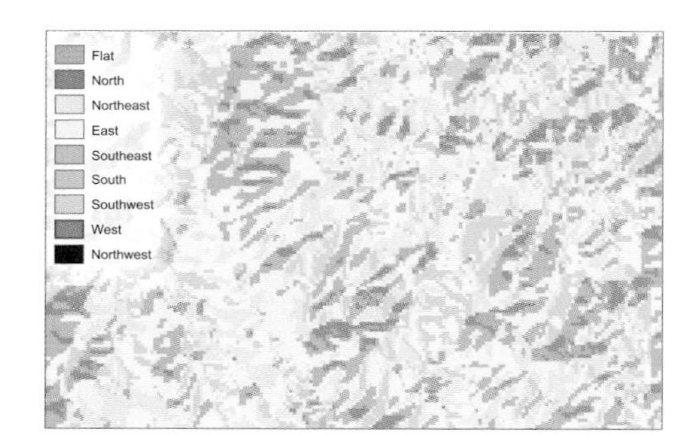

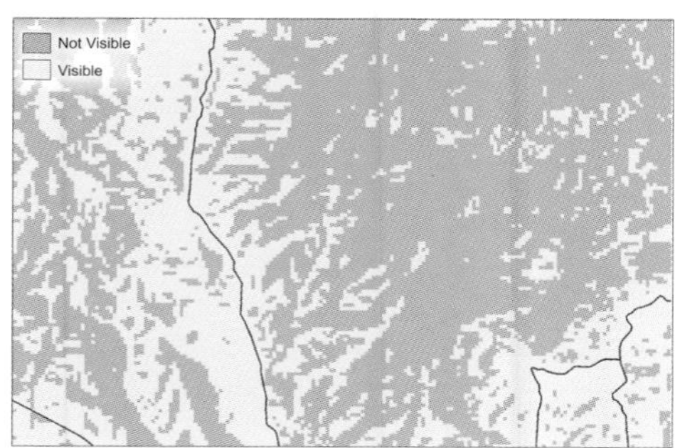

Terrain-based criteria layers such as aspect (left), solar radiation (center), and viewsheds (in this example, areas that can be seen from roads, shown as red lines) are derived from an elevation layer.

Suitability models often include distance criteria. For example, in a housing development model you may want to assign high suitability values to areas close to roads and far from streams. You can create distance rasters from points, lines, or polygons. The raster represents continuous distance from the features. A Euclidean distance raster measures straight-line distance.

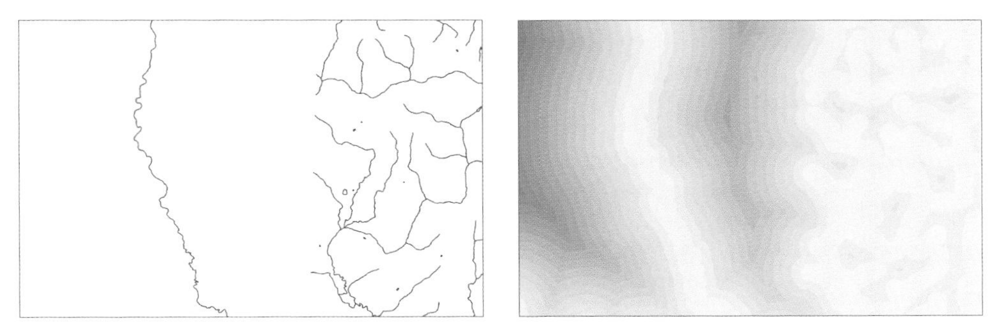

Streams (left) and raster layer of straight-line distance from streams (right)—the lighter the shade of blue, the closer that location is to a stream.

A cost distance raster represents the money, time, or effort involved in traversing the landscape along with the distance traveled. Cost distance rasters are discussed in chapter 4, "Modeling paths."

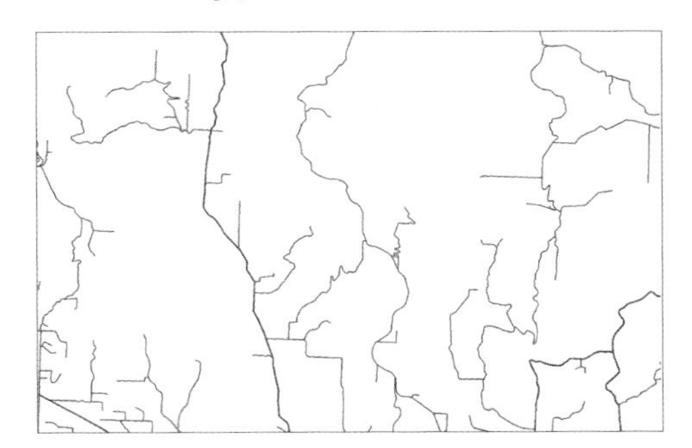

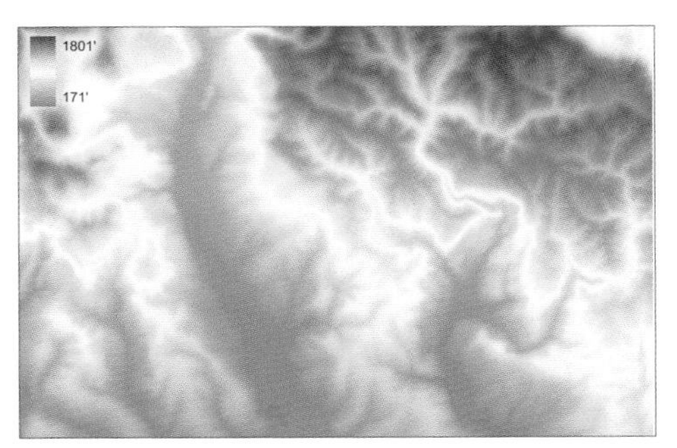

Roads (left) are combined with elevation (center) to create a cost distance raster showing the relative cost of building a road to each location (the steeper the slope the higher the cost of building a road)—light yellow areas represent low cost; dark orange areas high cost.

CREATE THE SUITABILITY LAYERS

Before combining the source layers, you need to assign relative suitability values to the values for each source layer, creating corresponding suitability layers.

The suitability values for each layer must use the same scale—for example, a scale of 1 to 5. This will ensure that the values are comparable (a value of 1 means low suitability for slope and equally low suitability for soils), and thus can be mathematically combined to get meaningful results.

The suitability values you assign are based on your knowledge of the subject, published research, or industry standards (as discussed in chapter 1, "Introducing GIS modeling"). For example, research on mule deer habitat may show which type of land cover the deer prefer (say forest over shrub). Similarly, building codes and regulations may specify which slopes and soil types are suitable for construction of housing.

To convert the source layers to suitability layers that use a common value scale, you first evaluate each source layer and reclassify it into relevant classes. You can then choose a suitability scale. Finally, you assign a suitability value to each class on each layer to create the suitability layers.

Evaluate the source layers and reclassify into relevant classes
The first step is to group the values on each source layer into classes based on how suitable for the use a value is relative to other values.

For source layers comprised of categories, you may be able to skip this step (you just assign the appropriate suitability value directly to each category). However, if you have many categories and some are equally suitable, you might want to generalize the source layer before assigning suitability values by reclassifying the layer into fewer categories. For example, if several categories of shrub in a land-cover layer are equally suitable, you might reclassify the categories into a single category. However, it limits flexibility in modifying your model, so you'd do this only if you're certain you won't want to assign different suitability values to the original categories at some future point (for example, if you need to create alternate scenarios).

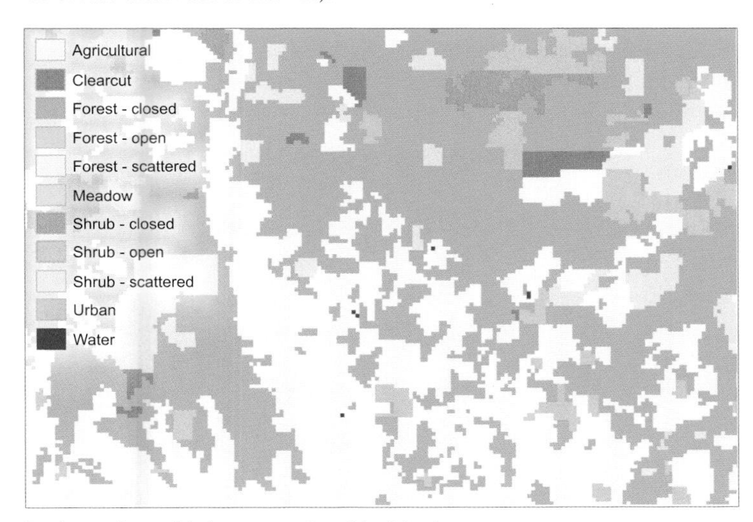

Land-cover layer with three categories of shrub land.

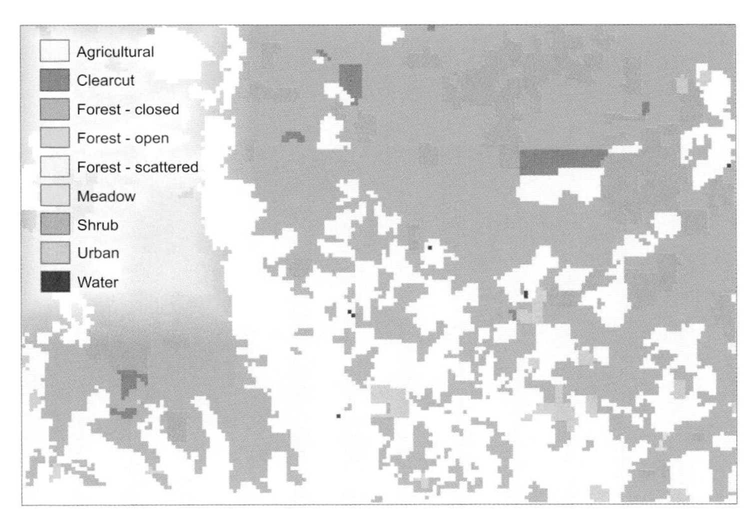

Land-cover layer with shrub land reclassified into a single category.

For source layers comprised of continuous values, such as elevation or distance from roads, you'll need to group the values into classes before assigning suitability values. In a raster having continuous values, each cell potentially has a unique value. At the very least there will be many different values in the raster, even if some cells share a value. Rather than assigning a suitability value to each individual value on the raster, it's much more efficient to group similar values into classes and then assign the appropriate suitability value to each class.

The classes you define should reflect the relative suitability of the source layer values and be based on meaningful class breaks. Published research, regulations, or accepted industry standards are often used as a guide when defining classes. For example, many local building regulations specify ranges of slope steepness, with steeper slopes having increasing restrictions on construction:

- 0% to 7%—no restrictions
- 7% to 15%—minor restrictions
- 15% to 30%—hillside restrictions in effect
- > 30%—limited construction

Assuming that the fewer the restrictions the more suitable the location, you'd reclassify the slope steepness layer into four classes to create a new layer with slopes matching these building regulations. When you assign suitability values (as discussed later in this section), the class with slopes of 0 percent to 7 percent would be assigned high suitability while the class with slopes > 30 percent would be assigned low suitability.

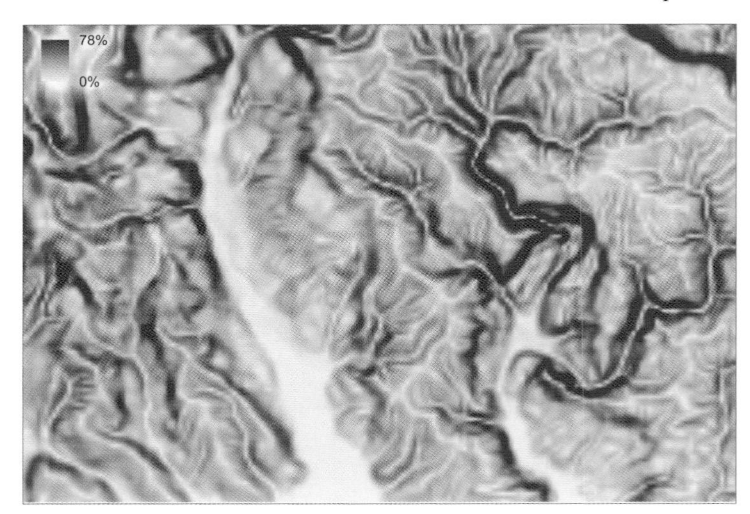

Original slope raster layer with continuous values.

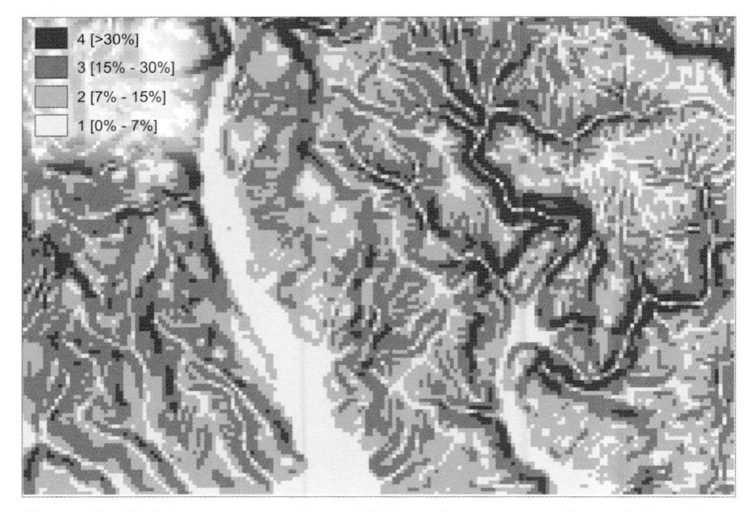

Slope reclassified into a new raster layer with four values corresponding to the four class ranges.

In some cases, you may have only two classes. For example, if regulations specify that there can be only limited development within 500 feet of streams, you'd reclassify the distance to streams layer into two classes: areas within 500 feet of a stream (which will be assigned low suitability) and all other areas (which will be assigned high suitability).

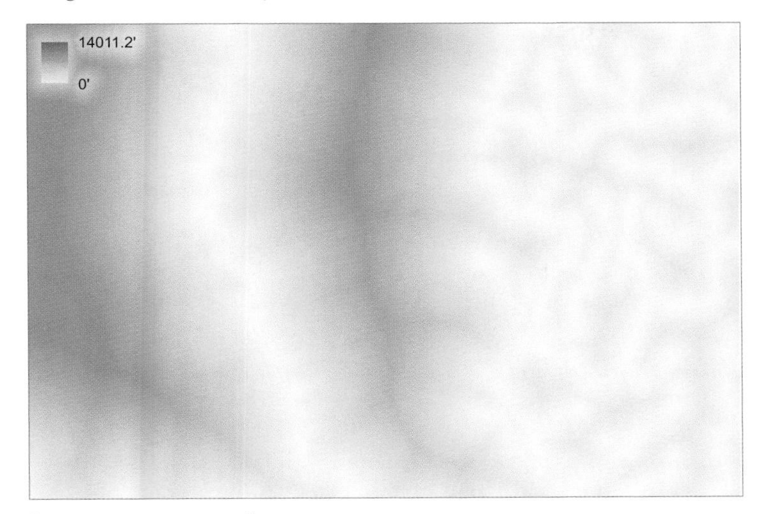

Distance to streams raster layer.

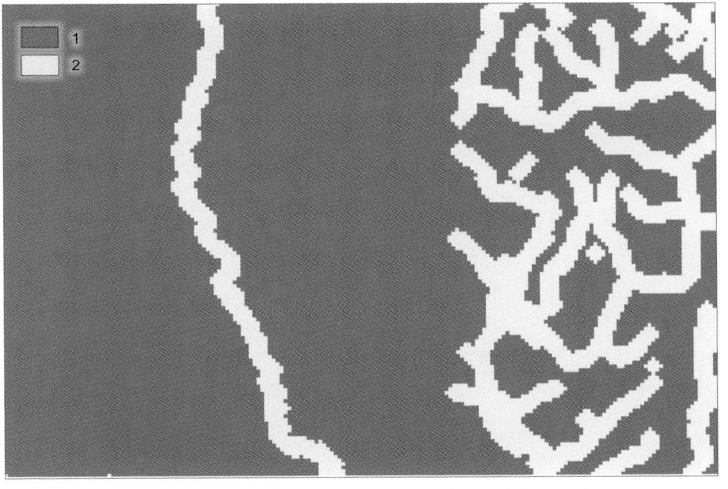

Distance to streams layer reclassified into a new raster layer with two values corresponding to areas within 500 feet of a stream and areas more than 500 feet from a stream.

In general, avoid the temptation to simply create classes using a predefined classification scheme, such as equal interval or Jenks' natural breaks. It's better to create classes that reflect your specific criteria.

However, there are times when using a standard classification scheme is appropriate. For example, one criterion might be that locations close to existing roads are more suitable for housing development (since that reduces the cost of constructing new roads). You might assume that—all other things being equal—the cost of building new roads increases at a constant rate as distance from existing roads increases. In this case you would in fact define classes of equal intervals—say quarter-mile increments (0 to 1,320 feet, 1,320 to 2,640 feet, and so on).

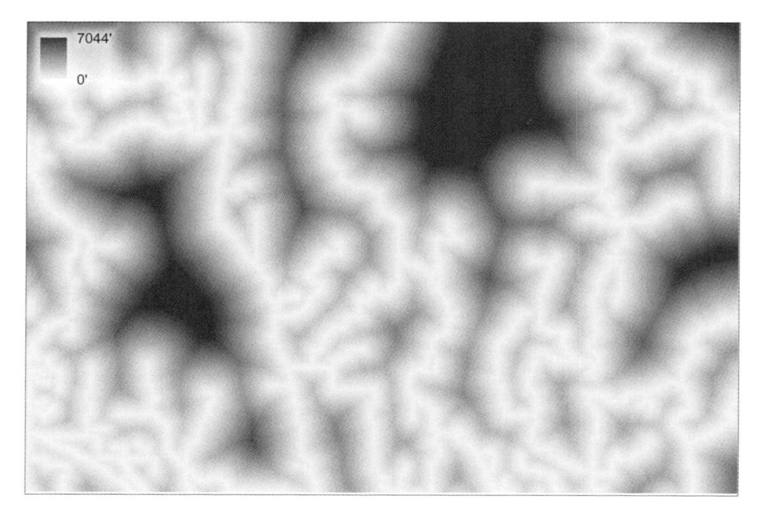

Distance to roads raster layer.

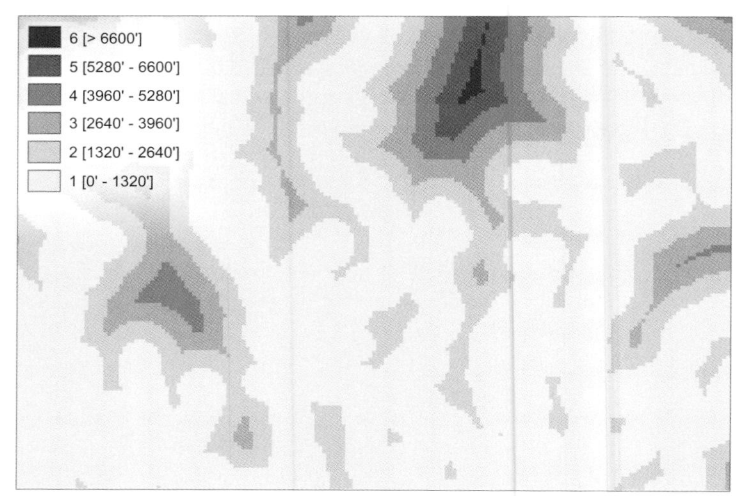

Distance to roads layer reclassified into a new raster layer using six equal interval classes.

In ArcGIS, it's not required to reclassify source layers having continuous values before assigning suitability values—you can instead assign suitability values directly to ranges of values at the same time you overlay the layers to create the overall suitability layer. However, grouping values into classes before assigning suitability values lets you easily change the suitability values for a class without having to start with the original continuous values raster each time you run the model. (For example, you might modify suitability values for classes when verifying the results of your model as described later in this section—for many models, you're more likely to change the suitability value assigned to a class than you are to change the class breaks themselves.)

Choose a suitability scale

After reclassifying the source layers as needed, you'll choose a suitability scale. The suitability scale is a way of comparing values across the source layers so there is a common standard. All source layer values are placed on the same scale with the same units. The values on the suitability scale should have the same meaning across the source layers—a 3 for land cover is as suitable as a 3 for soils. The same scale is used for all the individual suitability layers and for the final overall suitability layer.

Suitability values are assigned on an interval scale, which measures relative value (see chapter 1, "Introducing GIS modeling"). With an interval scale, you know, for example, that on a scale of 1 to 5, with 5 being very suitable, a location with a value of 4 is more suitable than a location with a value of 2, but you cannot state that it is twice as suitable.

By convention, suitability scales usually consist of positive values with a value of 1 representing the lowest suitability and the highest number on the scale representing the highest suitability. Often, scales use an odd number of values so that there is a middle value representing neutral.

You need to decide how many values your suitability scale has. The number of values on the scale should reflect the various criteria values, your knowledge of the subject, and the purpose of your model.

In practice, fewer than ten values is usually sufficient—it's difficult to distinguish more levels of relative suitability than this (which is why many things are rated "on a scale of 1 to 10" and not a scale of 1 to 20 or 1 to 50). And, in fact, research has shown that people can distinguish only about seven colors (and hence values) on a map.

Scales comprised of three, five, or nine levels of suitability are commonly used.

Using three levels (high, medium, and low) may be enough variation for many applications—it is easy to grasp the differences between the levels. However, it limits your ability to analyze trade-offs and compromises since the categories are broad.

SUITABILITY LEVEL	SCALE VALUE
High	*3*
Medium	*2*
Low	*1*

SUITABILITY LEVEL	SCALE VALUE
High	*5*
Moderately high	*4*
Medium	*3*
Moderately low	*2*
Low	*1*

When using five levels (high, moderately high, medium, moderately low, low), the differences between levels are still easy to grasp, but there is more differentiation. Five is a commonly used scale in every day experience so is familiar to people.

A scale with five levels of suitability gives you more flexibility in your decisions than does a scale with three levels. For example, if you're looking for a location to build a housing development and one location has an overall suitability of 5 (high) and another location has a suitability of 4, but the location with a value of 5 is much more expensive to purchase, you might decide that the location with a rating of 4 is good enough and locate your development there. (Or, you might decide to build in the location having a rating of 5 and pay the extra cost). If you were using a scale with only three levels and both locations were rated 3 (high) you would not know that the land you're paying less for is actually a little less suitable. Conversely, if the less expensive location had been assigned a rating of 2 (medium), you might reject it even though it's only slightly less suitable than the other location which was rated 3.

SUITABILITY LEVEL	SCALE VALUE
High +	*9*
High	*8*
High -	*7*
Medium +	*6*
Medium	*5*
Medium -	*4*
Low +	*3*
Low	*2*
Low -	*1*

You might use a scale with nine levels if one or more of your source layers has many values—for example, if one of your source layers is a vegetation layer with 40 categories and another is a layer of distance to roads that requires more than a few levels of suitability for your model. Having the additional levels on your scale will make it easier to condense the many values on the source layer into fewer values on the suitability scale. One way to approach assigning values on a scale with nine levels is to first classify the criteria values into high, medium, and low, and then assign the values on each of these three categories a rating of high, medium, or low within the category. So one value might be rated as a high high (or high+) while another might be rated a low high or (high-).

You can easily generalize the final overall suitability layer to three levels (high, medium, and low) for presentation if desired. You reclassify the layer to create a new layer with three levels (values of 9, 8, and 7 are assigned a value of 3, high, and so on). You still have the nine levels of suitability on the original overall suitability layer to use in further analysis, or for decision makers who want to consider second and third best options and compare the trade-offs when choosing the most suitable location.

If a particular source layer has fewer classes than the number of suitability values on the scale, not all suitability values will be used. In this example, using a scale of 1 to 5, slope classes might be assigned suitability values as shown in the table below.

SLOPE CLASS	VALUE RANGE	SUITABILITY LEVEL	SUITABILITY VALUE
1	*0% - 7%*	*High*	*5*
2	*7% - 15%*	*Medium*	*3*
3	*15% - 30%*	*Moderately low*	*2*
4	*> 30%*	*Low*	*1*

For source layers with more classes, you will need to combine classes (that is, assign the same suitability level to more than one class—for source layers with continuous values these will likely be adjacent classes). The distance to roads classes might be assigned suitability values as shown in the table below.

DISTANCE TO ROADS CLASS	VALUE RANGE	SUITABILITY LEVEL	SUITABILITY VALUE
1	*0' - 1320'*	*High*	*5*
2	*1320' - 2640'*	*Moderately high*	*4*
3	*2640' - 3960'*	*Medium*	*3*
4	*3960' - 5280'*	*Moderately low*	*2*
5	*5280' - 6600'*	*Low*	*1*
6	*> 6600'*	*Low*	*1*

Assign suitability values to source layers

Once you've chosen a suitability scale, you assign suitability values to each category or class on each source layer to create corresponding suitability layers.

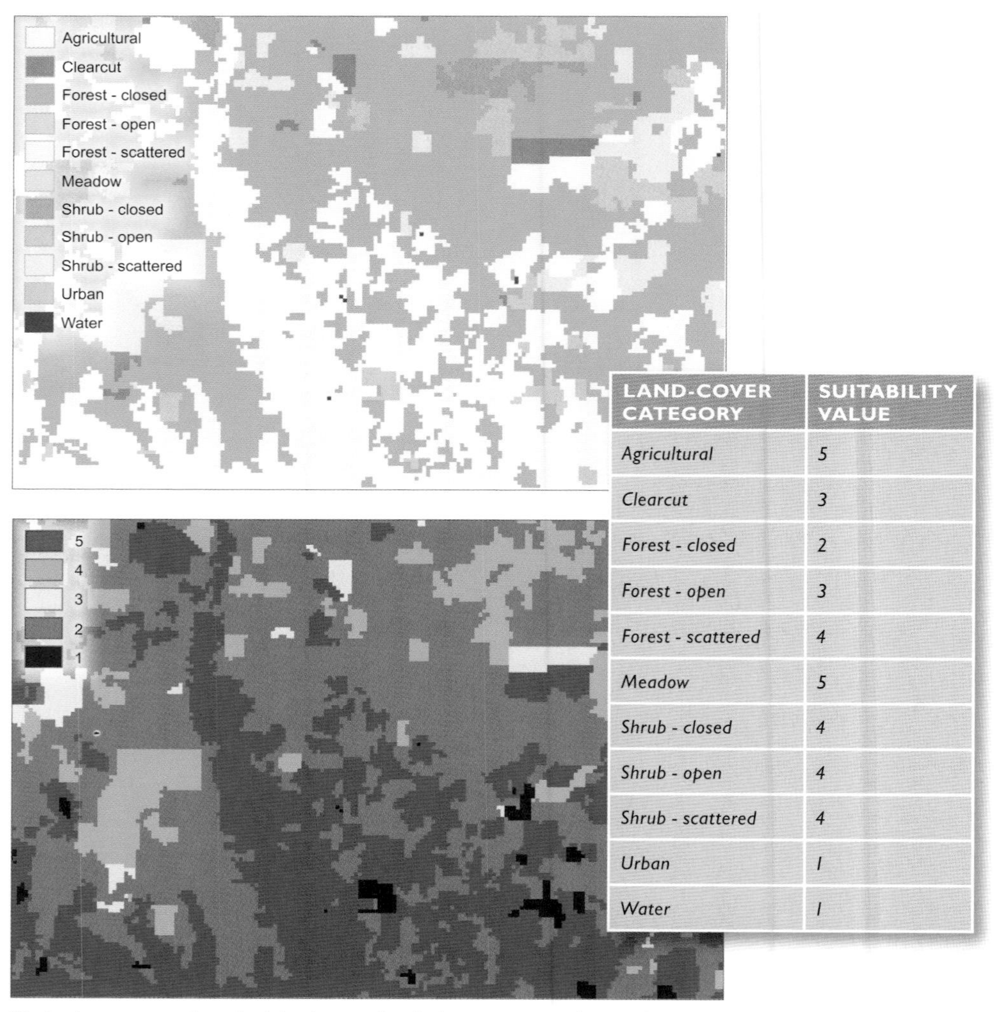

LAND-COVER CATEGORY	SUITABILITY VALUE
Agricultural	*5*
Clearcut	*3*
Forest - closed	*2*
Forest - open	*3*
Forest - scattered	*4*
Meadow	*5*
Shrub - closed	*4*
Shrub - open	*4*
Shrub - scattered	*4*
Urban	*1*
Water	*1*

The land-cover source layer (top) has been reclassified to create a new layer with suitability values (bottom) on a scale of 1 to 5 using the values listed in the table—each category is assigned a suitability value.

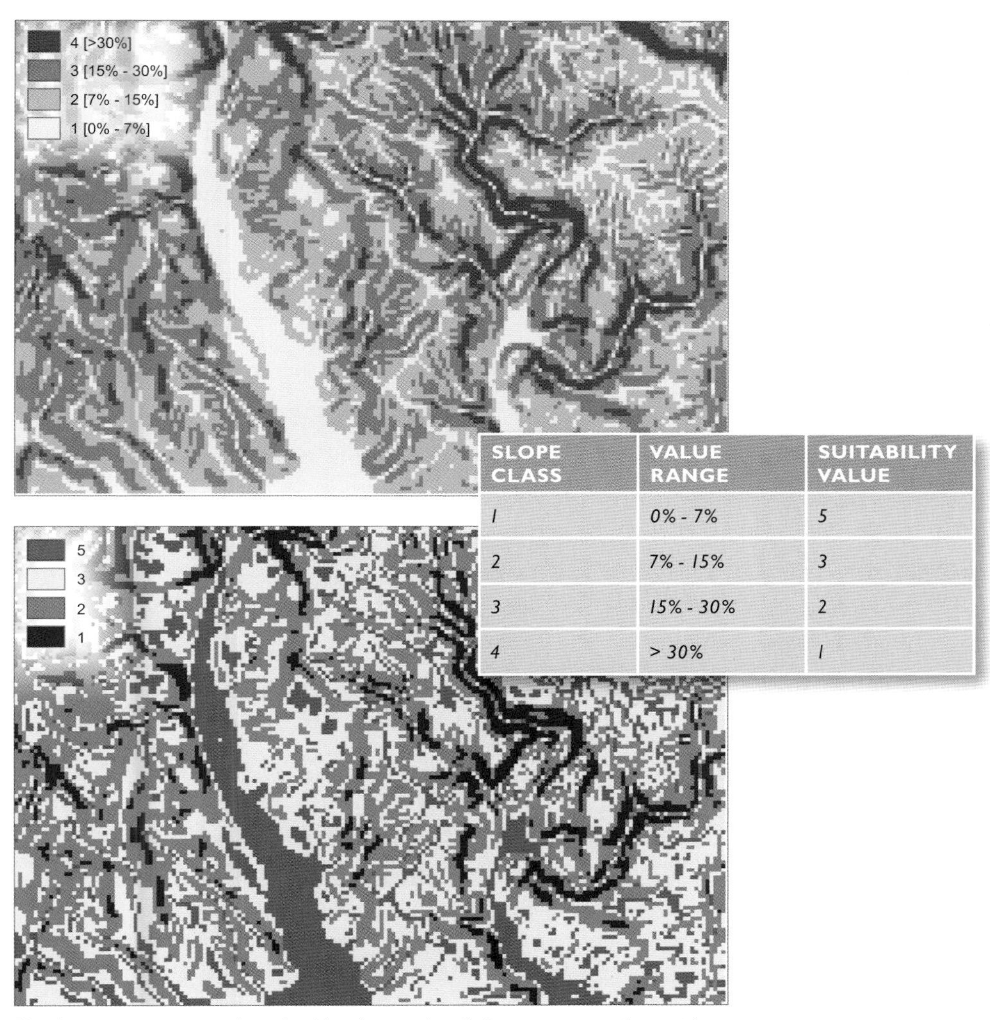

SLOPE CLASS	VALUE RANGE	SUITABILITY VALUE
1	0% - 7%	5
2	7% - 15%	3
3	15% - 30%	2
4	> 30%	1

The slope steepness source layer (top) has been reclassified to create a new layer with suitability values (bottom) on a scale of 1 to 5 using the values listed in the table—each of the four classes is assigned a suitability value. In this example, the category containing the flattest slopes was assigned the highest suitability value (5) since these slopes are much more desirable. Beyond a slope of 7 percent, suitability drops off, so the other classes were assigned the lowest suitability values (from 3 to 1).

Excluding unsuitable areas

You can exclude from consideration as suitable locations those areas that are completely unsuitable or are off limits. For example, when building a model for housing development, you might specify that urban areas and water bodies not be considered developable (the former because those locations are presumably already developed). You can also exclude areas beyond a threshold or cutoff value if your criteria require it. For example, regulations may specify that slopes greater than 30 percent cannot be developed.

When you combine the suitability layers to create the overall suitability layer (as discussed in the next section), you specify that the value or class that is off limits is "restricted." Those cells will receive a value in the overall suitability layer that indicates they are restricted (regardless of whether they have valid values on another input suitability layer).

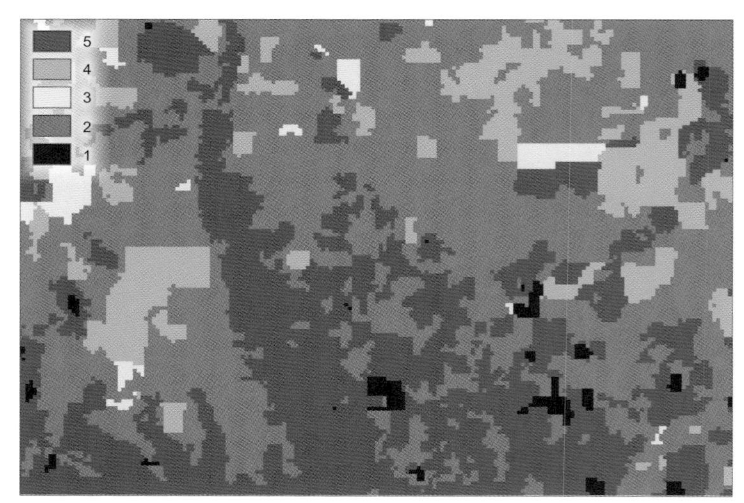

On the land-cover suitability layer, areas that are already developed or are water were assigned a value of 1. These cells will be specified as "Restricted" when the suitability layers are combined.

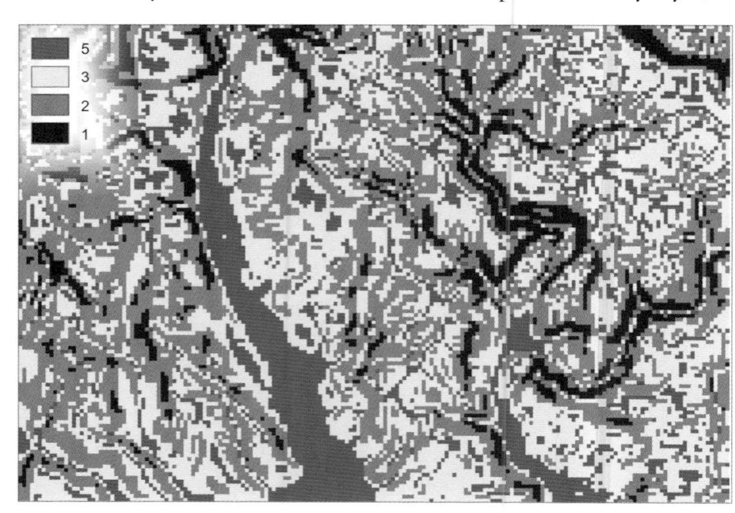

On the slope suitability layer, slopes greater than 30 percent were assigned a value of 1—they will also be specified as "Restricted" when the suitability layers are combined.

In ArcGIS, when the suitability layers are combined, a value of one less than the lowest value on the suitability scale is used to indicate that a cell is restricted. So for a scale of 1 to 5, with 1 being least suitable, any cell identified as restricted on any of the input suitability layers is assigned a value of 0 on the overall suitability layer. Using a value of "restricted" for areas that are off limits will allow you to distinguish these areas (on the overall suitability layer) from areas for which data is missing. (In ArcGIS, a value of NoData is used for the latter.)

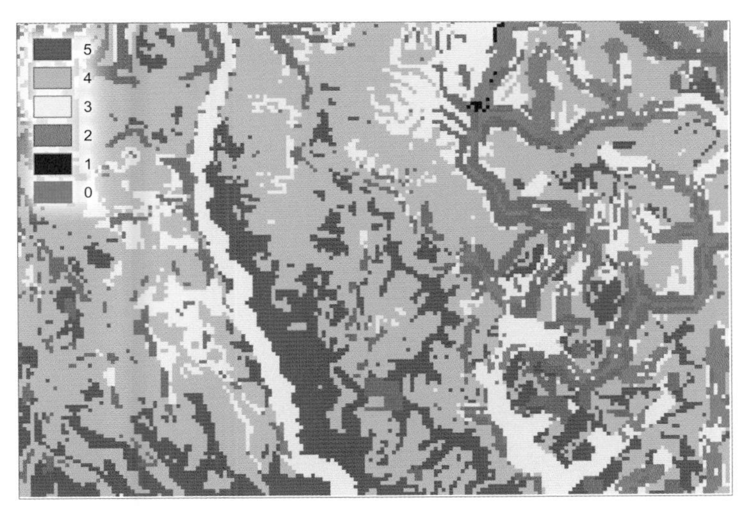

The results of this model (which also includes soils, stream distance, and road distance) show suitability values of 1 through 5. Areas that were specified as "Restricted" on the land cover and slopes suitability layers are assigned a value of 0 (shown in gray) on the output layer.

Creating alternative scenarios

Depending on your application you may need to create alternate scenarios. For example, when modeling suitable locations for a housing development, you may want to create one model that favors low development cost, one that favors preservation of agricultural land, and a third that favors protection of wildlife habitat. For each model, you'd assign suitability values to each source layer accordingly. Implementing your model as a ModelBuilder model document, as described in chapter 1, "Introducing GIS modeling," makes it relatively easy to create alternate scenarios—you simply copy and modify the model document to create a new model for each scenario.

For the housing development model that favors preservation of agricultural land, you'd assign the agricultural land category a value of 1 (low suitability) since you want agricultural land to have a low overall suitability rating for development, whereas for the model that favors low development cost you'd assign agricultural land a value of 5 (high suitability) since it costs relatively less to clear and build on. Similarly, if shrub land represents prime wildlife habitat, you'd assign it a low value when creating the housing development model that protects wildlife habitat. The values will be reflected in the overall suitability layer resulting from each model.

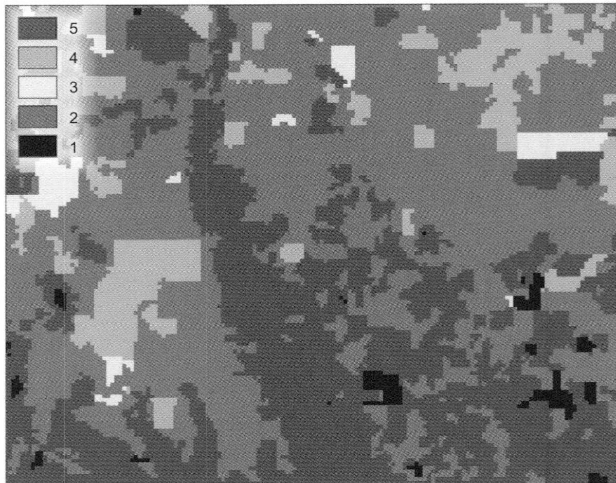

LAND-COVER CATEGORY	SUITABILITY VALUE
Agricultural	*5*
Clearcut	*3*
Forest - closed	*2*
Forest - open	*3*
Forest - scattered	*4*
Meadow	*5*
Shrub - closed	*4*
Shrub - open	*4*
Shrub - scattered	*4*
Urban	*1*
Water	*1*

Alternative A—Low development cost. Land-cover types are assigned suitability values based on development cost—agricultural land is assigned a value of 5 (low cost/high suitability); closed forest is assigned a value of 2 (high cost/low suitability).

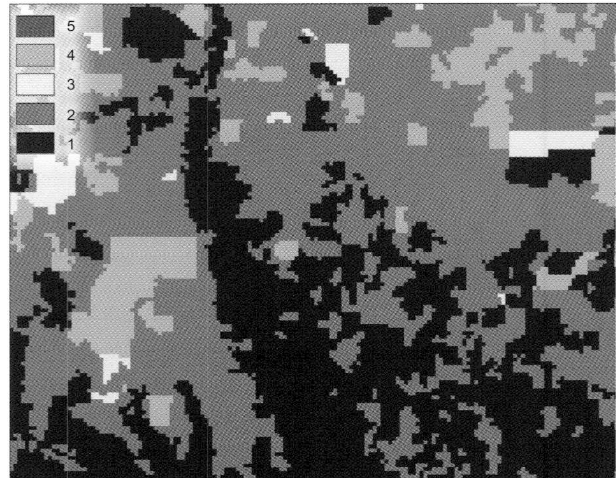

LAND-COVER CATEGORY	SUITABILITY VALUE
Agricultural	*1*
Clearcut	*3*
Forest - closed	*2*
Forest - open	*2*
Forest - scattered	*2*
Meadow	*5*
Shrub - closed	*4*
Shrub - open	*4*
Shrub - scattered	*4*
Urban	*1*
Water	*1*

Alternative B—Preserve agricultural land. Land-cover types are assigned suitability values based on preserving agricultural land—agricultural land cover is assigned a value of 1 (very low suitability) and forest a value of 2 (forest is protected to help conserve groundwater).

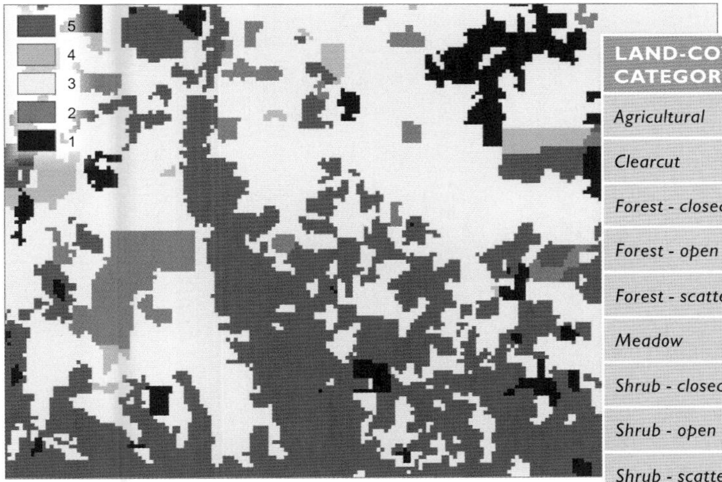

LAND-COVER CATEGORY	SUITABILITY VALUE
Agricultural	5
Clearcut	4
Forest - closed	3
Forest - open	3
Forest - scattered	3
Meadow	5
Shrub - closed	1
Shrub - open	1
Shrub - scattered	2
Urban	1
Water	1

Alternative C—Protect wildlife habitat. Land-cover types are assigned suitability values based on protecting wildlife habitat—shrub (representing sensitive habitat in this example) is assigned a value of 1 (closed/open shrub) or 2 (scattered shrub).

ASSIGN WEIGHTS AND COMBINE THE LAYERS

Once you've created the suitability layers, you combine them into a single layer to assign an overall suitability value to each cell.

In most models, some criteria are more important than others. Weighted overlay allows you to assign weights to the suitability layers to specify the extent to which suitability is dependent on each criterion. You specify the weights as a percentage—the weights for all the suitability layers sum to 100 percent. The suitability values assigned to the categories or classes in each layer are multiplied by the weight (represented as a decimal) resulting in more (or less) importance for that layer. When the suitability layers are summed, the results reflect this weighting.

The weights are ratio values which represent relative importance. So, in the example below, slope (with a weight of 40 percent) is twice as important as land cover (with a weight of 20 percent) and four times as important as distance to roads (with a weight of 10 percent).

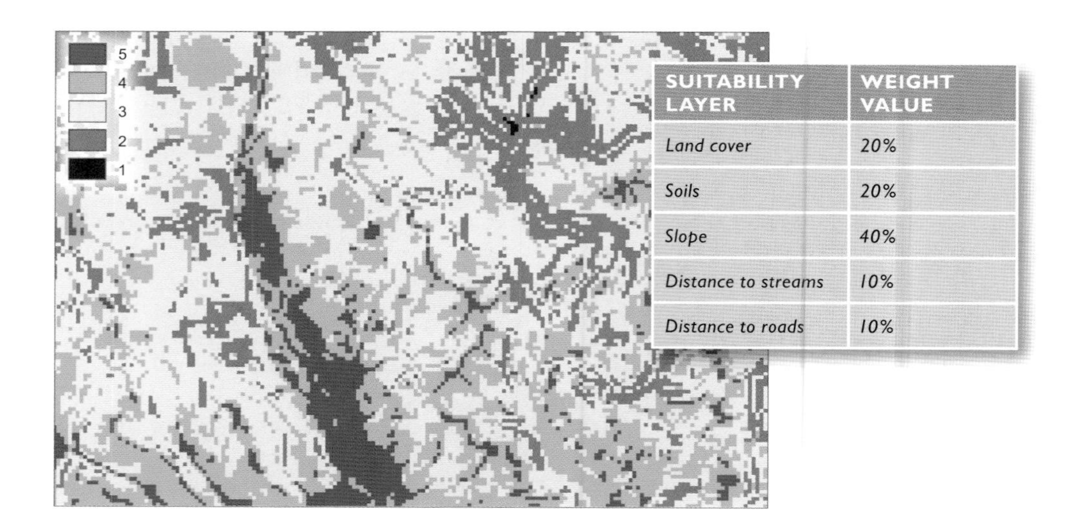

SUITABILITY LAYER	WEIGHT VALUE
Land cover	*20%*
Soils	*20%*
Slope	*40%*
Distance to streams	*10%*
Distance to roads	*10%*

With the slope criterion weighted more heavily, more suitable areas (5 and 4)are concentrated in areas with flat slopes while less suitable areas (1 and 2) are concentrated where slopes are steepest.

If all criteria are equally important (or if you're unable to determine which are more important), you assign equal weights to the layers. If you have five suitability layers, for example, each would receive a weight of 20 percent.

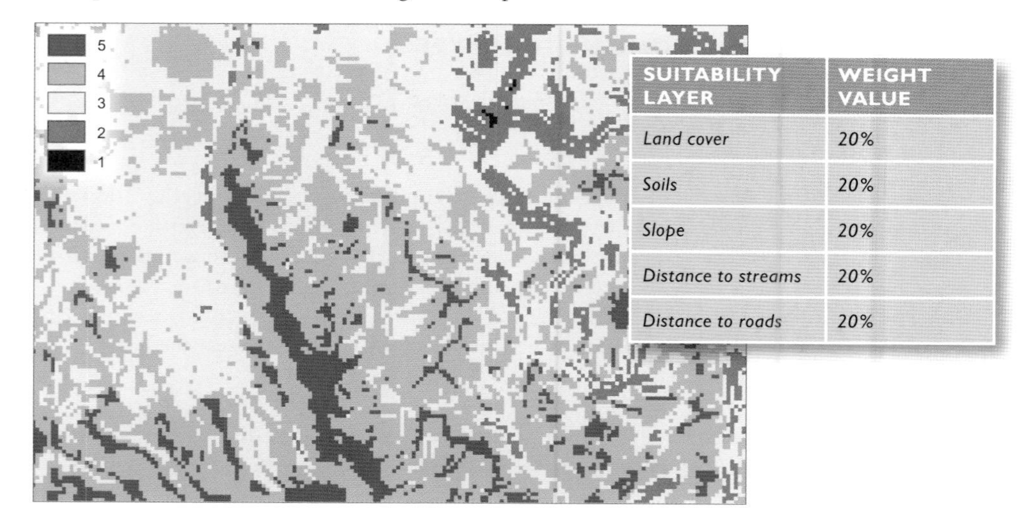

SUITABILITY LAYER	WEIGHT VALUE
Land cover	*20%*
Soils	*20%*
Slope	*20%*
Distance to streams	*20%*
Distance to roads	*20%*

With all criteria weighted equally, the influence of slopes seen in the previous map is diminished so that suitable locations are spread over a wider area.

With weighted overlay, then, the overall suitability layer is the result of two factors. One is the suitability value you assign to each value in a source layer (whether forest is highly suitable or not suitable, for example) when creating the suitability layers. The other is the weight you assign to each suitability layer when creating the overall suitability layer (for example, whether land cover is assigned a weight of 10 percent or 30 percent).

As with assigning suitability values to source layer values, the weight you assign to each suitability layer is dependent on the goal of your model as well as on your knowledge of the use you're modeling.

How weighted overlay works

The values in each suitability layer are multiplied by the weight for that layer, and these weighted values are summed. For example, if slope is four times more important than distance to roads, you'd assign a weight of 40 percent to the slope layer and a weight of 10 percent to the distance to roads layer. A cell in the slope layer with a value of 5 (gentle slope) will end up with a value of 2.0 (5*0.4); a cell in the distance to roads layer with a value of 5 (close to roads) will end up with a value of 0.5 (5*0.1). Two is four times 0.5, reflecting the fact that—in this model—slope is four times more important than distance to roads.

The weighted values for each suitability layer are added to derive the overall suitability rating for the cell. When using weighted overlay, the output scale is the same as the input suitability scale, since the values are multiplied by a decimal percentage (where 100 percent equals 1.0). For example—using the weight values in the table to the right—a cell with a suitability value of 5 for all five suitability layers would end up with an overall suitability value of 5 (1.0+1.0+2.0+0.5+0.5); and a cell with values of 1 for all five suitability layers would end up with a value of 1 (0.2+0.2+0.4+0.1+0.1).

SUITABILITY LAYER	WEIGHT VALUE
Land cover	20%
Soils	20%
Slope	40%
Distance to streams	10%
Distance to roads	10%

The suitability value of each cell is multiplied by the weight for that layer to create a weighted value. The weighted values are then summed to assign an overall suitability value to each cell.

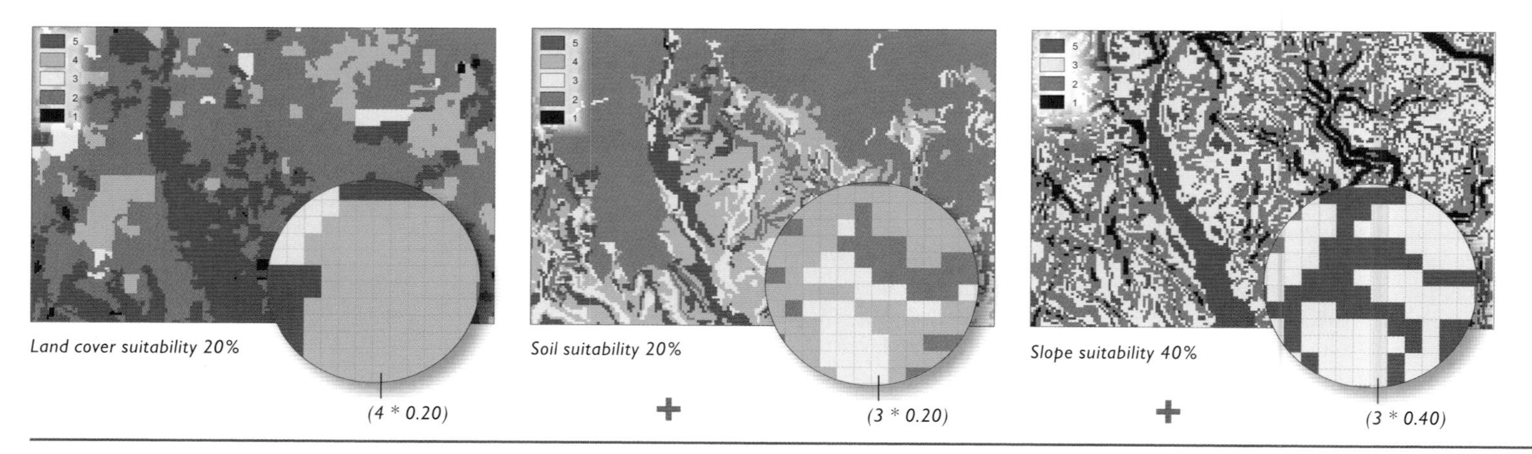

Of course, most cells will have different suitability values on each of the suitability layers and will end up with overall suitability values somewhere in the range between the high and low ends of the scale. For example, a cell with the input suitability values shown in the table below (and the diagram above) would end up with an overall suitability value of 3.6.

SUITABILITY LAYER	SUITABILITY VALUE	WEIGHT VALUE	WEIGHTED SUITABILITY VALUE
Land cover	4	20%	0.8
Soils	3	20%	0.6
Slope	3	40%	1.2
Distance to streams	5	10%	0.5
Distance to roads	5	10%	0.5
SUM		100%	3.6

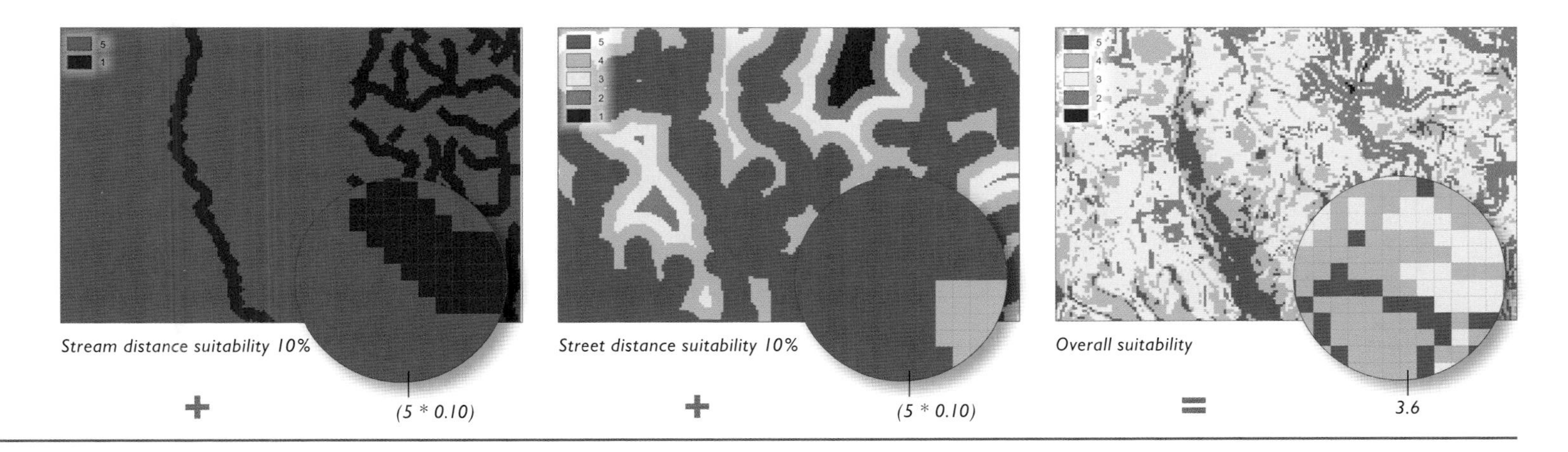

Stream distance suitability 10%

Street distance suitability 10%

Overall suitability

The summed weighted value is rounded up or down, resulting in an integer output raster—the overall suitability layer. That's because while the weights are ratio values, the values on the suitability layers are interval values. When you combine the suitability layers, you're adding interval values that happen to have been multiplied by a decimal value (representing a percentage) so they end up decimal values. However, they're still interval data values, so the rounding is done at the end to maintain the original interval scale. In the example, the cell with an overall suitability value of 3.6 is assigned a value of 4 on the output layer.

By weighting certain layers greater than others, you can emphasize one set of criteria over another. One scenario, for example, might emphasize the lowest direct development costs, while another might maximize environmental concerns. While neither of these alternatives may end up being the final option, they will at least highlight the constraints and possibilities for decision makers—for example, areas that are suitable under either scenario will be apparent.

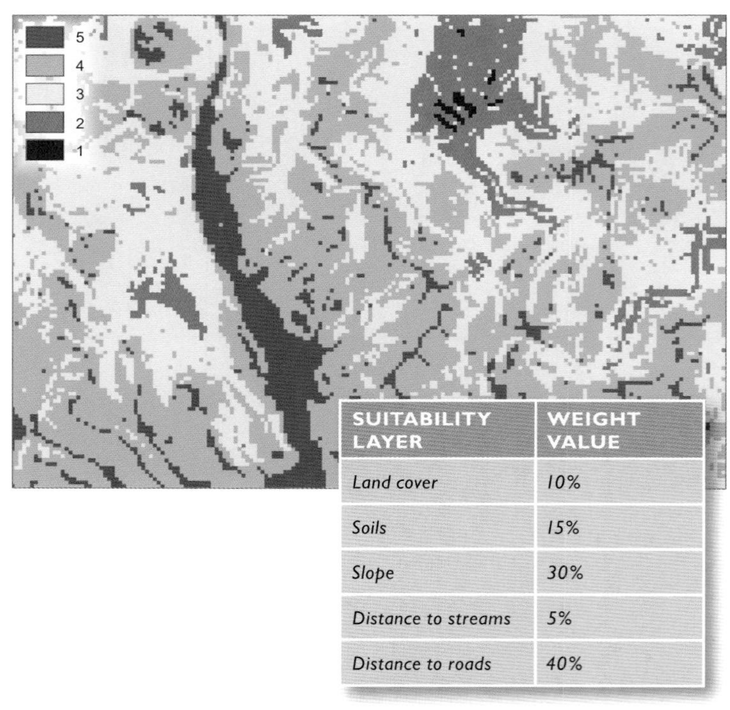

SUITABILITY LAYER	WEIGHT VALUE
Land cover	*10%*
Soils	*15%*
Slope	*30%*
Distance to streams	*5%*
Distance to roads	*40%*

In this scenario, which attempts to minimize development costs, the layers for slope steepness and distance to roads are weighted more heavily than land cover, soil suitability, or distance to streams. The result is that areas on gentle slopes and near roads are rated as more suitable.

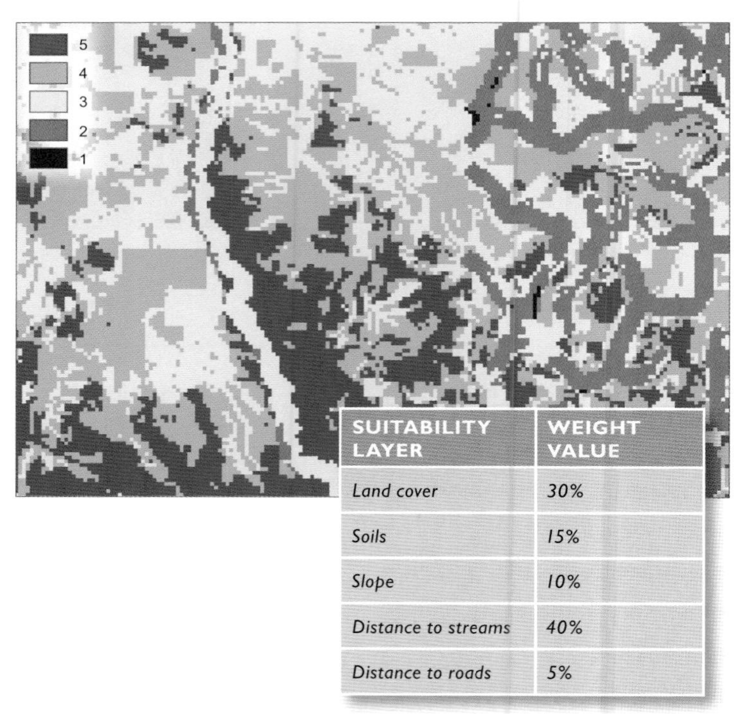

SUITABILITY LAYER	WEIGHT VALUE
Land cover	*30%*
Soils	*15%*
Slope	*10%*
Distance to streams	*40%*
Distance to roads	*5%*

In this scenario, which emphasizes protection of streams and forests, the layers for land cover and distance to streams are weighted more heavily.

EVALUATE THE RESULTS

Once you've created the overall suitability layer, you'll want to verify the results of your model. To start, visually compare the overall suitability layer to the input suitability layers to see if the pattern reflects the patterns on the suitability layers. If you weighted the various suitability layers differently, the patterns on the layers that were weighted higher should be more prominent on the overall suitability layer.

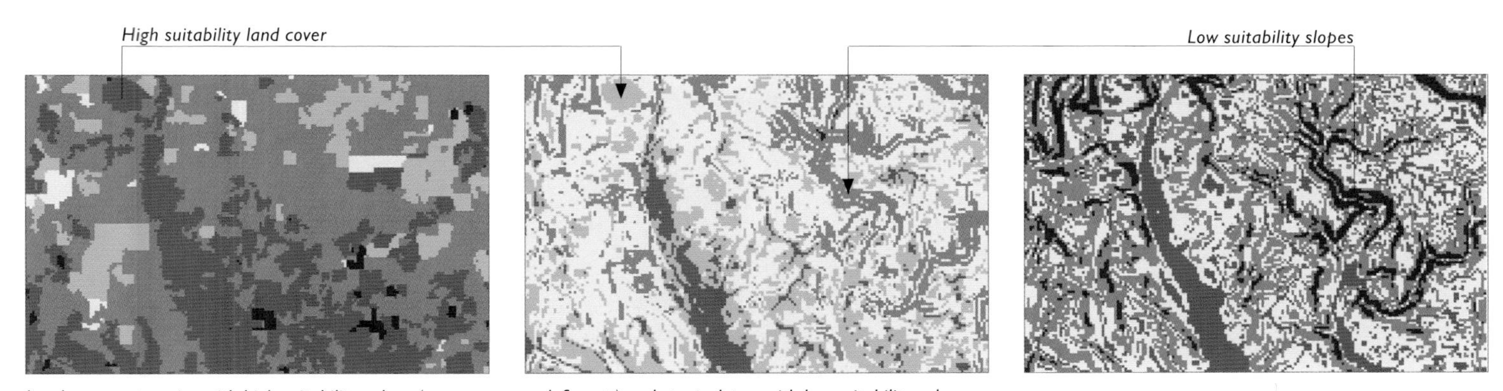

Land-cover categories with high suitability values (green areas on left map) and steep slopes with low suitability values (red areas on right map) are both reflected in the overall suitability layer (center).

Similarly, you can identify the values at different locations to see overall suitability value at that location, along with the corresponding source layer or suitability layer values.

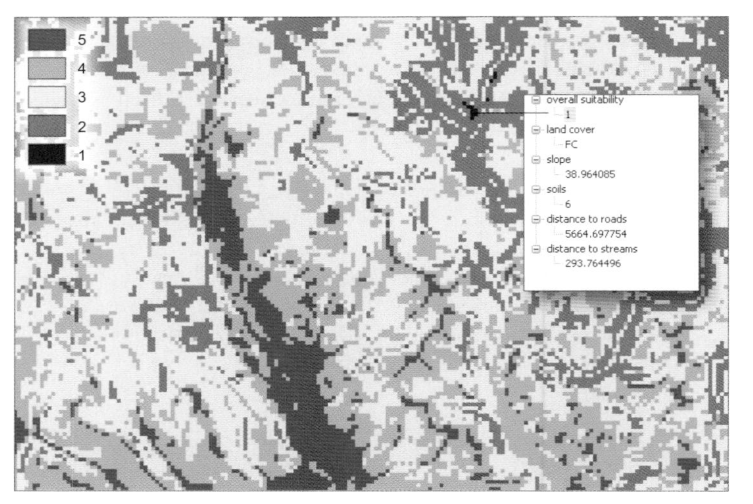

By identifying a location, you can see the values for that cell on the overall suitability layer, as well as on all of the source layers. The identified cell is covered by closed forest (FC), is on a steep slope (39 percent), on poor soils (class 6), over a mile from a road, and within 500 feet of a stream, all of which make it unsuitable for housing, as indicated by its overall suitability value of 1.

You'll want to make sure the results of your model correspond with your knowledge of the study area and the use you're modeling. If you're familiar with the study area, you may be able to see just by looking at the map if the results make sense. You may want to have colleagues or independent experts review the results. You can also check the results in the field and see if the suitable locations are in fact suitable.

You'll also want to analyze how robust the model is—that is, how much the results change if you modify the suitability values or the weights (which are in many cases based on expert opinion and are subjective). If shifting the values slightly has little effect on the results, you can have more confidence in your model. You might find that changing some criteria has little effect on the results, but changing other criteria does—the model is considered to be sensitive to these criteria. For the sensitive criteria, you'll want to take extra care to assign suitability values and weights that accurately reflect what is a suitable location for the particular use.

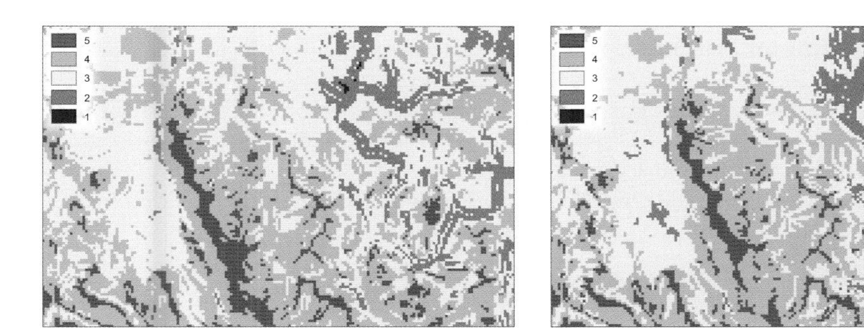

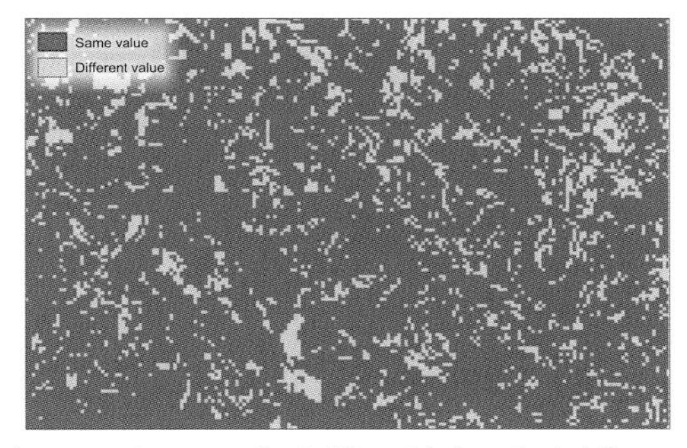

The map on the left shows overall suitability with all five input suitability layers weighted equally at 20 percent. The map in the center shows overall suitability with the soil suitability layer weighted more heavily at 40 percent, and the other four layers each weighted at 15 percent. The pattern is very similar to the map on the left (although the unsuitable—red—areas have expanded slightly), indicating that the model is not very sensitive to soil suitability. You can measure the difference between the maps by subtracting one from the other. The map on the right shows cells that have the same value on both maps in blue, and cells that have different values in orange. You can see that most of the cells (almost 85 percent) do in fact have the same value on both maps.

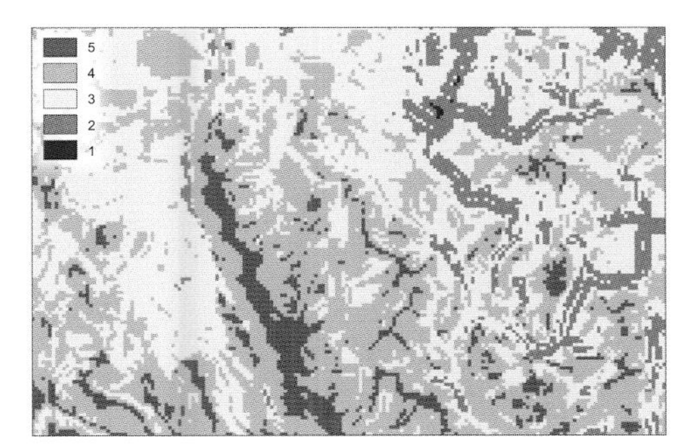

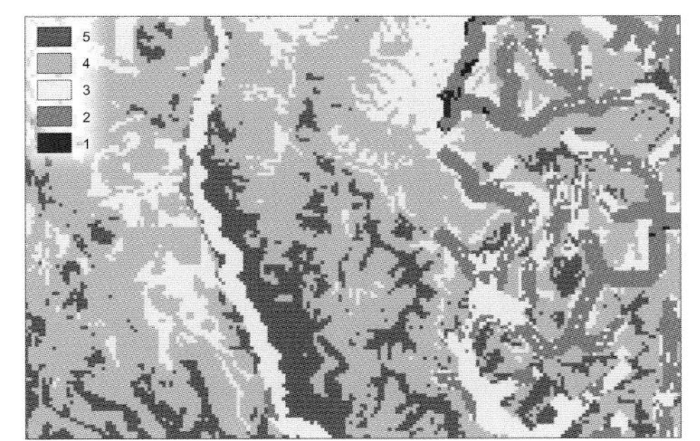

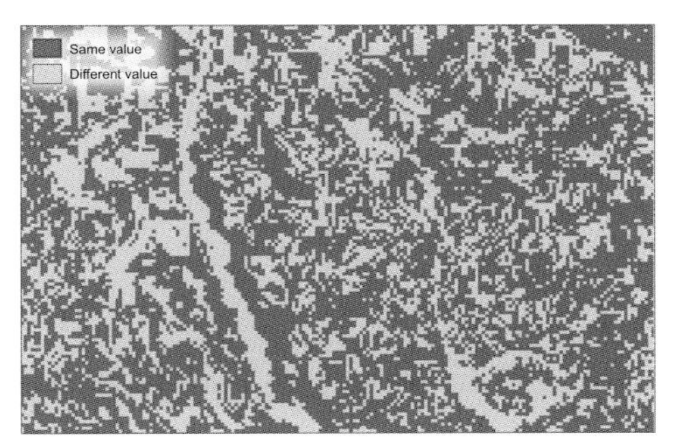

The map on the left again shows overall suitability with all five input suitability layers weighted equally at 20 percent. The map in the center shows overall suitability with the distance to streams layer weighted more heavily at 40 percent and the other four layers each weighed at 15 percent. The pattern is noticeably different—the suitable (green) areas are expanded and the influence of the distance to streams layer can be seen in the eastern portion of the map as unsuitable (red) areas. When the maps are subtracted, the result (right map) shows that fewer cells have the same value on both suitability maps (about 60 percent of cells). The fact that the pattern of suitability changes when the distance to streams layer is weighted more heavily indicates that the model is sensitive to distance to streams.

DISPLAY AND APPLY THE RESULTS

Once you're confident your model has generated useful results, you'll want to display the results and present them to others. Displaying the final suitability layer with additional layers, such as roads, streams, and terrain, will add context to the map and make it easier for people to understand the model results. You can also use the final suitability layer to derive statistics and other summary information.

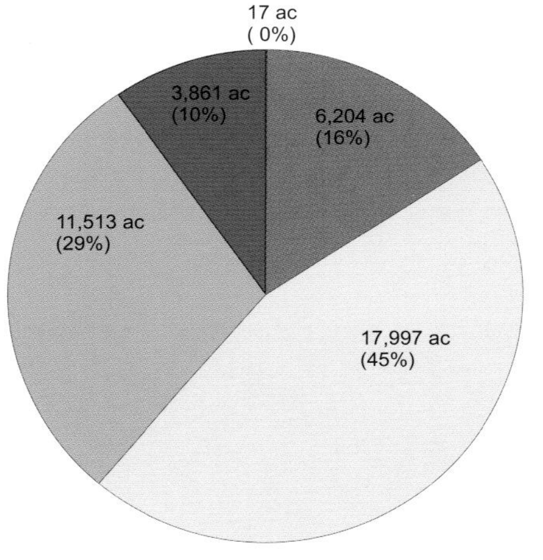

The chart shows the amount of land (and percent of the total) in each suitability category. There are, for example, 3,861 acres considered very suitable for housing development, about 10 percent of the study area, and only 17 acres considered very unsuitable (too little to be visible in the pie chart).

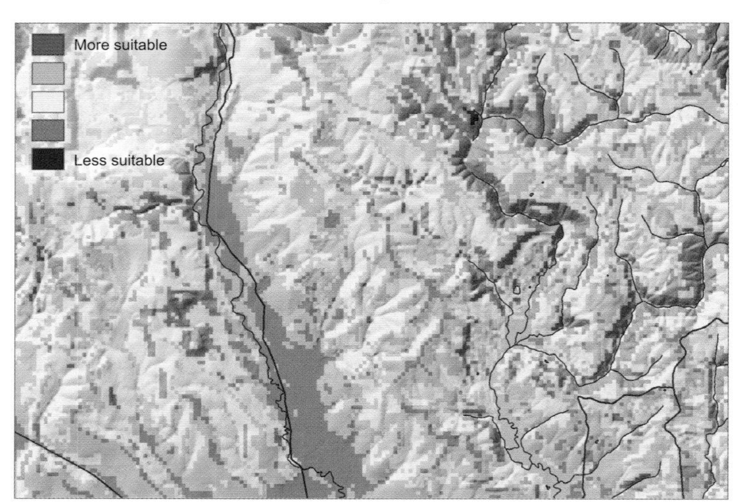

Suitability map for housing development showing major roads, streams, and terrain. It's clear that areas near roads and away from streams are more suitable. Including terrain shows that some suitable areas are along ridge lines (where slopes, which are heavily weighted in this model, are flat). These may or may not be viable locations for housing development—there may be regulations that prohibit building on ridges, the areas may be too difficult to build roads to, and the suitable areas may be too narrow in any case. Or, they may in fact be very suitable for some types of housing.

Including other related data layers on the map will assist decision makers. For example, for a housing development study, you might display suitable areas on top of land use. If you want to do additional analysis on the suitable areas, you may want to convert the suitability raster to a layer of polygons. This will make it easier, for example, to calculate the areal extent of each discrete suitable area. It will also be easier to perform overlay or selection operations to find, for example, the parcels that fall within the suitable area.

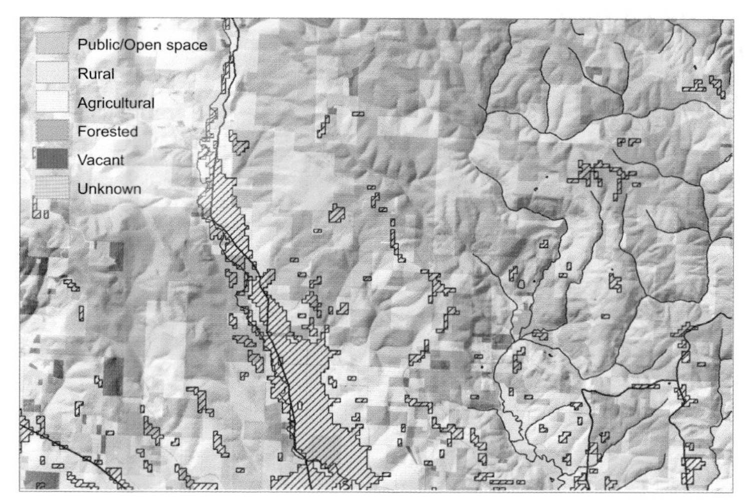

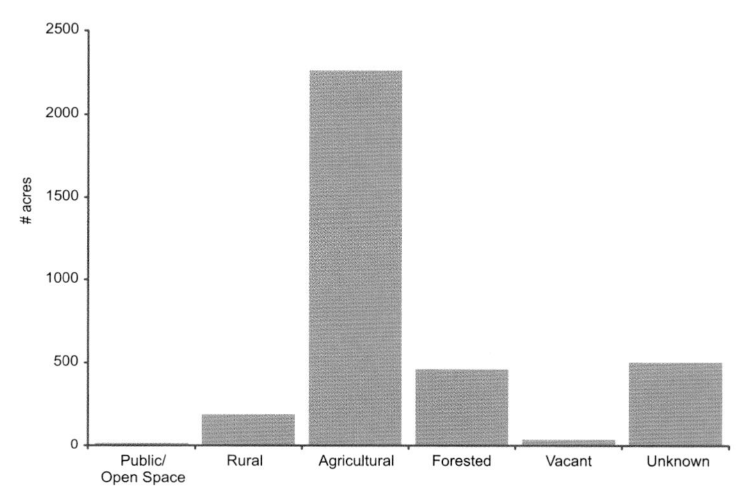

The highest suitability areas have been converted to polygons (diagonal shading) and displayed on top of parcels color-coded by land use. The chart shows the number of acres of suitable land in each land-use category.

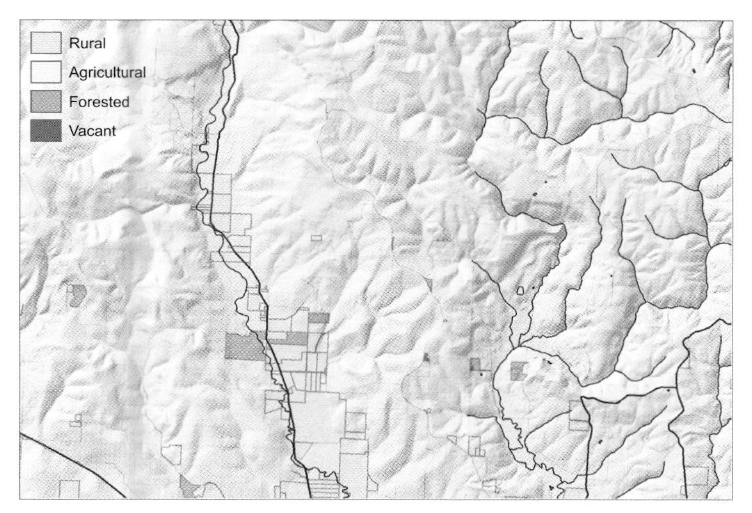

The suitable area polygons have been used to select the undeveloped parcels located in areas suitable for development.

If you use a scale of 1 to 9 to create the suitability layer, you may want to condense the number of suitability categories to make the information easier to interpret. You'd display the nine categories as three, for example.

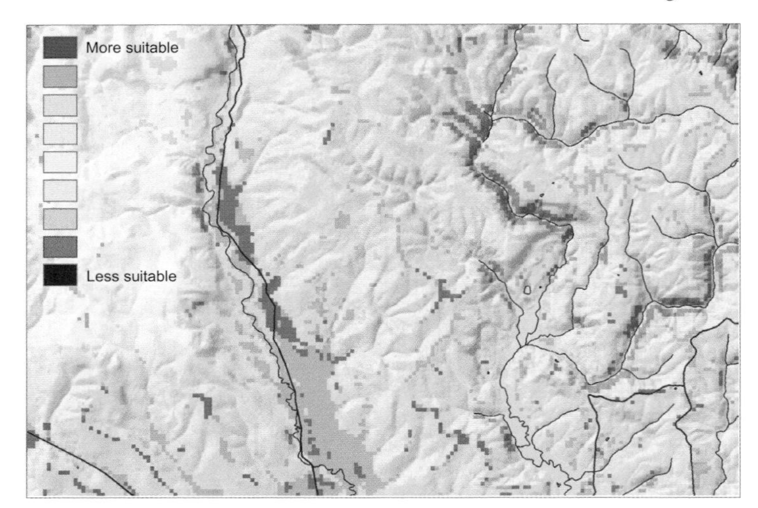

Suitability raster using nine categories—there is a fair amount of detail in the map but the various levels of suitability are hard to distinguish.

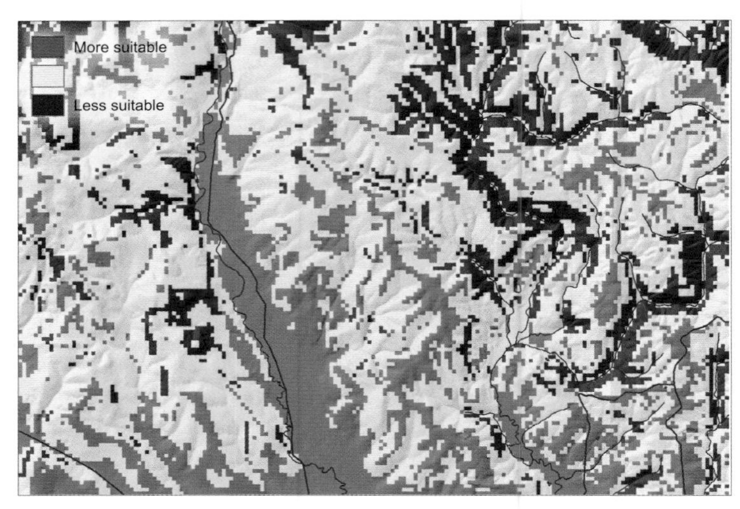

The suitability raster using nine categories has been reclassified into three categories, making the categories easy to distinguish, although some detail is lost.

When you create the map legend, use text to label the suitability categories (rather than the actual numeric values of the suitability scale). The more specific you are as to what the categories represent, the more meaningful the map will be for readers. For example, if your model shows locations at risk of wildfire, with three categories, you might specify legend labels of "Extreme wildfire risk," "Moderate wildfire risk," and "Slight wildfire risk." (In this case you might also reverse the colors so that the highest category is red and the lowest is green). Map readers will know that the lowest category represents some risk and not misinterpret the map as indicating that those locations have no risk.

If your analysis involved creating alternate scenarios, you can combine the results to show opportunities and constraints. For example, if you've modeled suitability for housing using one model that minimizes negative impacts to the environment and another model that minimizes development costs, you can create a map showing which areas are suitable for development under both scenarios, which are unsuitable under both, and which are suitable in one and not the other. This works best with suitability maps that use only three levels of suitability (otherwise, the various combinations between the scenarios make the map difficult to interpret).

This scenario shows suitability for land development based on environmental impacts. Areas in red are unsuitable due to high environmental impact. Roads (red lines) and streams (blue lines) are also displayed.

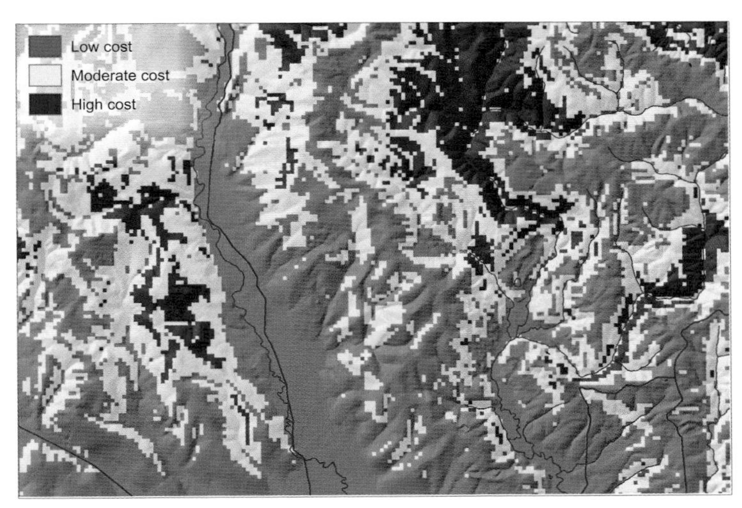

This scenario shows suitability based on development costs. Areas in red are unsuitable due to high development costs.

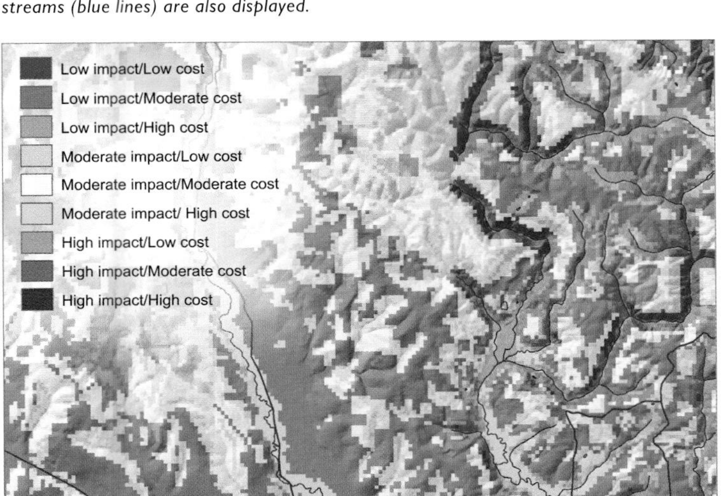

Combining the two suitability maps displays every combination of environmental impact and development cost. Areas with high impact and high development cost are unsuitable for development under both scenarios, while areas with low impact and low cost would be good candidates for development.

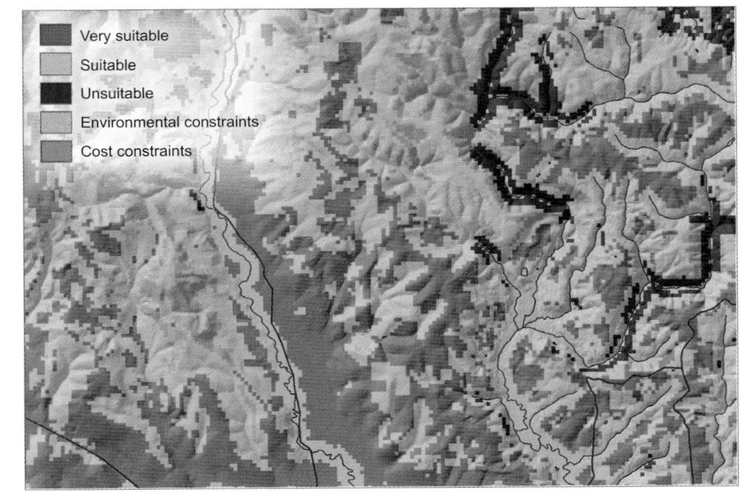

The combined suitability map can be classified into fewer categories to focus discussions. Areas in dark green are suitable under both scenarios. Areas in light orange have low development cost but high environmental impact. Areas in blue have low environmental impact but high development costs.

Finally, you will want to be able to explain how you obtained the results of your model. Displaying a ModelBuilder model document is a good way to present the criteria, the data layers that were used, and the process (see chapter 1, "Introducing GIS modeling"). It also serves as documentation for the model if you want to share it with others or need to revisit the model in the future. The document for the housing development suitability model shows that—using the weighted overlay method—after the source layers with continuous values were reclassified into meaningful classes, all the layers were assigned values on a suitability scale of 1 to 5, and the suitability layers combined to create the overall suitability layer using the assigned weights.

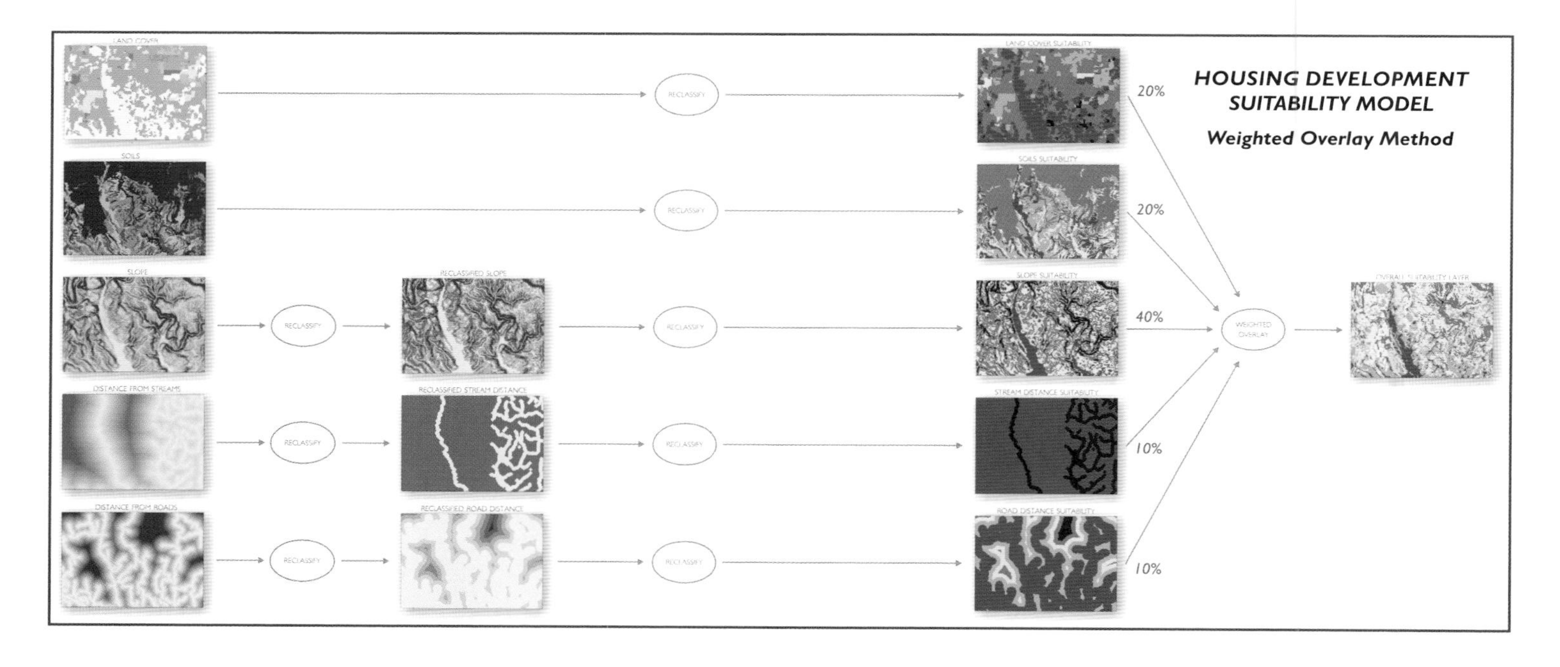

Use fuzzy overlay to rate locations based on their suitability when many of your criteria require source layers having continuous values and when the breakpoints between suitable values are not well defined.

For example, suppose you want to identify areas in a region that are suitable mule deer habitat (perhaps you are a researcher looking for the best places to study the deer). One site characteristic for mule deer habitat is low elevation. It may be difficult, though, to define the specific elevation ranges that are the most and least suitable (especially since the mule deer move around). You may only generally know that lower elevations are very suitable, high elevations are not suitable, and mid-range elevations are somewhat suitable. This is in contrast to, say, rating suitable locations for a particular forest type where the elevation ranges are fairly distinct (mountain hemlock might mainly occur between 4,500 and 5,400 feet, while alpine forest occurs above 6,000 feet).

Fuzzy overlay also provides a number of options for combining the individual suitability layers—beyond a weighted sum that the weighted overlay method provides—to create the overall suitability layer. You can, for example, specify that if one particular desired site characteristic is present at a location, then the location is considered suitable (relatively speaking) regardless of whether the other desired site characteristics are present. Having these options allows you to better capture how people think about what makes a location suitable.

Fuzzy overlay is based on the logic of set theory, in which, traditionally, you determine whether a value is a member of a set, or not. A variation on set theory, known as fuzzy logic, allows you to specify the *likelihood* that a given value is a member of the set (rather than merely specifying whether the value is either in or out of the set). In fuzzy logic, a numeric scale is used, with 1 representing full membership in the set and 0 representing nonmembership. Source layer values are assigned corresponding values on this continuous scale according to the likelihood they have membership in the set. These values are known as "fuzzy membership" values. In the mule deer habitat example, a very low elevation would get a fuzzy membership value of 1, while a mid-range elevation would get a value around 0.5.

The method as applied to suitability analysis has been researched and used extensively by geologists Graeme Bonham-Carter, Gary Raines, and others for finding likely locations to look for mineral deposits such as gold and tin. The goal was to capture—in a GIS—the knowledge of field geologists who had a sense of where to look for deposits based on their years of experience. These geologists could verbalize, but not necessarily quantify, the site characteristics they were looking for. A key element of the method early on was to show the results to the geologists, have them note where the results matched or differed from their knowledge of the study area, and then fine-tune the models accordingly.

In the GIS, the site characteristics are represented as cell values on raster layers (elevation values on one raster layer, land-cover categories on another, distance to streams on a third). By assigning fuzzy membership values to each layer and then combining the layers, you can calculate the likelihood that a given location has membership in the desired set on one or more of the input layers. The result is a layer showing the locations most likely and least likely to be suitable.

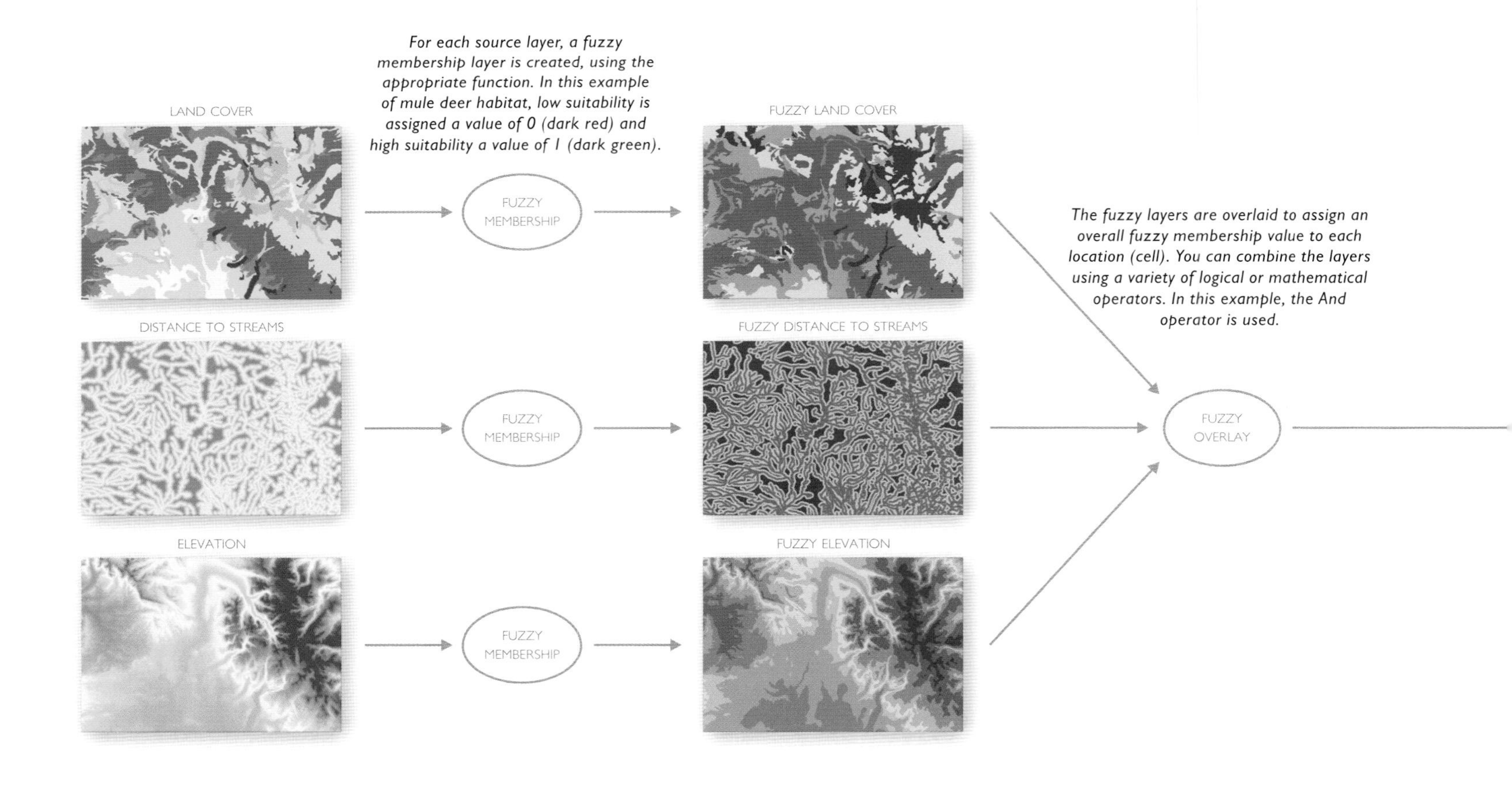

For each source layer, a fuzzy membership layer is created, using the appropriate function. In this example of mule deer habitat, low suitability is assigned a value of 0 (dark red) and high suitability a value of 1 (dark green).

The fuzzy layers are overlaid to assign an overall fuzzy membership value to each location (cell). You can combine the layers using a variety of logical or mathematical operators. In this example, the And operator is used.

The process for fuzzy overlay:

1. Collect criteria layers
2. Assign fuzzy membership values for each layer
3. Combine the fuzzy layers
4. Evaluate the results
5. Display and apply the results

The overall fuzzy membership layer has values from 0 (low suitability—shown in dark red) to 1 (high suitability—shown in dark green).

COLLECT THE SOURCE LAYERS

The source layers represent the site characteristics that are factors in determining which locations are suitable for the use you're modeling. Your own knowledge of the particular use or phenomenon, or a search of the literature, will guide you to likely source layers. For example, if you're modeling mule deer summer habitat, you might collect layers for land cover, distance to streams, and elevation. In the terminology of fuzzy overlay, the values on each layer—such as elevation in meters or land-cover type (riparian, meadow, and so on)—are referred to as "observed" values. (They are also sometimes referred to as "crisp" values, to distinguish them from the corresponding fuzzy membership values that will be assigned.)

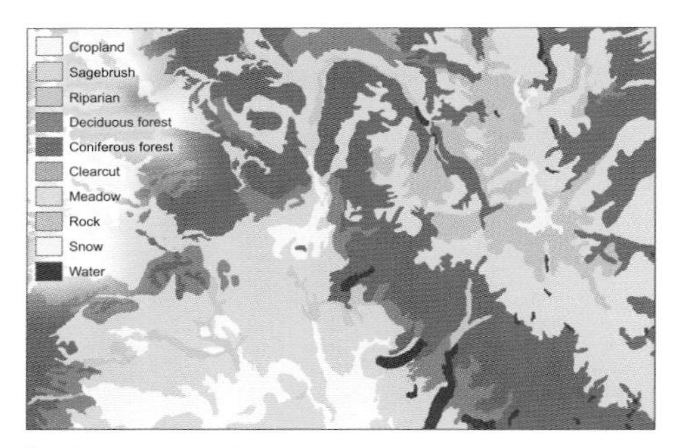

Land-cover categories.

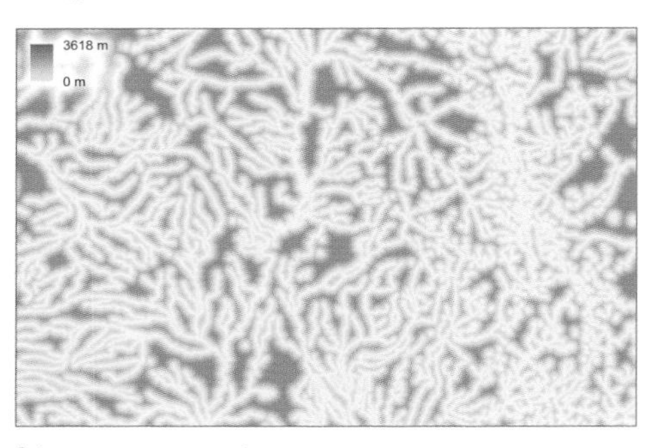

Distance to streams, in meters.

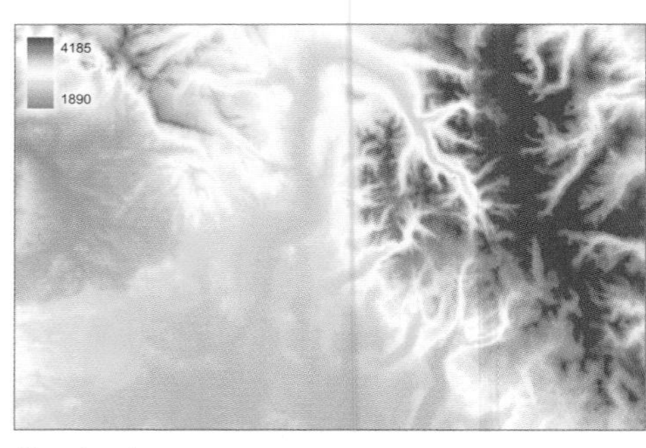

Elevation, in meters.

As with other suitability modeling methods, you start by listing your criteria. For example, for a model to rate areas that are likely mule deer summer habitat, your criteria statements might look like the ones in the following table.

CRITERIA	SOURCE LAYER
Areas that provide cover are more suitable	*Land cover*
Areas near streams are more suitable	*Distance to streams*
Lower elevations are more suitable	*Elevation*

ASSIGN FUZZY MEMBERSHIP VALUES TO OBSERVED VALUES

For each source layer, you specify the likelihood that each observed value is a member of the defined set of suitable locations, based on the values for that criterion. Likelihood is indicated by assigning a value on a scale of 1 (very likely) to 0 (not likely). For example, if one of your criteria is that low elevations are suitable and high elevations are not, you'd assign a value of 1 to the lowest elevations and 0 to the highest elevations. Elevations in between are assigned values between 1 and 0 accordingly. A new layer is created with corresponding fuzzy membership values ranging from 1 to 0. The assignment of these values is subjective—this is where expert knowledge comes into play. You (or other experts) define the relationship between observed values and fuzzy membership values that captures the best understanding of the phenomenon or use you're modeling (that is, how likely it is that a particular observed value is part of the defined set of suitable locations).

You may need to fine-tune the fuzzy membership values for each source layer by creating the fuzzy membership layer, reviewing the results with experts, adjusting the values, and re-creating the layer. After several iterations, the fuzzy membership layer will better reflect the best knowledge of the experts.

Assigning fuzzy membership to categories

To assign fuzzy membership values to a source layer having categorical data, you first determine the fuzzy membership value that will be assigned to each category, again, based on your knowledge of the phenomenon. You then create the fuzzy membership layer by assigning each category value the corresponding fuzzy membership value.

If you have a few categories, you may only need fuzzy membership values that use a single decimal place (0.0, 0.1, 0.2, 0.3, and so on, up to 1.0) That provides up to eleven unique fuzzy membership values to assign to the various categories. If you have many categories, you may need more than these eleven numbers to create a wider range of fuzzy membership values and capture a finer gradation of likely set membership. If so, you'd assign fuzzy membership values on a scale with two decimal places (0.01 to 0.99). For example, with forty categories of vegetation types, it may be that mountain hemlock has a higher likelihood of being a member of the set of suitable locations than does lodgepole pine (with a fuzzy membership value of 0.8) but less likelihood than subalpine fir (with a fuzzy membership of 0.9). With a scale that uses two decimal places, you could assign a fuzzy membership value of 0.85 to mountain hemlock.

The top map represents land-cover categories. The bottom map assigns fuzzy membership values to these categories for the set "land cover preferred by mule deer." In this example, five values were used, ranging from 0.9, indicating that a location has a high likelihood of membership in the set (that is, preferred land cover) to 0.1, indicating a low likelihood of membership (land cover avoided by mule deer). A value of 0.5 indicates neutral membership.

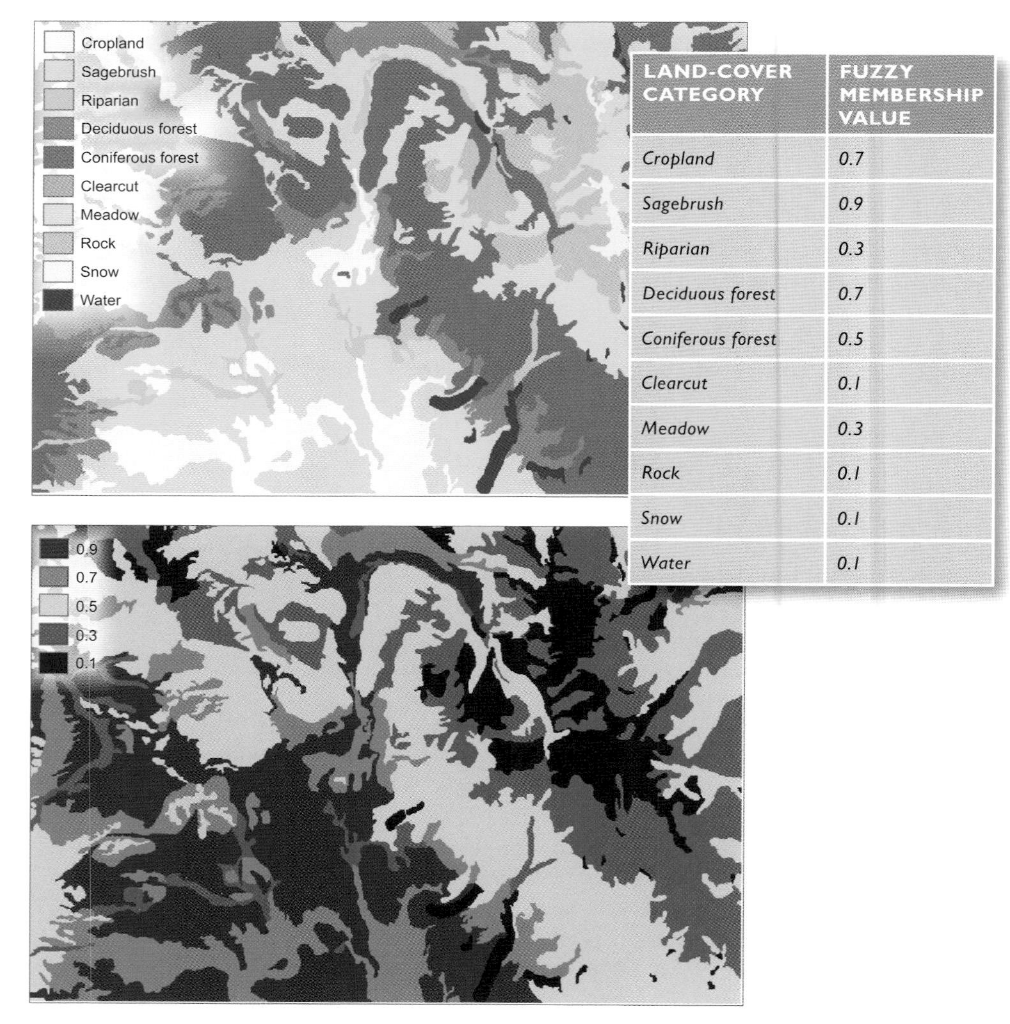

LAND-COVER CATEGORY	FUZZY MEMBERSHIP VALUE
Cropland	*0.7*
Sagebrush	*0.9*
Riparian	*0.3*
Deciduous forest	*0.7*
Coniferous forest	*0.5*
Clearcut	*0.1*
Meadow	*0.3*
Rock	*0.1*
Snow	*0.1*
Water	*0.1*

Some categorical source layers might represent essentially a single category—the presence or absence of the site characteristic indicates suitability. For example, a layer may show locations of a food source for a particular wildlife species. You'd assign two fuzzy membership values: locations where the characteristic is present are assigned a fuzzy membership value of 1; locations where it's known to be absent are assigned a value of 0.

DISTANCE TO STREAMS (METERS)	FUZZY MEMBERSHIP VALUE
0	1.000
95	0.967
134	0.954
190	0.935
212	0.927
269	0.907
285	0.902
301	0.896
343	0.882
380	0.869
392	0.865
403	0.861
. . .	. . .

Assigning fuzzy membership to continuous values

For continuous observed values such as elevation or distance to streams, assigning each observed value a fuzzy membership value would be tedious since there may be hundreds or thousands of observed values. Instead, you define the relationship between observed values and fuzzy membership. To do this, you select and run a mathematical function to assign the fuzzy membership values automatically. The function essentially plugs each observed value into an equation to calculate the corresponding fuzzy membership value. The result is a new raster layer with each cell assigned a fuzzy membership value based on its original, observed value. The results of the function can be displayed as a table of paired values, with each observed value assigned a corresponding fuzzy membership value.

Since there may be many pairs of values, a graph (rather than a table) is often used to express the relationship between observed and fuzzy membership values. Observed values are displayed on the horizontal axis and fuzzy membership values on the vertical axis. The pairs of values trace a line or curve on the graph.

To select the appropriate function, you first identify the general nature of the relationship—whether large observed values have high fuzzy membership (that is, are likely to be suitable locations); whether small observed values have high fuzzy membership; or whether observed values in the middle of the range have high fuzzy membership (with larger and smaller observed values having low fuzzy membership). You also need to consider the rate at which fuzzy membership values decrease as observed values decrease (or increase) and whether the rate of change is constant or varies. For each function, you define the nature of the relationship by specifying the appropriate parameters.

Assigning high fuzzy membership to large or small observed values

When high fuzzy membership is assigned to either large or small observed values, the rate of change can be constant or can vary. It may be—for the source layer you're assigning values to—that fuzzy membership increases at the same rate at which observed values increase. For example, you may know that mountain goats prefer higher elevations and that the likelihood of goats being found increases at a constant rate as elevation increases. The relationship between fuzzy membership and elevation is linear.

Elevation, in meters.

For the Linear function, you specify a minimum and maximum value to define the relationship. The minimum observed value is assigned a fuzzy membership value of 0. The maximum observed value is assigned a value of 1. Observed values between these extremes are assigned fuzzy membership values between 0 and 1, depending on where they fall. To assign high fuzzy membership to large observed values (for example, if high elevations are more suitable), set the maximum to be greater than the minimum—the slope of the line will be positive (fuzzy membership increases as observed values increase).

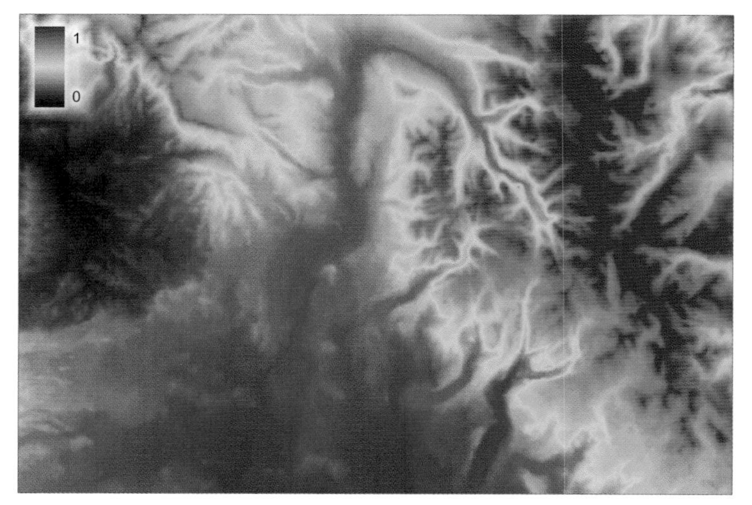

Fuzzy membership values for elevation using the Linear function with a minimum of 1,890 meters and a maximum of 4,185 meters. Locations at higher elevations are assigned a higher fuzzy membership (dark green on the map).

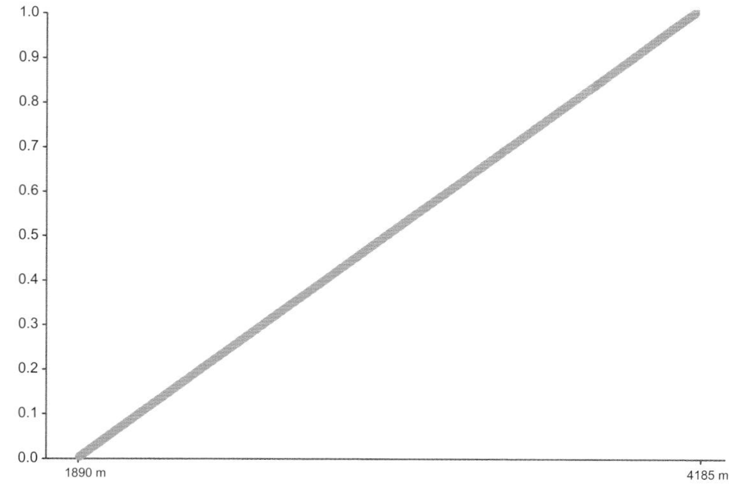

The relationship between elevation (horizontal axis) and fuzzy membership (vertical axis) can be shown on a graph. With the Linear function, fuzzy membership increases at a constant rate as elevation increases.

You can also use the maximum and minimum to establish threshold values above and below which observed values are definitely considered a member of the set (the maximum observed value) or definitely not a member of the set (the minimum observed value). Assume, for example, that for your model higher elevations are more suitable (more likely to have membership in the set of suitable locations). If the elevation range for the study area is 1,890 meters to 4,185 meters and you know that elevations greater than 3,500 meters are definitely suitable and elevations less than 2,300 meters are definitely not suitable, you'd specify 3,500 as the maximum and 2,300 as the minimum. All cells with elevations greater than or equal to 3,500 will be assigned a fuzzy membership value of 1. All cells with values less than or equal to 2,300 will be assigned a fuzzy membership value of 0. Cells with elevation values between 2,300 and 3,500 will be assigned fuzzy membership values between 1 and 0.

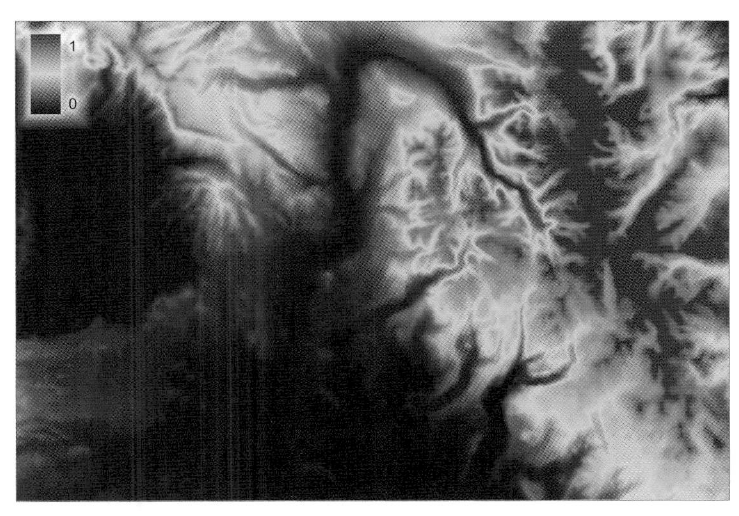

Fuzzy membership values for elevation using the Linear function with a minimum of 2,300 meters and a maximum of 3,500 meters. Locations below 2,300 meters elevation are assigned a fuzzy membership value of 0 (dark red on the map); locations above 3,500 meters are assigned a fuzzy membership value of 1 (dark green).

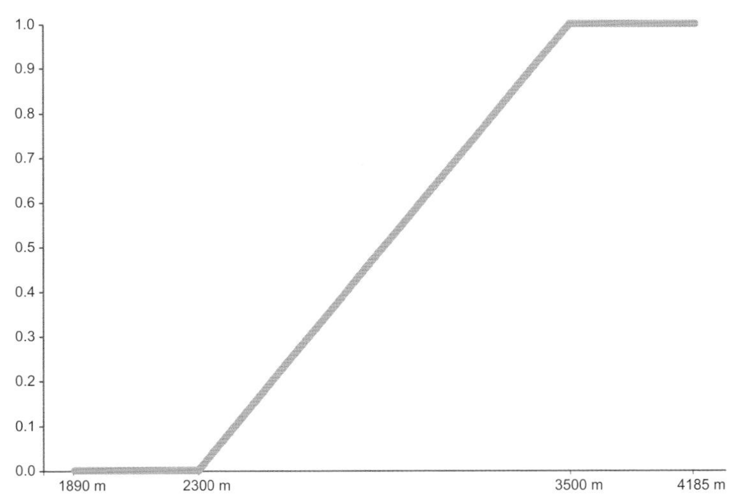

The relationship between elevation (horizontal axis) and fuzzy membership (vertical axis), using a minimum of 2,300 meters and a maximum of 3,500 meters.

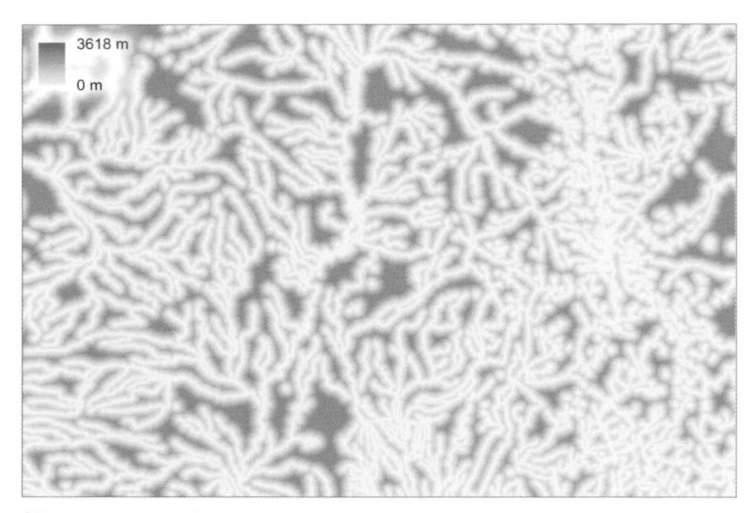

Distance to streams, in meters.

To assign high fuzzy membership to low observed values, set the maximum to be less than the minimum—the slope of the line will be negative (fuzzy membership decreases as observed values increase). For example, you may know that mule deer like to be near streams in the summer, and the likelihood of mule deer being found decreases at a constant rate as the distance from streams increases.

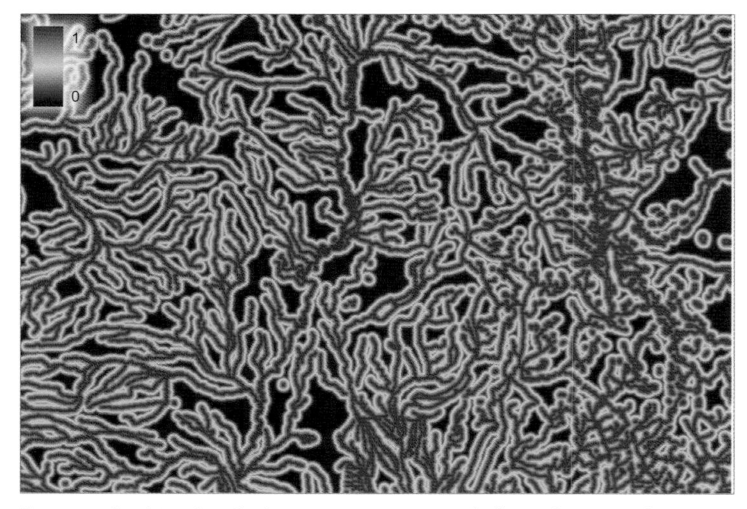

Fuzzy membership values for distance to streams using the Linear function with a minimum of 3,618 meters and a maximum of 0 meters. Locations near streams are assigned a higher fuzzy membership (dark green on the map). In this example, there are no threshold values—the fuzzy membership values decrease starting with the smallest distance (0 meters) and continuing to the largest (3,618 meters).

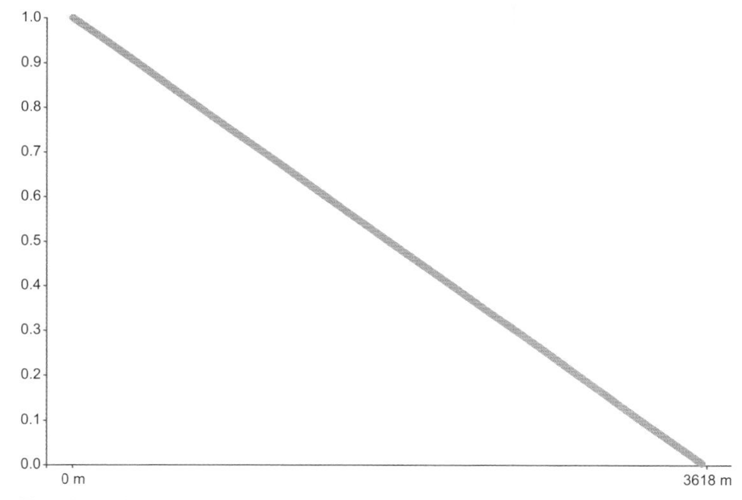

The relationship between distance to streams (horizontal axis) and fuzzy membership (vertical axis). The graph shows that fuzzy membership decreases as distance to streams increases.

In many cases, the relationship between observed values and fuzzy membership is not linear. Instead, membership in the defined set may decrease more or less rapidly as observed values increase. For example, within your study area, mule deer may prefer to be at elevations less than 2,600 meters, and the likelihood of mule deer being found at elevations higher than this decreases rapidly. To capture nonlinear relationships between observed values and fuzzy membership values, two other functions are used—Small and Large.

To assign high fuzzy membership values to small observed values, use the Small function. For example, if lower elevations are suitable habitat, use Small (small elevation values will have high fuzzy membership values). When using the Small function, you specify the midpoint—the observed value that is assigned a fuzzy membership value of 0.5. Observed values smaller than the midpoint value are assigned high fuzzy membership values, and observed values larger than the midpoint are assigned low fuzzy membership values. The midpoint can represent neutral fuzzy membership—for example, if mid-level elevations are sometimes used by mule deer and sometimes not. The midpoint can also represent uncertain fuzzy membership. For example, you may know that very low elevations are preferred by mule deer and very high elevations are avoided, but it's unclear whether mid-range elevations are desirable, or even which elevation is too high for the mule deer.

You also specify the rate of decrease of the fuzzy membership values. This parameter is known as the "spread." The spread defines how tightly the assigned fuzzy membership values cluster around the midpoint. If you specify a large spread value, the fuzzy values decrease rapidly from the midpoint, hence few observed values will be assigned fuzzy membership values in the middle of the scale (near 0.5). If you specify a small spread, the fuzzy membership values decrease more gradually and more observed values will be assigned fuzzy membership values in the middle of the scale. The spread is usually defined as a value between 1 and 10, although you can specify numbers beyond this range, if necessary, to capture the relationship between observed values and membership in the set.

Elevation, in meters.

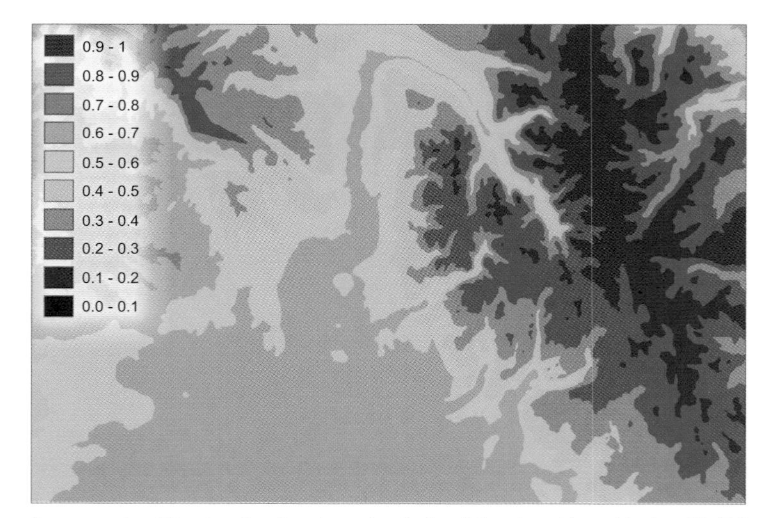

Fuzzy membership values for elevation, using the Small function with a midpoint of 2,600 meters and a spread of 5. Low elevation values are assigned high fuzzy membership values.

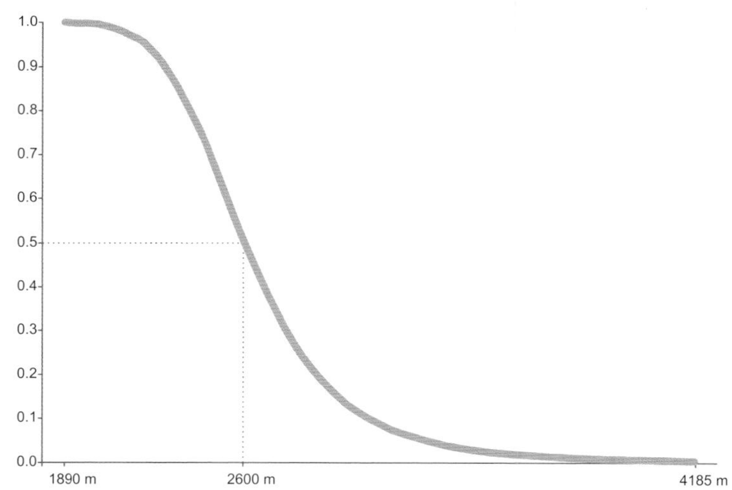

The graph allows you to visualize the relationship between observed values (elevation, in this example) and fuzzy membership values. The midpoint and the spread together define the relationship and the shape of the curve. The midpoint of 2,600 is assigned a fuzzy membership value of 0.5. With a spread of 5, the slope of the curve is moderately steep—fuzzy membership decreases fairly rapidly for locations with elevations higher than about 2,200 meters.

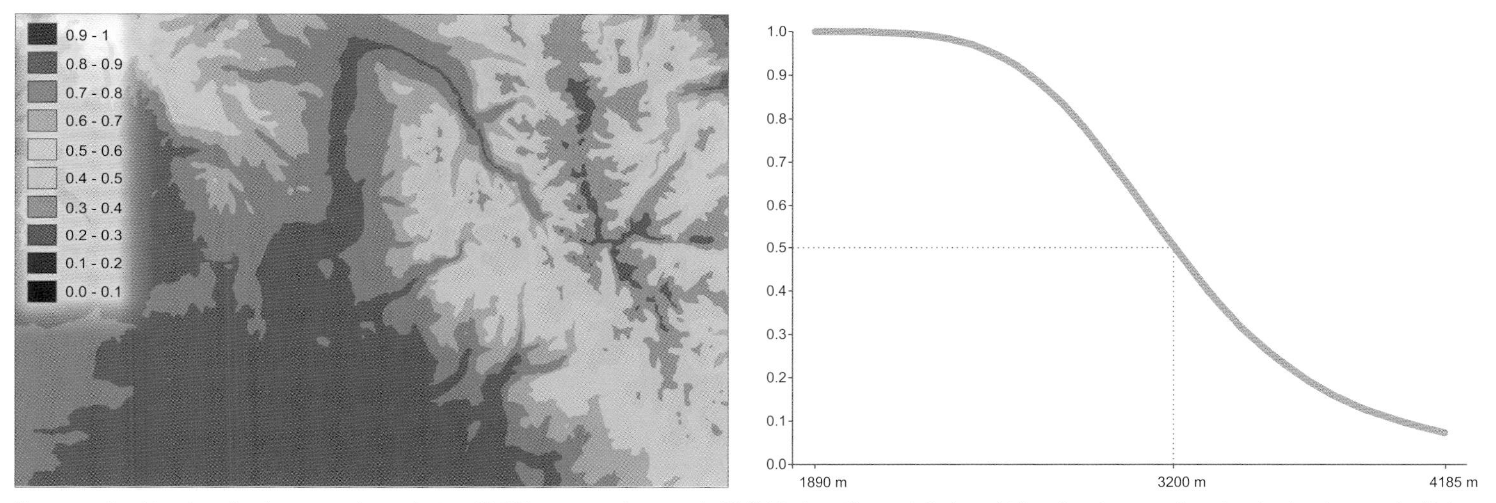

Fuzzy membership values for elevation, with a midpoint of 3,200 meters and a spread of 5. With the midpoint shifted to a higher elevation, more low-elevation locations receive high fuzzy membership values, and low fuzzy membership values are restricted to the highest elevations.

You choose the spread based on the best knowledge you have about the phenomenon you're modeling. For example, if one of your layers is elevation, and you know based on observation or research that almost all mule deer congregate at elevations below 2,600 meters (and above this level the number of mule deer drops off sharply), you'd use a large spread value to ensure fuzzy membership falls off rapidly. If, on the other hand, most mule deer are found below 2,600 meters elevation, but some are also found at higher elevations (although fewer and fewer as the elevation increases), you'd use a smaller spread value. Fuzzy membership for locations at higher elevations will decrease, but not as rapidly as with a large spread.

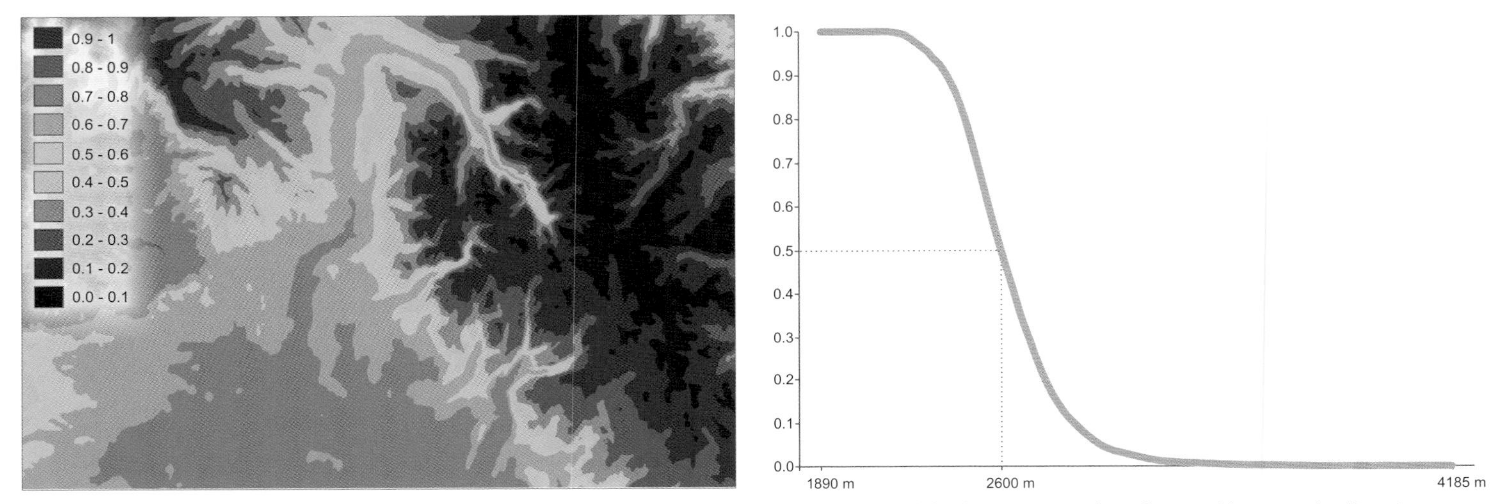

Fuzzy membership values for elevation, with a midpoint of 2,600 meters and a spread of 8. With a large spread, few locations are in the mid-range of fuzzy membership values—most locations have either very high or very low values, as can be seen on the map and the graph.

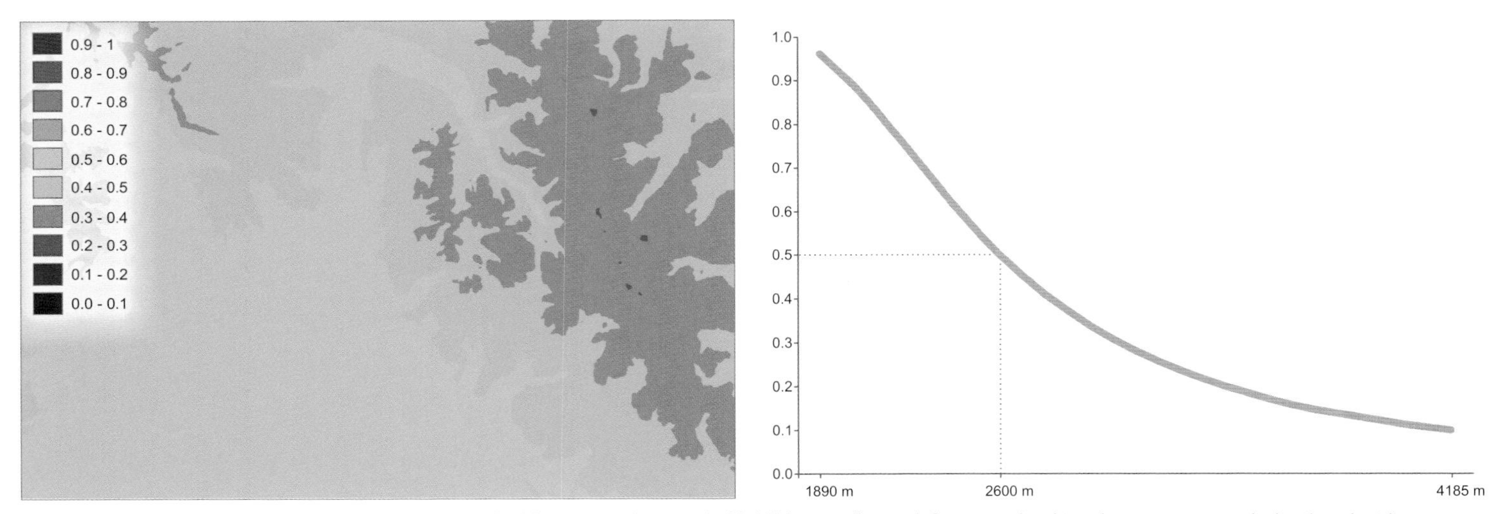

Fuzzy membership values for elevation, with a midpoint of 2,600 meters and a spread of 2. With a small spread, fuzzy membership values are more evenly distributed, with more locations in the mid-range of fuzzy membership values (light green, yellow, and light orange on the map).

To assign high fuzzy membership to large observed values, use the Large function. For example, if, for your particular model, high elevations are suitable (perhaps you're modeling mountain goat habitat), you'd use the Large function to assign high fuzzy membership values to high elevations. As with the Small function, the observed value used for the midpoint represents neutral fuzzy membership (0.5), and the spread controls the rate at which fuzzy membership increases from low to high.

Elevation, in meters.

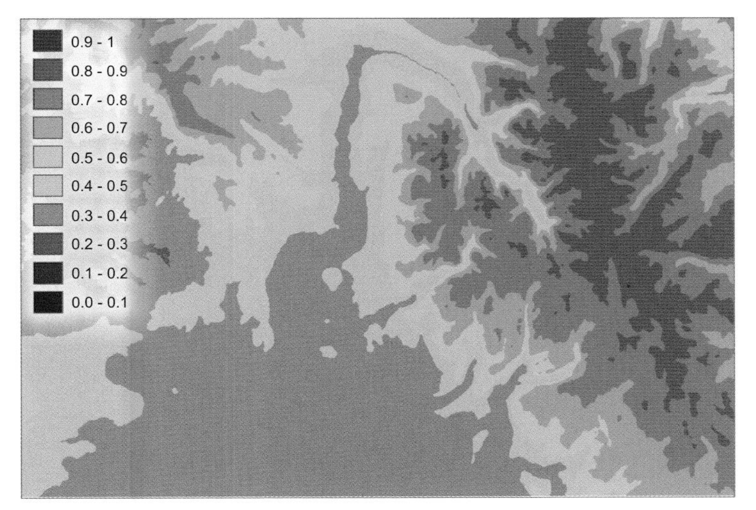

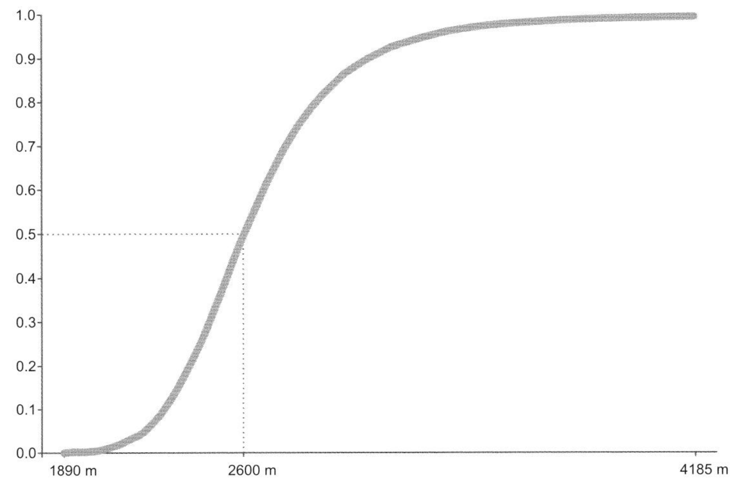

Fuzzy membership values for elevation, using the Large function with a midpoint of 2,600 meters and a spread of 5. High elevation values are assigned high fuzzy membership values (dark green on the map).

Assigning high or low fuzzy membership to observed values above the mean

In applications such as geology and biology, suitable locations are sometimes defined in part by high or low levels of something. For example, high concentrations of zinc in soil or rock can indicate the presence of tin deposits. Similarly, areas having extremely low annual rainfall could indicate suitable habitat for plants such as succulents. In such applications, scientists distinguish a normal "background" level from an extremely high or low level, termed an "anomaly." The background level is often defined as the mean value for the study area (for example, the mean of the average annual rainfall across the study area). The degree to which there is an anomaly is defined in terms of a number of standard deviations above or below the mean for a measurement at a given location.

The standard deviation is a measure of the extent to which values vary from the mean. The assumption is that most observed values cluster near the mean, with fewer values both higher and lower than the mean. In a normal distribution (bell curve), about 68 percent of observed values fall within one standard deviation above and below the mean value (about 95 percent of values are within two standard deviations). Geologists and other scientists looking for anomalies in the landscape often consider one standard deviation from the mean to be somewhat anomalous from the background value and two (or more) standard deviations very anomalous.

Average annual rainfall, in millimeters. The mean average annual rainfall for the study area is 635 mm, with a standard deviation of 169 mm.

Two functions are used to assign fuzzy membership to observed values greater or less than the mean. The functions are called MS Small and MS Large (MS standing for mean and standard deviation). Rather than requiring you to define the relationship between observed and fuzzy membership values (by specifying a midpoint and spread as with the Small and Large functions), these functions calculate the mean and standard deviation and use these values to define the relationship. They essentially define the relationship based on the distribution of observed values. The mean is calculated by adding the values of all cells on the map and dividing by the number of cells.

If low values indicate suitable locations, use MS Small. With MS Small, any observed values less than the mean are considered to be part of the set of suitable locations and are assigned fuzzy membership values of 1. MS Small assigns high fuzzy membership (but less than 1) to observed values greater than the mean, with decreasing fuzzy membership values as observed values increase. (For example, for a layer of average annual rainfall, if the mean is 635 mm with a standard deviation of 169 mm, a location with a value of 640 mm would be assigned a fuzzy membership value of 0.95, while a location with a value of 1,000 mm would be assigned a fuzzy membership value of 0.28.)

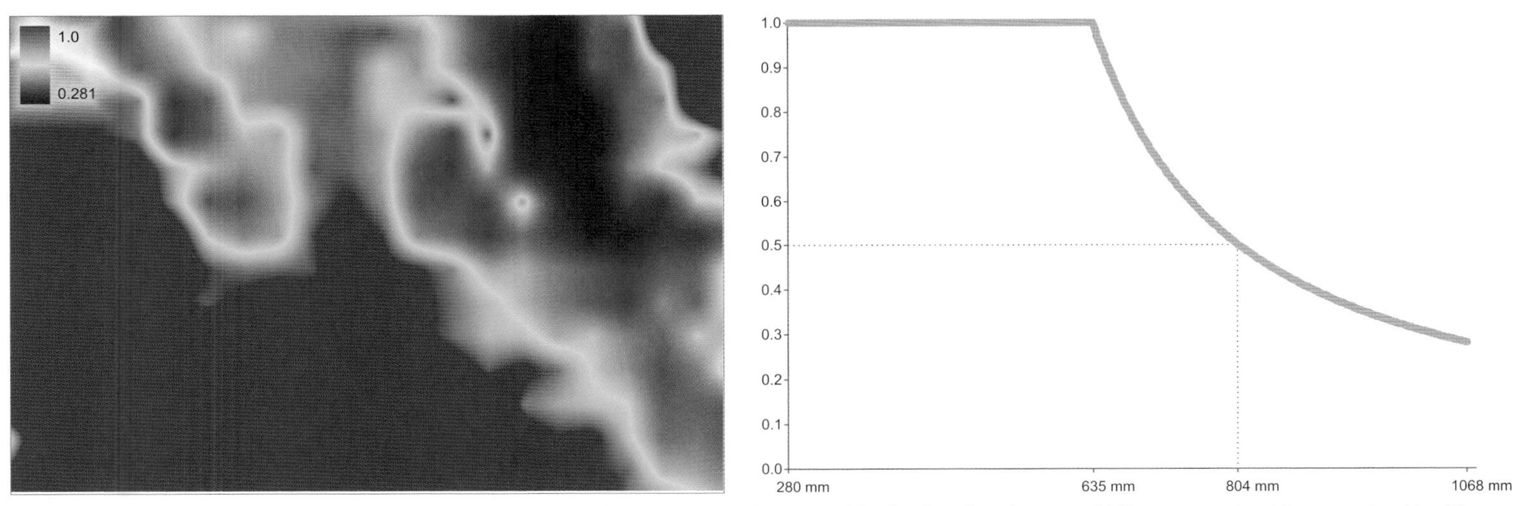

Fuzzy membership values for average annual rainfall, using the MS Small function. Locations having rainfall values less than the mean of 635 mm are assigned fuzzy membership of 1. Locations having rainfall values greater than the mean are assigned decreasing fuzzy membership values as shown by the curve. Locations having a rainfall value 1 standard deviation above the mean (804 mm) are assigned a fuzzy membership of 0.5.

If high values indicate suitable locations, use MS Large. You might use MS Large if, for example, you're modeling habitat for a species that is found where there are very high levels of rainfall within the study area. The difference between MS Large and MS Small is in how values less than the mean are treated—with MS Small, locations with values less than the mean are assigned a value of 1 (high fuzzy membership), while with MS Large, values less than the mean are assigned a value of 0. (MS Large does not, as you might think, assign a fuzzy membership value of 1 to values *greater* than the mean.) MS Large assigns low fuzzy membership to observed values just above the mean, increasing toward 1 as observed values increase.

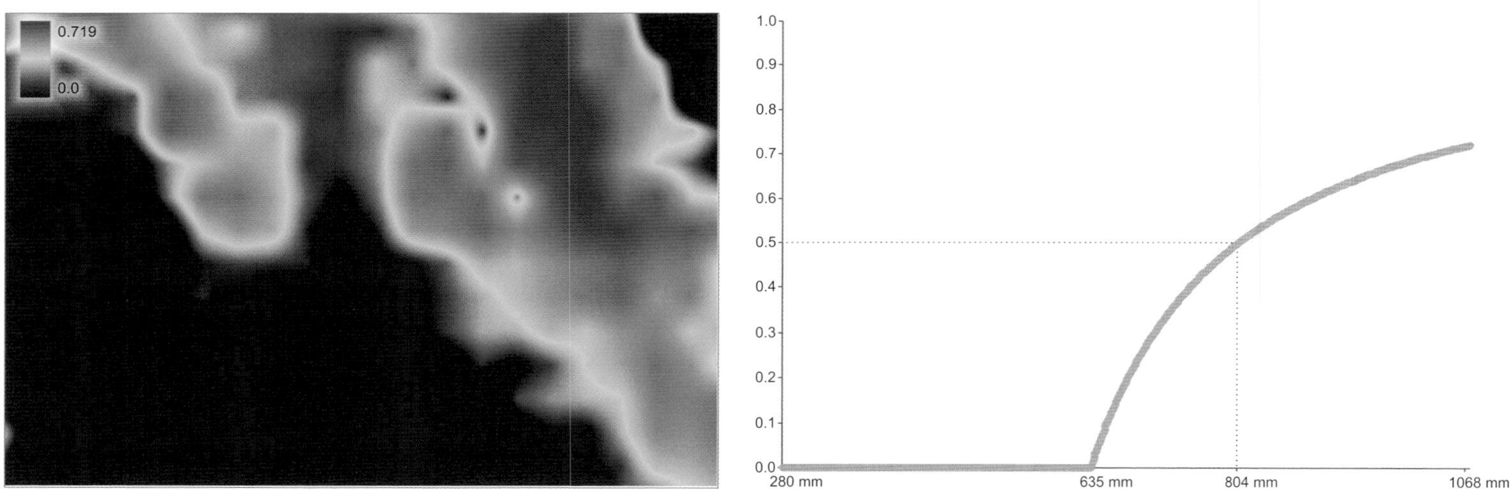

Fuzzy membership values for average annual rainfall, using the MS Large function. Locations having rainfall values less than the mean of 635 mm are assigned fuzzy membership of 0. Locations having rainfall values greater than the mean are assigned increasing fuzzy membership values as shown by the curve.

With both MS Small and MS Large, you can specify a multiplier to adjust the base value below or above which a fuzzy membership value of 0 is assigned. For example, if the mean for a rainfall layer is 635 mm, using a multiplier of 0.5 would set the base value to 317.5 mm. If you're using MS Large, all cells with a value of 317.5 or less would be assigned a fuzzy membership value of 0. This allows you to adjust the fuzzy membership based on the characteristics of the phenomenon you're modeling. For example, suitable habitat for a particular species might be found at half the mean rainfall and above.

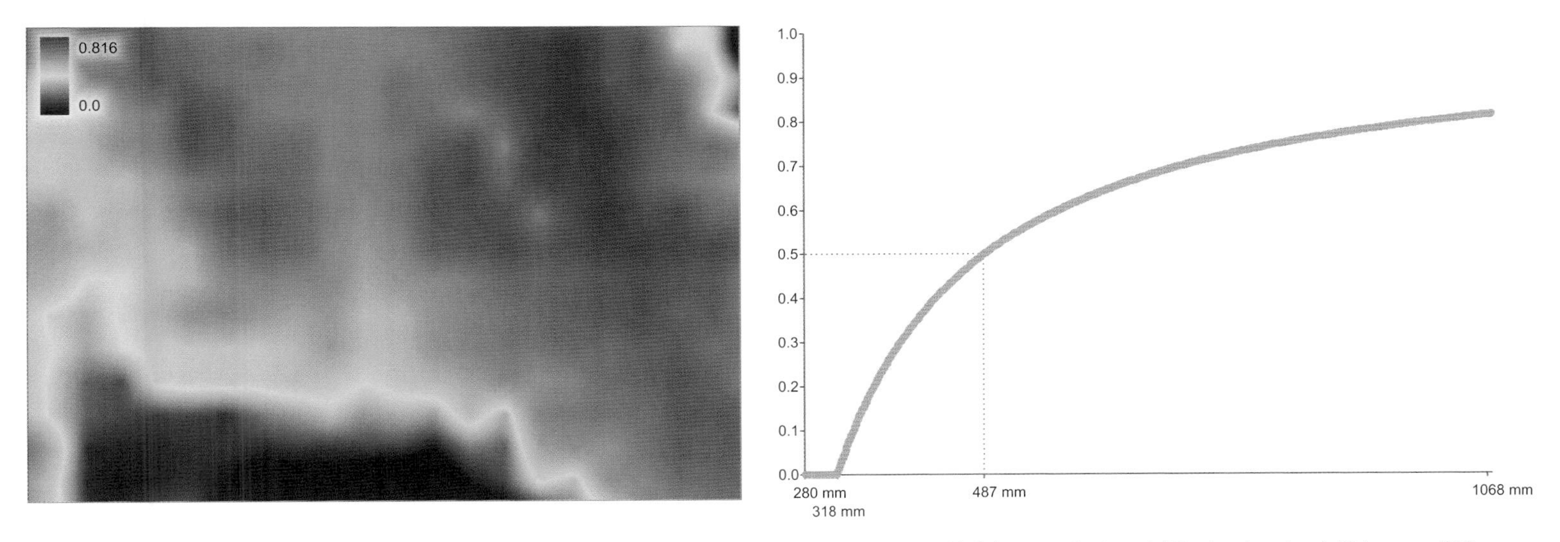

Fuzzy membership values for average annual rainfall, using the MS Large function, with a multiplier for the mean of 0.5. Locations having rainfall values less than half the mean (318 mm, rounded up) are assigned fuzzy membership of 0. Locations having rainfall values greater than this value are assigned increasing fuzzy membership values. The map shows that using the multiplier, fewer locations have low fuzzy membership (red). The standard deviation of 169 mm is added to the base value (318 mm) to calculate the observed value assigned a fuzzy membership value of 0.5 (487 mm).

You also use a multiplier to specify the observed value that will be assigned a fuzzy membership value of 0.5 (the midpoint or neutral value). Rather than specifying the actual observed value (as with the Small and Large functions), you specify the number of standard deviations above the mean. A multiplier of 1 (the default) specifies that the observed value one standard deviation above the mean will be assigned a fuzzy membership value of 0.5. A multiplier of 2 specifies that an observed value two standard deviations above the mean will be assigned a fuzzy membership value of 0.5. For example, with a mean average annual rainfall of 635 mm and a standard deviation of 169 mm, using a multiplier of 1 would assign a fuzzy membership value of 0.5 to cells with an average annual rainfall of 804 mm. Using a multiplier of 2 would assign a fuzzy membership value of 0.5 to cells with an annual average rainfall of 973 mm: (169 * 2) + 635.

With a mean multiplier of 0.5 and standard deviation multiplier of 2 (meaning an anomaly is considered any value more than two standard deviations above half the mean rainfall for the study area), a fuzzy membership value of 0 is assigned to observed values up to 318 mm, and a fuzzy membership value of 0.5 is assigned to an observed value of 656 mm: (169 * 2) + 318.

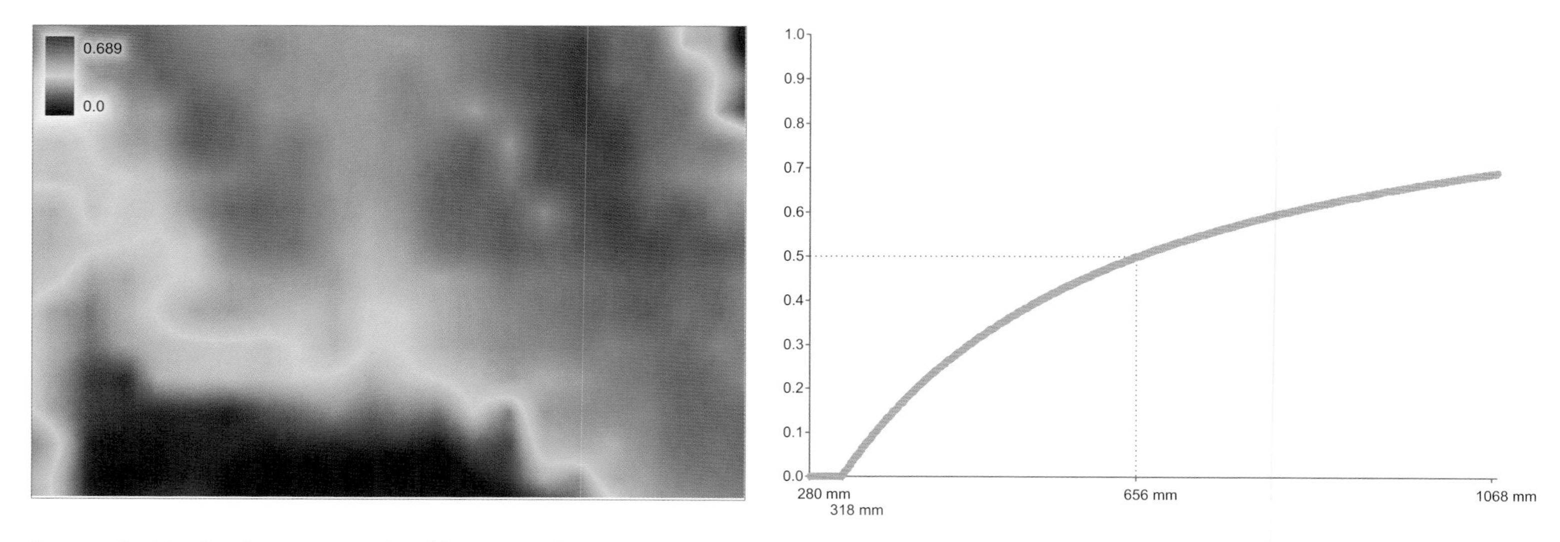

Fuzzy membership values for average annual rainfall, using the MS Large function, with a multiplier of 0.5 for the mean, and a multiplier of 2 for the midpoint (2 standard deviations). The curve is flatter than when using 1 standard deviation, and more locations are assigned fuzzy membership values in the middle of the range (light green, yellow, and light red on the map).

Assigning high fuzzy membership to mid-range observed values

If an observed value in the middle of the range of values represents high membership in the set, use the Near function. With Near, the midpoint you specify represents the highest fuzzy membership (unlike with the Small and Large functions, where the midpoint represents neutral fuzzy membership). As values drop away from the midpoint value (both higher and lower), fuzzy membership values decrease. For example, in a mule deer habitat model, suppose one of your layers is elevation with a range of 1,890 meters to 4,185 meters. If mule deer are likely to be found between 2,400 and 2,800 meters, you'd use Near with an elevation of 2,600 meters as the midpoint.

As with the Small and Large functions, you use the spread parameter to control how rapidly fuzzy membership values decrease. Unlike with these functions, however, the spread values for Near may be outside the typical range of 1 to 10.

Elevation, in meters.

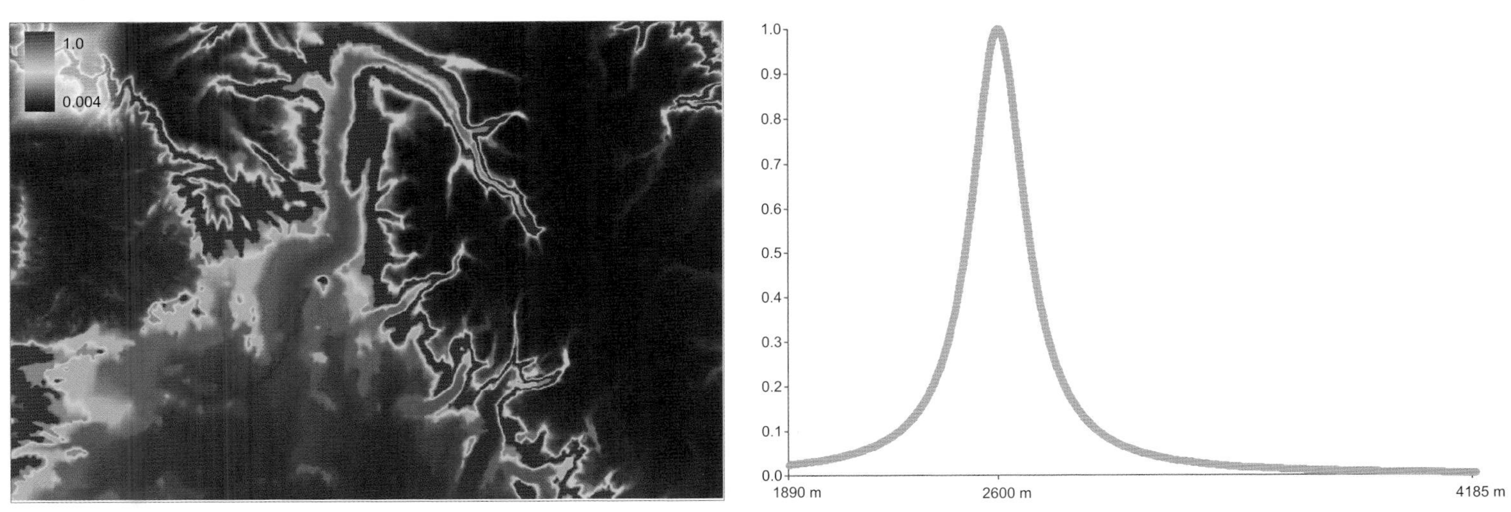

Fuzzy membership values for elevation, using the Near function with a midpoint of 2,600 meters and a spread of 0.0001. Locations with elevation values near 2,600 meters are assigned high fuzzy membership (dark green on the map), with decreasing fuzzy membership for locations at higher and lower elevations.

If your source layer has many unique values—more than 100, or so—you may need to specify a very small spread value (as shown in the map on the previous page). This is often the case with continuous data, such as elevation, where each cell may have a unique value. Very few cells will have the exact midpoint value you specify. These cells will have high fuzzy membership values, while nearby cells with somewhat lower and higher observed values will be assigned lower fuzzy membership values, creating a result which likely will not be useful in your model.

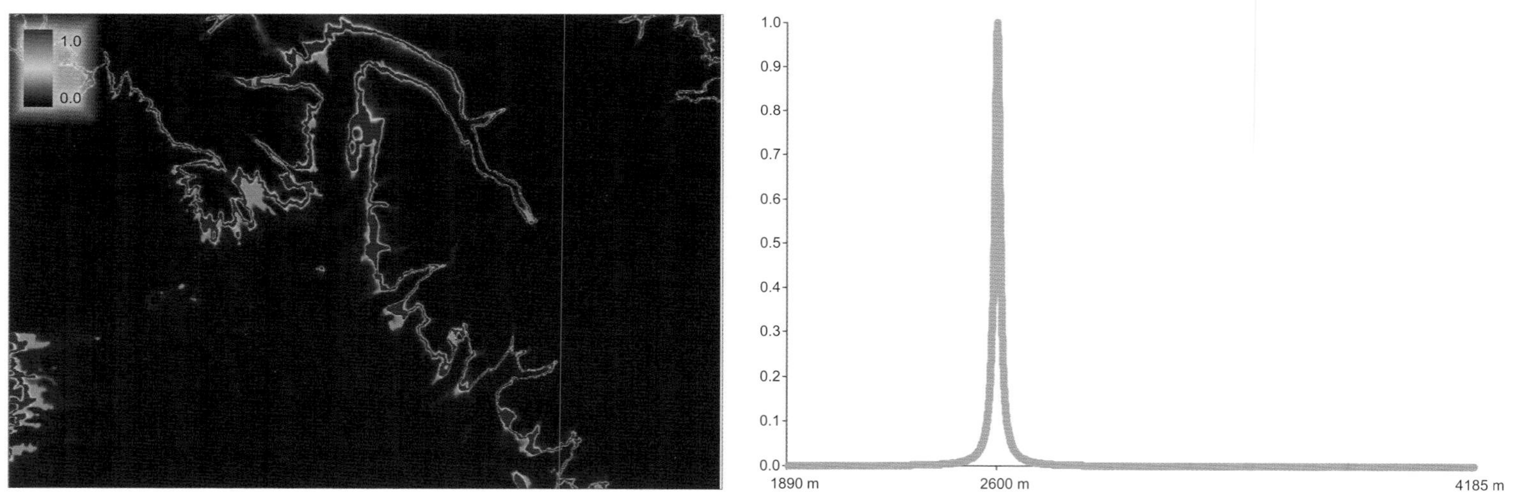

Fuzzy membership values for elevation, using the Near function with a midpoint of 2,600 meters and a spread of 0.1. With this spread, only those equal to or very near the midpoint value are assigned high fuzzy membership—most cells are assigned very low fuzzy membership.

In these cases, the spread value you ultimately use may require some experimentation. Alternatively, you can reclassify the source layer to create ranges of observed values prior to creating the fuzzy membership layer. When you reclassify the source layer, cells within the same range of values are assigned to the same class, so the class you specify as the midpoint will have enough cells (as will the other classes) to create a useful result.

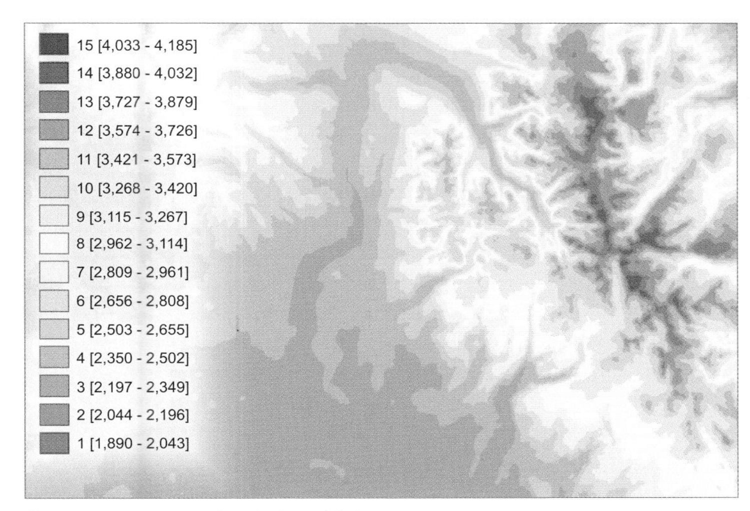

Elevation, in meters, reclassified into 15 classes.

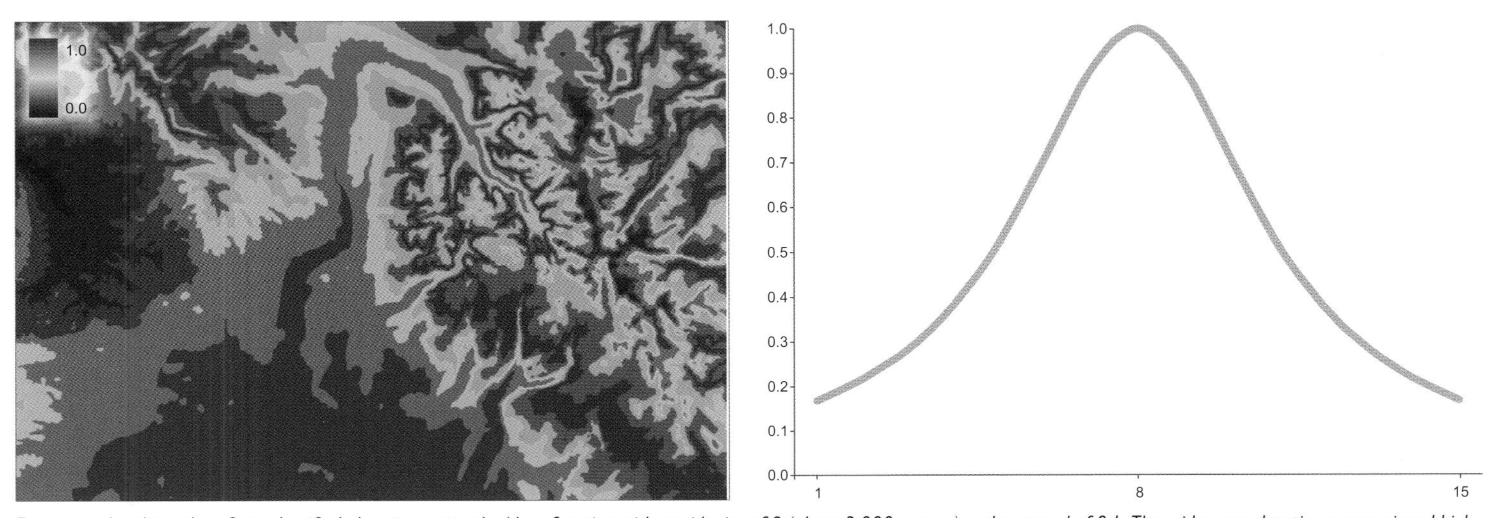

Fuzzy membership values for reclassified elevation, using the Near function with a midpoint of 8 (about 3,000 meters) and a spread of 0.1. The mid-range elevations are assigned high fuzzy membership values, while lower and higher elevations are assigned lower fuzzy membership values. Using a classified raster allows you to assign high fuzzy membership to a wider range of elevation values.

Using a hedge factor

All of the fuzzy membership functions allow you to specify a hedge factor of either "very" or "somewhat." These give you an additional way to capture expert opinion by translating a verbal statement into a mathematical function. For example, someone with knowledge about the phenomenon being modeled might say that locations with "very" low elevation values are suitable, or that locations with "somewhat" high elevation values are suitable.

With the "very" factor, more observed values are assigned high or low fuzzy membership values (with fewer assigned mid-range fuzzy membership values) than would be the case without the hedge factor. It is calculated as the square of the fuzzy membership function it's used with. The "somewhat" factor results in more observed values being assigned mid-range fuzzy membership values—only the very highest or lowest observed values are assigned fuzzy membership values near 1 or 0. It is calculated as the square root of the fuzzy membership function it is used with.

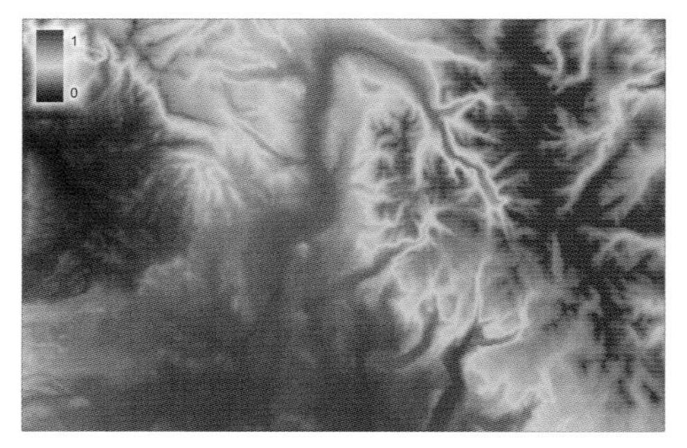

Fuzzy membership layer for elevation using the Linear function and no hedge factor (high elevations are considered more suitable).

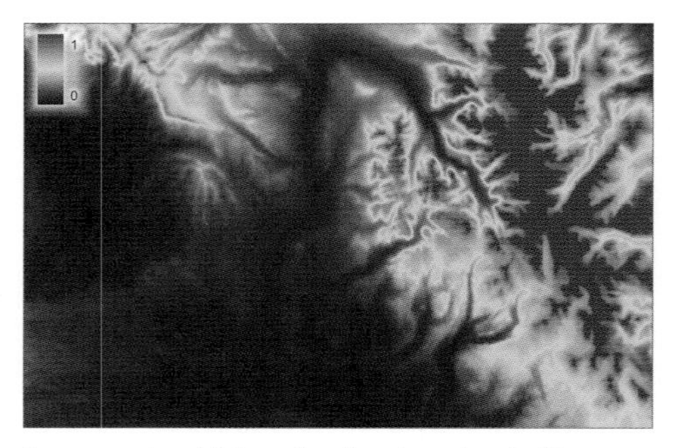

Fuzzy membership layer for elevation using the Linear function with the "very" hedge factor—more of the lower elevations get low fuzzy membership values, and more of the highest elevations get a high fuzzy membership value.

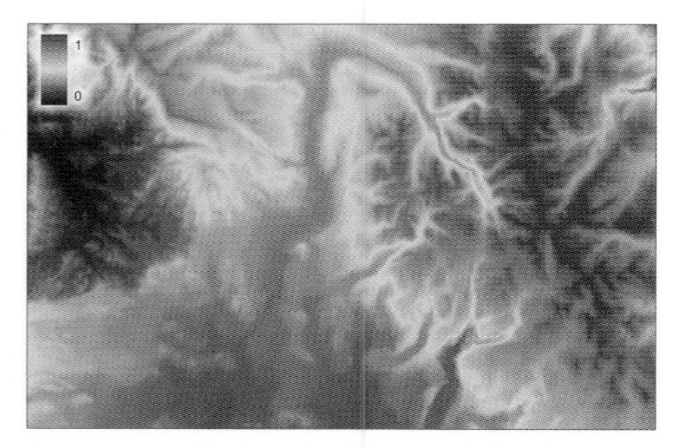

Fuzzy membership layer for elevation using the Linear function with the "somewhat" hedge factor—the lowest elevations get a low fuzzy membership value, and the highest elevations get a high fuzzy membership value. Most elevations get a mid-range fuzzy membership value.

Summary of fuzzy membership functions

For most suitability models, you'll likely use the Small, Large, and Near functions for continuous data, as well as assigning membership for categorical data. The other functions are mainly used in specialized cases.

FUNCTION	ASSIGNS HIGH FUZZY MEMBERSHIP TO:	WHAT YOU SPECIFY	SAMPLE GRAPH
Assign membership	*Specified categories*	*•Fuzzy membership for each category*	
Linear	*Large or small observed values, with fuzzy membership decreasing at a constant rate*	*•Observed value to be assigned fuzzy membership of 0 ('minimum')* *•Observed value to be assigned fuzzy membership of 1 ('maximum')*	
Small	*Small observed values*	*•Observed value to be assigned fuzzy membership of 0.5 ('midpoint')* *•Rate of decrease from high to low fuzzy membership ('spread')*	
Large	*Large observed values*	*•Observed value to be assigned fuzzy membership of 0.5 ('midpoint')* *•Rate of increase from low to high fuzzy membership ('spread')*	
MS Small	*Observed values less than the mean*	*•Optional multiplier to the mean to increase or decrease the observed value below which fuzzy membership of 1 is assigned* *•Observed value to be assigned fuzzy membership of 0.5 (in standard deviations above the mean or modified mean)*	
MS Large	*Observed values greater than the mean*	*•Optional multiplier to the mean to increase or decrease the observed value below which fuzzy membership of 0 is assigned* *•Observed value to be assigned fuzzy membership of 0.5 (in standard deviations above the mean or modified mean)*	
Near	*Mid-range observed values*	*•Observed value to be assigned fuzzy membership of 1 ('midpoint')* *•Rate of decrease from high to low fuzzy membership ('spread')*	

COMBINE THE FUZZY MEMBERSHIP LAYERS

Once you've created fuzzy membership layers for all your source layers, you combine them to create a layer showing the overall likelihood of membership in each of the defined sets for each location (that is, each cell in the raster). Generally, locations having high membership values in all or most of the sets are the most suitable. As with the fuzzy membership layers, the output scale ranges from 0 to 1.

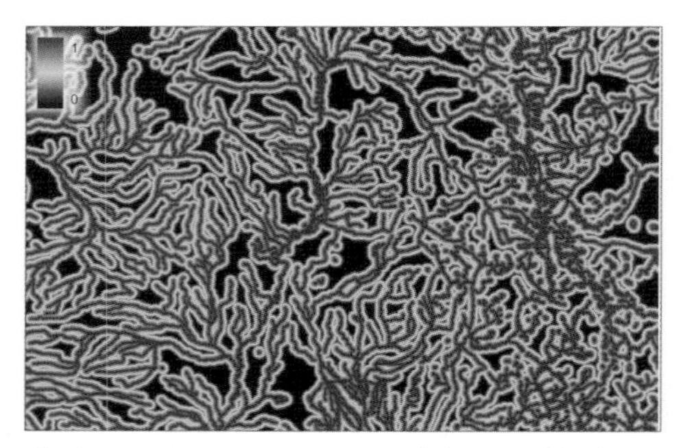

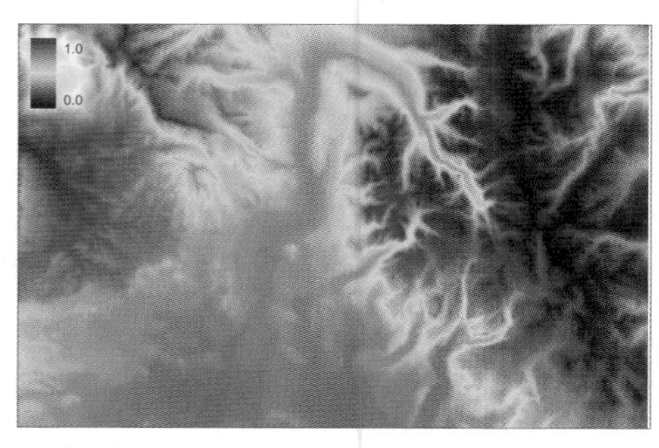

The fuzzy membership layers representing mule deer habitat—land cover, distance to streams, and elevation. The layers are combined to create a layer of overall fuzzy membership values (and hence suitable and unsuitable locations).

There are several different operators you can use to combine the fuzzy layers, depending on the requirements of your model. The operators give you a great deal of flexibility in combining the site characteristics to make your model better match reality. For a wildlife habitat model, it may be that a food source must be present for a location to be considered minimally suitable—if the food source is present, then the presence of other site characteristics (such as vegetation cover or proximity to water) make that location even more suitable. Or it may be that the presence of two (or more) particular characteristics in a location increases the likelihood of the location being suitable to a degree that is greater than if simply any two desirable characteristics are present. This is sometimes the case in mineral exploration, for example, when the presence of particular minerals or rock types in a location greatly increases the likelihood that the mineral you're modeling will be found at that location. You can use various fuzzy overlay operators alone or in conjunction to combine fuzzy membership layers to model these and other scenarios. The operators also let you combine layers to create a more inclusive model asking, for example, where are all the possibly suitable locations, or a more exclusive model that determines only the most suitable locations.

Both logical and mathematical operators can be used to combine fuzzy layers.

Combining layers using logical operators

Two logical operators—And and Or—allow you to combine layers to find locations that meet all the criteria or those that meet any of the criteria, respectively.

When *all* site characteristics must be present for the location to be considered suitable, use the And operator. For suitability analysis, this is commonly the result you want, so the And operator is often used.

With the And operator, the output cell is assigned the minimum value from all the input fuzzy membership layers. If the fuzzy membership value for a particular cell is 0.300 on one layer, 0.952 on another, and 0.659 on the third, the value assigned to the cell on the output layer is 0.300. This produces a more conservative (or exclusive) result with smaller overall membership values, since to have high membership in the output, the cell must have high membership on all the input layers. You therefore know that cells with high output values are more likely to meet all the criteria.

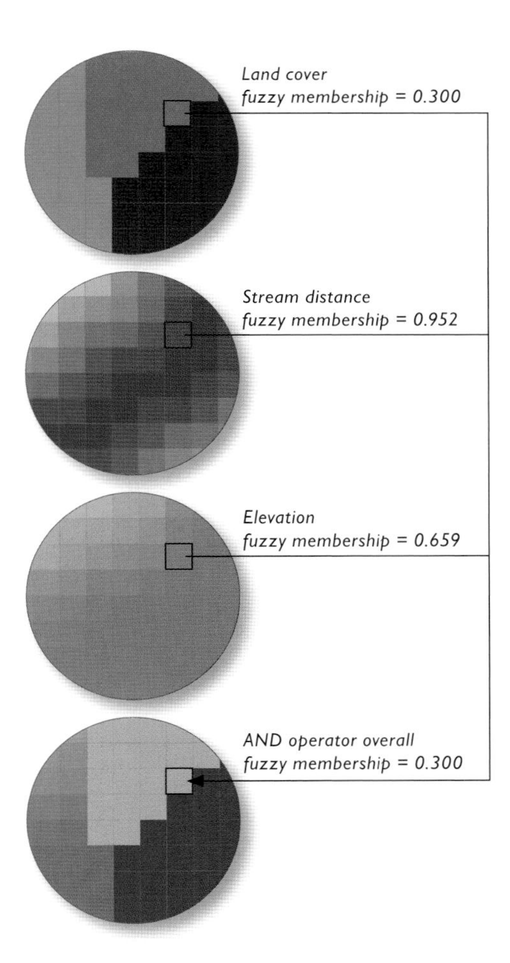

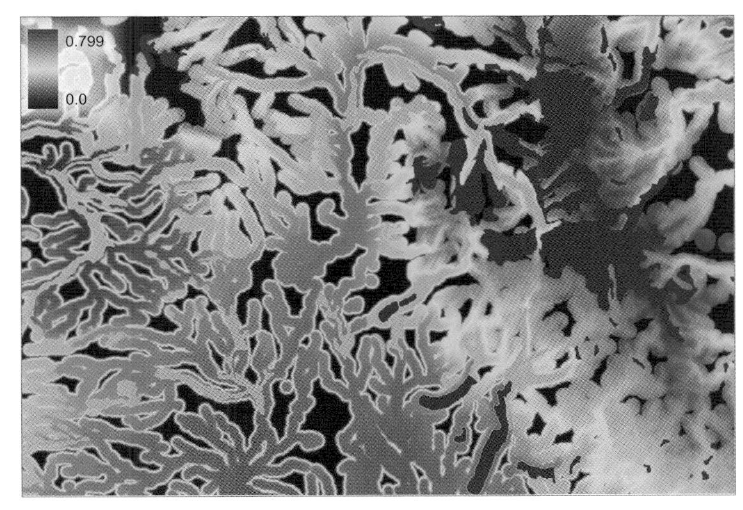

Land cover, distance to streams, and elevation combined using the And operator. The result emphasizes the most likely habitat (dark green).

If *any* of the desired site characteristics can be present for a location to be considered suitable, use the Or operator. With the Or operator, the maximum value for a cell on any of the layers is carried through to the output layer. This produces a less conservative (and more inclusive) result, since to have a high value on the output layer, a cell need only have high membership on any one of the input layers. This operator is useful in situations where the presence of even one desirable characteristic is enough to consider the location suitable. You'd use the Or operator, for example, if you're identifying wildlife habitat for an animal that is dependent on at least one of several plant species (as is the case with some butterflies). If the locations of each plant species are on different layers, using the Or operator to combine the layers would ensure that a location where one of the plants is present gets a high value in the overall fuzzy membership layer, regardless of whether other desired site characteristics (including the other plant species) are present.

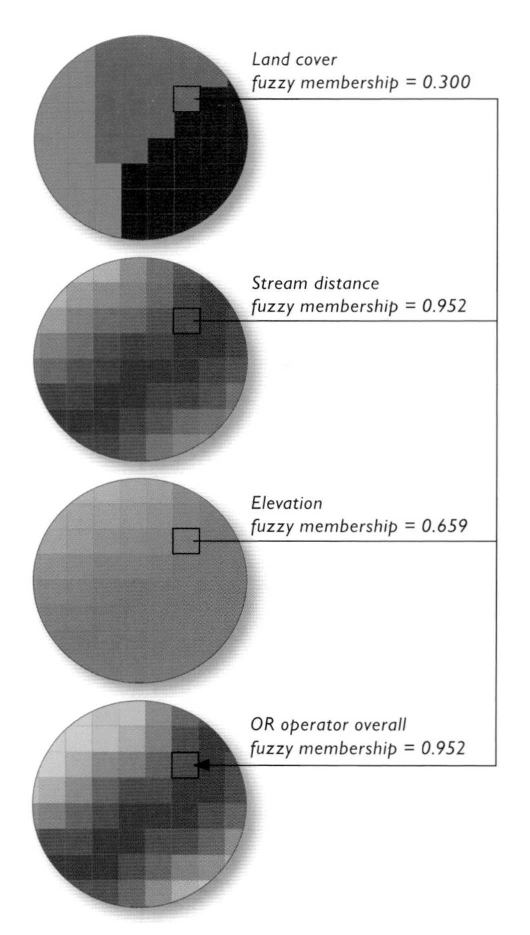

Land cover, distance to streams, and elevation combined using the Or operator. Only the least suitable locations—those having unsuitable land cover for mule deer, far from streams, and at the highest elevations are given a low fuzzy membership value.

Combining layers using mathematical operators

Another set of operators combine the values on the fuzzy membership layers mathematically—the output layer reflects the combined input of all the source layers. (This is unlike the And and Or operators, where the minimum or maximum fuzzy membership value from only one input layer is carried through to the results layer—values on the other layers are essentially discarded and don't influence the results.) The mathematical operators include Product, Sum, and Gamma.

With the Product operator, fuzzy membership values on the input layers are multiplied. The operator decreases the output values. For example, if a cell in one input layer has a fuzzy membership value of 0.8 and the same cell in another layer has a fuzzy membership value of 0.9, the result for the cell on the output layer is 0.72, which is less than either of the input values. The Product operator also tends to increase the range of fuzzy membership values in the output layer, so cells that have high fuzzy membership values on all the input layers retain high (but slightly reduced) values on the output layer, but cells that have low fuzzy membership values on all the input layers will have greatly reduced values on the output layer (if, for example, the input values are 0.06 and 0.02, the output cell will have a value of 0.0012). The output values at the low end of the range can become very small if you're combining more than two or three input layers. The Product operator is, in fact, not often used in suitability analysis.

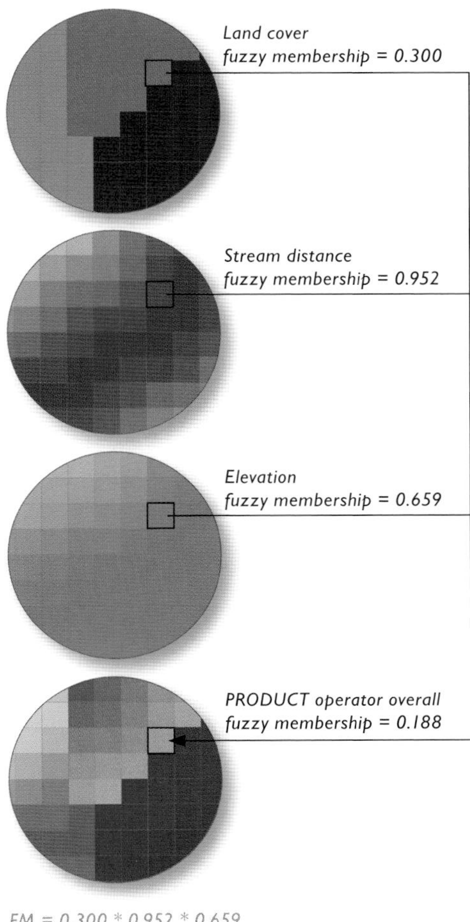

FM = 0.300 * 0.952 * 0.659
= 0.188

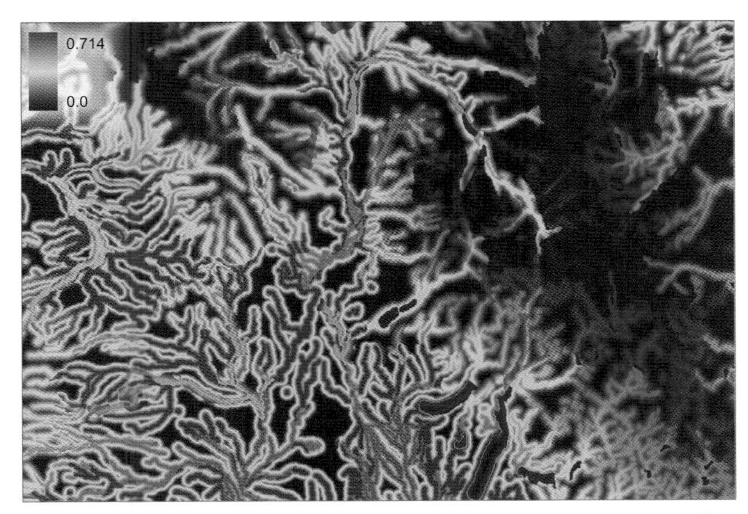

Land cover, distance to streams, and elevation combined using the Product operator. The result emphasizes the extremes of the value range—locations are considered either very suitable or not suitable.

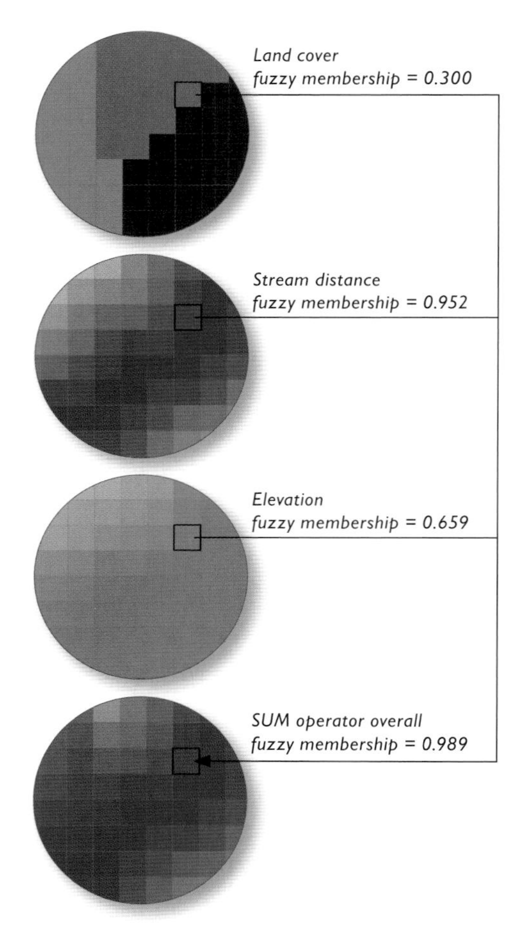

With the Sum operator, fuzzy membership values on the input layers are added together. This increases the output values, with the results being slightly higher than any of the input values. The effect is to emphasize high fuzzy membership since high values for a particular cell on two (or more) input layers will result in an even higher value for that cell in the output layer. The Sum operator also tends to decrease the range of values on the output layer since low values being added increases the output value for a cell having several low input values. You'd use Sum if high fuzzy membership on two (or more) input layers increases the suitability of the location beyond the highest membership value on any of the input layers. In the example, you can see that the overall fuzzy membership value of 0.989 is greater than the fuzzy membership value for that cell on any of the input layers.

To calculate Sum, the input values for a cell are each subtracted from 1, the results multiplied, and the product subtracted from 1 to get the output value (as shown in the diagram to the left). This calculation is used to ensure the output value remains within the scale of 0 to 1.

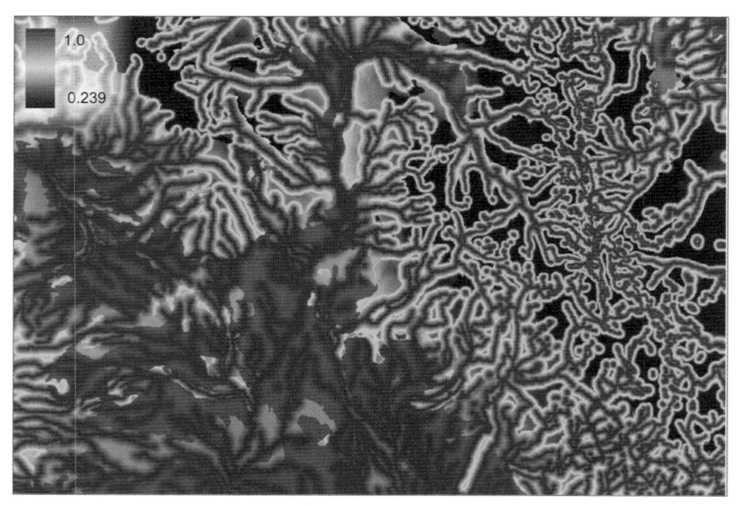

Land cover, distance to streams, and elevation combined using the Sum operator. The result emphasizes the high end of the value range—locations with high fuzzy membership on two or more layers receive even higher fuzzy membership scores in the output layer.

The Gamma operator allows you to achieve a result for the output layer between what you would get using either the Sum or Product operators. You specify a factor (called a "gamma" value) to control how much influence is from Sum and how much from Product. It's calculated as the result of Sum raised to the power of the gamma factor, multiplied by the result of Product raised to 1 minus the gamma factor:

$$\text{COMBINED FUZZY MEMBERSHIP} = (\text{SUM})^{G} * (\text{PRODUCT})^{1-G}$$

The Gamma operator allows you additional control over how the input fuzzy membership layers are combined. With a factor of 1, the result is the same as using the Sum operator; with a factor of 0, the result is the same as using the Product operator. Gamma is particularly useful for fine-tuning your model—you (or other experts) can review the results of the initial Gamma operation, then rerun the operation using a different gamma factor to better reflect your understanding of the phenomenon you're modeling. You can continue to modify the factor through several iterations of the model, if necessary.

Land cover, distance to streams, and elevation combined using the Gamma operator, with a gamma value of 0.6. In this example, the result is similar to the results using the Product or And operators, but with fewer high fuzzy membership locations than when using Product and more nuance than when using And.

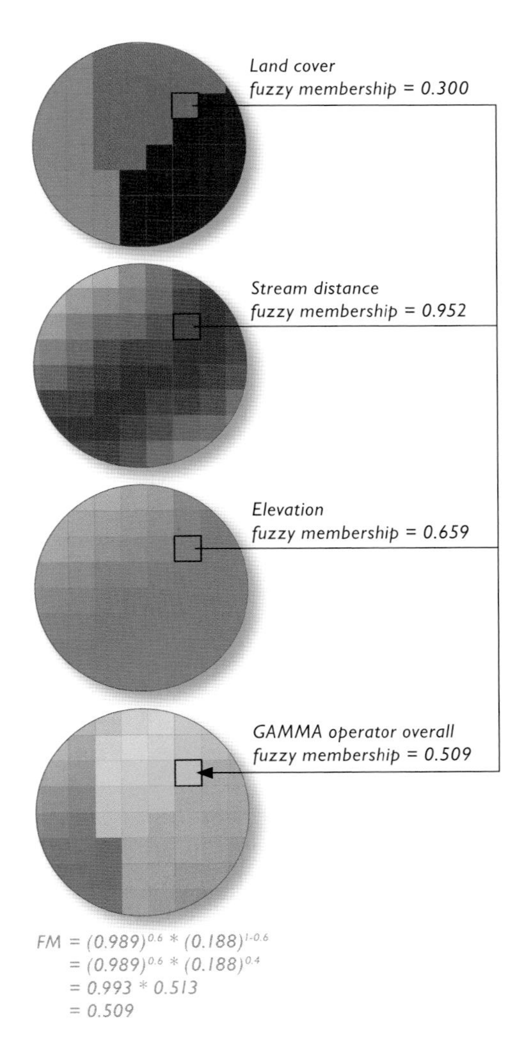

$$\begin{aligned} FM &= (0.989)^{0.6} * (0.188)^{1-0.6} \\ &= (0.989)^{0.6} * (0.188)^{0.4} \\ &= 0.993 * 0.513 \\ &= 0.509 \end{aligned}$$

The Gamma operator is also useful for combining several submodels—created using the other operators—to create the overall fuzzy membership layer. Using Gamma with a high factor (closer to 1) will emphasize the high output values, which will highlight the most suitable areas—this is likely what you want as the final output of your model.

Which operator should I use?

The operator you use depends on the requirements of your analysis. If all characteristics have to be present for a site to be considered suitable, use And. If any one characteristic being present makes a site suitable, use Or. If having two or more characteristics together increases suitability, use Sum.

If two or more of your source layers are correlated, avoid using the mathematical operators to combine layers. Since Sum, Product, and Gamma combine the values on the various input layers, if the layers are correlated, the site characteristic they represent may have too much influence on the result. For example, elevation and slope steepness are often correlated, with steep slopes occurring at higher elevations. Similarly, vegetation types and elevation are also often correlated, with one type of forest occurring at high elevations and another type at lower elevations. If your model includes elevation, slope, and vegetation source layers, the effect of elevation is likely to be magnified in the overall fuzzy membership layer when using one of the mathematical operators. This is not the case with the logical operators And and Or, where, for any given cell, the values from only one layer are retained. (The values for the other layers are essentially discarded, for a given cell, so the result reflects the influence of only one of the input layers for that cell.) This decreases the possibility that correlated layers will result in a particular site characteristic having undue influence.

One advantage of the fuzzy overlay method is that it's straightforward to use the various operators in conjunction to create logical statements, similar to the selection statements described in chapter 2, "Finding suitable locations." This allows you flexibility in modeling the phenomenon. For example, suppose the criteria for good mule deer habitat include areas that are at low elevations and are either near streams or have suitable land cover (that is, it isn't necessary for the location to both be near a stream and have suitable land cover—either will do). You'd first use the Or operator to combine the land-cover layer and the distance to streams layer. You'd then combine the result with the elevation layer using the And operator.

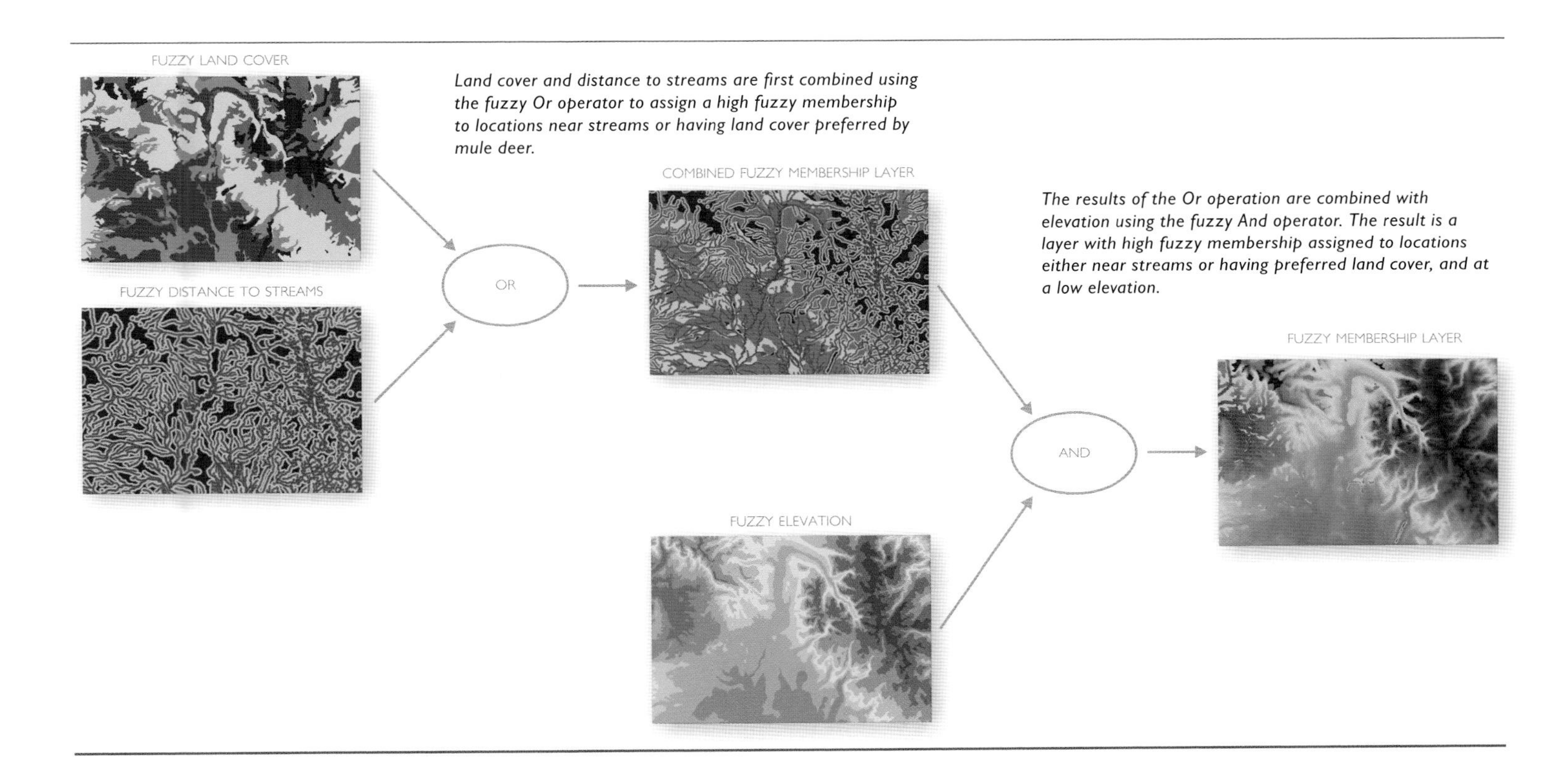

Land cover and distance to streams are first combined using the fuzzy Or operator to assign a high fuzzy membership to locations near streams or having land cover preferred by mule deer.

The results of the Or operation are combined with elevation using the fuzzy And operator. The result is a layer with high fuzzy membership assigned to locations either near streams or having preferred land cover, and at a low elevation.

Similarly, if—in a wildlife habitat model—the presence of a food source is required while either vegetation cover or proximity to water increases the suitability of the location, you'd first combine the vegetation cover and distance to streams layers using the Sum operator (since locations with both good cover and close to water are even more suitable). You'd then combine the results of the Sum with the food source layer using the And operator to ensure the high membership cells on the output layer are ones that have a food source.

Summary of operators

OPERATOR	OUTPUT VALUE	TENDENCY	BEST USE
And	*Minimum of input values*	*Exclusive—locations must have high fuzzy membership on all input layers to receive high fuzzy membership on output layer*	*Finding the suitable locations that meet all the criteria*
Or	*Maximum of input values*	*Inclusive—locations can have high fuzzy membership on any one input layer and receive high fuzzy membership on output layer*	*Finding the suitable locations that meet any of the criteria*
Product	*Product of input values*	*Decreasive—output value for a location is lower than any of the input values for the location; highlights locations with highest fuzzy membership on all input layers*	*Finding the most suitable locations using combined input fuzzy membership values*
Sum	*Sum of input values, transformed to a scale of 0 to 1*	*Increasive—output value for a location is higher than any of the input values for the location; minimizes difference between high and low output values*	*Finding all potentially suitable locations using combined input fuzzy membership values*
Gamma	*Combination of Sum and Product values*	*Increasive with large gamma factor; decreasive with small gamma factor*	*Fine-tuning models based on expert knowledge; combining submodels*

EVALUATE THE RESULTS

Once you've created the overall fuzzy membership layer, analyze the results of your model to determine how valid it is and whether it needs to be fine-tuned or otherwise corrected.

One way to confirm that your model is valid is to check the results with existing research or with data from the field. For example, a wildlife biologist may have already done research and analysis on mule deer habitat. You could map the data collected in the field (such as mule deer sightings) with the results of your model and see how well they match.

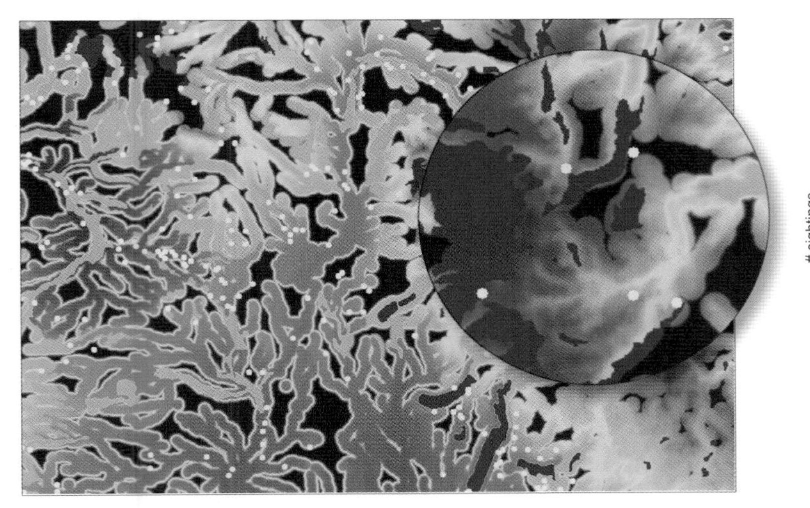

The map for mule deer summer range (created using the And operator), with recorded mule deer observations (blue dots). While many observations are in the areas you'd expect (green), some are in areas considered least likely (orange or red) as shown in the close-up view.

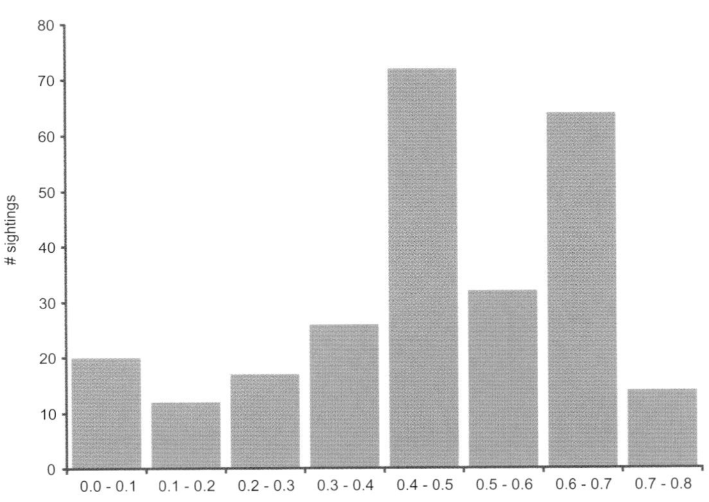

The graph shows the number of sightings in each overall fuzzy membership range. It confirms what the map shows—while many of the sightings occur in locations with fuzzy membership values at the high end of the scale (where you'd expect), quite a few occur in the middle of the scale, and some at the low end of the scale. By tuning the fuzzy membership values for the elevation and land-cover layers, you may be able to create a better model. (Alternatively, you may not have included all the relevant input layers.)

Another way to check the validity of your model is to run several versions using different parameters (such as midpoint and spread) for the various input fuzzy membership layers, as well as various operators to combine the layers. You or other experts can review the results and choose the version that best fits your knowledge of the phenomenon being modeled.

The overall goal is to reflect the best expert knowledge of the phenomenon. Fuzzy overlay is usually an iterative process in which you may modify the fuzzy membership values of the input layers, as well as use different combinations of operators, to achieve the best result.

DISPLAY AND APPLY THE RESULTS

Once you've verified the results of your model, you can display the final suitability layer symbolized by fuzzy membership values to show the most and least suitable locations. Geologist Gary Raines and his fellow researchers have found that using an equal interval classification for the final map works well. That way, a value of 0.5, for example, is always displayed using the same color, making it easy to compare the results of different iterations of the model.

To make the map more readable for nonexperts you may want to reclassify the values into three or five classes, using labels such as "high, medium, low" or "most suitable" and "least suitable"). Including reference layers—such as streams, roads, towns, or a hillshade layer for shaded relief—will also make it easier for map readers to orient themselves.

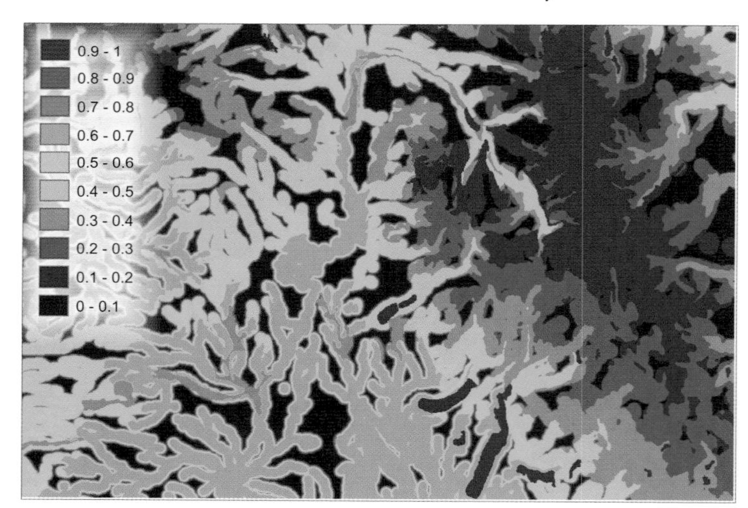

The final map reclassified into ten equal interval classes of likely mule deer summer habitat.

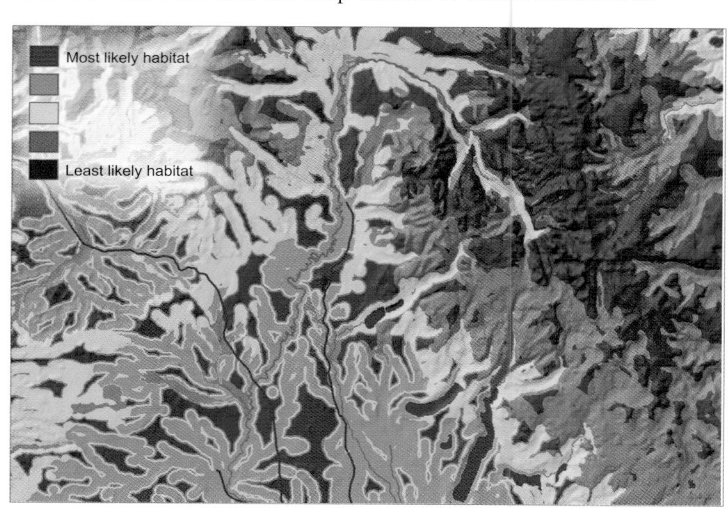

The final map reclassified into five classes of likely mule deer summer habitat, along with terrain, major streams, and roads. Using five classes makes the map easier to interpret.

You can use the results of your model to perform additional analysis. For example, if your research involves analyzing mule deer habitat by watershed, you can overlay watershed boundaries with the final reclassified results layer and identify which watersheds mainly contain likely mule deer habitat and which mainly contain unlikely habitat.

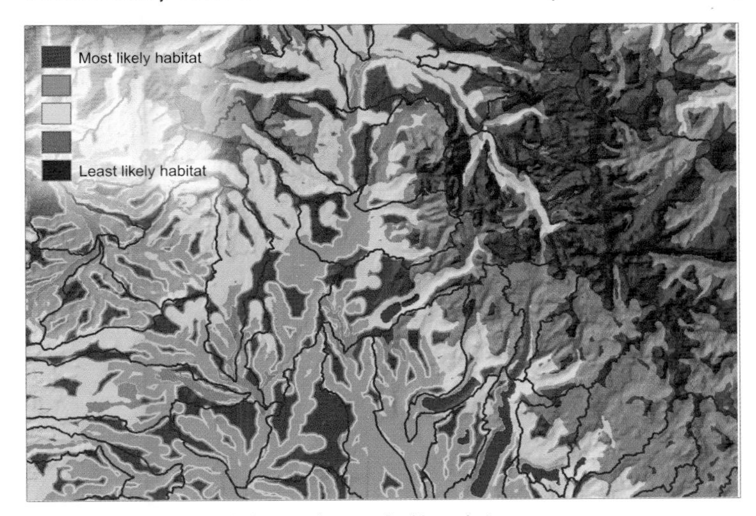

Likely mule deer summer habitat with watershed boundaries.

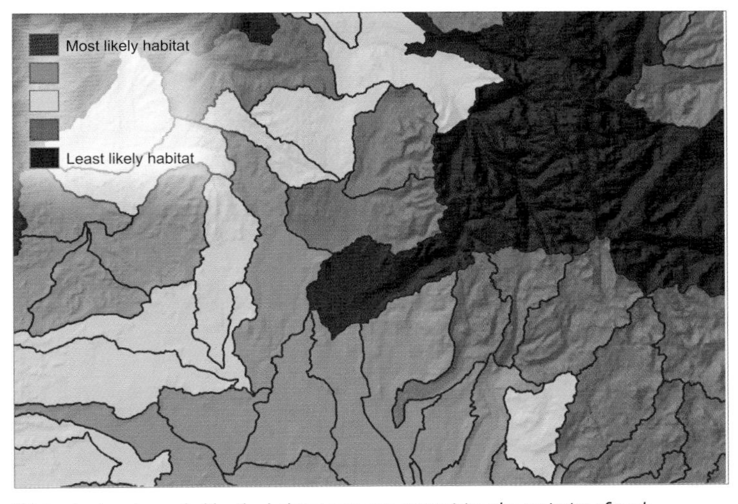

Watersheds color-coded by the habitat category comprising the majority of each watershed. The majority of cells in the dark red watersheds, for example, are in the "Least likely habitat" category.

As with a weighted overlay model (or any other suitability model), presenting the ModelBuilder model document will allow you to show others the criteria for your model and the process used to achieve the result. The document for the mule deer summer habitat model shows that, using the fuzzy overlay method, the source layers were assigned fuzzy membership values and then these layers were combined using the And operator to create the overall fuzzy membership layer.

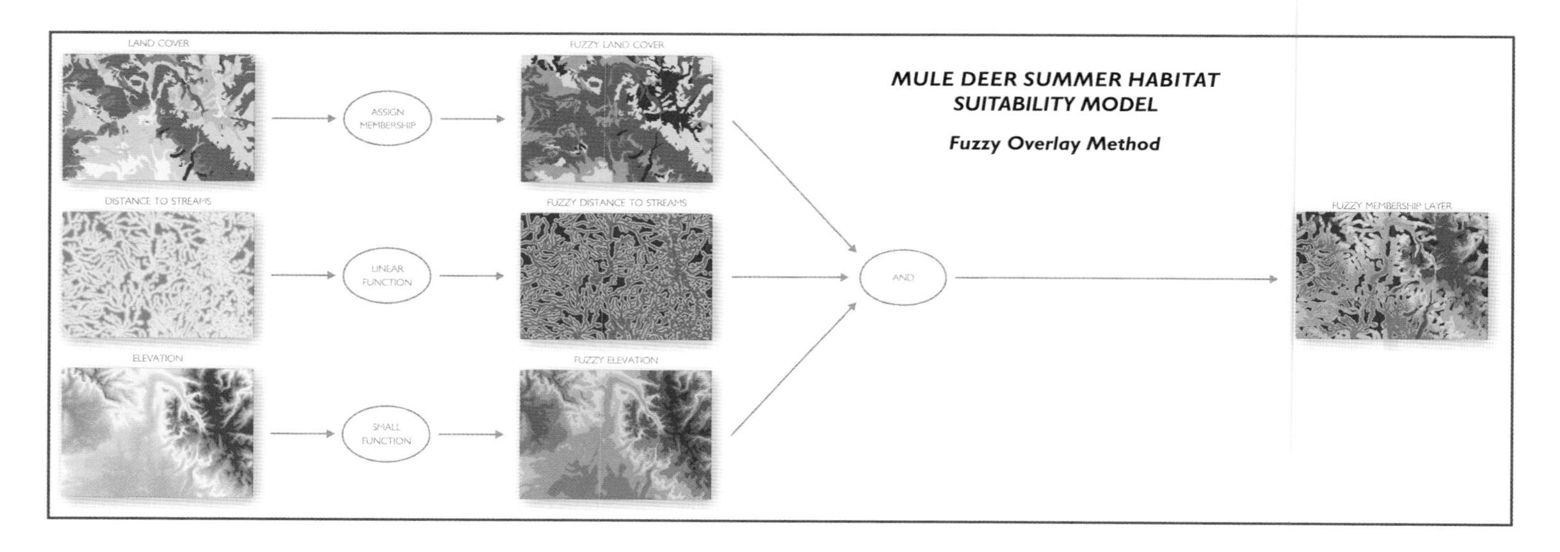

Bonham-Carter, Graeme F. 1994. *Geographic Information Systems for Geoscientists: Modelling with GIS*. Oxford, UK; New York, NY: Pergamon. Chapter on analyzing multiple maps includes discussion of fuzzy logic methods and provides examples of geologic applications.

Malczewski, Jacek. 1999. *GIS and Multicriteria Decision Analysis*. New York, NY: Wiley. Malczewski presents the process for decision making with GIS and describes approaches to selecting evaluation criteria.

McHarg, Ian. 1969. *Design with Nature*. Garden City, NY: Doubleday/Natural History Press. A seminal work that describes the overlay methodology for suitability analysis. Through the course of the book, McHarg lays out an approach to landscape design (the actual purpose of the work).

Raines, G.L. and G.F. Bonham-Carter. 2006. "Exploratory Spatial Modelling Demonstration for Carlin- type deposits, Central Nevada, USA, using Arc-SDM." In *GIS Applications in Earth Sciences: Special Publication,* edited by J.R. Harris. Geological Association of Canada, Special Publication 44. Describes the applications of a fuzzy overlay model to predict mineral deposits and compares the results to other modeling techniques.

Tomlin, C. Dana. 1990. *Geographic Information Systems and Cartographic Modeling*. Englewood Cliffs, NJ: Prentice Hall. Classic work on raster analysis. Tomlin also describes the approach to modeling map information.

Modeling paths

People model paths to find the best way to move people or goods from place to place. You can use GIS to model paths in several ways. One is to model the best path for transporting people or objects over existing infrastructure. For example, you can create the best path for a truck making deliveries to markets based on travel time along streets, the capacity of the truck, and the delivery times promised to customers.

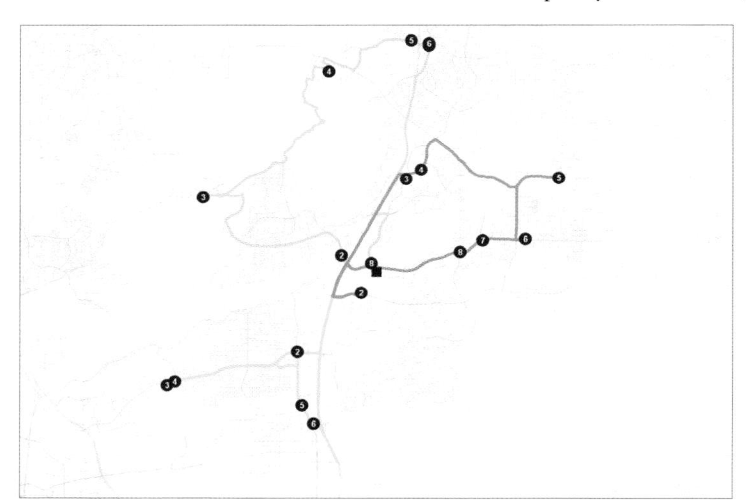

Orders

RouteName	Sequence	Name	ArriveTime	DepartTime	Delivery Quantities	Pickup Quantities
Truck 1	2	9320 MIRA MESA	4/4/2010 9:49:52 AM	4/4/2010 10:29:52 AM	16	4
Truck 1	3	6695 MIRA MESA	4/4/2010 10:36:04 AM	4/4/2010 11:16:04 AM	16	4
Truck 1	4	6795 MIRA MESA	4/4/2010 11:16:23 AM	4/4/2010 11:56:23 AM	16	4
Truck 1	5	9720 CARROLL CENTRE	4/4/2010 12:03:34 PM	4/4/2010 12:43:34 PM	16	4
Truck 1	6	9393 KEARNY MESA	4/4/2010 12:45:03 PM	4/4/2010 1:25:03 PM	16	4
Truck 2	2	12860 RANCHO PENASQUITOS	4/4/2010 10:17:06 AM	4/4/2010 10:57:06 AM	16	4
Truck 2	3	13985 TORREY DEL MAR	4/4/2010 11:05:28 AM	4/4/2010 11:45:28 AM	16	4
Truck 2	4	16629 DOVE CANYON	4/4/2010 11:58:26 AM	4/4/2010 12:38:26 PM	16	4
Truck 2	5	17011 BERNARDO	4/4/2010 12:43:59 PM	4/4/2010 1:23:59 PM	16	4
Truck 2	6	11898 RANCHO BERNARDO	4/4/2010 1:25:06 PM	4/4/2010 2:05:06 PM	16	4
Truck 2	7	11891 RANCHO BERNARDO	4/4/2010 2:05:11 PM	4/4/2010 2:45:11 PM	16	4
Truck 2	8	12610 SABRE SPRINGS	4/4/2010 2:52:55 PM	4/4/2010 3:32:55 PM	16	4
Truck 3	2	12033 SCRIPPS SUMMIT	4/4/2010 10:03:40 AM	4/4/2010 10:43:40 AM	16	4
Truck 3	3	11030 RANCHO CARMEL	4/4/2010 10:49:02 AM	4/4/2010 11:29:02 AM	16	4
Truck 3	4	11815 CARMEL MOUNTAIN	4/4/2010 11:29:55 AM	4/4/2010 12:09:55 PM	16	4
Truck 3	5	14147 TWIN PEAKS	4/4/2010 12:16:58 PM	4/4/2010 12:56:58 PM	16	4

Page 1 of 2

Three routes for deliveries to service station convenience markets, color-coded by route.

You can also use GIS to select the location for a path to be constructed. For example, you can determine the best path for a pipeline based on steepness of slope, land cover, soils, the location of protected natural areas, and other factors.

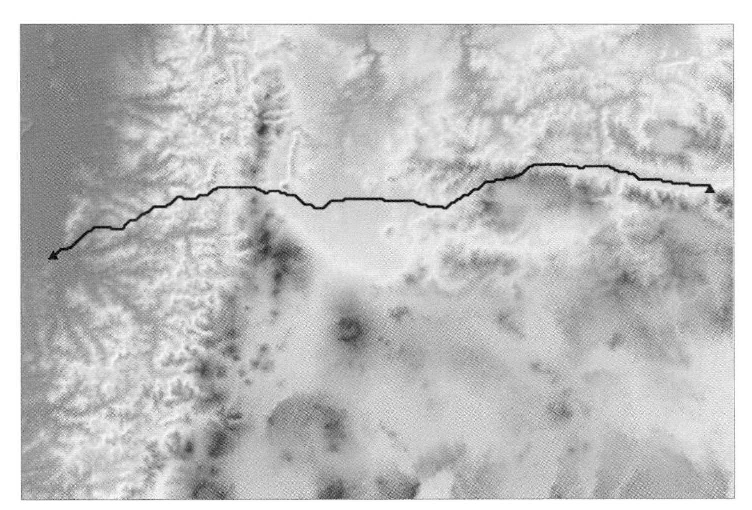

Path for a pipeline.

Similarly, you can model the path people or animals are likely to take between two points. For example, you could model the path elk would take from summer to winter feeding grounds, based on vegetation, terrain, location of water, and so on.

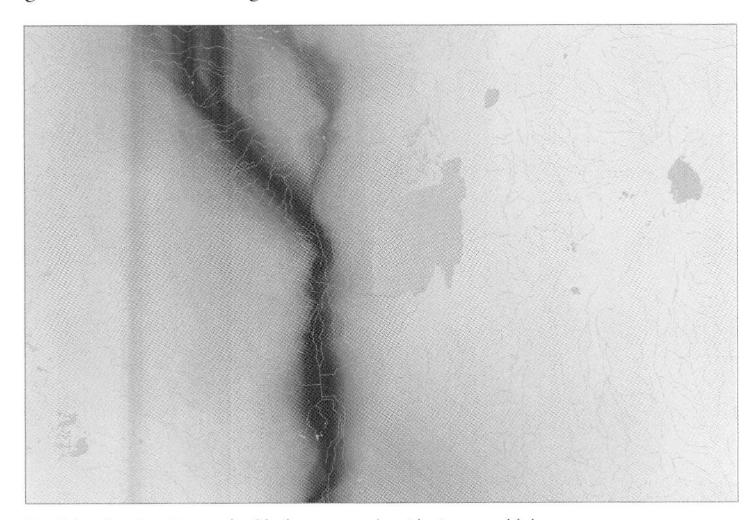

Possible elk migration paths (dark tan areas), with rivers and lakes.

In all these cases, the goal is to find the path with the least cost. Cost can be measured in a number of ways, including distance, money, and time. Often there are trade-offs between these costs. For example, if you're modeling the path for a new road, you may find that the route that requires the least time for motorists to travel costs the most to build if it is more direct but requires bridges. Path models are particularly useful for quantifying and presenting the costs associated with alternate paths.

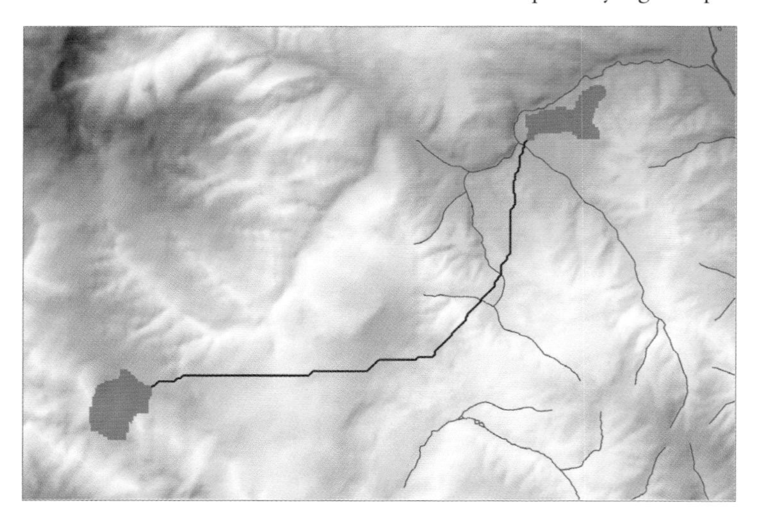

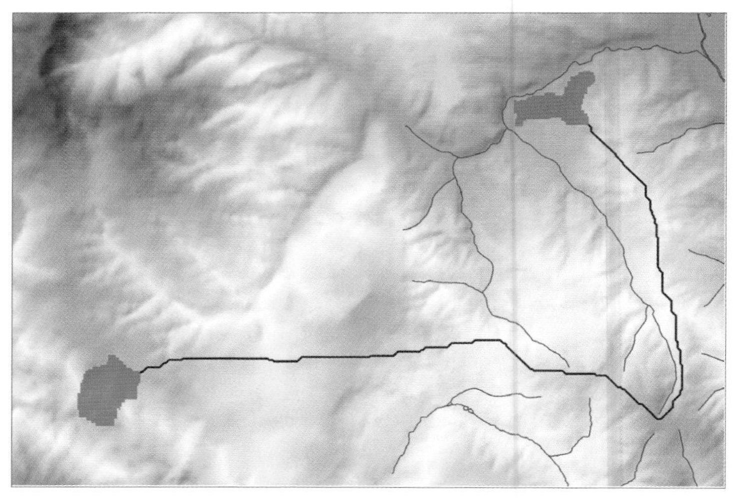

Alternate paths for a road connecting two campgrounds. The path on the left is more direct, but may cost more to build since it crosses several rivers.

Paths generally fall into two categories: network paths (such as a delivery route) and overland paths (such as the path for a new pipeline).

Network paths

In a GIS, a network is used to represent a fixed infrastructure such as roads, stormwater pipes, or cables. Objects can move only along the lines that comprise the network. The two most common networks are utility networks and transportation networks.

In a utility network, you model how water, gas, electricity, or data flows through the pipes, wires, or cables comprising the network. This flow is often from one point to many (such as with electricity) or many locations to one (such as with sewage) along multiple paths. Tracing flow over a utility network is discussed in chapter 5, "Modeling flow."

In a transportation network, such as a street network, you model the paths that cars, trucks, and other vehicles take as they travel through, or over, the network. Examples of modeling paths over a transportation network include finding the best path for a truck from the fire station to a fire or the most efficient route for a truck making deliveries to stores. Of course, the best path may not be the shortest one—depending on the time of day and amount of traffic on the streets, a slightly longer path may be the quicker one. The concept is the same as that found in navigation systems used in cars and trucks. The ability to find the best path has been a fundamental part of GIS for decades.

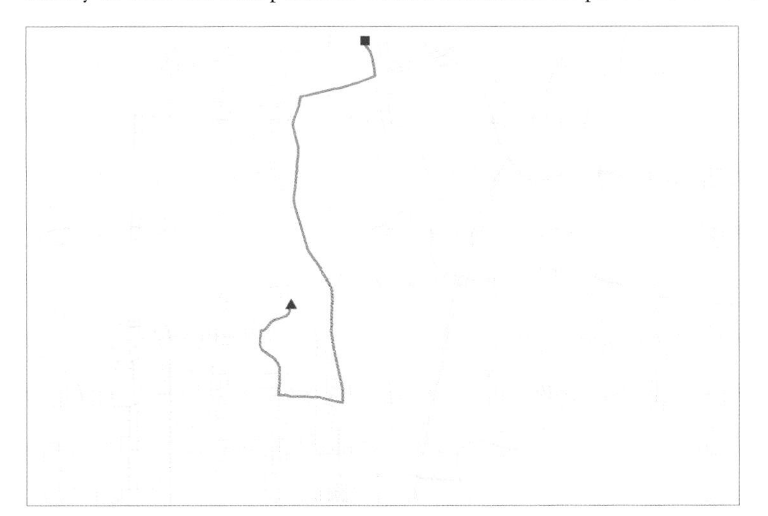

Quickest path from the fire station (square) to the fire (triangle), based on travel time along streets

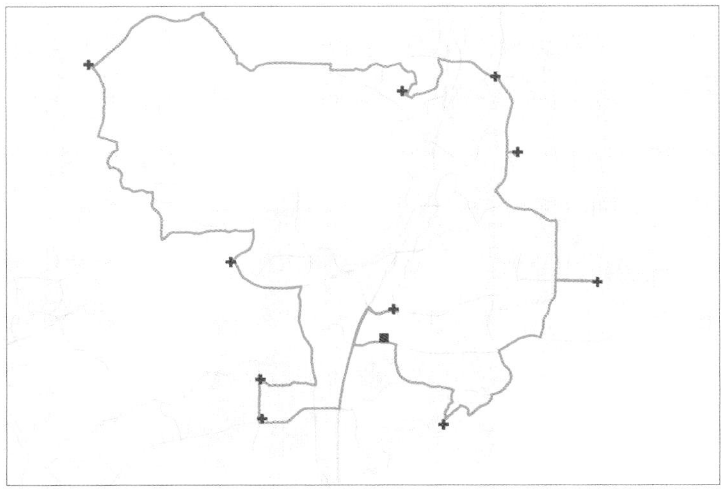

This path for a salesperson making visits to doctors' offices includes many stops.

A path over a transportation network can simply connect two points or can include stops along the way. The parameters for paths that include stops can become quite complex. You may need to drop off or pick up goods or people, or the path may need to cover a certain territory, as with the path for a snow plow or a meter reader. Your model can include the capacity of the vehicle, the amount of goods or people to be picked up or dropped off, and time windows in which the stops need to be made.

Paths may be more or less variable. Some routes change daily (such as with a furniture delivery van), some rarely change (such as with a school bus route), and some vary within a fixed framework (such as for a courier service where the route remains primarily the same from day to day, but the actual stops may vary depending on who is getting a delivery or shipping a package).

The process for modeling a path over a transportation network is discussed in this chapter.

Overland paths

Overland paths are useful for modeling the movement of objects that don't travel over a fixed infrastructure. For example, you could model the route elk are likely to travel from a breeding ground to a forage area or find potential wildlife corridors between two protected natural areas. Modeling overland paths is also useful when you want to find the best route for new infrastructure, such as highways, pipelines, or power lines. The same type of path model is used to predict the spread of some phenomenon from an origin location, such as with a wildfire. In this case, you are essentially modeling the path in all directions from the origin.

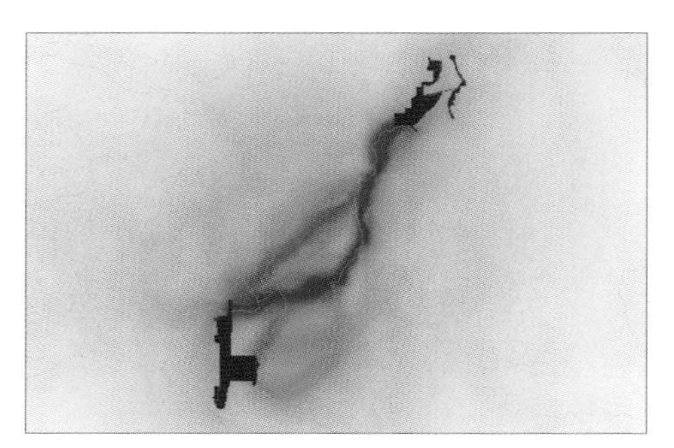

Possible locations for a wildlife corridor connecting two protected natural areas. The dark tan paths are the ones most likely to be used by wildlife. Streams are also shown (blue lines).

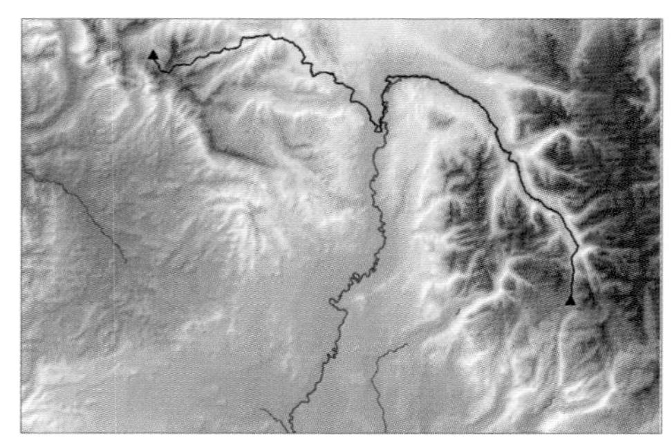

Path for the construction of a road between two towns. The model ensures the path avoids steep slopes.

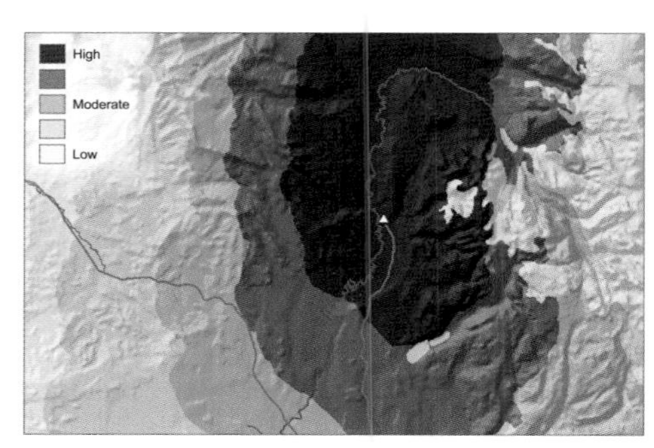

Wildfire spread surface classified into five classes to show areas at high, moderate, and low risk for a fire originating at a particular campground (white triangle).

To design a path model, you define the problem you want to solve, and then define the parameters of the path—how cost is measured, the direction of the path (whether it runs both directions or is one way), and whether there are any barriers that affect the location of the path.

DEFINE THE PROBLEM

Framing the problem as a question is a good technique. Being specific when you define the problem will make it easier to determine the parameters for your model.

GENERAL	SPECIFIC
What is the fastest route between the fire station and the fire?	*What is the fastest route between the fire station and the fire during morning rush hour, while avoiding residential streets, and taking into account one-way streets and traffic lights?*
What is the best corridor for building a new highway between two cities?	*What is the best corridor for building a new highway between two cities, considering existing roads, the economic impact on towns along the route, the locations of sensitive wildlife habitats and cultural sites, and the cost of construction?*

Once you've defined the problem, you need to define the parameters to be included in your model and decide how each will be measured. A good way to do this is to list the parameters along with their corresponding source data layers and measurement standards. The list for the route from the fire station to the fire might look like this:

PARAMETER	SOURCE LAYER	MEASUREMENT
Shortest travel time	*Streets*	*Total time in minutes*
Time period 6-9 a.m.	*Streets*	*Impedance based on traffic counts by time of day*
Travel on secondary streets or higher	*Streets*	*Street type*
Minimize travel against traffic	*Streets*	*Street direction (one-way street)*
Account for slowing at traffic lights	*Traffic lights*	*Average wait time*

The list for the corridor for a new highway might look like this:

PARAMETER	SOURCE LAYER	MEASUREMENT
Avoid sensitive wildlife habitats	*Habitats*	*Habitats rated by sensitivity*
Avoid archaeological/historical sites	*Cultural sites*	*Distance to cultural site*
Minimize construction costs	*Land cover; elevation; soils; geology*	*Dollars per linear foot*

DEFINE THE PARAMETERS

The most common parameters people use when modeling paths are cost, direction, and barriers.

Cost

Cost is often expressed as distance, time, or money. Your parameters may include more than one of these costs and, in some cases, there may be trade-offs between them. For example, when creating the path for an oil pipeline, you'd probably want to create the most direct path, while minimizing construction costs. However, a longer route (meaning a longer time for the oil to arrive at its destination) may have lower construction costs—if the path detours around a mountain range, for example. You need to consider the trade-offs between time and money.

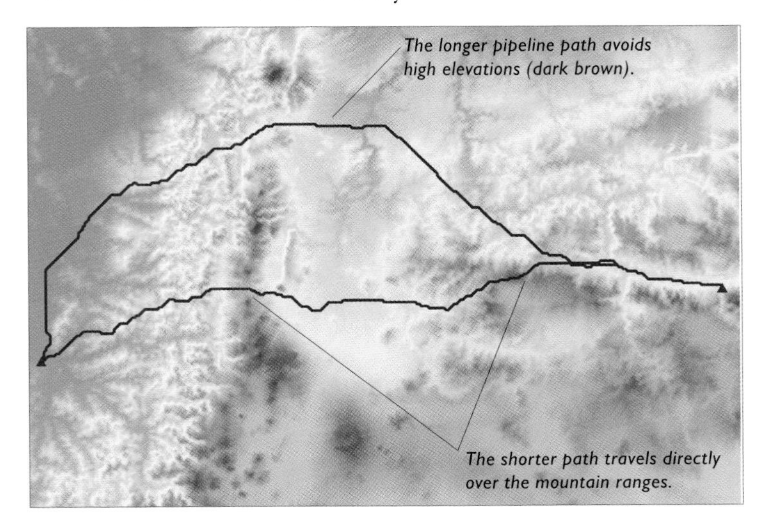

Other costs, such as social, environmental, or political costs, may impact the final decision about the path. These costs can be difficult to quantify. However, you can include data that takes these costs into account to some extent. For example, if you're building a new highway, you can consider environmental costs by including the locations of sensitive wildlife habitats so these areas can be avoided. Similarly, social costs can be considered, to some extent, by including a layer of landownership—building the highway on public lands would have higher social costs than building it on privately owned lands (although likely lower monetary costs).

For a path over a network, costs are associated with network edges and with intersections and turns. For an overland path, cost is represented by the cell values in each raster layer.

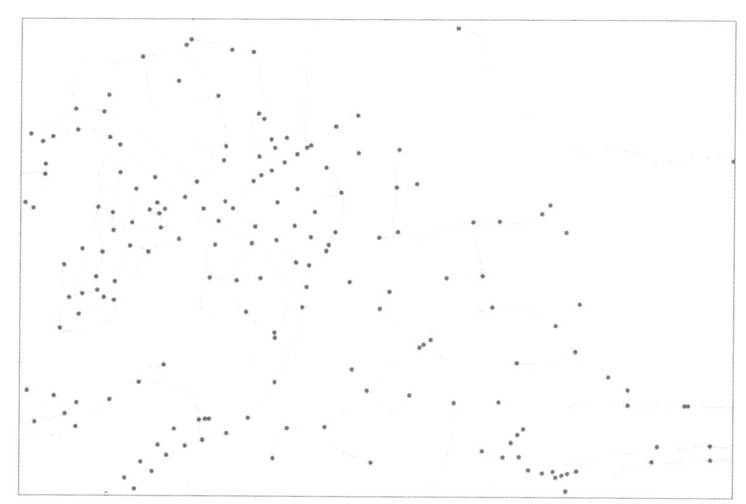

A network layer—costs are associated with edges (lines) and intersections (dots).

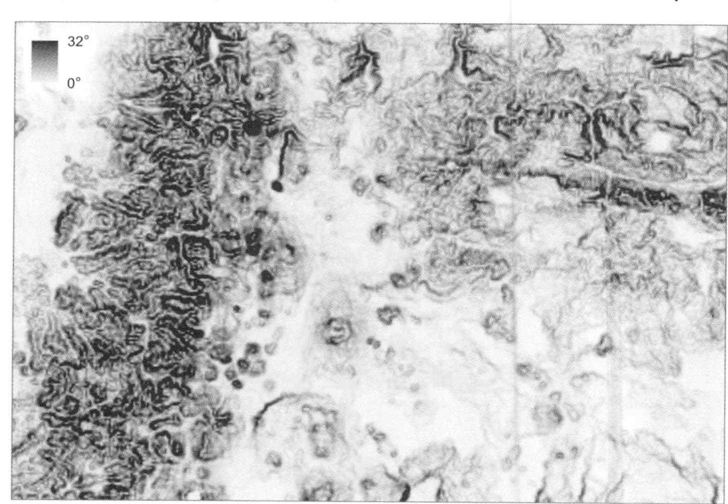

A raster layer of slope representing one of the costs for an overland path.

Cost usually varies along the path. For a highway corridor, the land cover may change from forest to grassland or the terrain change from steep slopes to gradual ones. Similarly, for a fire truck traveling along a street network, the travel time for each street may vary according to the average amount of traffic on the street and the number of lanes.

Cost measures can be either direct or indirect. For a street network, for example, if cost is measured in terms of time, a direct measure would be the actual time required to travel each street segment. If you don't know the actual travel time along each street, you could use the speed limit and the length of the street to estimate the travel time—an indirect measure.

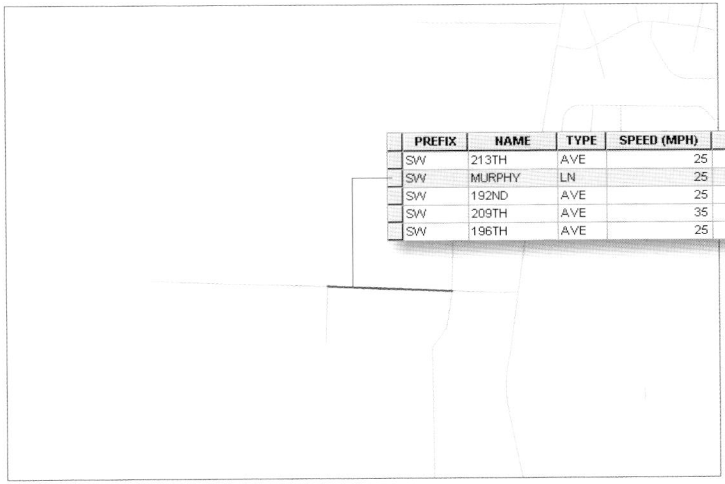

PREFIX	NAME	TYPE	SPEED (MPH)	LENGTH (FT)	MINUTES
SW	213TH	AVE	25	493.1	0.22
SW	MURPHY	LN	25	1050.4	0.48
SW	192ND	AVE	25	230.3	0.1
SW	209TH	AVE	35	492	0.16
SW	196TH	AVE	25	265.9	0.12

Travel time over a street network, calculated from street length and speed limit, is an indirect measure.

Once you've defined the measures, you can determine if the data you have supports the model you want to build. You may need to add data (either collect it in the field or get it from a commercial data supplier or another GIS user). If the data is not available or is too expensive to collect, you'll need to use a different measure—perhaps an indirect one—or eliminate that parameter from the model.

Path direction

You'll want to consider whether objects can move over the path in both directions or in only one direction. For example, cars travel on a highway in both directions, while oil in a pipeline flows in one direction (from where it's drilled to where it's refined).

Some networks limit travel to one direction—for example, a street network may have one-way streets. If this is the case, you'll need to set up the transportation network to indicate the direction of travel on restricted edges.

For an overland path, travel may be more difficult in one direction than the other. For example, elk crossing a ridge to travel between summer and winter forage range may face a steeper climb in one direction than the other. In this case, you'd run two separate models—one for the path to winter forage and another for the return path.

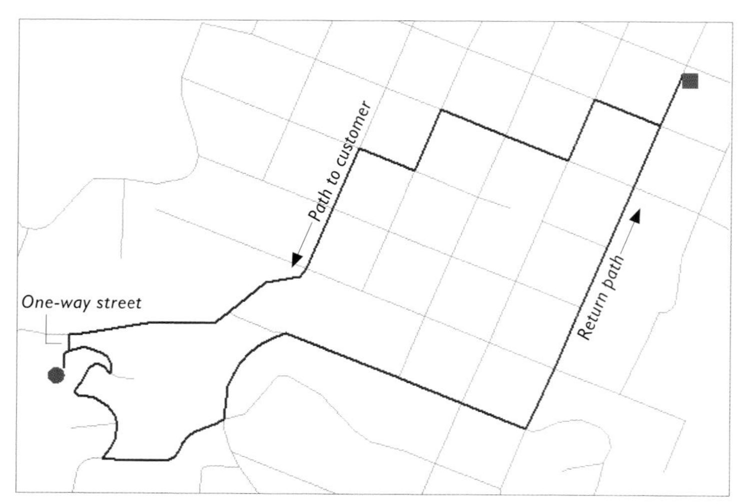

Because the delivery location is off a one-way street, the shortest return path to the store (blue square) is different than the path from the store to the customer (blue dot).

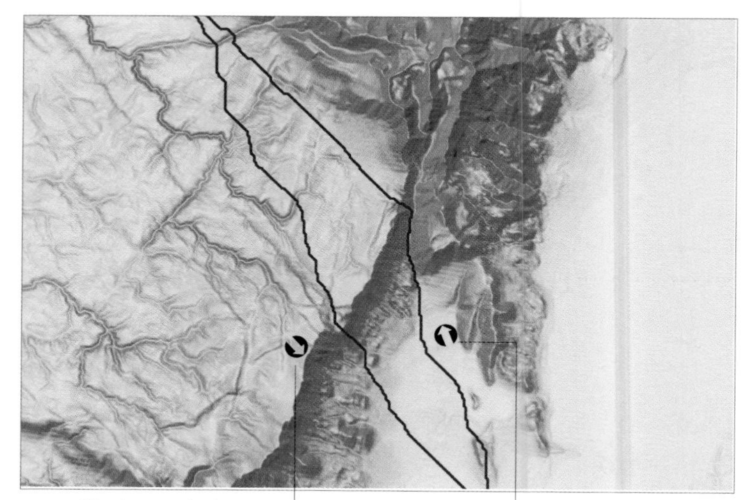

Heading south, the path travels straight downhill over the slope face (dark red indicates steeper slopes).

Heading north, the path passes through a gap in the ridge, traveling over a gradual slope that requires less effort.

Barriers

Barriers let you limit where a path can go. Examples of barriers for a path over a street network would be a washed-out bridge, a road closed for repairs, or an accident blocking an intersection.

The map on the left shows the fastest route from a fire station (square) to the fire when no barriers are present. With an accident blocking the circled intersection, the GIS finds the alternate path shown in the map on the right.

For overland paths, barriers are often linear features that can't be crossed. For example, the path for a wildlife corridor cannot cross a highway. Barriers can also be area features that the path can't travel through. For example, a pipeline may be prohibited from crossing a protected natural area, a lake, or a town.

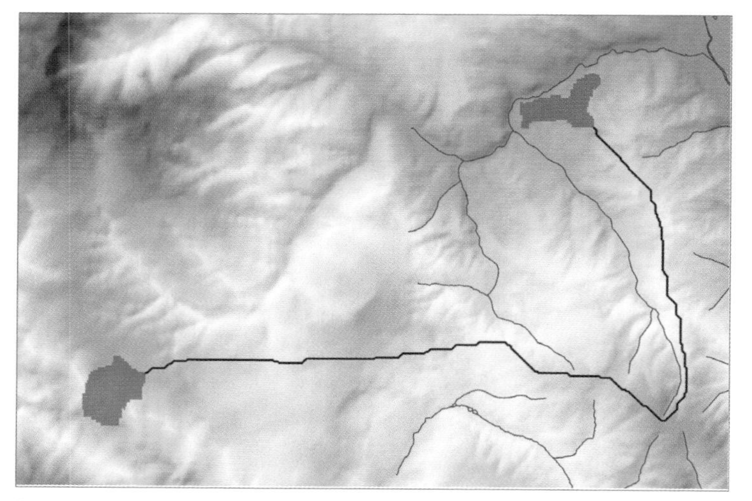

For this proposed path for a road between two campgrounds, rivers are used as a barrier, forcing the path to travel around them.

A path over a transportation network may lead directly from one point to another or it may include a number of stops along the way. The stops may involve time windows and making deliveries to drop off (or pick up) people or goods.

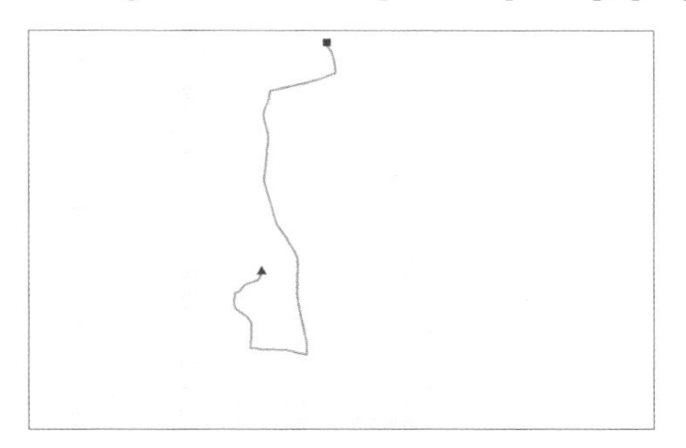

Quickest path from the fire station (square) to the fire (triangle).

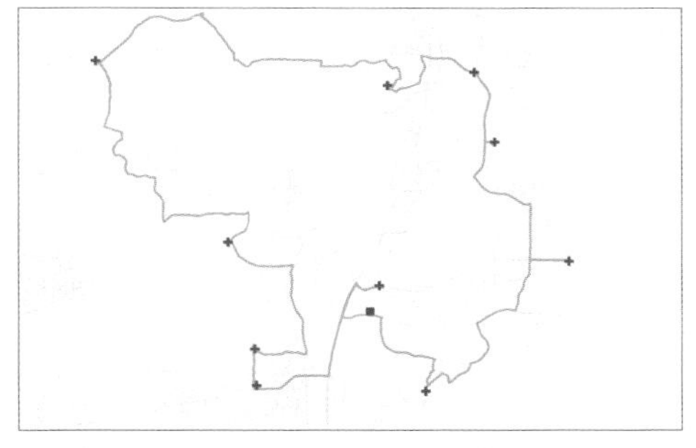

Path for a salesperson visiting doctors' offices (crosses). The path starts from and returns to the sales office (square).

To model the least-cost path, you need the locations of the stops (including the start and end locations) and the time window within which each stop will be visited. In addition, if the stops include deliveries, you need information about each delivery (the amount to drop off or pick up and how long the delivery itself takes) and about the vehicles making the deliveries (how much they can carry and how much they cost to operate). In all cases, the first step is to create or obtain a network layer representing the transportation infrastructure.

CREATE OR OBTAIN THE NETWORK LAYER

A GIS network layer is a collection of edges and junctions representing the physical infrastructure. An edge is the line that runs between two junctions. In a street network, an edge would represent a street segment between two intersections (represented by junctions). The GIS knows which edges connect at each junction, so it can trace a path through the network. You can use the GIS to create a network layer from an existing layer of linear features such as streets or rail lines.

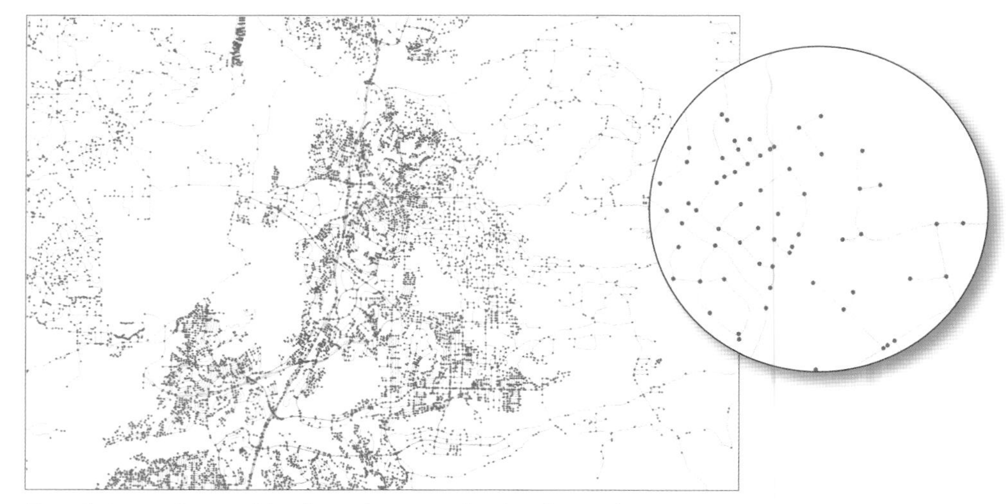

Network dataset of streets. Edges (streets) connect at junctions (intersections—gray circles). Junctions also occur at ends of edges and may occur along an edge.

Networks can include different modes of transportation. These networks are referred to as "multimodal." For example, if you need to move new cars from a port to a city in another state using trucks and trains, you'd create the least-cost path using a multimodal network that includes both highways and rail lines. You control where the different modes connect (these become transfer points). For example, the highways and rail lines would connect at rail yards, but not at highway overpasses that cross rail lines.

In a multimodal network, the transfer points between modes have to be included as features along with the network edges. For example, in a network used for modeling pedestrian paths over streets and subways, the network would include streets, subway entrances, subway stations, and subway lines. The transfer points have to be located on network edges and linked to each other. So the subway entrances are located on streets, and the subway stations are located on subway lines. Additional network edge features link entrances to stations and link transfer stations to each other.

When the path is created, it traces over all these features: along a street to a subway entrance, along the link from the entrance to the connected station, and then along the subway line. At subway transfer stations, the path traces along the link from station to station.

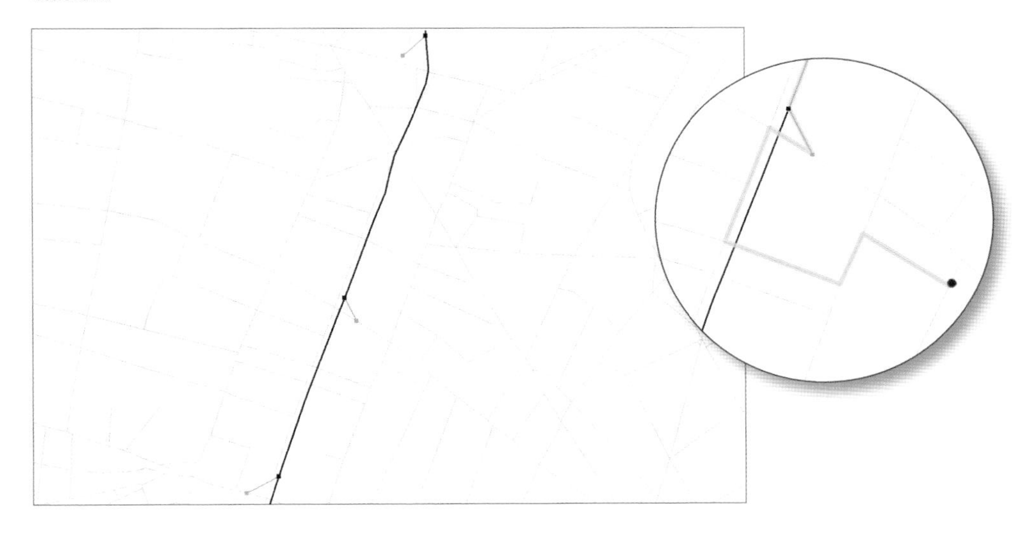

In this multimodal network for modeling a pedestrian path, the streets are in gray and subway lines in dark blue. The street-level subway entrances are shown as light blue squares, and the subway stations as dark blue squares. Network edge features link entrances to stations (light blue lines). The close-up shows how the path (yellow line) traces the connected network features— it starts at the origin (brown dot), traces along streets, enters at a subway entrance, traces along the link to the station, and then continues north on the subway line.

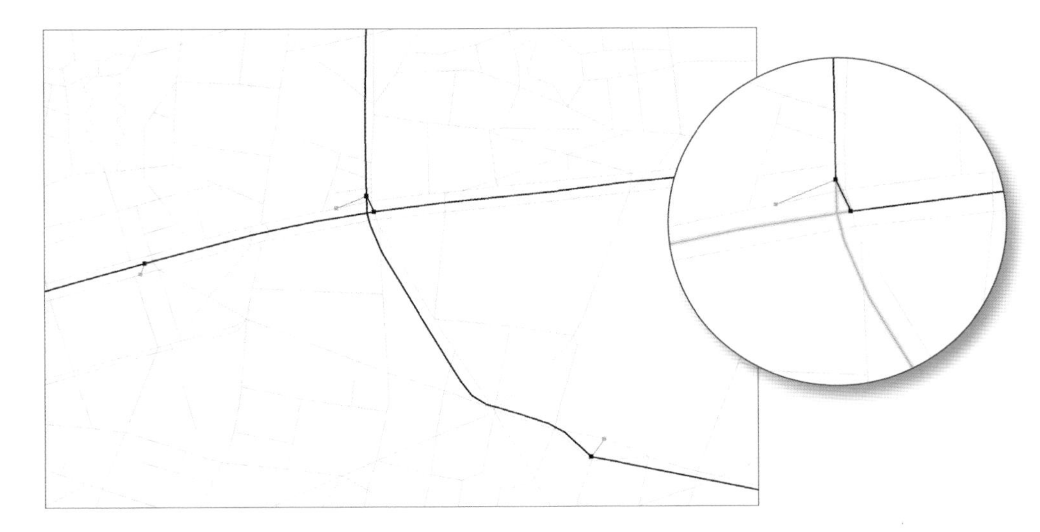

Network edge features also link transfer stations to each other (thin blue line). The close-up shows how the path follows one subway line north, exits at a station connected to the line, traces along the transfer link to the station located on the east–west line, and then continues west on the new subway line.

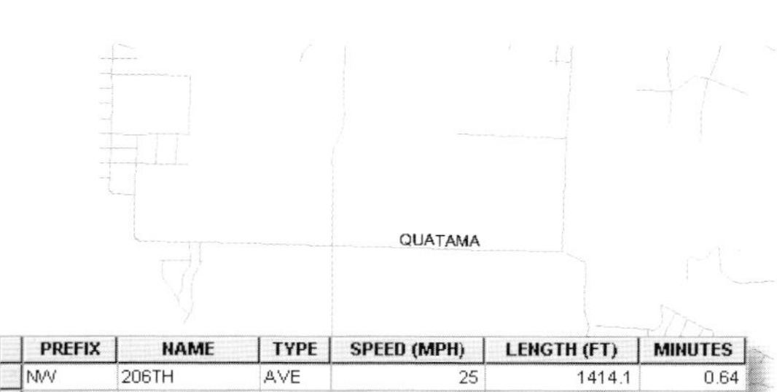

PREFIX	NAME	TYPE	SPEED (MPH)	LENGTH (FT)	MINUTES
NW	206TH	AVE	25	1414.1	0.64
NW	QUATAMA	RD	25	2770.8	1.26
SE	8TH	AVE	25	235.6	0.11
SE	WASHINGTON	ST	25	442.5	0.2
NE	18TH	AVE	25	356.1	0.16

The GIS stores the length of each edge by default. You can assign other costs, such as travel time.

Assign costs to edges and junctions

The GIS finds the least-cost path by calculating the cumulative cost along the network. To allow the GIS to do this, you need to assign the relevant cost values to each edge and, optionally, to junctions.

Cost is usually measured in terms of distance, time, or money. In a transportation network, time or money is often the most important cost. For example, the least-cost path from a fire station to a fire will be the quickest path, not the shortest one. A delivery van company would want to calculate the least expensive path based on time (labor costs), fuel, maintenance, tolls, and the additional fuel and labor costs of being stalled in traffic or at intersections.

The network always stores the length of each edge, so you can create the shortest path without adding any costs to the database. If you want to use costs other than distance in your model, you'll need to add them to the network by assigning the cost values to each edge.

Many path models use travel time as the cost. If you have the actual measured travel time for each edge, you can include those values when you build the network. You may even have different travel times for different times of day, such as mid-morning and rush hour. (These would be stored as separate attributes—you'd specify which travel time to use when you create the path.) The more detailed your time measurements, the more accurate the results of the model will be.

As an alternative, you can calculate approximate travel times from the speed of travel for each edge—this could be the assigned speed limit for the edge or the measured average traffic speed. (Speed is more common in network datasets than measured travel times.)

PREFIX	NAME	TYPE	SPEED (MPH)	LENGTH (FT)	MINUTES
NE	ELAM YOUNG	PKY	25	1024.5	0.47
NW	WALKER	RD	35	646.2	0.21
NE	QUEENS	LN	25	295.4	0.13
NE	BELKNAP	CT	25	821.5	0.37
NE	DONELSON	ST	25	262.9	0.12
NW	WALKER	RD	35	84.8	0.03
NE	3RD	AVE	25	490	0.22
NE	DONELSON	ST	25	247.9	0.11
NE	6TH	AVE	25	363.3	0.17

The Minutes attribute shows the time required to travel the length of the segment. It was calculated from the Length and Speed values.

To calculate travel time from the speed and length of each edge, convert the speed (usually miles or kilometers per hour) to the same unit of measure as the street segment length (usually feet or meters) and to a shorter time unit (minutes or seconds) appropriate for the short distances traveled along most street segments. For example, you might convert miles per hour to feet per minute. You then divide this value—for each segment—by the length of the segment to get travel time.

You can do these calculations in a single operation. For example, to calculate travel in minutes from length in feet and speed in miles per hour, you can use the following formula:

```
MINUTES = LENGTH / ((SPEED * 5280) / 60)
```

You can also use money as a cost value. This is stored as an amount for traversing each edge, calculated from the per mile or per kilometer cost of travel and the length of the edge. You'd use monetary cost, for example, when creating the least-cost path for a furniture delivery truck from a warehouse to a store. The cost per mile value would include the cost of labor (the driver), fuel, and truck maintenance.

In addition to assigning costs to edges, you can assign costs to junctions. For example, you might assign street intersections the average wait time at a stop sign or traffic light, as well as the time required to make a left or right turn. Turns are stored as separate features in the network. Each turn is represented as a pair of edges that are connected at the junction, and each has its own cost. For example, going straight through an intersection may have a cost of one minute while making a left turn at the intersection may have a cost of one and one-half minutes.

Assign restrictions

You can assign restrictions to network features, limiting where the path can go. This will help the model reflect the situation in the real world. Restrictions include:

- blocked edges, such as streets closed for repair
- edges limited in one direction, such as one-way streets
- blocked junctions, such as flooded street intersections
- restricted turns, such as intersections at which no left turn is allowed

Restrictions can be temporary, such as a flooded intersection, or permanent, such as a one-way street. You assign restrictions either by selecting and tagging network features (often used for temporary restrictions) or by setting an attribute value in the edge feature attribute table.

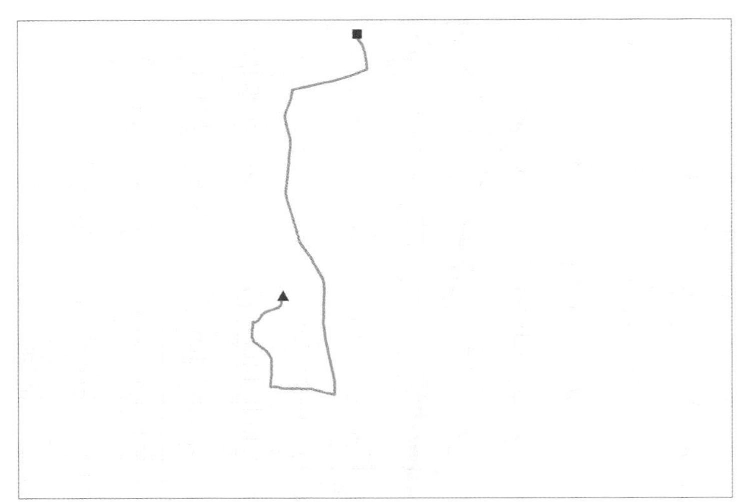

The quickest path from the fire station (square) to the fire (triangle)—7.6 minutes.

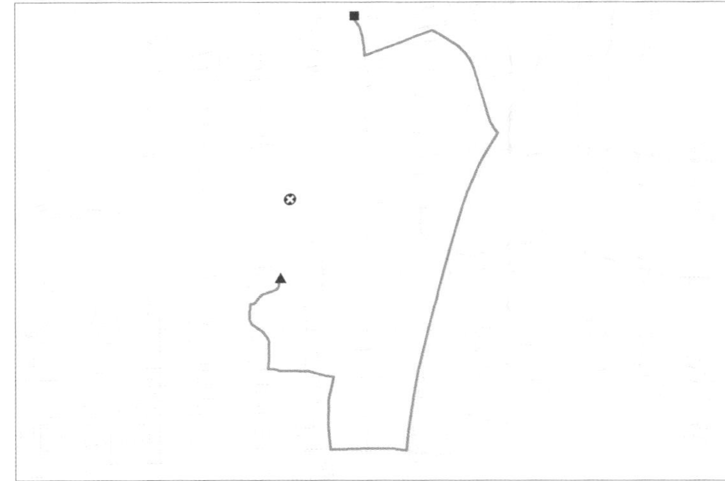

The quickest path from the fire station to the fire with a street closed for repair (indicated by the X)—9.7 minutes.

MODEL VISITS TO LOCATIONS

You can model the least-cost path for making stops at a set of locations. The simplest path has two stops—an origin and a destination—and the path shows the route between the two. For example, a fire department dispatcher might want to find the path with the shortest travel time from the station to the scene of a fire. You can also create a path that has many stops and returns to the origin. For example, a realtor showing her client five homes for sale could find the shortest path that starts at the realty office, visits all five houses, and returns to the office.

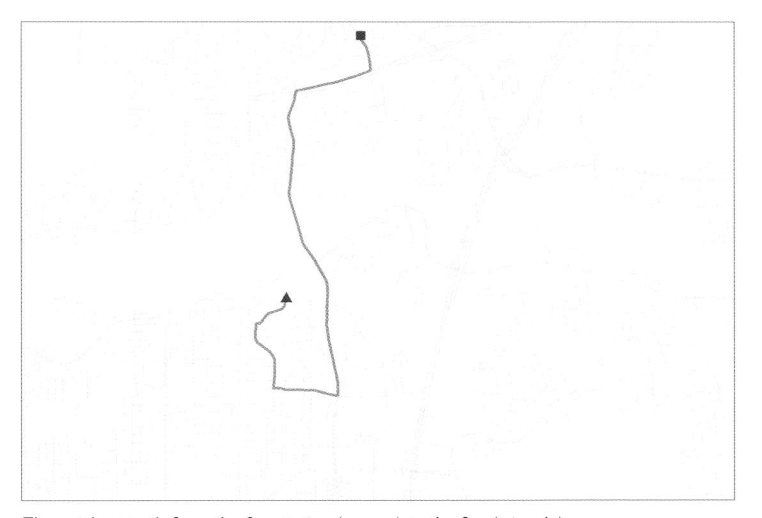

The quickest path from the fire station (square) to the fire (triangle).

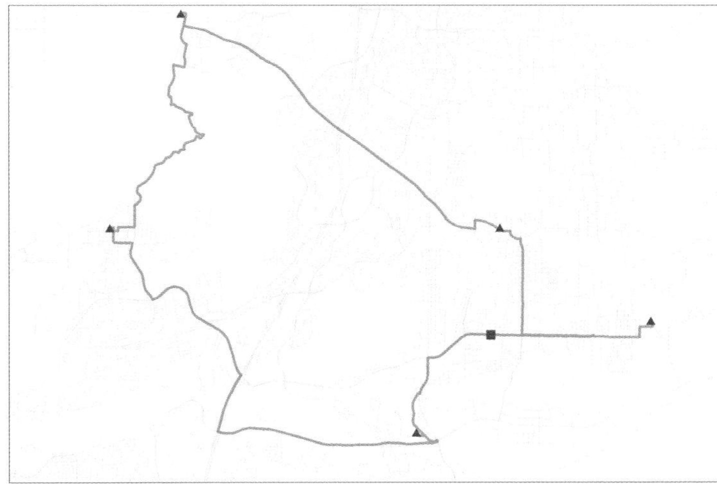

The shortest path starting at the realty office (square), visiting five homes (triangles), and returning to the office.

The process for modeling visits to locations:

1. Create the route layer
2. Specify the location and time window for each stop
3. Create the path
4. Evaluate the results
5. Display and apply the results

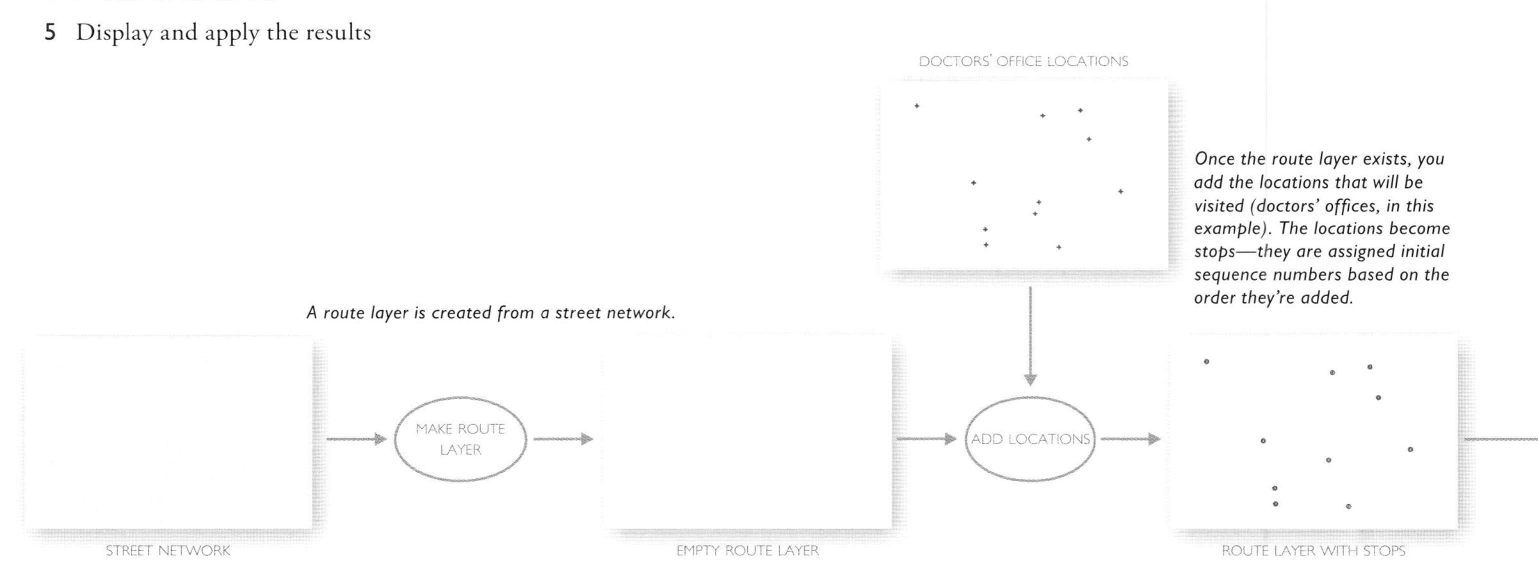

Create the route layer

The first step is to create a route layer that holds the elements of the route, such as the stops and barriers. You create a route layer from a network dataset. While the route layer doesn't include the network itself, the underpinning parameters of the layer are derived from the network dataset—the impedance values (distance, time, or cost), whether one-way streets are included, and so on. In addition, any locations (such as stops or barriers) added to the route layer are located using the network dataset. The route layer is initially empty until you add the stops and, optionally, barriers. After running the model, the route layer contains the input elements as well as the newly created path.

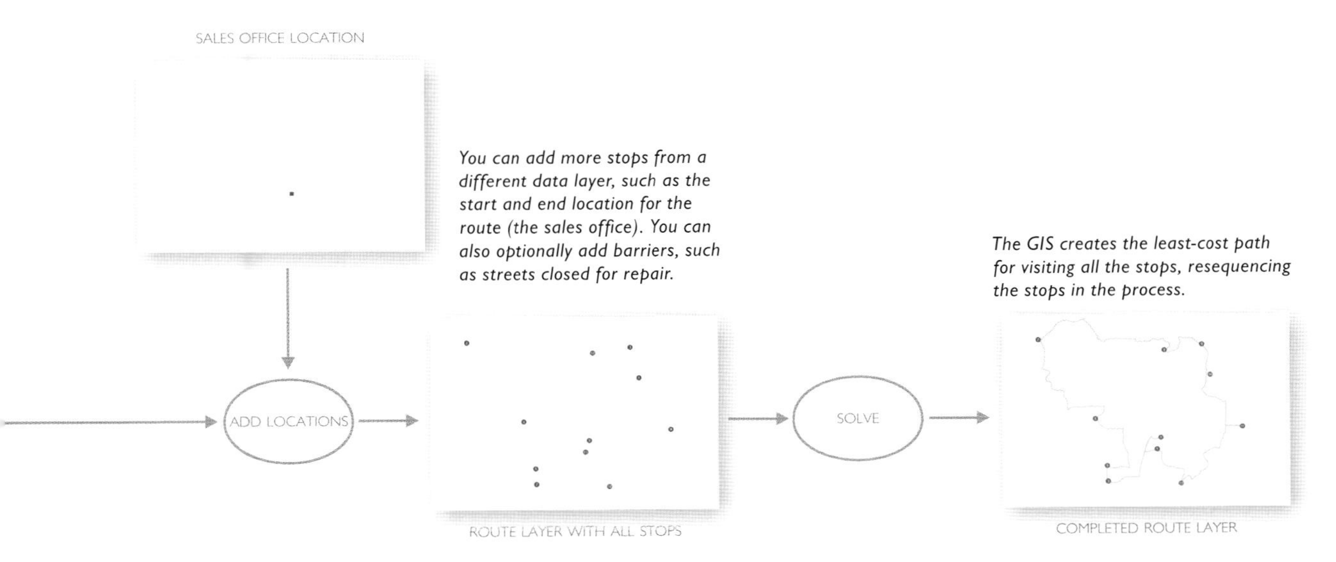

Specify the stops

Next you specify the locations of the stops. In addition, you can specify the time window within which each stop should be visited.

Specifying the location

You locate the stops by placing markers at those locations on the network. You can do this interactively by adding locations on the displayed network layer. More commonly, you'll specify stops by loading a dataset of locations that have already been assigned geographic coordinates. For example, a salesperson would load stops from a dataset of customer locations.

You also add the origin of the route as a stop. The GIS assigns stops an initial sequence number in the route based on the order they're added.

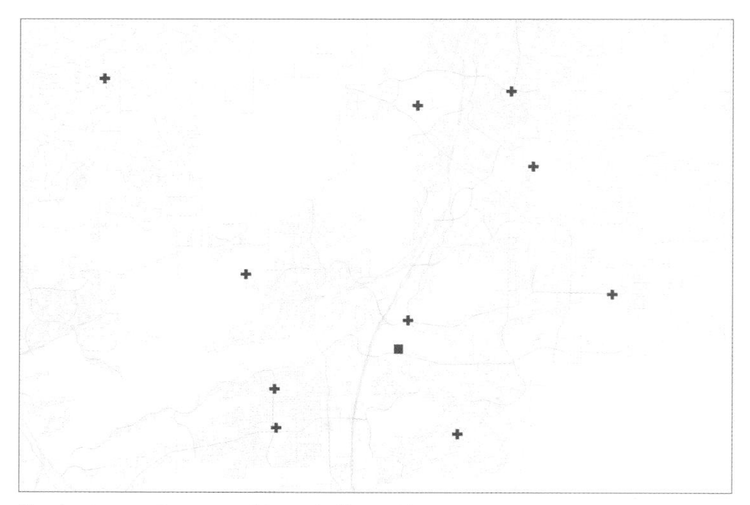

The destinations (locations of doctors' offices—blue crosses) are stored in one dataset, and the origin (the location of the medical equipment sales office—blue square) is stored in another. The street network is also shown.

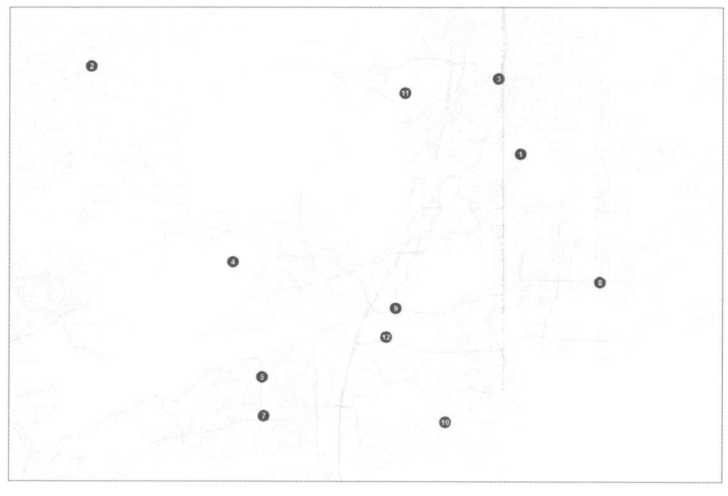

When loaded into the model, the eleven doctors' office locations become numbered stops, in the order they're stored in the dataset. The sales office was loaded after the doctors' offices and becomes the twelfth stop.

If you know the sequence that the stops should be visited you can modify the sequence numbers (you do this by reordering the stops in the GIS). More likely, you'll let the GIS reassign the sequence of stops later when it creates the route—that allows it to find the least-cost path. However, you'll at least want to specify that the origin is the first stop. Similarly, if you want the route to end at a specific stop, specify it as the last stop. When the GIS creates the route, you specify that the first and/or last stops should be preserved. If you want the route to return to the origin, copy the first stop and specify the copy as the last stop.

You can also use street addresses to locate stops. A realtor might do this to create a route between several houses that are for sale. The GIS term for this is geocoding (since you're coding each location with a geographic coordinate, usually based on a street address or intersection). Some GIS software does this at the same time it creates the path. You can enter addresses interactively, or load stops from a dataset that has street addresses.

You may end up loading stops using more than one method. For example, you might load houses for sale by entering street addresses, but add the origin—the real estate office—by pointing to the location interactively.

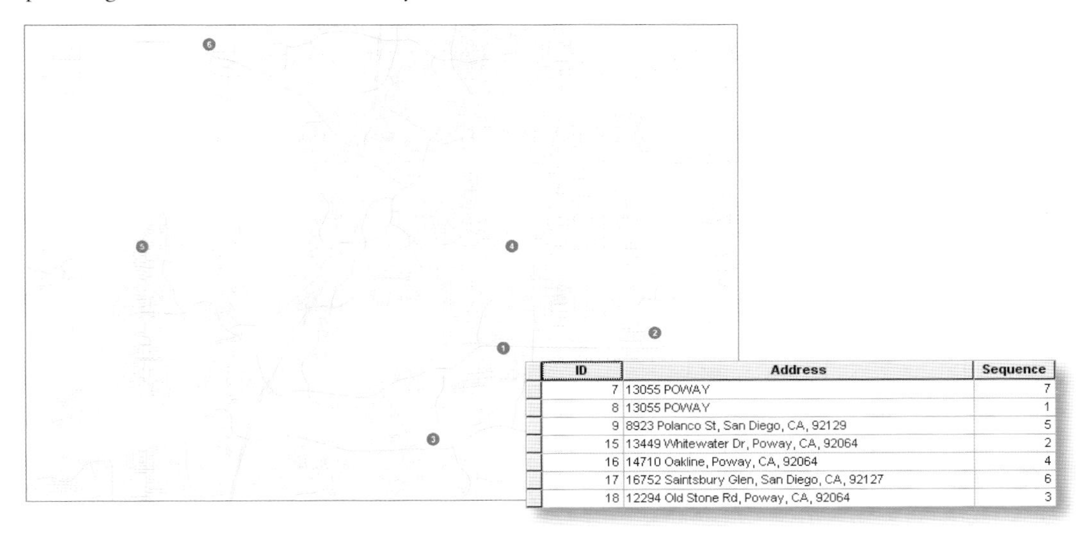

ID	Address	Sequence
7	13055 POWAY	7
8	13055 POWAY	1
9	8923 Polanco St, San Diego, CA, 92129	5
15	13449 Whitewater Dr, Poway, CA, 92064	2
16	14710 Oakline, Poway, CA, 92064	4
17	16752 Saintsbury Glen, San Diego, CA, 92127	6
18	12294 Old Stone Rd, Poway, CA, 92064	3

The five houses for sale were loaded as stops by entering the street addresses. The sequence number indicates the order they were entered. The location of the real estate office was loaded as a stop by pointing to the location on the map. The stop was then copied and the two stops specified as the first and seventh stops to ensure the route starts and ends at the office. The other stops will be resequenced when the GIS creates the route.

Specifying time windows

If there are more than two stops, you can specify a time window for each stop by assigning an arrival, service, and departure time for each. Some paths do not require time windows for the stops—the stop can be visited at any time. For example, a salesperson at a pharmaceutical company might create a route for visiting a set of doctors' offices and dropping off samples—she can visit any time during business hours, so no time windows are needed. Other routes require time windows—a salesperson who has set up appointments with clients will want to create a route that includes the appointment times and durations. You can also create routes where some stops have time windows and others don't. A realtor visiting both occupied and vacant houses with a prospective buyer would likely set up a route that included time windows for the occupied houses (so the owner would know when to expect the visit), but no time windows for the vacant ones since they could be visited at any time.

You create the window for each stop by specifying the earliest allowed arrival time (time window start), the latest allowed arrival time (time window end), and the expected amount of time spent at the stop (service time). The service time might consume the entire time between the time window start and time window end, but does not necessarily have to. (That allows for some leeway in creating the route.) For example, a salesperson visiting clients at set appointment times would ensure the time window start is the time of the appointment and the service time is the length of the appointment. On the other hand, a realtor visiting houses for sale might specify that they will view a specific house within an hour and a half window (say 11:00 a.m. to 12:30 p.m.) and spend no more than thirty minutes (so the time window start would be 11:00 a.m., the time window end 12:00 p.m., and the service time thirty minutes).

The time window start, time window end, and service time are stored as attributes in the attribute table for the stops.

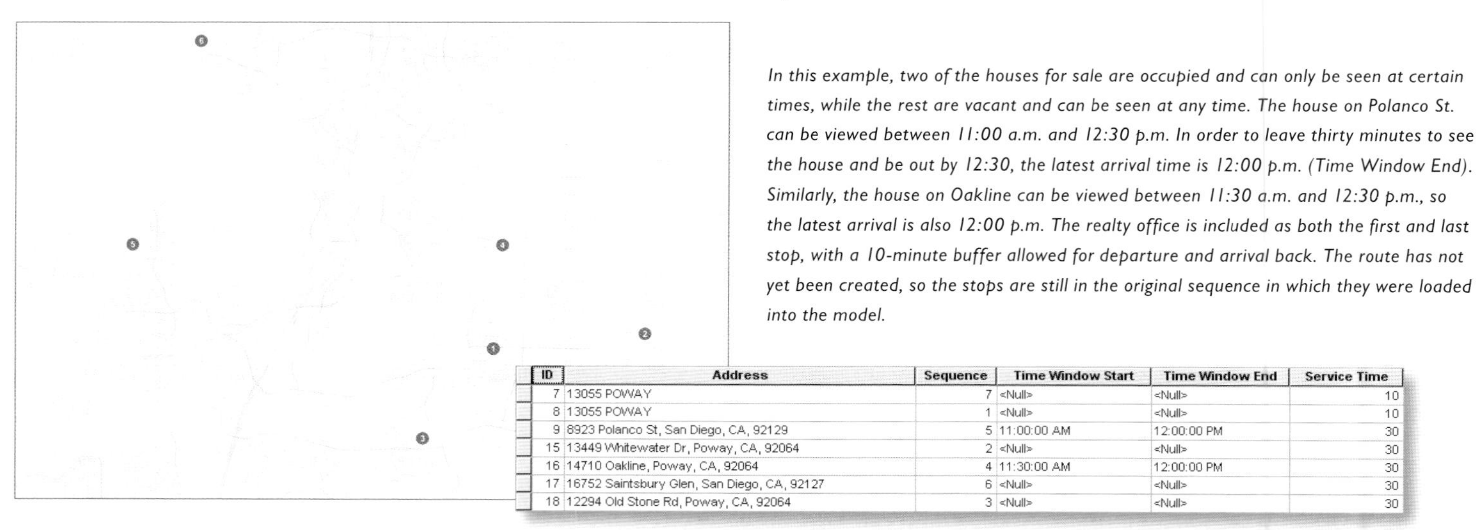

ID	Address	Sequence	Time Window Start	Time Window End	Service Time
7	13055 POWAY	7	<Null>	<Null>	10
8	13055 POWAY	1	<Null>	<Null>	10
9	8923 Polanco St, San Diego, CA, 92129	5	11:00:00 AM	12:00:00 PM	30
15	13449 Whitewater Dr, Poway, CA, 92064	2	<Null>	<Null>	30
16	14710 Oakline, Poway, CA, 92064	4	11:30:00 AM	12:00:00 PM	30
17	16752 Saintsbury Glen, San Diego, CA, 92127	6	<Null>	<Null>	30
18	12294 Old Stone Rd, Poway, CA, 92064	3	<Null>	<Null>	30

In this example, two of the houses for sale are occupied and can only be seen at certain times, while the rest are vacant and can be seen at any time. The house on Polanco St. can be viewed between 11:00 a.m. and 12:30 p.m. In order to leave thirty minutes to see the house and be out by 12:30, the latest arrival time is 12:00 p.m. (Time Window End). Similarly, the house on Oakline can be viewed between 11:30 a.m. and 12:30 p.m., so the latest arrival is also 12:00 p.m. The realty office is included as both the first and last stop, with a 10-minute buffer allowed for departure and arrival back. The route has not yet been created, so the stops are still in the original sequence in which they were loaded into the model.

The GIS will attempt to create the least-cost route given the time windows and the distance between stops. In some cases, the arrival time for a particular stop may be before the start of the time window for that stop. In these cases, there will be a wait time added to the route. Similarly, if a stop cannot be visited until after the end of the time window, the stop will still be included in the route, but there will be a "violation." For example, if the specified time window end is 3:15 p.m., but the stop will not be reached until 3:45 p.m., there is a thirty minute violation. This additional time is included in the routing for later stops on the route. The GIS attempts to create the least-cost path without any wait time or violations. If a violation is not acceptable, you can remove it by removing a stop or by changing the time window. (For example, the realtor could reschedule the visit to that house for another day, or ask the owner if her clients can arrive a half-hour late.)

Create the path

Once you've specified the network to use, assigned the costs and restrictions, and specified the locations of the stops—and, optionally, the time windows—you have the GIS calculate the least-cost path.

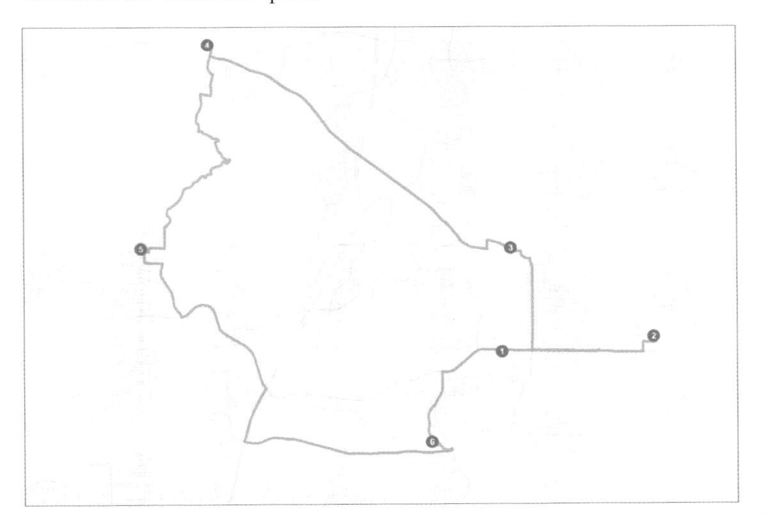

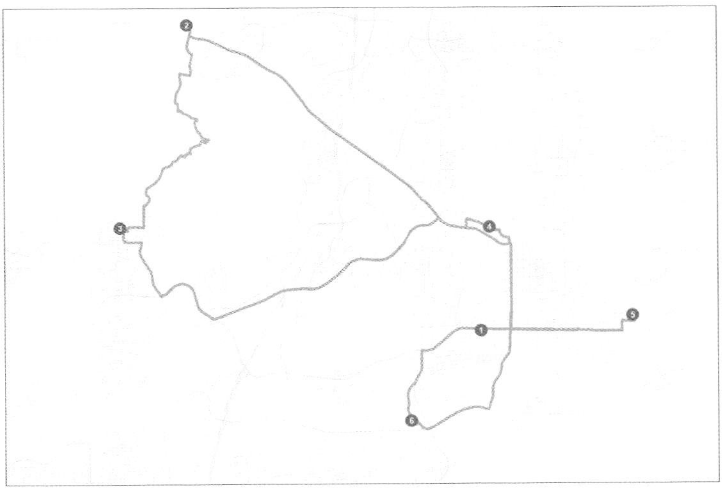

Two routes for visiting houses for sale, starting at the realty office (stop 1) and returning to the office. The route on the left was created without time windows, while the one on the right was created with time windows for two stops (as shown in the previous map and table). In both cases, a new sequence of stops was assigned and the quickest path between the stops created.

Checking every possible path between the stops to find out which one has the least cost would take an inordinate amount of time. Instead, the GIS uses an efficient approach to explore the network and find the least-cost path. One of the more common approaches is Dijkstra's algorithm, developed by Dutch computer scientist Edsger Dijkstra. (See "References and further reading" at the end of this chapter.)

Evaluate the results

The route is only as good as the network data it's based on. Common problems include streets that are missing, dirt roads or trails shown as traversable streets, and one-way streets not identified as such. To the extent your network contains any of these errors, or others, the path will be less reliable. Many organizations start with commercially available data and improve it based on field checks and current local information, such as new subdivision plans.

Many routes are based on average costs, such as average travel time for each block. Hence, the path the GIS creates may not, in fact, be the least cost. The path is merely a best guess which should be used in conjunction with your knowledge of a place and situation. This is especially true for paths that can't be field-checked, such as in an emergency response situation.

Display and apply the results

The model creates a new route feature that is displayed on the map as a line connecting the stops. Information about the route, as well as the stops, is contained in the attribute tables for the features.

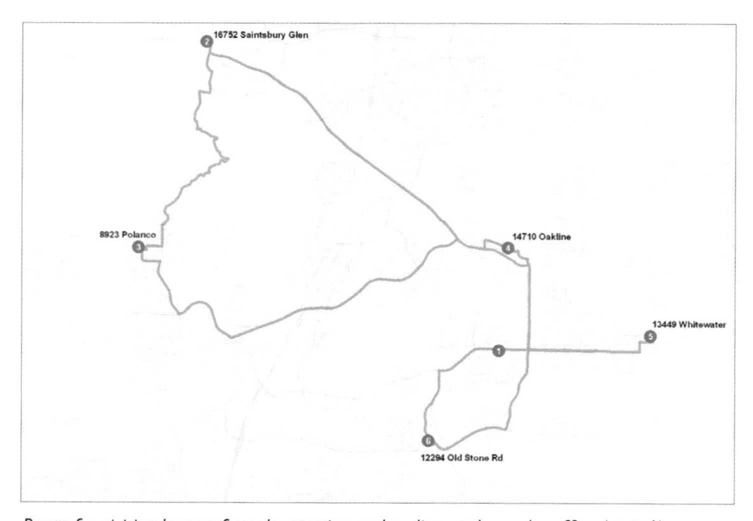

Route for visiting houses for sale, starting and ending at the realty office (stop 1).

The table for the route identifies the beginning and ending stops and lists the total number of stops in the route. It also includes the start and end times for the route and the total impedance (travel time, distance, or cost), along with the total wait and violation time, if any.

First Stop ID	Last Stop ID	Stop Count	Start Time	End Time	Total Minutes	Total Wait	Total Violation
1	24	7	4/6/2010 10:00:00 AM	4/6/2010 1:47:35 PM	227.58	0	0

The table for the stops includes, in addition to the input attributes you specified, the sequence of the stops (which may be different from the sequence you specified if you allowed the GIS to resequence the stops), the arrival and departure time for each stop, and the wait time or time violation for the stop (if any). It also includes the cumulative impedance, cumulative wait time, and cumulative time violations from the beginning of the route up to and including that stop—that is, including travel time and time allotted to all the previous stops.

ID	Address	Sequence	Time Window Start	Time Window End	Service Time	Arrive Time	Depart Time	Wait	Violation	Cumul Minutes	Cumul Wait	Cumul Violation
1	13055 POWAY	1	<Null>	<Null>	10	4/6/2010 10:00:00 AM	4/6/2010 10:10:00 AM	0	0	10.00	0	0
23	16752 Saintsbury Glen, San Diego, CA, 92127	2	<Null>	<Null>	30	4/6/2010 10:24:06 AM	4/6/2010 10:54:06 AM	0	0	54.09	0	0
19	8923 Polanco St, San Diego, CA, 92129	3	11:00:00 AM	12:00:00 PM	30	4/6/2010 11:05:16 AM	4/6/2010 11:35:16 AM	0	0	95.27	0	0
22	14710 Oakline, Poway, CA, 92064	4	11:30:00 AM	12:00:00 PM	30	4/6/2010 11:46:55 AM	4/6/2010 12:16:55 PM	0	0	136.91	0	0
21	13449 Whitewater Dr, Poway, CA, 92064	5	<Null>	<Null>	30	4/6/2010 12:24:22 PM	4/6/2010 12:54:22 PM	0	0	174.37	0	0
20	12294 Old Stone Rd, Poway, CA, 92064	6	<Null>	<Null>	30	4/6/2010 1:03:03 PM	4/6/2010 1:33:03 PM	0	0	213.05	0	0
26	13055 POWAY	7	<Null>	<Null>	10	4/6/2010 1:37:35 PM	4/6/2010 1:47:35 PM	0	0	227.58	0	0

Attribute table for stops for a route visiting houses for sale. There are no wait times or violations for this route—the two houses with time windows are visited within the windows.

In addition to maps showing the path of the route and the stops, you can create a set of directions with the distance or time for each segment of the route and a corresponding detailed map.

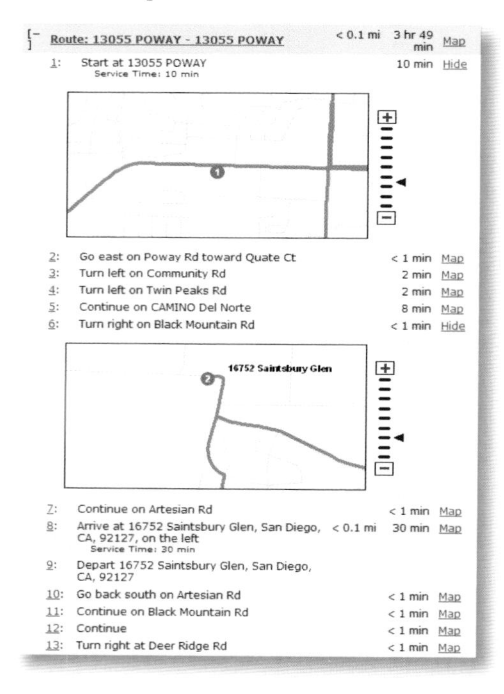

Creating alternate paths

By changing the impedance values, you can create alternate paths for a particular use. For example, you might create alternate paths when finding the best hazardous materials route through a city. One path may be the shortest, one the quickest, and a third pass the fewest houses. (In the last case, the cost would be the number of houses on each street segment.)

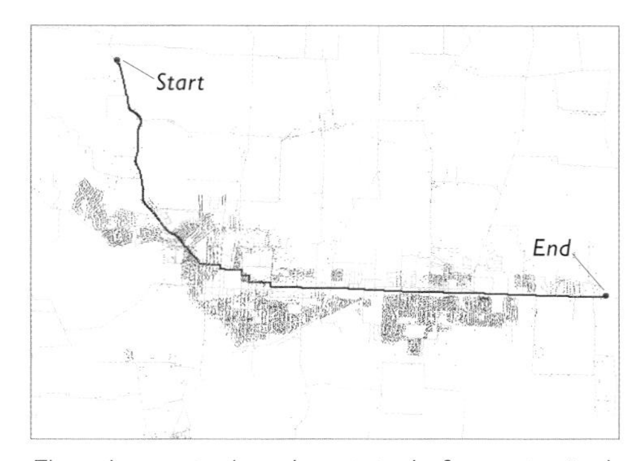

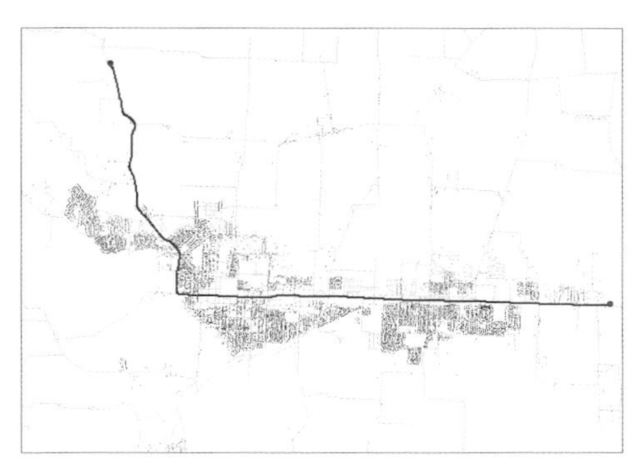
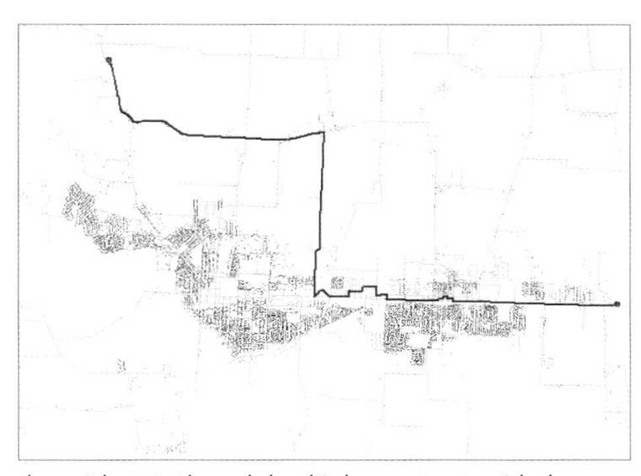

These three maps show alternate paths for transporting hazardous materials. The first is the shortest distance, the second is the quickest path, and the third uses streets with the fewest houses (the gray dots). In all three cases, the second half of the route is essentially the same, so discussion of alternatives could focus on the first half of the route.

If you've added several different cost attributes to each edge, it is fairly easy to create alternate paths using the different costs (or combinations of costs). If, for example, you're creating a path for shipping cars by truck and rail, you could run the analysis using various combinations of truck and rail carriers, including the corresponding shipping rates and travel times. Since the GIS calculates the total cost of each path, it's easy to compare alternates.

You can also create alternate paths by adding or removing impediments associated with the network edges or junctions to create "what-if" scenarios.

MODEL DELIVERIES

You can use GIS to model the most efficient path for delivering goods or people to various locations, and, optionally, for picking up goods or people, as well. The underlying methodology is similar to modeling visits to stops, but there are more parameters.

The locations you are delivering to (or picking up from) are known as orders. They are analogous to the stops you specify when modeling visits, with the major difference being that orders have an associated quantity to be delivered or picked up.

When you model deliveries, a route—instead of being simply a path along which you travel—is associated with a vehicle that has characteristics such as a capacity and an operating cost. Your model can have multiple routes, each with its own characteristics.

Routes start at, and often return to, a depot. You can have several routes starting from one or more depots.

The GIS assigns orders to routes and creates paths so that each route fulfills as many orders as possible at the least cost, given constraints such as the delivery time window for each order, the capacity of the truck, and the maximum time or distance allowed for the route.

One type of model creates the most efficient routes for deliveries from a depot to the order locations and back. This is the classic model for a delivery route, such as a truck delivering groceries from a distribution center to stores or an airport shuttle picking up people and delivering them to an airport.

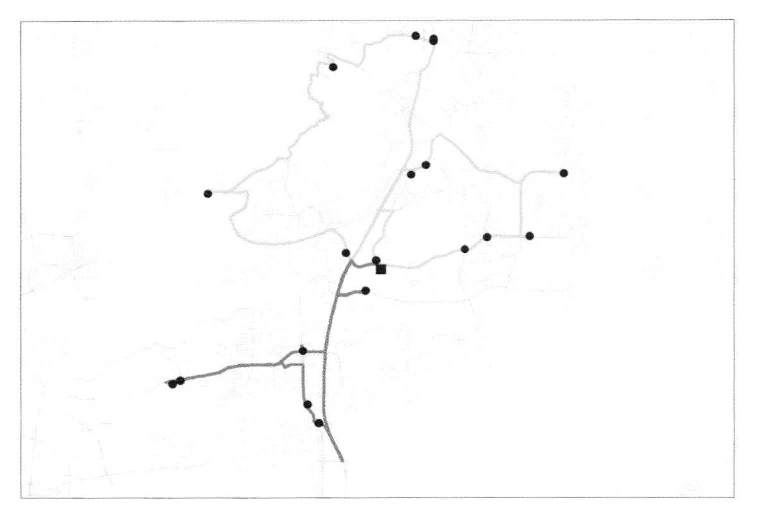

Three routes for delivery vans delivering goods to service station convenience markets (circles). Each route starts and ends at the distribution center (square).

Another similar type of model is used for decentralized deliveries, where people (usually) are picked up at various locations and delivered to various other locations. This type of model is used, for example, for a paratransit van that picks up people at their homes and drops them off at doctors' offices, hospitals, or stores. At any given time, the van may be carrying people going to different locations. For this type of model, you create what are known as "order pairs," which contain the pickup and drop-off locations for each order (person). All the pickup and drop-off locations are included as orders. Additionally, a list of order pairs is specified—the list matches each pickup location with a drop-off location.

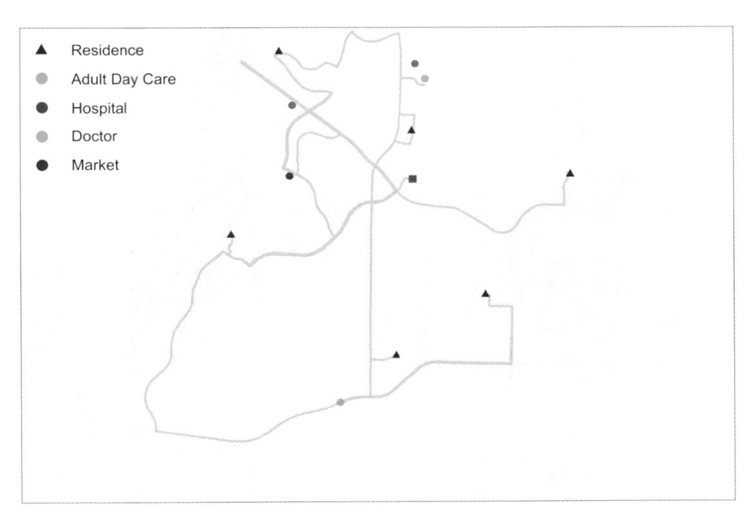

Route for a paratransit van picking up people at their homes, delivering them to various locations, and then picking up and returning them home. Each rider is matched with a destination to create an order pair. A single route is shown here, but you can create multiple routes for order pairs, in which case the GIS assigns the pairs to the best route.

With either type of model you have the option of assigning very specific characteristics to each route. You can limit the geographic area a route covers, specify breaks for drivers (and whether they're paid or unpaid breaks), specify whether overtime is allowed and what the overtime pay rate is, and specify whether a route has a special capability such as a wheelchair lift or bike rack. (Orders requiring these specialties will only be assigned to those routes.)

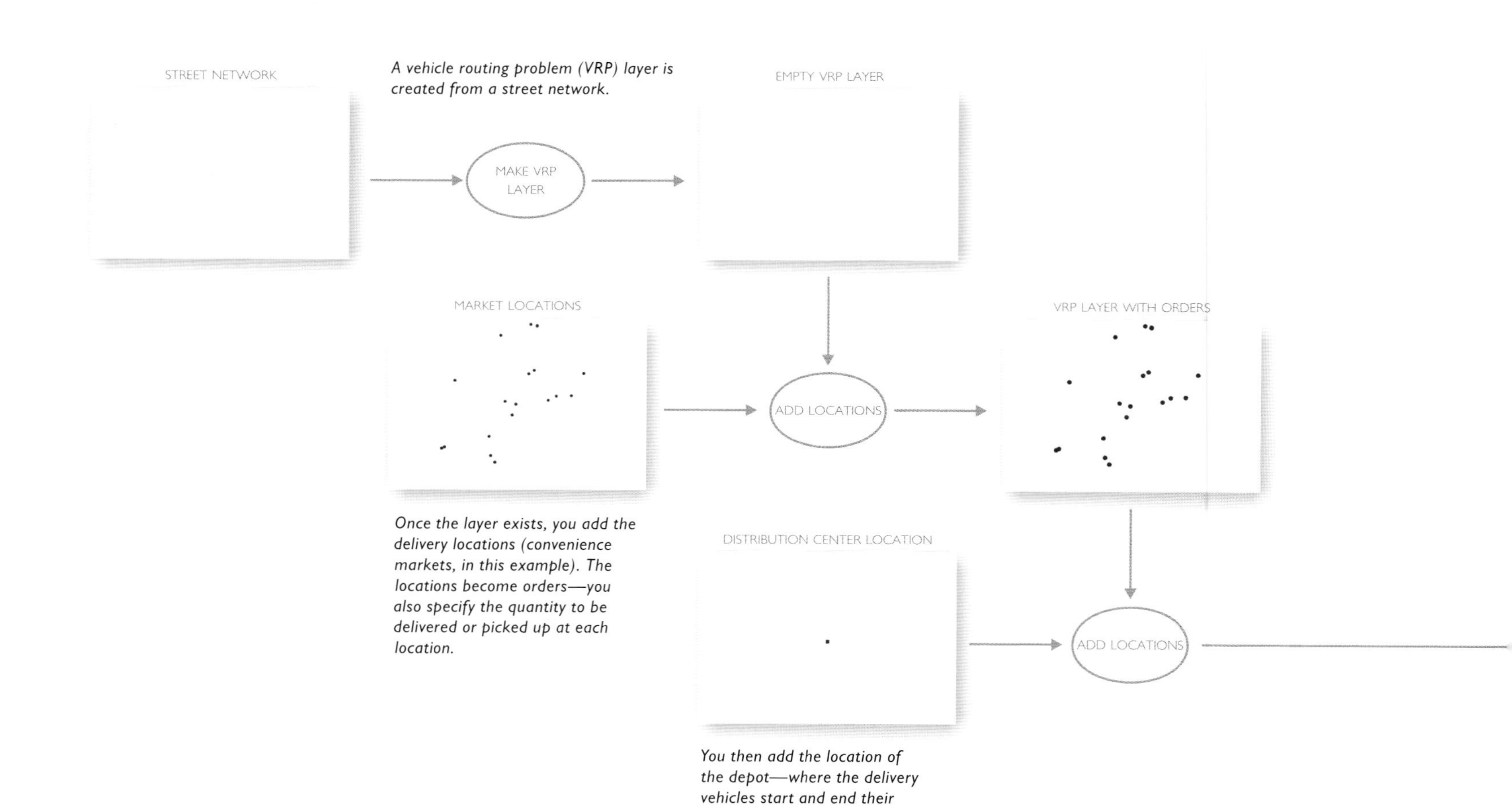

A vehicle routing problem (VRP) layer is created from a street network.

Once the layer exists, you add the delivery locations (convenience markets, in this example). The locations become orders—you also specify the quantity to be delivered or picked up at each location.

You then add the location of the depot—where the delivery vehicles start and end their routes.

ROUTE ATTRIBUTES

Route	Capacities	Cost Per Minute	Max Order
Truck 1	90	0.25	12
Truck 2	140	0.25	18
Truck 3	120	0.25	15

Next you add the characteristics of each route—the capacity of each truck, the labor costs, and so on. These are stored as attributes, since there is no geographic representation of the route at this point.

The process for modeling deliveries is:

1. Create the vehicle routing problem layer
2. Specify the orders (and, optionally, order pairs)
3. Specify the depots
4. Specify the route characteristics
5. Solve the routing problem
6. Evaluate the results
7. Display and apply the results

Create the vehicle routing problem layer

As when you model visits to stops, the first step is to create a layer to contain the various elements of the delivery routes. These include the orders themselves, the depots the trucks depart from and return to, and the characteristics of the delivery vehicles. You can specify additional parameters of the routes before running the model—the zones routes operate within, breaks for drivers—along with any barriers on the network, such as flooded intersections. The GIS adds the results of the model to the layer (the route each order is assigned to, the least-cost paths between the orders, the total time and cost for each route, and so on).

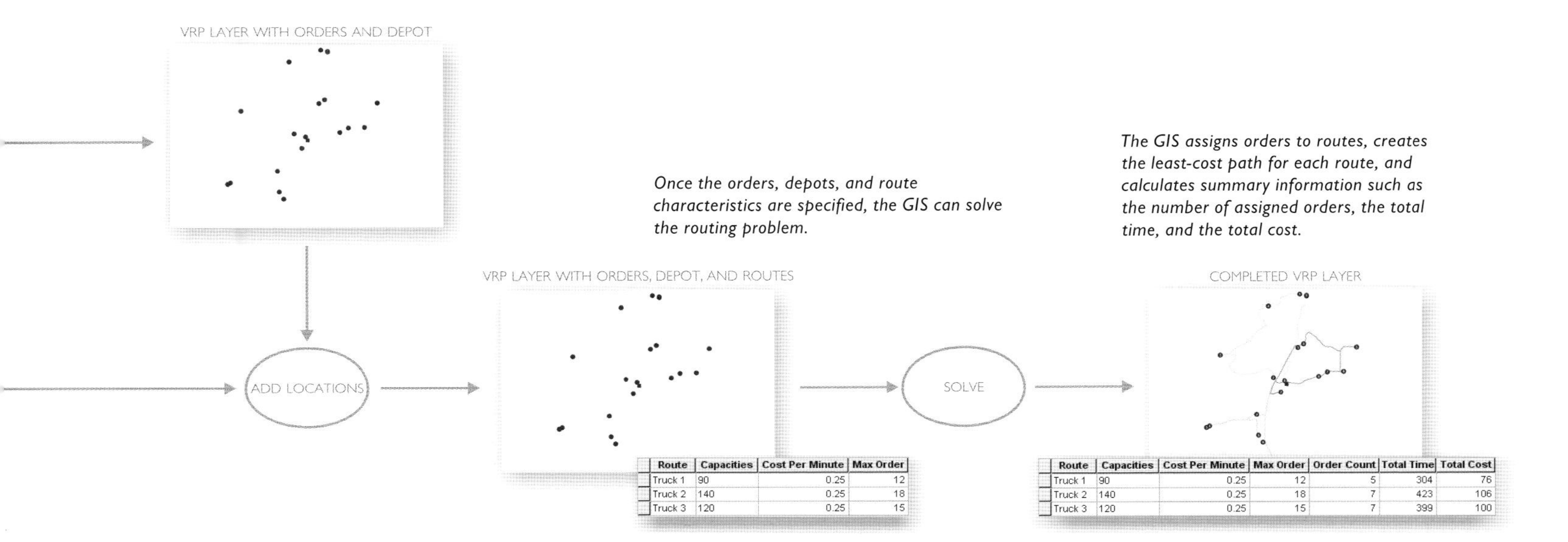

Route	Capacities	Cost Per Minute	Max Order
Truck 1	90	0.25	12
Truck 2	140	0.25	18
Truck 3	120	0.25	15

Route	Capacities	Cost Per Minute	Max Order	Order Count	Total Time	Total Cost
Truck 1	90	0.25	12	5	304	76
Truck 2	140	0.25	18	7	423	106
Truck 3	120	0.25	15	7	399	100

Once the orders, depots, and route characteristics are specified, the GIS can solve the routing problem.

The GIS assigns orders to routes, creates the least-cost path for each route, and calculates summary information such as the number of assigned orders, the total time, and the total cost.

Specify the orders

As with specifying stops, you specify the locations of orders by pointing to locations on the screen, by loading existing point features from a dataset, or by using street addresses to assign locations. Again, as with stops, once the orders are located, you specify the time window for each order—the starting and ending arrival times, the time required to load or unload (the service time), and the latest allowed arrival (the maximum time violation).

In addition, you specify, for each order, the amount or quantity to be picked up or delivered. This could be the number of people or objects, or the weight or volume of goods. For example, three students may board at one bus stop and five at another. Or, one customer may receive a dishwasher weighing 80 lbs. and another a refrigerator weighing 240 lbs. The GIS ensures the capacity of the vehicle is not exceeded.

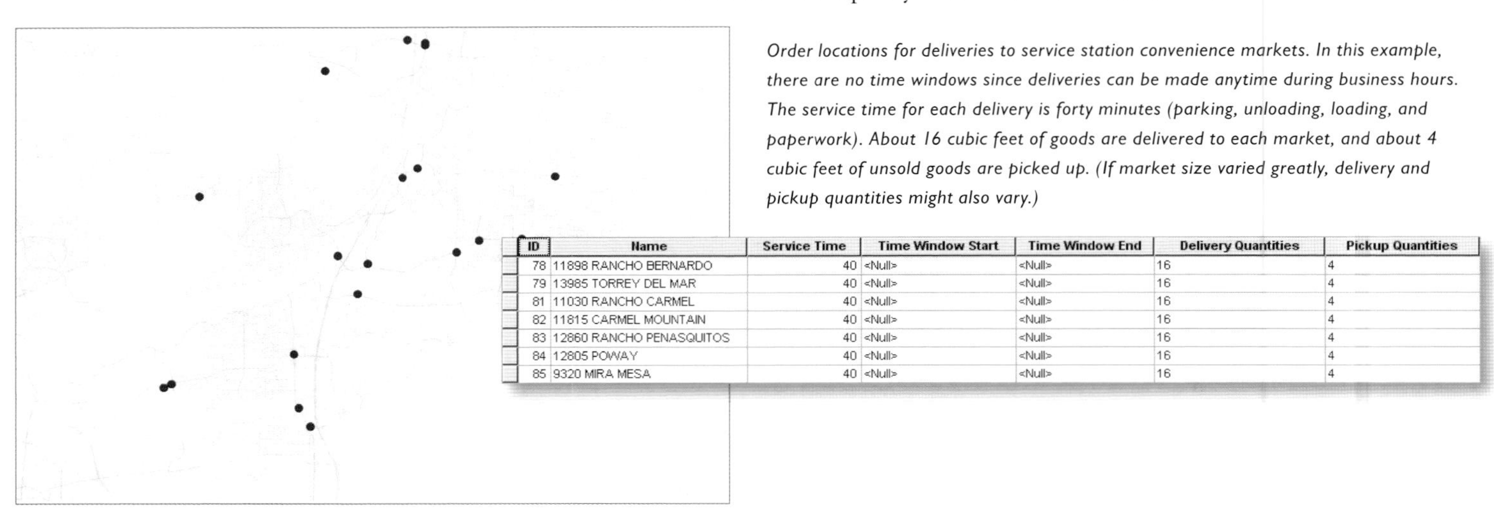

ID	Name	Service Time	Time Window Start	Time Window End	Delivery Quantities	Pickup Quantities
78	11898 RANCHO BERNARDO	40	<Null>	<Null>	16	4
79	13985 TORREY DEL MAR	40	<Null>	<Null>	16	4
81	11030 RANCHO CARMEL	40	<Null>	<Null>	16	4
82	11815 CARMEL MOUNTAIN	40	<Null>	<Null>	16	4
83	12860 RANCHO PENASQUITOS	40	<Null>	<Null>	16	4
84	12605 POWAY	40	<Null>	<Null>	16	4
85	9320 MIRA MESA	40	<Null>	<Null>	16	4

Order locations for deliveries to service station convenience markets. In this example, there are no time windows since deliveries can be made anytime during business hours. The service time for each delivery is forty minutes (parking, unloading, loading, and paperwork). About 16 cubic feet of goods are delivered to each market, and about 4 cubic feet of unsold goods are picked up. (If market size varied greatly, delivery and pickup quantities might also vary.)

You can optionally specify any special requirements for the order, such as the need for a vehicle with a wheelchair lift or a bicycle rack. These specialties are defined in a separate table and then assigned as characteristics of both orders and routes, so that an order will be assigned to a route that can handle the special request.

If your route involves order pairs, each pickup and each drop-off is included as an order. So, for example, if a person is picked up at home, dropped at a hospital, then later picked up at the hospital and delivered back home, that generates four orders: the pickup at the house, the drop-off at the hospital, the pickup at the hospital, and the drop-off at the house. So you may, in fact, have several orders associated with each location. (In this example, there are two orders associated with the house and two with the hospital.) Similarly, if two people are picked up at different locations and delivered to the same location, two drop-off orders are generated for that location.

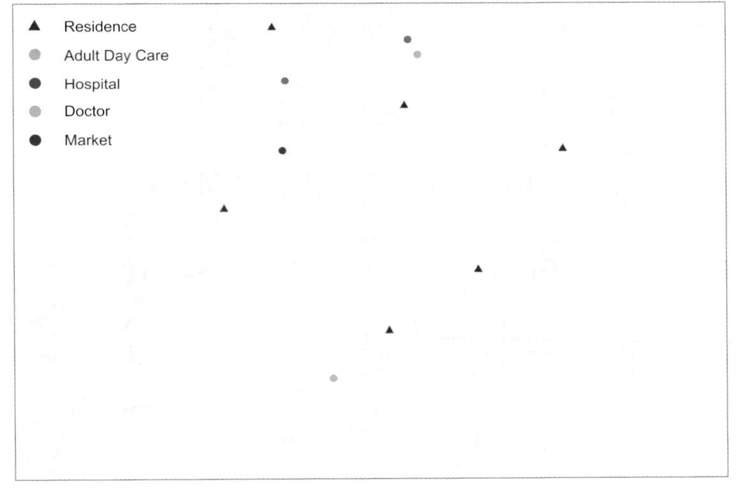

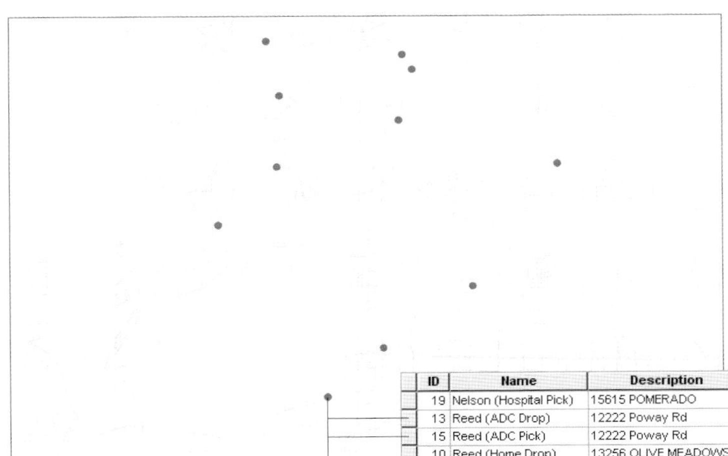

The top map shows the residences where people will be picked up and the various locations where they will be dropped off (and later picked up from). The bottom map shows the locations after they have been loaded as orders—pickup and drop-off locations both become orders. In fact, since, in this example, people are delivered back to their homes, each location is both a pickup and a drop-off order, as shown in the orders attribute table. While each location appears as a single dot, there may be, in fact, several orders associated with each. For example, the adult day-care center has four orders associated with it—dropping off the rider named Reed, later picking her up, dropping off the rider named Sanders, and later picking him up. Orders have time windows (the window within which the rider must be picked up or dropped off). Drop-off orders have delivery quantities (the number of people being dropped) while pickup orders have pickup quantities. The service time for each order is three minutes (the time required for riders to board the van, assuming they are mainly elderly and may have walkers or wheelchairs).

ID	Name	Description	ServiceTime	Time Window Start	Time Window End	Delivery Quantities	Pickup Quantities
19	Nelson (Hospital Pick)	15615 POMERADO	3	11:30:00 AM	12:00:00 PM	<Null>	1
13	Reed (ADC Drop)	12222 Poway Rd	3	9:30:00 AM	10:00:00 AM	1	<Null>
15	Reed (ADC Pick)	12222 Poway Rd	3	4:00:00 PM	4:30:00 PM	<Null>	1
10	Reed (Home Drop)	13256 OLIVE MEADOWS	3	<Null>	<Null>	1	<Null>
4	Reed (Home Pick)	13256 OLIVE MEADOWS	3	<Null>	<Null>	<Null>	1
14	Sanders (ADC Drop)	12222 Poway Rd	3	11:15:00 AM	11:45:00 AM	1	<Null>
16	Sanders (ADC Pick)	12222 Poway Rd	3	1:30:00 PM	2:00:00 PM	<Null>	1
7	Sanders (Home Drop)	11505 WINDCREST	3	<Null>	<Null>	1	<Null>
1	Sanders (Home Pick)	11505 WINDCREST	3	<Null>	<Null>	<Null>	1

Once the orders are loaded, you load the table of order pairs to specify which drop-off order each pickup order is linked to and the maximum time the person (or items being delivered) can be en route for that delivery. The table can be created within the GIS or loaded from other software, such as Microsoft Excel.

ID	Pick Order	Drop Order	Max Transit Time
1	Reed (Home Pick)	Reed (ADC Drop)	25
2	Sanders (Home Pick)	Sanders (ADC Drop)	25
3	Cruz (Home Pick)	Cruz (Hospital Drop)	25
4	Nelson (Home Pick)	Nelson (Hospital Drop)	25
5	Collins (Home Pick)	Collins (Doctor Drop)	20
6	Murray (Home Pick)	Murray (Market Drop)	30
7	Reed (ADC Pick)	Reed (Home Drop)	25
8	Sanders (ADC Pick)	Sanders (Home Drop)	25
9	Cruz (Hospital Pick)	Cruz (Home Drop)	25
10	Nelson (Hospital Pick)	Nelson (Home Drop)	25
11	Collins (Doctor Pick)	Collins (Home Drop)	20
12	Murray (Market Pick)	Murray (Home Drop)	30

Order pairs table for the paratransit route. The table links each pickup order in the orders table with the corresponding drop-off order. The GIS uses the information to create the required paths between the pickup and drop-off locations, given the time window and maximum transit time constraints.

Specify the depots and renewal locations

Depots are where goods to be delivered are stored and loaded onto vehicles. It is where a route starts and ends. As with orders, the locations of depots can be specified interactively on the screen, by loading locations from an existing dataset, or by using street addresses. Depots can have a start (open) and end (close) time, so that routes have to operate between those hours.

You can optionally specify satellite depots where vehicles can pick up additional goods while out on deliveries. These are known as "renewal locations." You don't schedule or require the vehicle to visit them, but they can be added to the route, if necessary, by the GIS, so the vehicle can complete its deliveries without having to return to the depot.

Depot for deliveries to service station convenience markets. In this example, the depot is a distribution center, and all routes start and end at this depot. You can have more than one depot in your model.

Specify the route characteristics

When you're modeling visits to a set of stops (as discussed in the previous section), the route does not exist until you run the model. The route that is created is simply the path that you take to visit all the stops. Essentially, the term "route" is synonymous with "path."

When you're modeling deliveries, however, the term "route" takes on a slightly different meaning. It denotes not only a path, but elements including a vehicle and a delivery person (such as with a newspaper "route" or a postal "route"). Consequently, before running the model you define these characteristics and create the route. After the model is run, the sequence of orders and the path between them are added to the route, along with other characteristics (such as the arrival and departure time for each order).

Your model can include a single route or multiple ones. If there is more than one route, the characteristics can be the same for all routes, or they can vary based on the vehicles and drivers assigned to each. In most cases, some characteristics will be consistent across all routes (such as labor cost or the start time) and some will vary (such as the capacity of each truck).

The key characteristic in modeling deliveries is the capacity of the vehicle. The capacity is the amount, weight, or volume the vehicle assigned to the route can hold. For example, a school bus may have seats for thirty students or a delivery truck may carry up to 2,000 lbs. Orders that are near each other will be assigned to a route until the capacity is reached. If another route has excess capacity, orders will be assigned to that route. The GIS attempts to balance the orders between routes so that each route is efficient (that is, can handle the most orders at the lowest cost).

In addition to the capacity of the route, you specify the parameters for the route. These include which depot it starts and returns to, the earliest and latest start time, and the time required to load or unload. To ensure that the routes are at least roughly balanced, you can specify the maximum duration, distance, cost, and number of orders for each route. For example, you might specify that all trucks make 10 to 12 stops or that all drivers take between 7 and 9 hours to finish their routes.

You can also specify how the costs for the route will be calculated in terms of any fixed costs and per-unit costs. You can specify unit costs for labor by using a cost per unit of time. For example, if the driver is paid $15 per hour, and the route is calculated using minute increments, the per-minute cost is $0.25 per minute. You can also specify a different rate if the route runs into overtime. Similarly, a cost per unit of distance can be used to account for fuel and maintenance for the vehicle (say, $0.30 per mile). The costs are accumulated over the entire route to provide the overall cost.

ID	Route	Start Depot	End Depot	Start Service Time	End Service Time	Earliest Start	Latest Start	Capacities	Cost Per Minute	Cost Per Mile	Max Order	Max Time
1	Truck 1	10931 CREEKBRIDGE	10931 CREEKBRIDGE	45	30	7:30:00 AM	9:00:00 AM	90	0.25	0.3	12	480
2	Truck 2	10931 CREEKBRIDGE	10931 CREEKBRIDGE	75	30	7:30:00 AM	9:00:00 AM	140	0.25	0.35	18	480
3	Truck 3	10931 CREEKBRIDGE	10931 CREEKBRIDGE	60	30	7:30:00 AM	9:00:00 AM	120	0.25	0.3	15	480

Routes for deliveries to service station convenience markets. The capacity of each truck is shown in cubic feet. The larger the truck, the more orders it can handle, but the longer it takes to load (start service time).

Breaks for drivers can also be assigned to routes. Including scheduled breaks will ensure that time windows are met. You define each break—the time window within which it can occur, its duration, and whether or not the driver is paid for the break (so the GIS knows whether to include the pay period in the total cost for the route). You then assign the break to a specific route. If all drivers get the same break, you can assign the single break to several routes.

If you want a particular route to cover a specific geographic area, you can delineate a zone within which the route must be contained. You can do this by interactively drawing a polygon on the screen or by loading predefined zones. You'd use zones if, for example, you regularly deliver to or pick up from a set of customers that have already been assigned to market areas. Alternatively, you can specify a seed location for a route—the GIS will attempt to create the route so it is in the vicinity of the seed location. You might use a seed location if a driver is familiar with a particular part of town—you pick a seed location centrally located among orders in that part of town and assign it to the driver's route. You'd most likely use zones or seed locations when the delivery areas are consistent, but the specific delivery locations may change daily. This is the case with parcel delivery or paratransit routes, as opposed to routes where the specific delivery locations don't change, as with deliveries to grocery stores.

Finally, you assign any specialties that have been defined (as described earlier) to the route. Orders that require a specialty, such as a wheelchair lift, will only be assigned to routes that have that specialty.

Solve the routing problem

Once you've loaded the orders and depots and specified the characteristics of the routes, you have the GIS assign orders to the routes, sequence the orders, and create the path for each route.

If you're creating several routes at one time, the GIS first finds the orders that are nearest each other. To do this, it calculates the distance between each pair of orders, using all the characteristics of the network—the distance along network edges, the costs associated with edges and turns, any barriers on the network, one-way streets, and so on. It then groups orders that are closest to each other until it reaches the capacity of the route. It also makes sure the delivery time windows on a given route don't overlap.

Next, the GIS assigns the sequence of the orders. The GIS attempts to minimize back-tracking and the amount of time the vehicle has to wait between orders in order to meet the assigned time windows. It then begins swapping orders between routes to see if it can reduce the overall cost of the routes. Trying all possible combinations of orders and routes to find the guaranteed optimum solution would take an inordinate amount of time. So the GIS uses an intelligent search method to quickly find a solution that is most likely the optimum one.

Finally, the GIS creates the least-cost path between each order in the specified sequence, as described in the previous section for visiting stops.

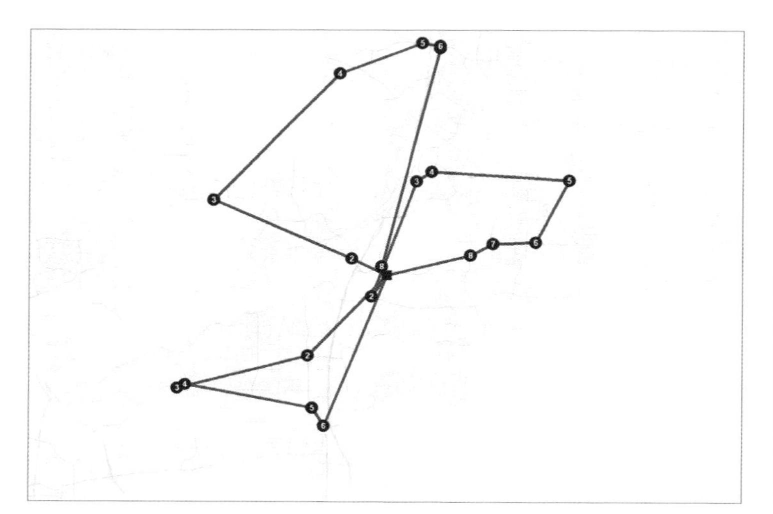

The GIS assigns orders to routes and assigns the sequence of the orders. The first sequence number for each route is the depot itself, so the first delivery is labeled 2 in the sequence.

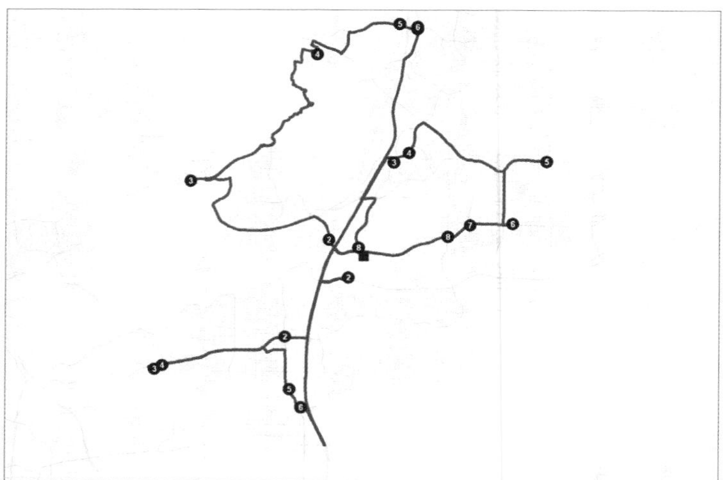

The GIS then calculates the least-cost path between orders, along streets.

Evaluate the results

You'll want to check the assignment of orders to routes, the sequence of orders, the path between the orders, and the various time and cost values to see if these make sense based on your knowledge of the orders, the routes, and the street network. There are many parameters for the various components of the route. If the routes do not look valid, check to see if incorrect values were entered for vehicle capacity, quantity for an order, service time, or other parameters.

As with any network analysis, the quality of the street network will determine how accurate, and useful, your results are. Network features that don't connect where they should will affect the results. Similarly, incorrect impedance for edges or turns, barriers in the wrong location, and edges incorrectly identified as one way, or having the wrong direction, will all affect the validity of the route.

You may find that making all the stops within the specified time windows increases the cost of the path. For example, in order to meet the required time windows, you may need to add a vehicle or extend the time a vehicle is on the road. You need to decide whether meeting the time windows is important enough to justify the additional cost.

Display and apply the results

You can color-code the routes to make it easier to distinguish them.

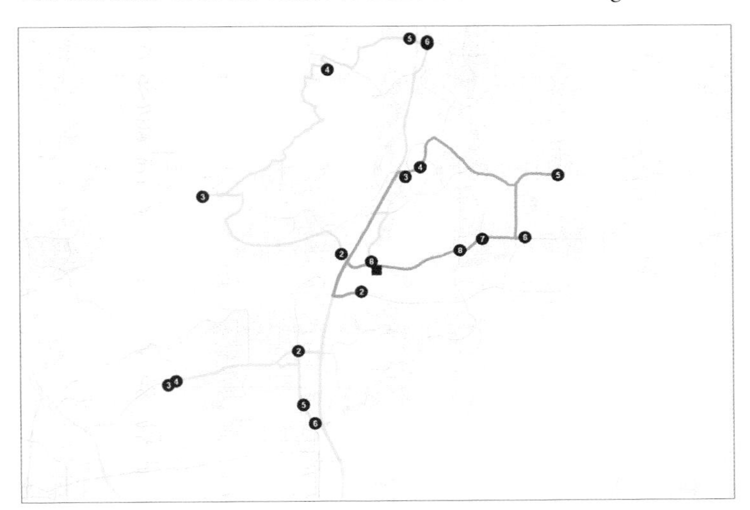

Three routes for deliveries to service station convenience markets, color-coded by route.

Or, you can create a separate map for each route.

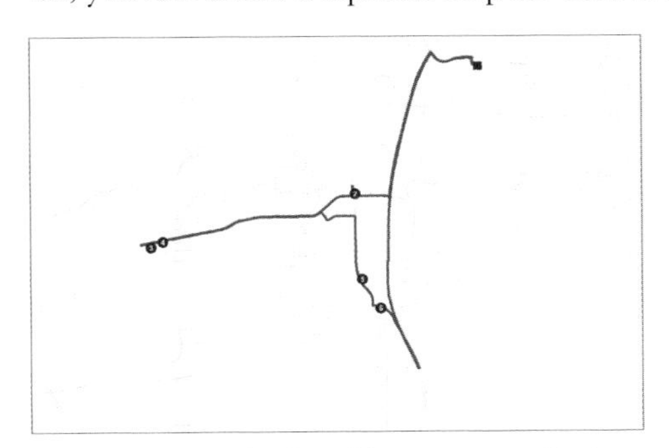

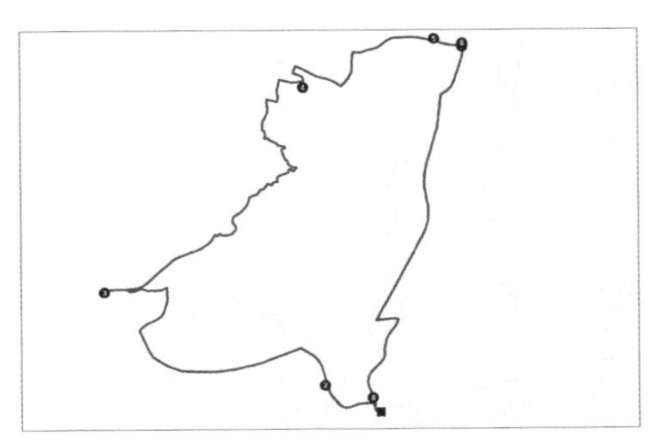

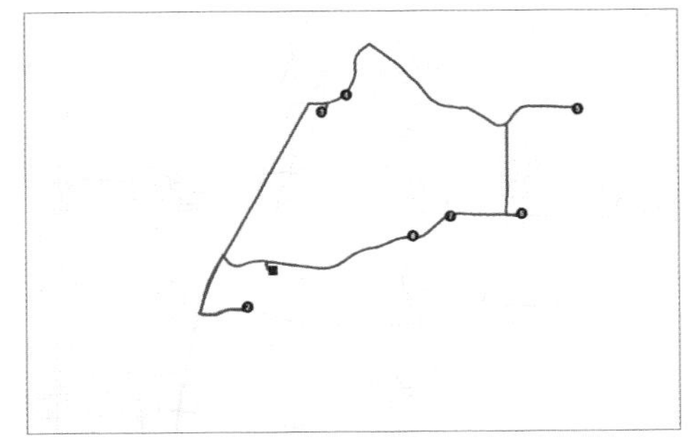

Individual maps for the three delivery routes can be given to the driver for each route.

Optionally, you can have the GIS create street directions for each order, with detailed maps for each.

Information about the orders and the route is contained in the attribute tables associated with the orders, depots, and routes.

The attribute table for the orders includes, for each order, the route the order is assigned to, the sequence number of the order on that route, and the arrival and departure time. It also includes any wait time (early arrival) or time violation (late arrival), as well as the cumulative wait time and time violations for the route up to that point. Plus, it lists the time and distance from the previous order location (so, once the vehicle leaves an order location, you can notify people waiting for the next delivery approximately how long it will take to get there). Finally, it lists the cumulative time, distance, and cost to that point in the route, so you know how long it will take to reach the order location once the vehicle leaves the depot.

ID	Name	Service Time	Delivery Quantities	Pickup Quantities	Route	Sequence	Arrive Time	Depart Time	Cumul Travel Time	Cumul Time
85	9320 MIRA MESA	40	16	4	Truck 1	2	4/6/2010 9:49:53 AM	4/6/2010 10:29:53 AM	5	90
93	6695 MIRA MESA	40	16	4	Truck 1	3	4/6/2010 10:36:05 AM	4/6/2010 11:16:05 AM	11	136
89	6795 MIRA MESA	40	16	4	Truck 1	4	4/6/2010 11:16:24 AM	4/6/2010 11:56:24 AM	11	176
90	9720 CARROLL CENTRE	40	16	4	Truck 1	5	4/6/2010 12:03:34 PM	4/6/2010 12:43:34 PM	19	224
86	9393 KEARNY MESA	40	16	4	Truck 1	6	4/6/2010 12:45:03 PM	4/6/2010 1:25:03 PM	20	265
83	12860 RANCHO PENASQUITOS	40	16	4	Truck 2	2	4/6/2010 10:17:07 AM	4/6/2010 10:57:07 AM	2	117
79	13985 TORREY DEL MAR	40	16	4	Truck 2	3	4/6/2010 11:05:28 AM	4/6/2010 11:45:28 AM	10	165
97	16629 DOVE CANYON	40	16	4	Truck 2	4	4/6/2010 11:58:27 AM	4/6/2010 12:38:27 PM	23	218
76	17011 BERNARDO	40	16	4	Truck 2	5	4/6/2010 12:43:59 PM	4/6/2010 1:23:59 PM	29	264

Portion of the orders attribute table for deliveries to service station convenience markets. In addition to the input order characteristics (including service times for each and delivery and pickup quantities), the table shows some of the attributes added as a result of running the model. These include the route each order is assigned to, the sequence in that route, the arrival and departure times for the order, the cumulative travel time up to that order (time the truck is on the road), and the cumulative total time to that order (including service time for the depot and all previous orders).

The attribute table for routes contains a row for each route showing the number of orders assigned to each and the total distance traveled. The table also contains information on the time and costs associated with the route. It lists the calculated start and end times for each route and the total time required for the route. Time is also broken out by total travel time, total service time, and total break time. Similarly, the total cost for the route is listed, along with cost broken out by regular time cost, overtime cost, and distance unit cost. It also lists the total wait time or violation time, if any, for the route.

ID	Route	Start Service Time	End Service Time	Capacities	Cost Per Minute	Max Order	Order Count	Total Order Service Time	Total Travel Time	Total Time	Total Cost	Start Time	End Time
1	Truck 1	45	30	90	0.25	12	5	200	29	304	76	4/6/2010 9:00:00 AM	4/6/2010 2:03:48 PM
2	Truck 2	75	30	140	0.25	18	7	280	38	423	106	4/6/2010 9:00:00 AM	4/6/2010 4:03:28 PM
3	Truck 3	60	30	120	0.25	15	7	280	29	399	100	4/6/2010 9:00:00 AM	4/6/2010 3:38:30 PM

The attribute table for the three convenience market delivery routes. The table shows some of the input route characteristics, including start and end service times, truck capacities, cost per minute, and maximum allowed order count. It also shows some of the attributes added after running the model, including the actual number of orders assigned to each route (Order Count), the total service time for orders for each, the total travel time (time on the road), and the total time for each route, including the start and end service times in minutes. The total cost is calculated by multiplying the total time by the cost per minute. The calculated start and end time for each route is also shown. You can see that the smaller truck (Truck 1) was assigned fewer orders.

The attribute table for depot visits contains a log of each time a vehicle assigned to a route visits a depot (usually at the beginning and ending of the route, but sometimes also in between). Perhaps most useful is that it lists the quantity of goods the truck loads or unloads each time it visits the depot, so you can see how much each truck is carrying.

ID	Depot	Route	Visit Type	Service Time	Arrive Time	Depart Time	Loaded Quantities	UnloadedQuantities
33	10931 CREEKBRIDGE	Truck 1	Start	45	4/6/2010 9:00:00 AM	4/6/2010 9:45:00 AM	80.000000	0.000000
34	10931 CREEKBRIDGE	Truck 1	End	30	4/6/2010 1:33:48 PM	4/6/2010 2:03:48 PM	0.000000	20.000000
35	10931 CREEKBRIDGE	Truck 2	Start	75	4/6/2010 9:00:00 AM	4/6/2010 10:15:00 AM	112.000000	0.000000
36	10931 CREEKBRIDGE	Truck 2	End	30	4/6/2010 3:33:28 PM	4/6/2010 4:03:28 PM	0.000000	28.000000
37	10931 CREEKBRIDGE	Truck 3	Start	60	4/6/2010 9:00:00 AM	4/6/2010 10:00:00 AM	112.000000	0.000000
38	10931 CREEKBRIDGE	Truck 3	End	30	4/6/2010 3:08:30 PM	4/6/2010 3:38:30 PM	0.000000	28.000000

The attribute table for depot visits. Each route has two visits to the depot—at the start and end of the route. Start visits have a loaded quantities value while end visits have an unloaded quantities value. You can see that the smaller truck (Truck 1) loaded and delivered a smaller quantity than the other two larger trucks.

From the tables, you can create a report for the driver listing each order, the address, the arrival and departure time, the quantity to deliver and pick up, and so on.

Orders

RouteName	Sequence	Name	ArriveTime	DepartTime	Delivery Quantities	Pickup Quantities
Truck 1	2	9320 MIRA MESA	4/4/2010 9:49:52 AM	4/4/2010 10:29:52 AM	16	4
Truck 1	3	6695 MIRA MESA	4/4/2010 10:36:04 AM	4/4/2010 11:16:04 AM	16	4
Truck 1	4	6795 MIRA MESA	4/4/2010 11:16:23 AM	4/4/2010 11:56:23 AM	16	4
Truck 1	5	9720 CARROLL CENTRE	4/4/2010 12:03:34 PM	4/4/2010 12:43:34 PM	16	4
Truck 1	6	9393 KEARNY MESA	4/4/2010 12:45:03 PM	4/4/2010 1:25:03 PM	16	4
Truck 2	2	12860 RANCHO PENASQUITOS	4/4/2010 10:17:06 AM	4/4/2010 10:57:06 AM	16	4
Truck 2	3	13985 TORREY DEL MAR	4/4/2010 11:05:28 AM	4/4/2010 11:45:28 AM	16	4
Truck 2	4	16629 DOVE CANYON	4/4/2010 11:58:26 AM	4/4/2010 12:38:26 PM	16	4
Truck 2	5	17011 BERNARDO	4/4/2010 12:43:59 PM	4/4/2010 1:23:59 PM	16	4
Truck 2	6	11898 RANCHO BERNARDO	4/4/2010 1:25:06 PM	4/4/2010 2:05:06 PM	16	4
Truck 2	7	11891 RANCHO BERNARDO	4/4/2010 2:05:11 PM	4/4/2010 2:45:11 PM	16	4
Truck 2	8	12610 SABRE SPRINGS	4/4/2010 2:52:55 PM	4/4/2010 3:32:55 PM	16	4
Truck 3	2	12033 SCRIPPS SUMMIT	4/4/2010 10:03:40 AM	4/4/2010 10:43:40 AM	16	4
Truck 3	3	11030 RANCHO CARMEL	4/4/2010 10:49:02 AM	4/4/2010 11:29:02 AM	16	4
Truck 3	4	11815 CARMEL MOUNTAIN	4/4/2010 11:29:55 AM	4/4/2010 12:09:55 PM	16	4
Truck 3	5	14147 TWIN PEAKS	4/4/2010 12:16:58 PM	4/4/2010 12:56:58 PM	16	4

Page 1 of 2

Modeling an overland path lets you create a least-cost path where no fixed infrastructure is in place. You can create a single least-cost path or a corridor that presents multiple potential paths between the locations. You can use the same method to create a layer showing the spread of a phenomenon (such as a wildfire or insect infestation)—the model essentially creates the least-cost path across the surface in all directions from a point of origin.

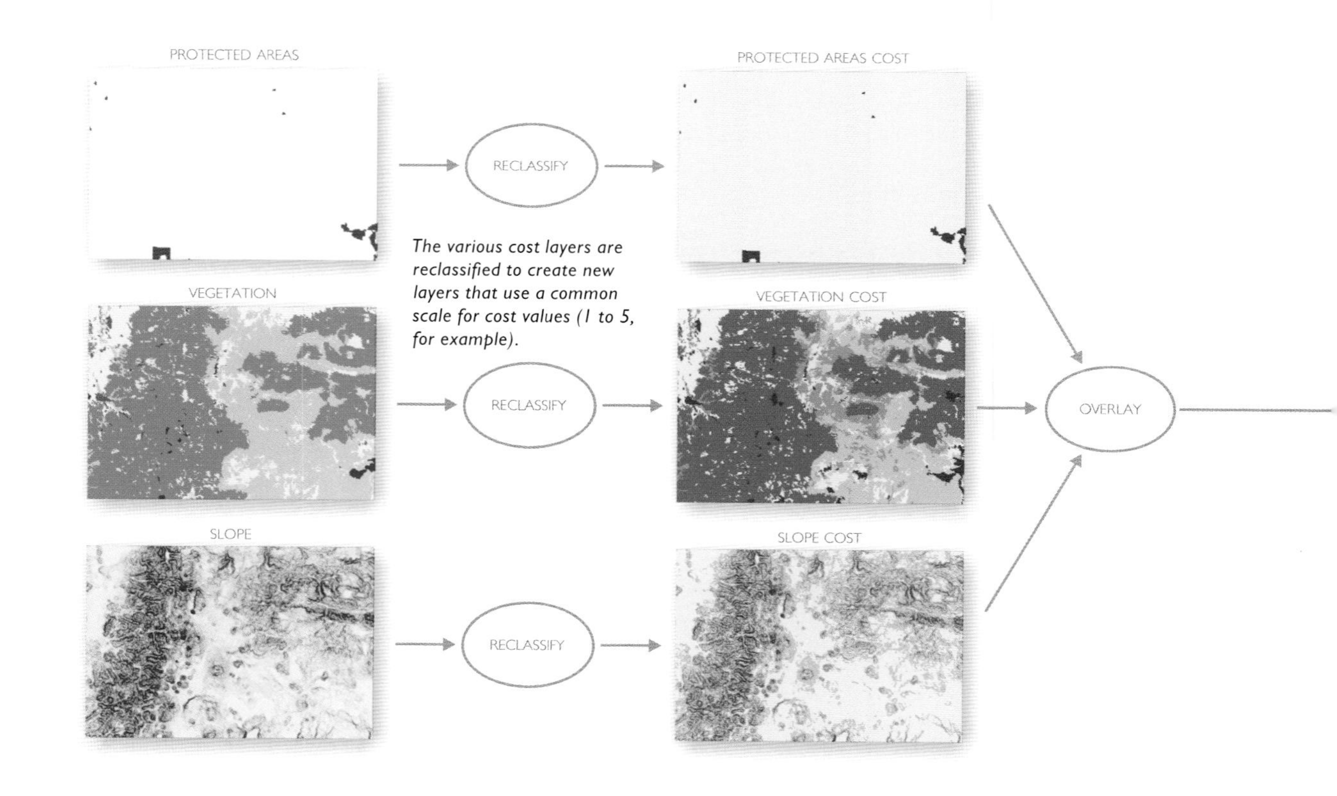

The various cost layers are reclassified to create new layers that use a common scale for cost values (1 to 5, for example).

Steps for modeling an overland path:

1 Specify the origin
2 Choose the method for calculating distance
3 Create the cost surface layer
4 Specify the movement factors
5 Create the cost distance surface
6 Create the spread surface, path, or corridor
7 Evaluate the results
8 Display and apply the results

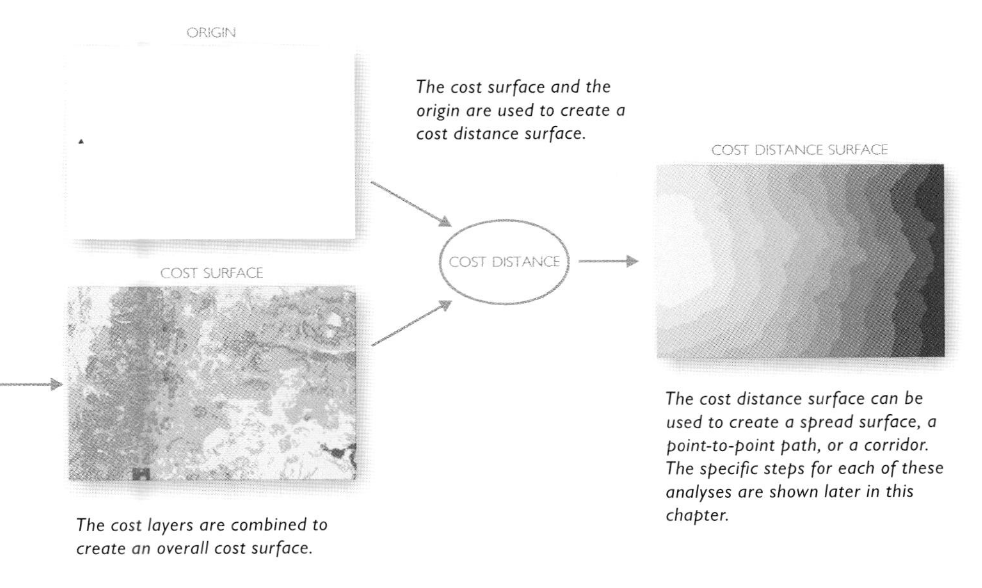

The cost layers are combined to create an overall cost surface.

The cost surface and the origin are used to create a cost distance surface.

The cost distance surface can be used to create a spread surface, a point-to-point path, or a corridor. The specific steps for each of these analyses are shown later in this chapter.

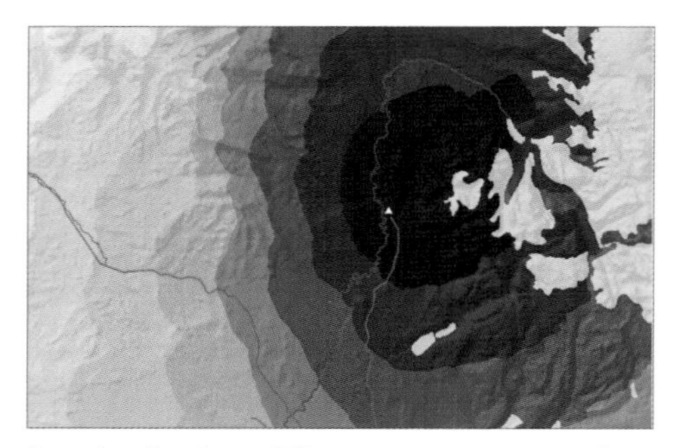

Spread surface for a wildfire igniting at a campground (white triangle). The surface shows the areas the fire would potentially reach first (dark red).

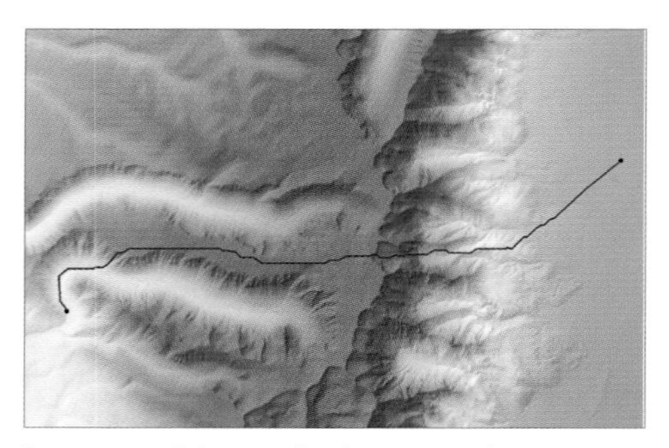

Least-cost path for a pipeline between two locations.

Potential paths (tan) for a wildlife corridor between two protected areas. Roads (red) and rivers (blue) are also shown.

Modeling movement is based on the idea that there is a cost (usually in the form of distance, time, or money) involved in moving across the surface. Surface features such as land cover can impede movement. Features such as a lake or a river can block travel. Moving over hills and valleys increases the actual distance traveled, while moving uphill can speed up movement (in the case of a wildfire) or slow it down (in the case of water through a pipeline). And factors such as wind can influence the direction of movement.

Depending on your application, you may not need to include all of these factors. However, if they're applicable, and if you have the data, including them will make the results of your model more accurate and useful.

In the following sections, some examples use a single factor by itself, in order to more easily present the concepts. In practice, several factors are normally used together to create a more realistic model.

Each of the factors is implemented in the model as a data layer along with parameters you specify. Since the path or corridor can potentially be located anywhere within your study area, it is best modeled using raster data—which represents geographic phenomena as a continuous surface. Similarly, a spread surface is by definition a continuous surface and hence modeled as a raster layer. Any input layers that are stored as vector data (such as land-cover polygons or river line features) will need to be converted to raster data format.

SPECIFY THE ORIGIN

The basis of your path model is a distance surface created from an origin location. You can modify the distance surface using various cost and movement factors to get a more realistic result from your model, as you'll see in the following sections. The modified distance surface is referred to as a cost distance surface. It essentially shows the cumulative cost of movement to each cell from the origin location.

Calculating cost distance starts with a layer that contains only the feature (or features) representing the source locations. In this example, the point (triangle) represents the origin location of a new highway.

The origin is usually either a point feature, such as a town or transmission tower, or an area, such as a timber stand or protected natural area. You can also use a line feature, such as a road or stream, as an origin.

If you are creating a spread surface, the cost distance surface is the end result of your model (that is, the cost distance surface *is* the spread surface). If you're creating a corridor between two locations, you create a cost distance surface from each location (you essentially have two origins) and combine them to identify the cells having the least cost to both locations. And if you're creating a point-to-point path, you specify an origin to create the cost distance surface, and you then specify a destination location to find the least-cost path from origin to destination.

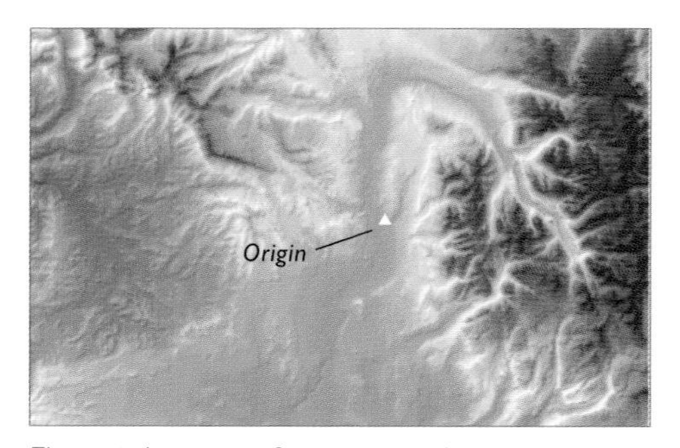

The origin location is often a point, such as a campground at which a wildfire starts.

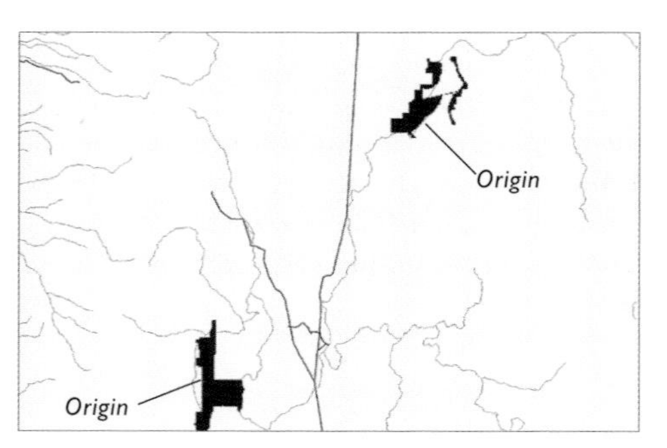

Two protected natural areas serve as the origin locations for creating a wildlife corridor—the origins are area features.

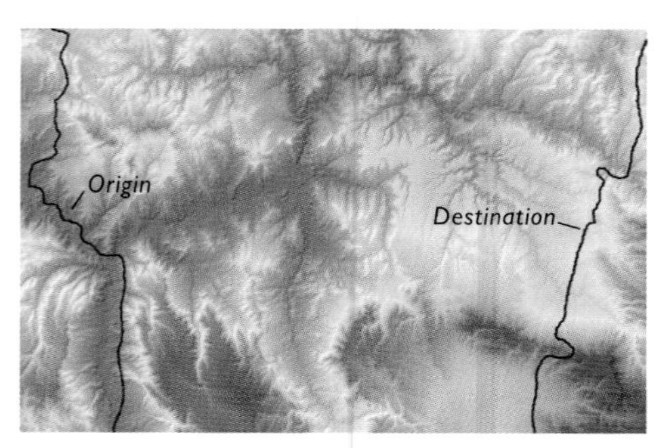

Two existing highways become the origin and destination for a path for a new highway connecting them—the origin and destination are line features.

CHOOSE THE METHOD FOR CALCULATING DISTANCE

In this step, you decide how the distance surface will be calculated—either over a plane or uphill and downhill over terrain. You specify which method to use when you create the cost distance surface.

Calculating distance over a plane
In its simplest form, distance is measured assuming a flat surface, or plane. You'd use this method if the path travels over something where there is little variation in the surface (such as a highway over a flat landscape or the spread of oil on the surface of a lake) or when changes in elevation would not significantly influence the results.

When calculating distance for a path model, the distance from the origin is measured as a cumulative distance, from cell to cell. (This is different from a simple distance surface, in which the Pythagorean theorem is used to calculate the straight line, or Euclidean, distance directly from the center of the origin cell to the center of each cell in the raster.)

The cell-to-cell distance is used because various factors involved in moving from one cell to the next are used to modify the distance (as you'll see in the next sections). These factors vary across the surface, thus the distance calculation requires measuring distance on a cell-to-cell basis. This approach is known as creating a cost distance surface, as opposed to a simple Euclidean distance surface.

The GIS calculates distance from cell center to cell center, based on the cell size. If the distance being calculated is on the diagonal, the GIS multiplies the distance by 1.41421. If the cells were one unit in size (for example, one meter or one foot), the distance from cell center to cell center on a diagonal would be 1.41421—the square root of 1^2 plus 1^2, or the square root of 2. Since the cells in the raster are square, the GIS can use this value as a factor to calculate diagonal distance. So when calculating the distance diagonally between cells, it simply multiplies the cell size by 1.41421.

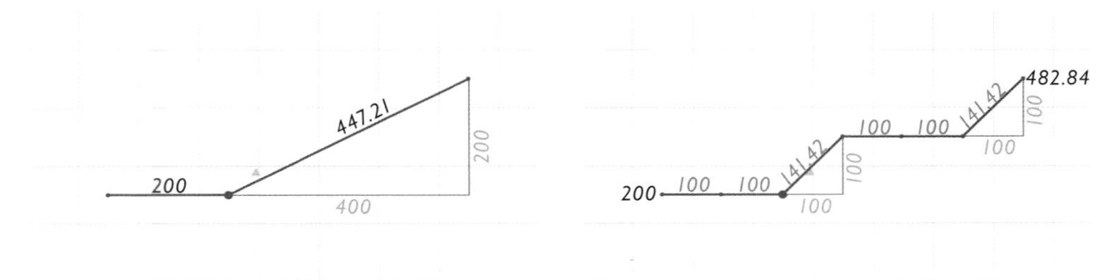

Euclidean distance method. The green triangle is the origin feature location and the large dot is the center of the origin cell. With a cell size of 100 m, the distance to the cell west of the origin cell is 200 m; the distance to the cell to the northeast is 447.21 m.

Cost distance method. The distance to the cell to the west is still 200 m, but the distance to the cell to the northeast is 482.84 m—farther than when using Euclidean distance.

To create the cost distance surface, the distance is essentially measured from the cell containing the origin location to each surrounding cell and then accumulated outward from cell to cell.

The result is that while with Euclidean distance the distance pattern appears circular, with the cost distance method the pattern appears octagonal. This is because for cells that are not on a straight horizontal, vertical, or diagonal line from the source cell, the distance is calculated in a stair-step pattern, increasing the calculated cumulative distance.

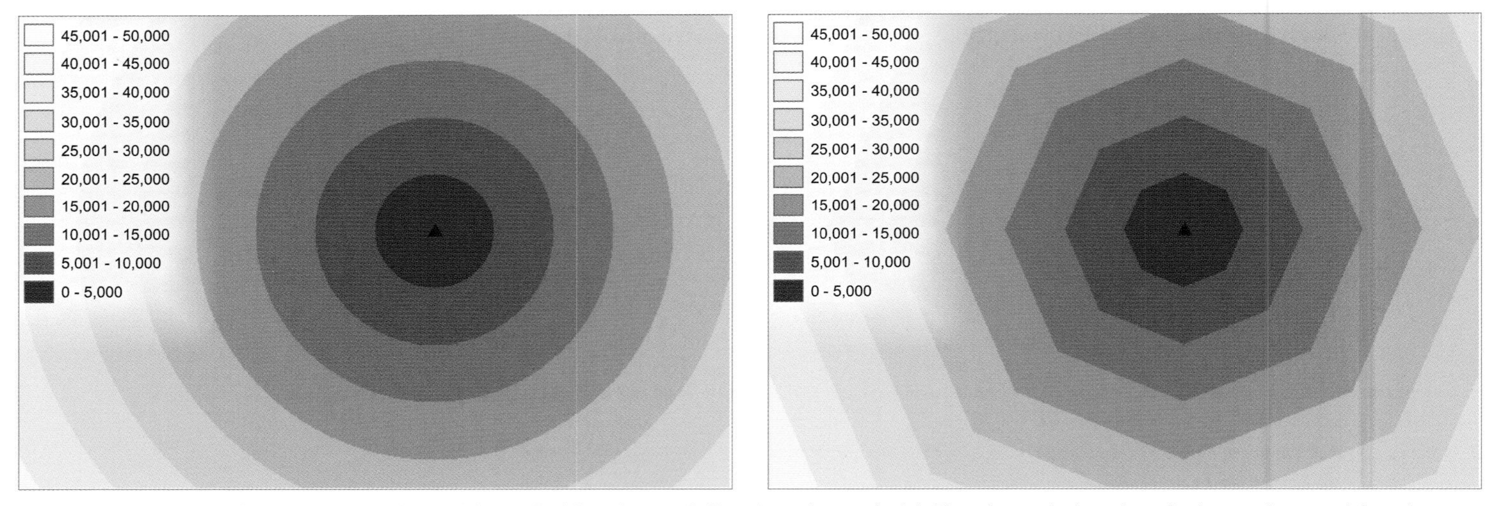

A distance surface created from the origin (black triangle) using Euclidean distance (left) and cost distance (right). The values in the legend are the distance (in meters) from the origin.

For cells that are on a straight horizontal, vertical, or diagonal line, the distance calculated using the cost distance method equals the Euclidean distance.

Calculating distance over terrain

Rather than calculating distance over a plane, you can measure the distance traveled over hills and valleys. This measure is termed "surface distance." Using surface distance provides a measure of actual distance traveled, hence making your model more accurate. For example, if you're creating a path for a water pipeline, you'd want to include the distance traveled uphill and downhill for an accurate measure of the total length of the pipeline.

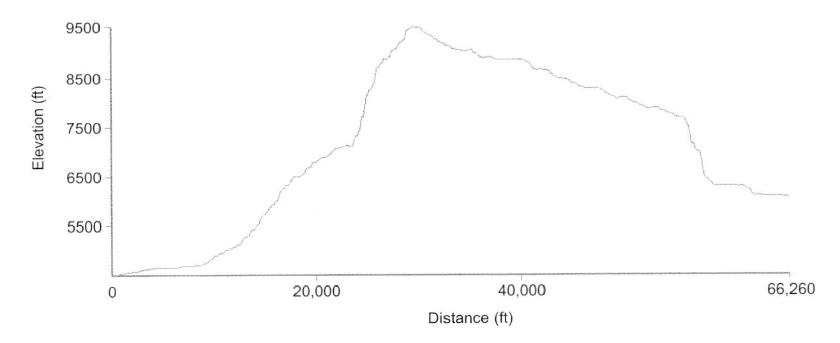

A path that travels up and down slopes covers a longer distance than one traveling over a flat surface (represented by the x-axis).

Unless the topography is especially varied, including the distance traveled over terrain may not greatly affect the actual location of the path or corridor. However, measuring distance over terrain will increase the total cost of the path. Thus it is useful when comparing alternate paths as well as for calculating construction costs (on a cost-per-mile basis).

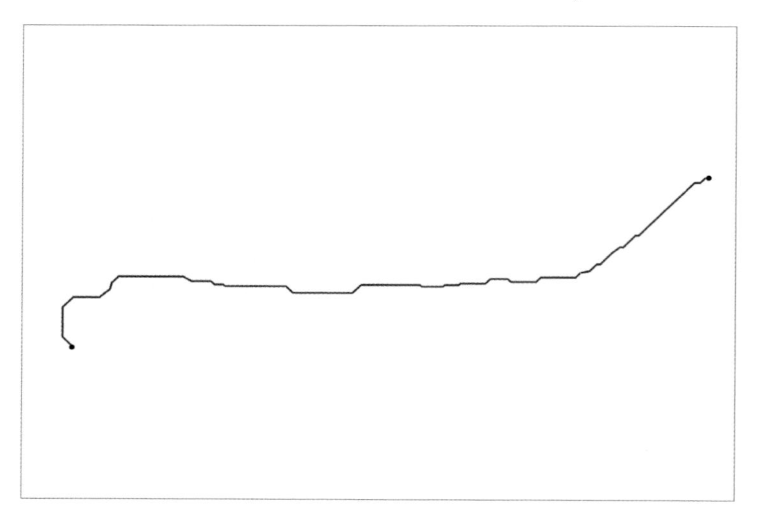

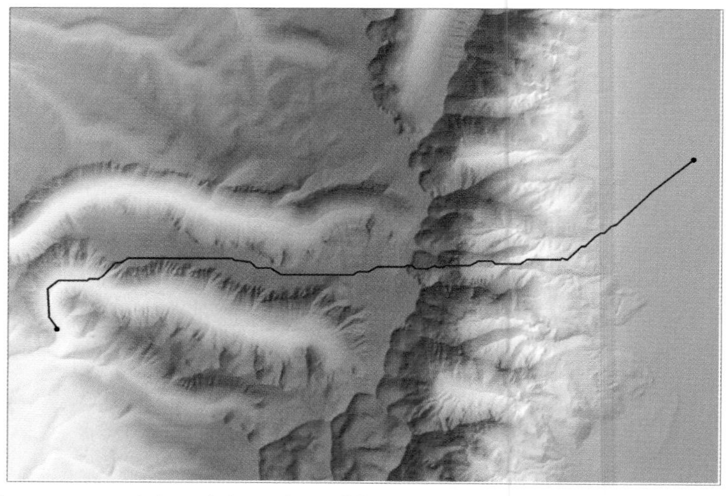

Path calculated over an assumed plane (left) and over terrain (right). In this example, when distance over terrain is used, the total cost of the path is 6 percent higher—on a $25 million project, calculating the path over a plane would underestimate the construction cost by $1.5 million.

To measure the distance traveled over terrain, you include a surface layer representing the terrain when creating the cost distance surface (this is usually an elevation layer).

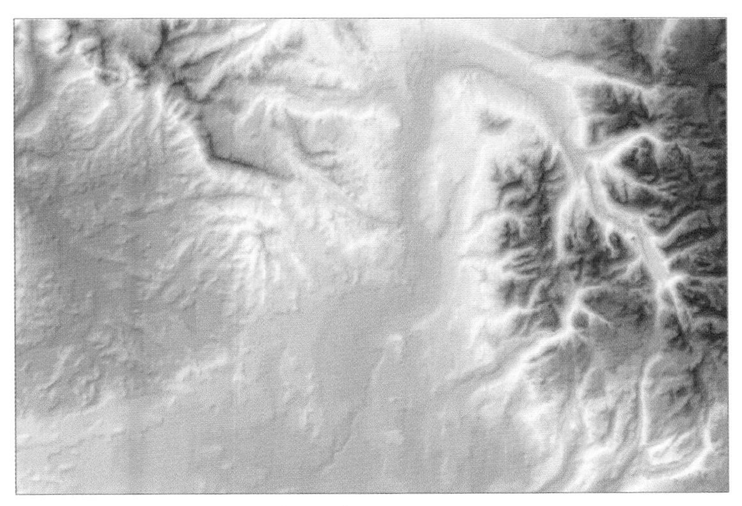

An elevation surface is used to calculate distance over terrain.

When you calculate distance over a plane, it's assumed that all cells are at the same elevation, and there are no hills and valleys.

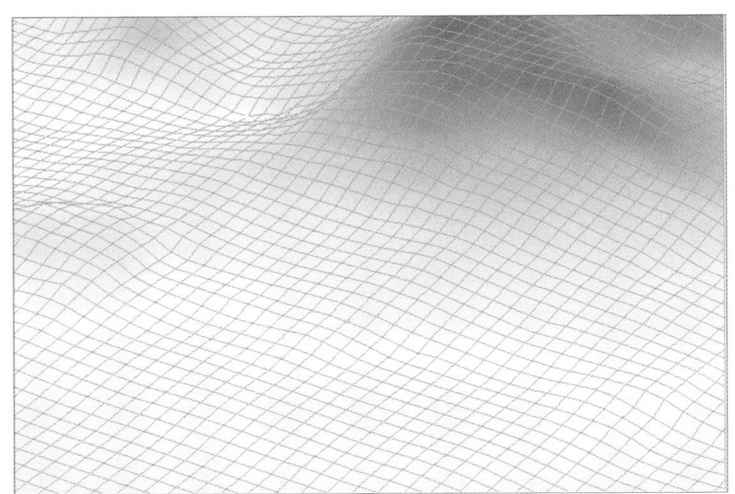

When you calculate distance over terrain, the cell-to-cell distance is calculated using the difference in elevation between adjacent cells.

The GIS calculates the distance from cell center to cell center, taking into account the difference in elevation between the two. This time, the Pythagorean theorem *is* used, to calculate the vertical dimension.

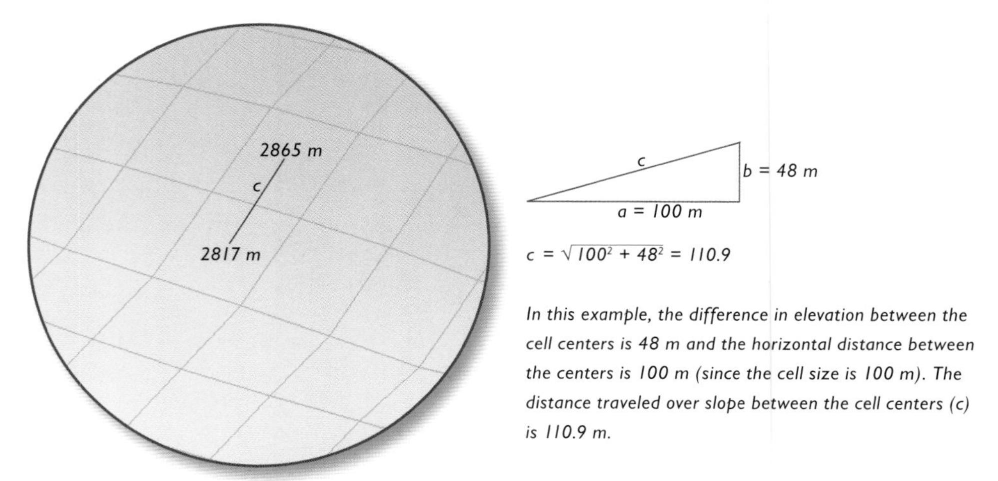

In this example, the difference in elevation between the cell centers is 48 m and the horizontal distance between the centers is 100 m (since the cell size is 100 m). The distance traveled over slope between the cell centers (c) is 110.9 m.

If the travel is diagonal, the horizontal distance ("a" in the equation) is multiplied by 1.41421 (as described earlier) and then squared.

To ensure the calculations are correct, the units used to measure elevation must match those of the map coordinates (if the map coordinates are in feet, elevation must also be in feet). If necessary, you can convert the elevation units by multiplying the existing elevation raster by a conversion factor to create a new raster with elevation values in the correct units. Some GIS tools that process elevation rasters allow you to specify a conversion factor for elevation units so that the output raster values are calculated correctly on the fly (thus making it unnecessary to convert the elevation raster first).

CREATE THE COST SURFACE LAYER

What is on the surface can affect the location of the least-cost path. A wildfire will burn faster through vegetation that is dense and dry—such as sagebrush—than through sparse or damp vegetation—such as a wetland. The ease (or difficulty) of movement is termed the "cost" of travel. To capture these costs, you create a cost surface layer and include it when you create the cost distance surface.

Assigning cost values

The cost surface layer represents the cost of crossing each cell. The costs are based on the criteria for your model. For example, the requirement for a path for a new road might be that the path travel over the flattest slopes—the slope layer would be the cost surface layer. Typically, several input layers are combined to make the cost surface layer.

For the cost surface to be accurate, and hence the results of your path model valid, the cost surface values should use a ratio scale; that is, the scale the values are measured on has a meaningful and absolute zero. You then know that a value twice that of another value represents twice the cost. A slope layer, for example, uses a ratio scale. A value of 0 for a cell means that cell has no slope, or is flat—so the scale has a meaningful and absolute zero. Hence a slope of 20°, for example, is twice as steep as a slope of 10°.

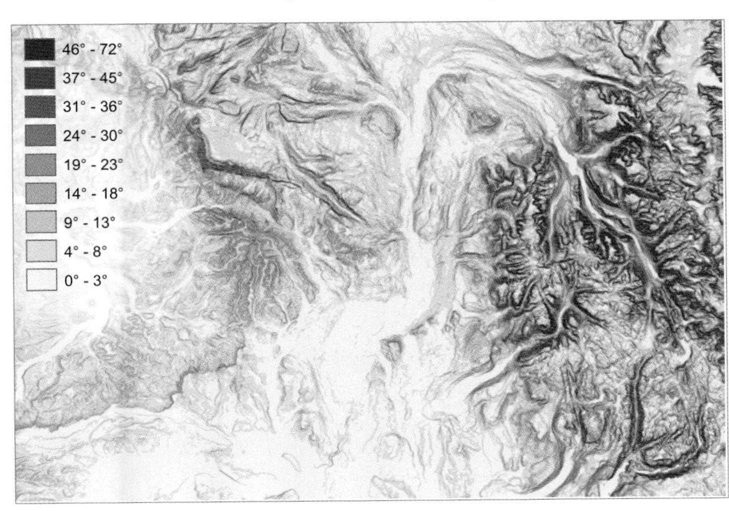

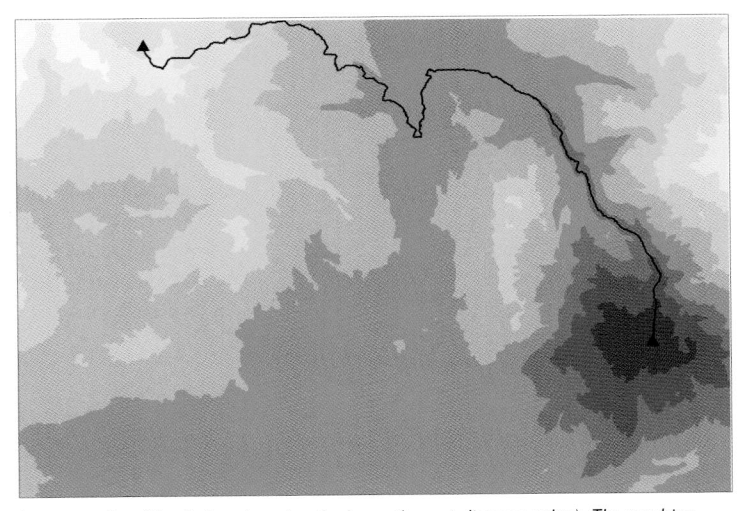

The left map shows slope. The map on the right shows a cost distance surface using slope as the cost surface (the darker the color, the lower the cost distance value). The resulting path from the origin (lower right) to the destination is also shown. The path attempts to follow the flattest slopes.

If the values on the source layer are categorical (or other non ratio values), you can covert them to values on a ratio scale. The key is assigning realistic and meaningful cost values to the categories in the source layer. For example, your requirement might be that the path should pass through land that is easy to clear for construction. If you know the cost of clearing land for each land-cover category, you'd create a new layer, assigning the corresponding monetary cost to each category. For example, it might cost $25 per foot to clear forest, and $5 per foot to clear meadow. (Money, like slope, is valued on a ratio scale).

LAND-COVER CATEGORY	$COST TO CLEAR/FT
Cropland	*$5*
Sagebrush	*$10*
Riparian	*$40*
Forest	*$25*
Clearcut	*$15*
Meadow	*$5*

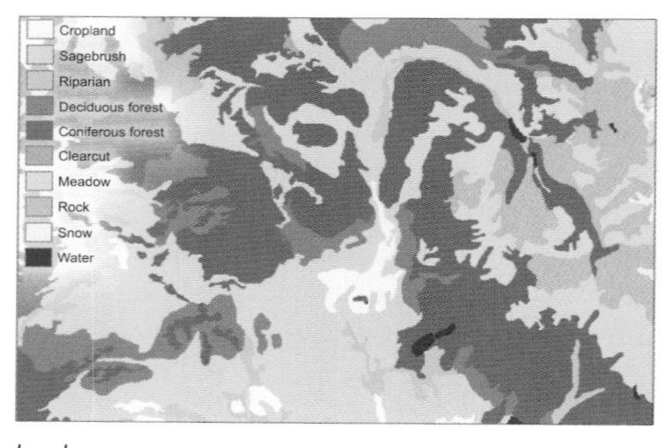

Land cover.

The reclassified map assigns monetary values to each land-cover category. (Rock, snow, and water are restricted—the cost to construct the path through these areas is prohibitive.)

LAND-COVER CATEGORY	COST VALUE
Cropland	*1*
Sagebrush	*2*
Riparian	*5*
Forest	*4*
Clearcut	*3*
Meadow	*1*

If you don't know the actual cost, you can assign values on a ratio scale that you define, such as a scale of 1 to 5 (with 1 being low cost and 5 being high cost). For example, you might not know how much money per foot it costs to clear forest, meadow, and so on—only that clearing forest is more expensive than clearing meadow. You'd assign forest a high value on the scale and meadow a low value. (If you think clearing forest is four times as costly, you'd assign forest a value of 4 and meadow a value of 1.) However, to the extent that you are estimating the costs, your model will be an approximation.

In many cases, several factors contribute to the overall cost of the surface. You combine the corresponding layers to create an overall cost surface. If the cost factors use different units of measure (dollars for clearing land and degrees for slope), you assign new values to the original source layers, using a common scale such as 1 to 5 or 1 to 9. (Again, a ratio scale is used.) Cost values should be consistent between cost layers—so a cost value of 8 for land cover is comparable to a cost value of 8 for slopes.

LAND-COVER CATEGORY	$COST TO CLEAR/FT	COST VALUE
	$45	9
Riparian	$40	8
	$35	7
	$30	6
Forest	$25	5
	$20	4
Clearcut	$15	3
Sagebrush	$10	2
Cropland/Meadow	$5	1

Land-cover categories, with a known cost of clearing, are assigned cost values on a scale of 1 (low cost) to 9 (high cost), before combining with other cost layers to create an overall cost surface.

SLOPE CATEGORY	COST VALUE
46° - 72°	9
37° - 45°	8
31° - 36°	7
24° - 30°	6
19° - 23°	5
14° - 18°	4
9° - 13°	3
4°- 8°	2
0° - 3°	1

Slope value ranges are assigned values on the same scale. In this example, there are nine classes of slope, so all nine cost values are used. Depending on your application, you might not use all the cost values on the scale (if only five slope class values are needed, for example).

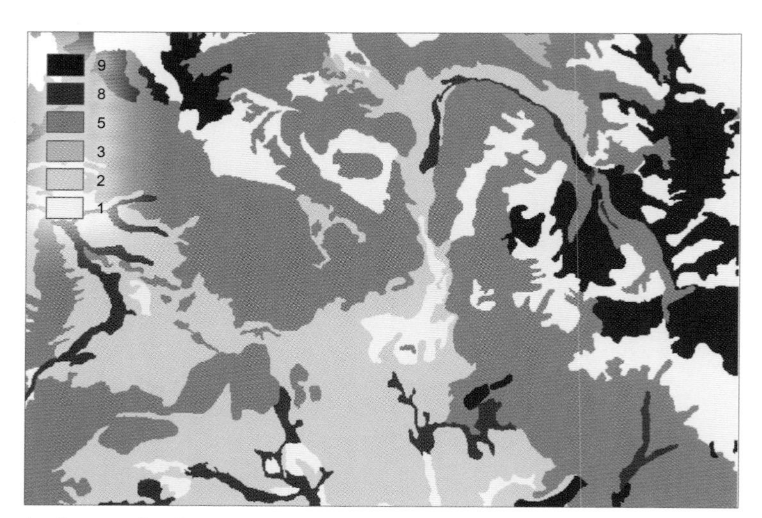

Land cover reclassified into a cost surface for a highway path. Cropland and meadow have the lowest cost (1). Rock, snow, and water are assigned a very high cost (9).

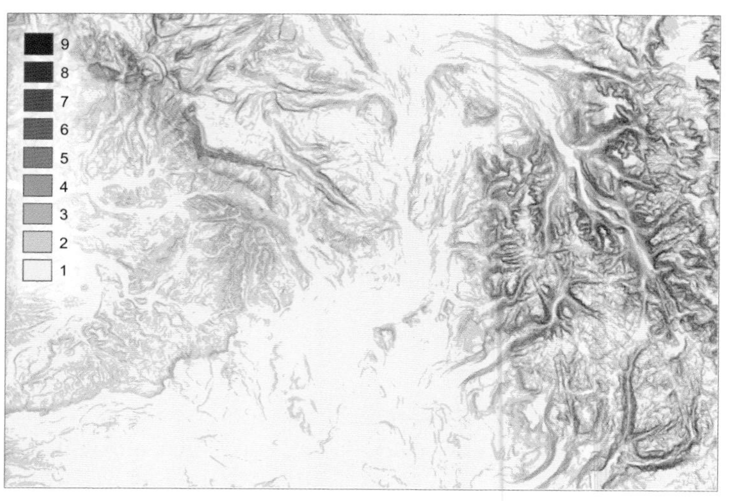

Slopes have been classified and assigned cost values of 1 (gentle slopes) to 9 (steep slopes).

In some cases, there may be costs for traversing certain features or areas that are so high that placing a path or corridor in those areas would be prohibitive. For example, the cost of building a bridge may prohibit a road from taking a path that crosses a river. If the cost makes traversing the feature or area difficult (but not impossible), you'd assign these areas or features a very high value on the scale (a 9, if your scale is 1 to 9). If travel over these areas is not permitted, you'd assign these areas a value of "restricted" when you combine the layers (as described later in this section).

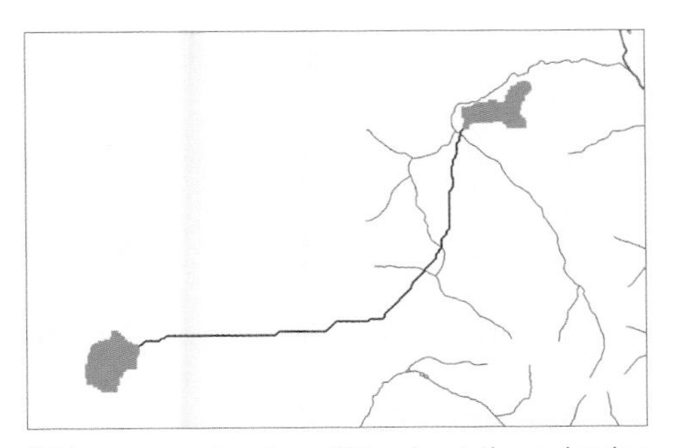
With streams assigned no additional cost, the road makes several crossings.

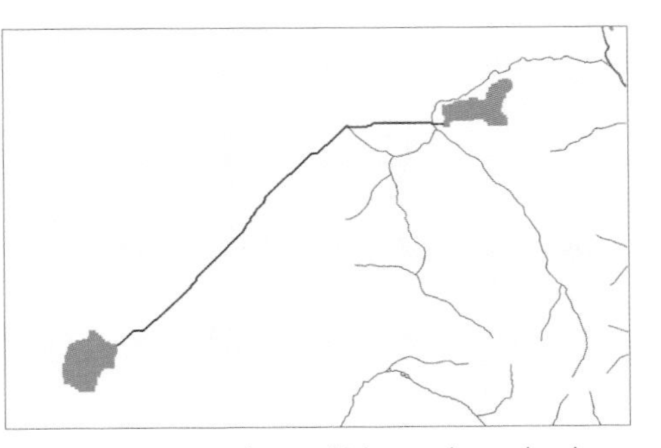
With streams assigned a very high cost, the road makes only one crossing.

With streams assigned a value of "Restricted," the road makes no crossings.

Creating the overall cost surface layer

Once you've created the individual cost layers, you combine them to create the overall cost surface layer. The process for creating a cost surface is very similar to the process for creating a suitability layer using the weighted overlay method, described in chapter 3, "Rating suitable locations" (with cost values corresponding to suitability values). And, in fact, using weighted overlay is an efficient way to create a cost surface.

When creating the overall cost surface, you can specify that all input factors are equally important by assigning equal weights to the input cost layers. Or, you can specify that some factors are more important than others.

For example, if the cost layers for creating a highway are land cover, slope, and rivers, and land cover is three times as important as the other two, you'd assign it a weight of 60 percent and the other two weights of 20 percent each (the weights must sum to 100 percent). The values on each cost layer are multiplied by the weight for the layer when the layers are combined.

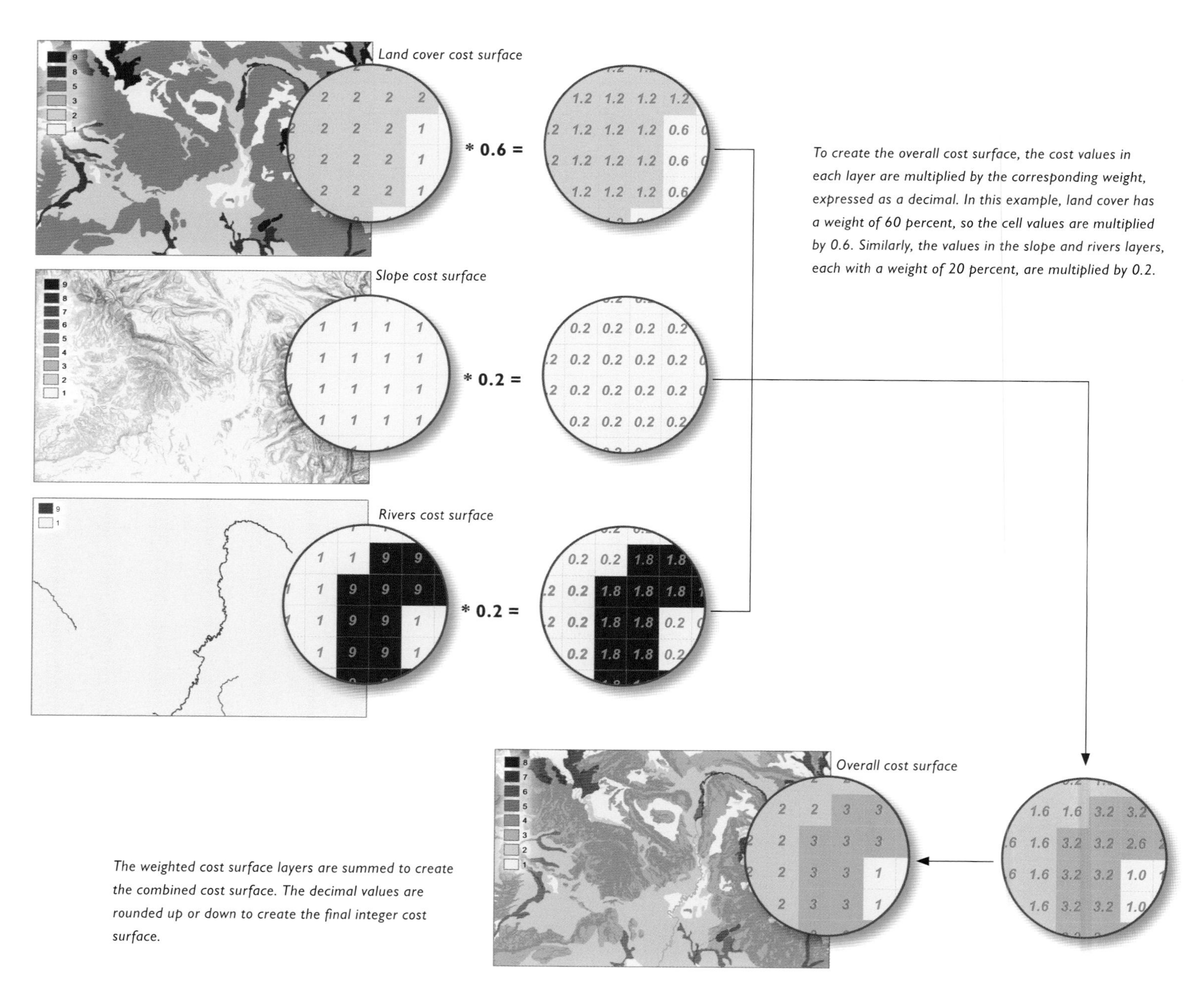

To create the overall cost surface, the cost values in each layer are multiplied by the corresponding weight, expressed as a decimal. In this example, land cover has a weight of 60 percent, so the cell values are multiplied by 0.6. Similarly, the values in the slope and rivers layers, each with a weight of 20 percent, are multiplied by 0.2.

The weighted cost surface layers are summed to create the combined cost surface. The decimal values are rounded up or down to create the final integer cost surface.

SPECIFY THE MOVEMENT FACTORS

In addition to the cost of traveling over the surface, you can include the ease of moving uphill and downhill over terrain and factors (such as prevailing wind) that influence the direction of travel.

Modeling ease of movement over terrain

When you model travel over slope, there are two issues to consider: the cost of constructing the path and the cost of transporting whatever is moving over the path. In some cases, such as for an electrical transmission line, the only issue is the construction cost—the transmission of the electricity, once the line is built, is not affected by the slope of the terrain. In other cases, both the construction cost and the transportation cost are issues. For example, for a water pipeline, the slope of the terrain will impact both the cost of constructing the pipeline and the cost of moving water through it. The relationship between slope and cost may not be the same for these factors. Construction costs will be higher on both steep uphill and steep downhill slopes, while the cost of moving the water through the pipeline will be high for uphill slopes (where a pump is required) but low for downhill slopes (where gravity can be used). The least-cost path will be the one that minimizes construction cost and the cost of transporting water from the origin to the destination.

The cost of construction (where uphill and downhill costs are typically the same) is generally included in the model by combining slope with any other layers when you create the cost surface.

The cost of transporting something over a particular slope, however, is included in the model by specifying a slope weight. The cost of each cell is multiplied by the slope weight, which is in turn derived from the slope angle. You specify a function (which can be represented as a line on a graph), and the GIS uses the function to look up the weight for each angle. In the model, this function is referred to as a vertical factor. Using a vertical factor gives you a great deal of control over the relationship between slope and cost. The function lets you specify the pace at which costs increase as slope increases, and an angle above which travel or transport becomes infeasible.

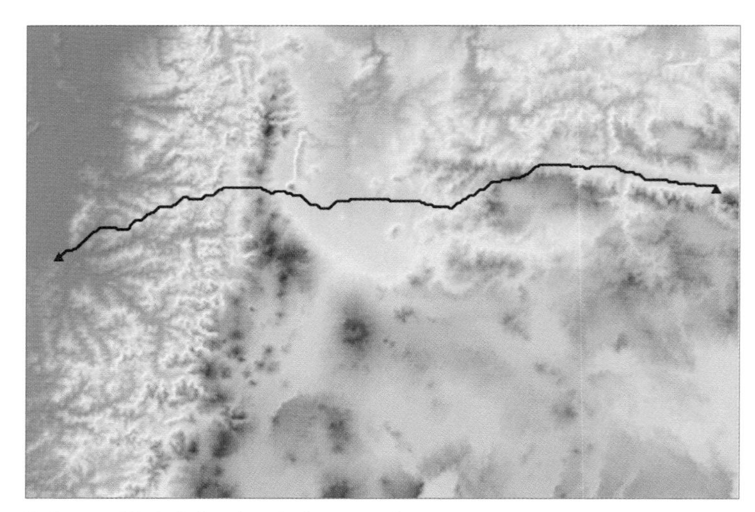

Path created by including slope in the cost surface.

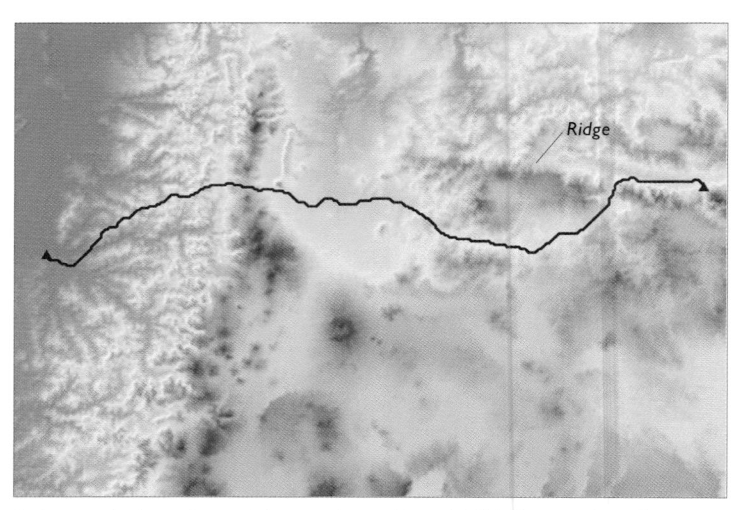

Path created using a slope weight curve (vertical factor). With the cost of traveling over slopes included, a portion of the path takes a more southerly route through flatter terrain, avoiding a mountain ridge.

How the GIS calculates slope weight

The GIS first calculates the slope angle between each cell pair based on the elevation assigned to each, using the following formula:

SLOPE = RISE / RUN

Since, when it creates the cost distance surface, the GIS knows the direction of travel across the surface outward from the origin, it also knows the direction of travel from each cell to the next (that is, which is the "From" cell and which is the "To" cell). It calculates the rise by subtracting the z-value (elevation) of the From cell from the z-value of the To cell. It then divides this by the defined cell size (this is the run—the distance between cell centers). If the cells are diagonal to each other, it multiplies the cell size by 1.41421 to account for the extra distance when traveling diagonally, and then divides the rise by this value. To get the slope angle, it takes the arctangent of the result (and, optionally, converts the resulting radians into degrees of slope).

In this example, the cell size is 50 feet, the elevation of the From cell is 162.0 feet and the elevation of the To cell is 175.5 feet.

SLOPE = (175.5 – 162.0) / 50 = 0.27

SLOPE ANGLE = ATAN(0.27) = 0.26 RADIANS = 15.1 DEGREES

If the To cell has a higher elevation than the From cell, travel is uphill. The difference in elevation is positive and the slope angle has a positive value. If the To cell has a lower elevation than the From cell, travel is downhill and the slope angle has a negative value. In the example above, if travel was in the opposite direction, the difference in elevation would be a negative value, and the slope angle would be -15.1°. (The angle is the same, only the sign is reversed.)

The GIS looks up the angle for the From cell using the function you specify to find the weight. When the least-cost path is calculated, the weight is applied to the cost for the corresponding cell in the cost surface layer (that is, the cost is multiplied by the weight). So, a flat surface may have a weight of 1 (basically, no additional cost) while a steep slope may have a weight of 3 (tripling the cost).

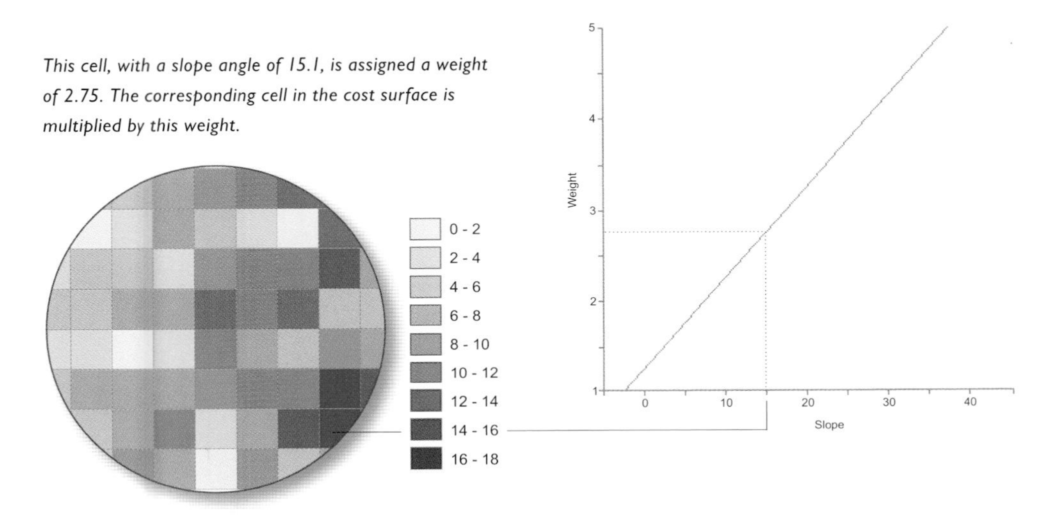

This cell, with a slope angle of 15.1, is assigned a weight of 2.75. The corresponding cell in the cost surface is multiplied by this weight.

Defining a slope weight function

It can be difficult to establish an exact relationship between a given slope angle and a weight. For a water pipeline, you might know that for each degree rise in slope, the cost increases by some monetary amount based on the size of the pump required and the energy needed to run it. This might not be a linear relationship—the cost may increase more rapidly as the slope gets steeper. Once you've established this relationship, you use it to define the function. You do this by creating a table containing the cost factor associated with each of several angles, using as many pairs as needed to adequately define the function (the GIS interpolates between the points). If you're using a function in which downhill and uphill weights are not the same, the downhill weights are fractions, so the cost is reduced rather than increased. For example, a downhill slope with an angle of –10, might have a weight of 0.8—when multiplied by the weight, a cell in the cost surface that has a slope of -10 would have a cost value equal to 80 percent of its original value.

SLOPE ANGLE	VERTICAL FACTOR
-30°	0.4
-20°	0.6
-10°	0.8
0°	1.0
10°	1.2
20°	1.4
30°	1.6

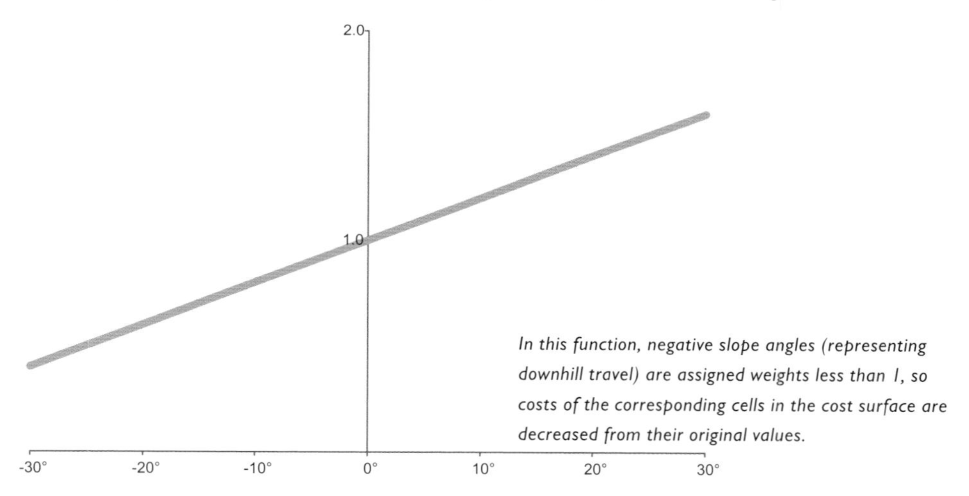

In this function, negative slope angles (representing downhill travel) are assigned weights less than 1, so costs of the corresponding cells in the cost surface are decreased from their original values.

Instead of using a table of slope angle and vertical factor pairs, you can use a mathematical function that defines the relationship between cost and slope steepness (and which can be expressed as a curve on a graph). ArcGIS provides several commonly used functions that allow you to specify whether traveling up slope is easier or harder than traveling downslope, as well as the rate at which the difficulty increases.

You define the relationship by specifying a particular function to use (for example, linear) and setting some parameters that define its slope or the shape of its curve.

You first specify the vertical factor to use for flat slopes (a slope angle of 0). This factor (known as the "zero factor") defines where the curve crosses the y-axis (that is, the intercept). By default, a zero factor of 1 is used, meaning that cost distance is not modified when movement is across flat slopes (the measured distance is multiplied by 1). Increasing the zero factor will not necessarily change the pattern of the cost distance surface. Rather, it increases the cost distance value assigned to each cell. You would use a zero factor greater than 1 if you want to increase the influence of the vertical factor layer when it is combined with the other factors to create the cost distance surface.

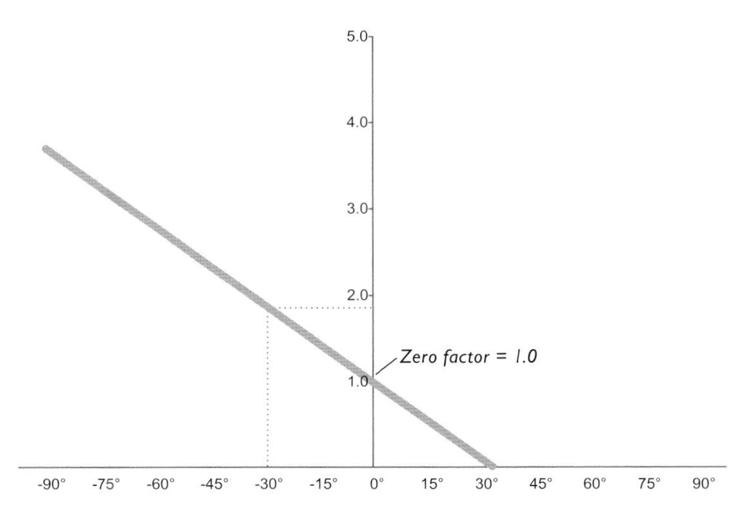

This graph shows a linear vertical factor function with a zero factor of 1. The x-axis represents the cell-to-cell slope of the surface, in degrees, and the y-axis represents the vertical factor. If, for example, the cell-to-cell slope is -30° (downhill), the vertical factor is 1.9—downhill movement is more difficult than uphill movement.

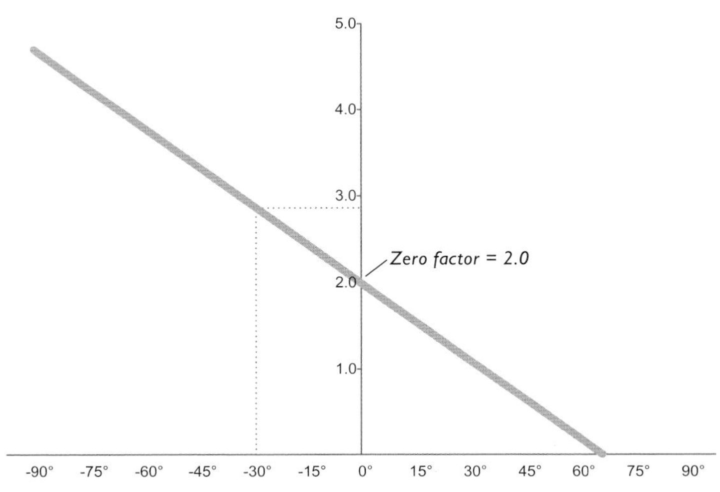

This graph shows the same function, but with a zero factor of 2.0. With this function, a slope of -30° has a vertical factor of 2.9.

The line on the graph does not extend past the x-axis because there cannot be negative vertical factor values. If the angle is such that the vertical factor—according to the function and the graph—would be negative, a vertical factor of 0 is used. When the accumulated cost distance is calculated, the value doesn't increment for the To cell. Rather, the To cell is assigned the cost distance value of the From cell.

You can also specify a slope angle beyond which the steepness is so great that travel is essentially impossible (sort of like a barrier on the cost surface). This is known as a cut angle. You can specify different cut angles for uphill and downhill slopes. If the angle of the cell-to-cell slope is greater than the cut angle, a cost value for the To cell can't be calculated. The To cell is assigned a value of NoData on the output map.

While there are a number of functions available to define the slope-weight relationship, there are a few common situations in which one or another of the functions is applicable.

If the movement is easier uphill and more difficult downhill (such as with a wildfire), use the linear function with a negative slope value. The linear function assumes that the ease of movement increases at a constant rate as the surface slope angle increases. Since wildfire can move downhill—just not as rapidly as uphill—you'll want to use a low value for the low cut angle and a high value for the high cut angle.

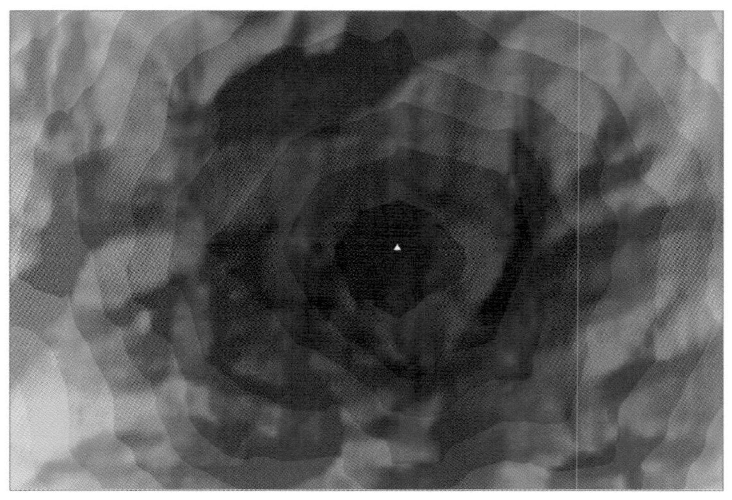

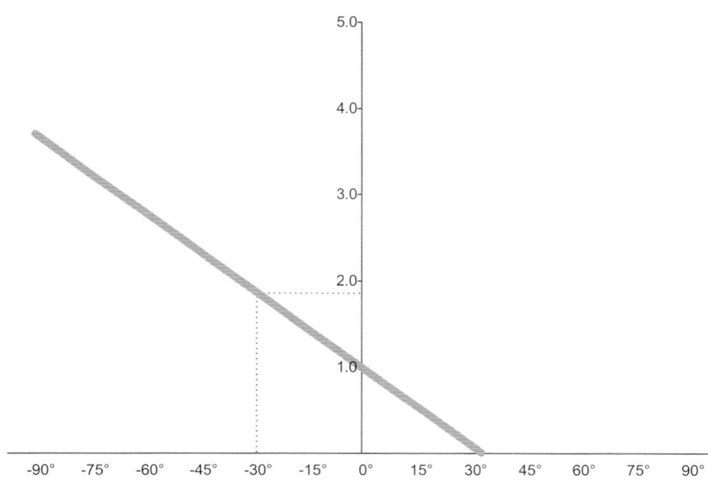

Spread surface from a point (white triangle), created using the linear function with a zero factor of 1, a slope value of -0.03, a low cut angle of -90°, and a high cut angle of 90°. A slope of -30° has a vertical factor of 1.9. Terrain is also shown.

Specifying a larger slope value for the function will create a steeper line. Greater downhill slope angles will have larger vertical factors. In other words, the difficulty of travel increases sharply with the downhill slope. (Likewise, the ease of travel increases sharply with the uphill slope).

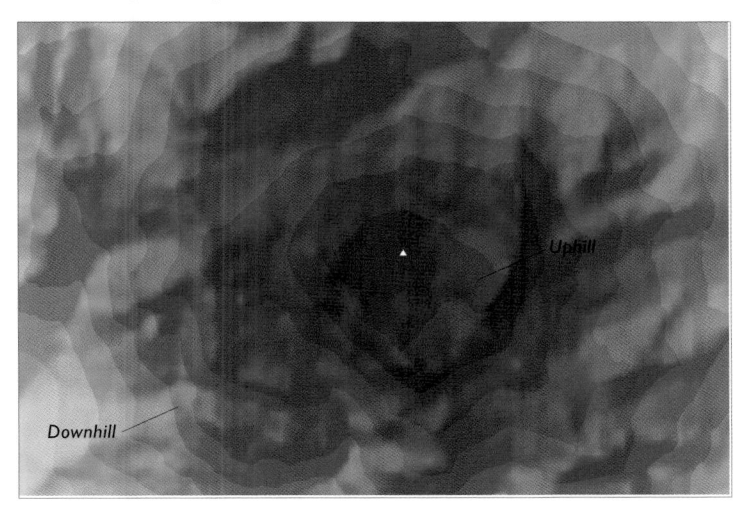

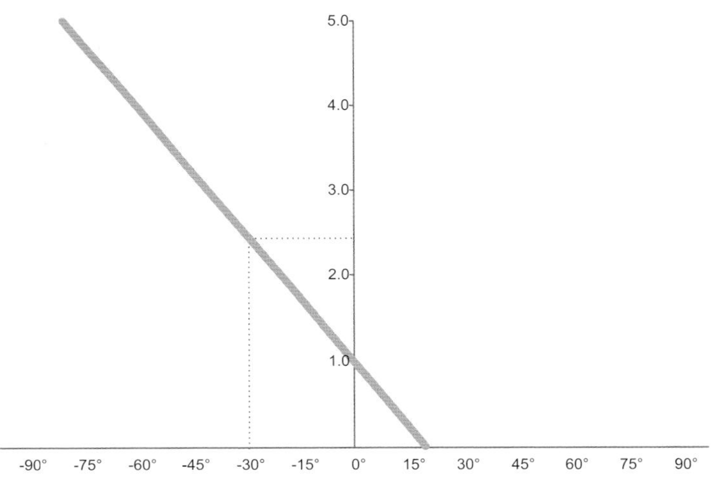

Spread surface created using the linear function with a zero factor of 1, a slope value of -0.05, a low cut angle of -90°, and a high cut angle of 90°. A slope of -30° has a vertical factor of 2.5. With a slope value of -0.05, the fire spreads more easily uphill and less easily downhill than with a value of -0.03 (map and graph on previous page).

Other mathematical functions, such as symmetric linear, secant, and cosine-secant, let you model various scenarios for traveling over slopes.

If moving over gentle slopes is easy and over steep slopes (either uphill or downhill) is difficult, use the symmetric linear function. A symmetric linear relationship assumes that the difficulty of movement (and the slope weight value) increases at a constant rate as the surface slope angle increases. You would use this option, for example, if you're modeling the migratory path of an elk herd—it would be relatively easier for elk to walk over gentle uphill or downhill slopes, but difficult for them to travel over steep uphill or downhill slopes.

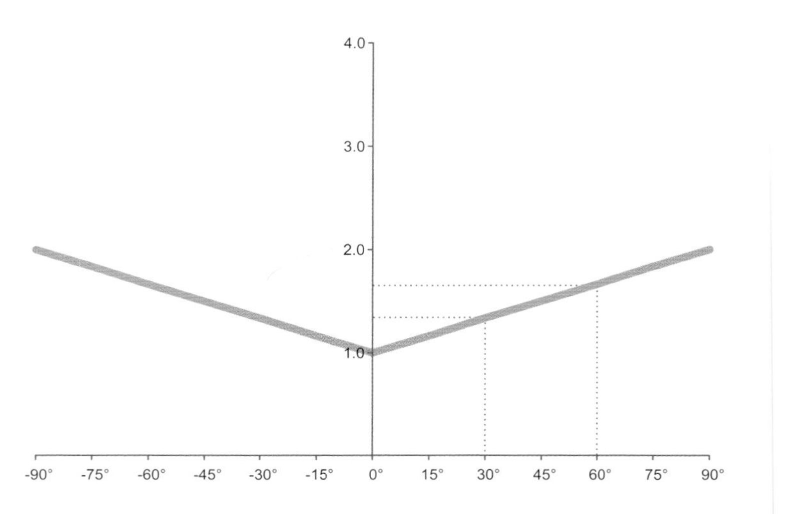

Symmetric linear function with a slope value of 0.01, a zero factor of 1.0, and cut angles of -90° and 90°. The slope weight increases for both uphill (positive) and downhill (negative) slopes as the slope steepness increases. A slope of 30° (or -30°) has a weight of 1.33, while a slope of 60° (or -60°) has a weight of 1.67.

If movement over very steep uphill or downhill slopes is very difficult, but movement over moderate slopes is relatively easy, use the secant function. The secant function assumes the difficulty of movement (and the slope weight value) increases at a faster rate as the slope angle increases. You might use this function for a highway path model, where the goal is to locate a path that allows cars and trucks to maintain a constant speed (and avoid having to accelerate uphill and brake downhill).

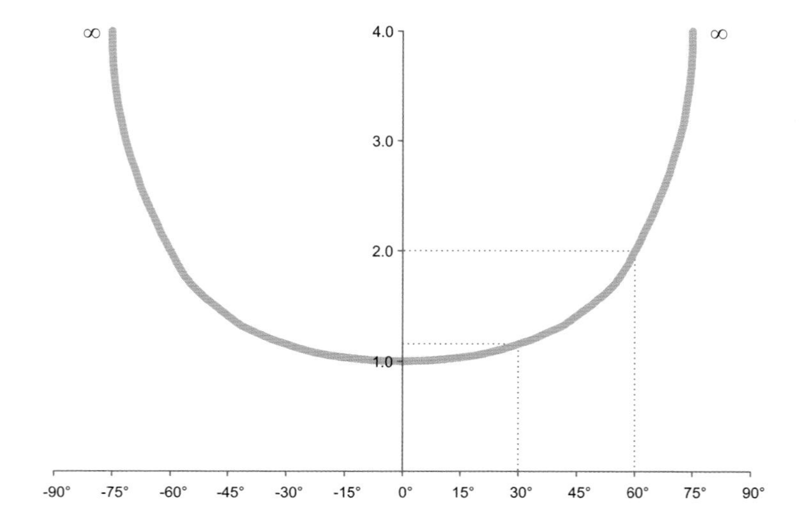

Secant function with a power of 1 and cut angles of -90° and 90°. The slope weight increases for both uphill (positive) and downhill (negative) slopes at a faster rate as the slope steepness increases. A slope of 30° (or -30°) has a weight of 1.15, while a slope of 60° (or -60°) has a weight of 2.0—the difference between the weights for these slopes is greater than with the symmetric linear function.

If movement over very steep downhill slopes is easy, and movement over very steep uphill slopes is difficult, use the cosine-secant function. You might use this function in a model for a water pipeline path, for example.

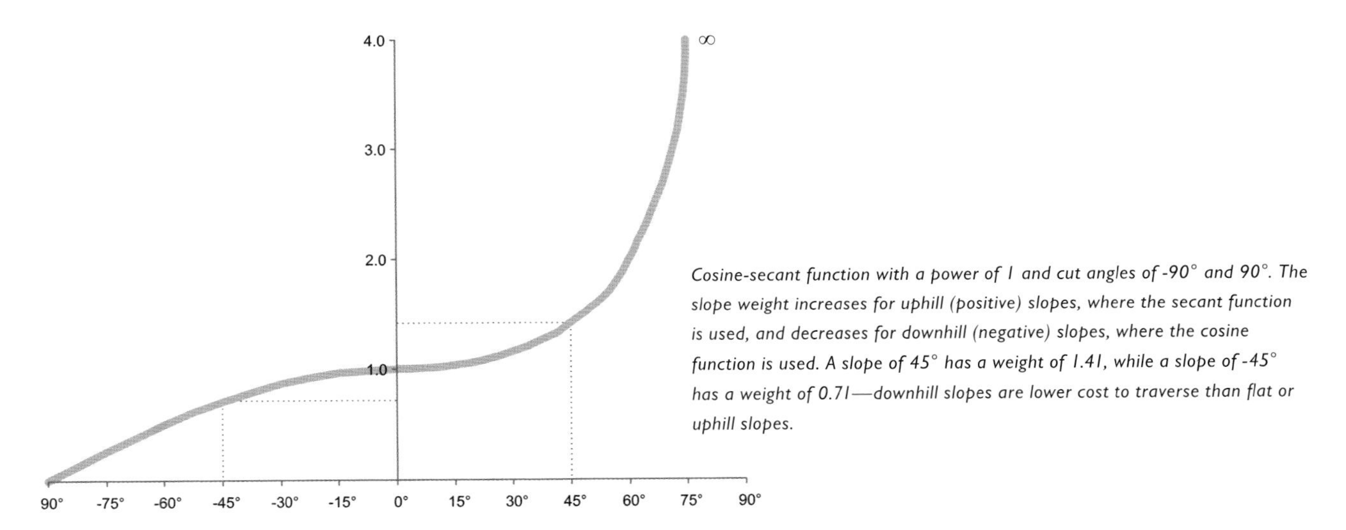

Cosine-secant function with a power of 1 and cut angles of -90° and 90°. The slope weight increases for uphill (positive) slopes, where the secant function is used, and decreases for downhill (negative) slopes, where the cosine function is used. A slope of 45° has a weight of 1.41, while a slope of -45° has a weight of 0.71—downhill slopes are lower cost to traverse than flat or uphill slopes.

Modeling directional influence on movement

In addition to vertical factors, such as slope, you can include factors in your model that may impact the forward momentum of travel or movement over the path. For example, a strong head wind will slow the progress of a truck and add cost, but a tail wind could decrease cost.

Horizontal factors such as wind are mainly useful when you're modeling the spread of a phenomenon, such as the direction in which a wildfire will burn. Generally, horizontal factors are not an issue for finding the least-cost point-to-point path or corridor, especially for pipelines or transmission lines. (Wind direction doesn't really impact the movement of electricity over a transmission line or water through a pipeline.) However, if wind or other horizontal factors influence whatever is traveling over the path, and hence the location of the least-cost path, you can include them in your model.

Increased cost of movement can be incurred from traveling against or perpendicular to an external force (such as prevailing wind). Likewise, the cost may decrease for movement parallel to the direction of the force.

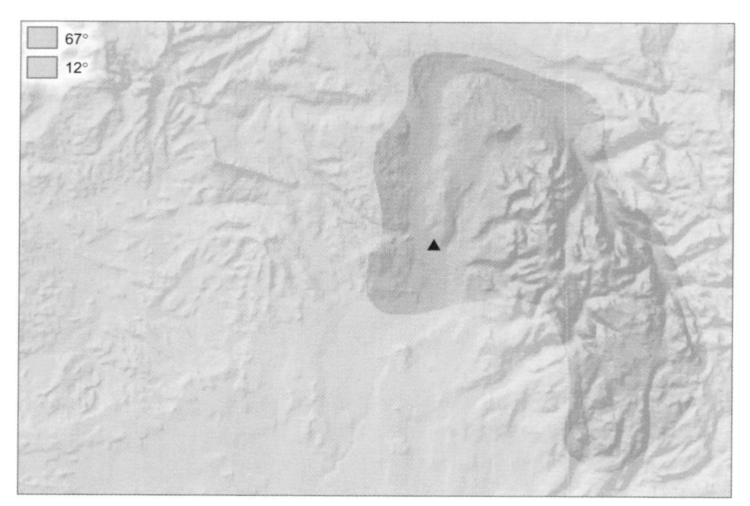

This layer shows the direction of the prevailing wind in the study area (along with terrain). For most of the area, it's toward east–northeast, which is 67° clockwise from north (0°). The prevailing wind direction for the area in darker blue is north–northeast, or 12° from north, due to the north–south trending mountain ranges. The origin location (triangle) is also shown.

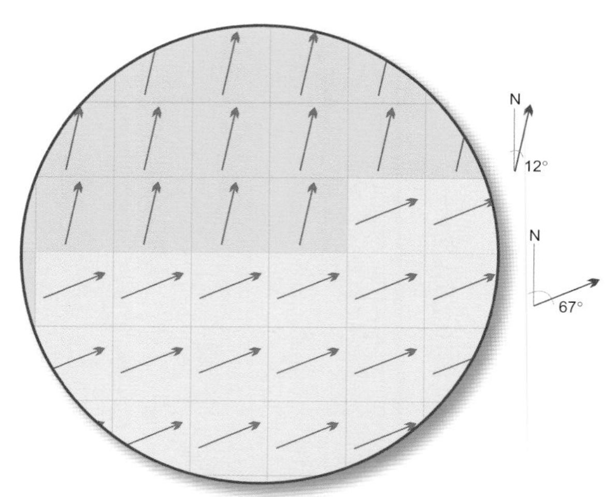

This close-up illustrates the prevailing wind direction for the horizontal factor layer. The arrow in each cell is the wind direction, which forms an angle of 12° from north (darker blue cells) or 67° from north (lighter blue cells). Hence, each cell is assigned a corresponding value of 12 or 67.

In practice, horizontal factors can be hard to quantify, but if you have the data they can greatly increase the accuracy of your model (the spread of wildfire, for example, is greatly influenced by wind direction and speed).

Specifying a horizontal factor is similar to specifying a vertical factor. You define the relationship between the direction and strength of the force and the associated cost. As with a vertical factor, you can do this by creating a table of paired values or by using a mathematical function.

Either way, you essentially specify the cost of traveling at a given angle to the direction of the force (remembering that the movement across the surface is from cell center to cell center horizontally, vertically, or diagonally). If the movement is directly opposite to the direction of the force (180 degrees), the cost is greatest. If the movement is in the same direction as the force (0 degrees) the cost is least, and in fact, may be less than 1 (the force actually assists the movement). If the force is at an angle to the direction of the movement between these extremes, the cost is increased or decreased accordingly. For example, in a wildfire spread model, the cost of moving from one cell to a neighboring cell in the direction of the prevailing wind will be less than 1—the wind actually makes movement easier than if there were no wind. The cost of moving from one cell to a neighboring cell in the opposite direction to the force will be very high, and may well block movement in that direction.

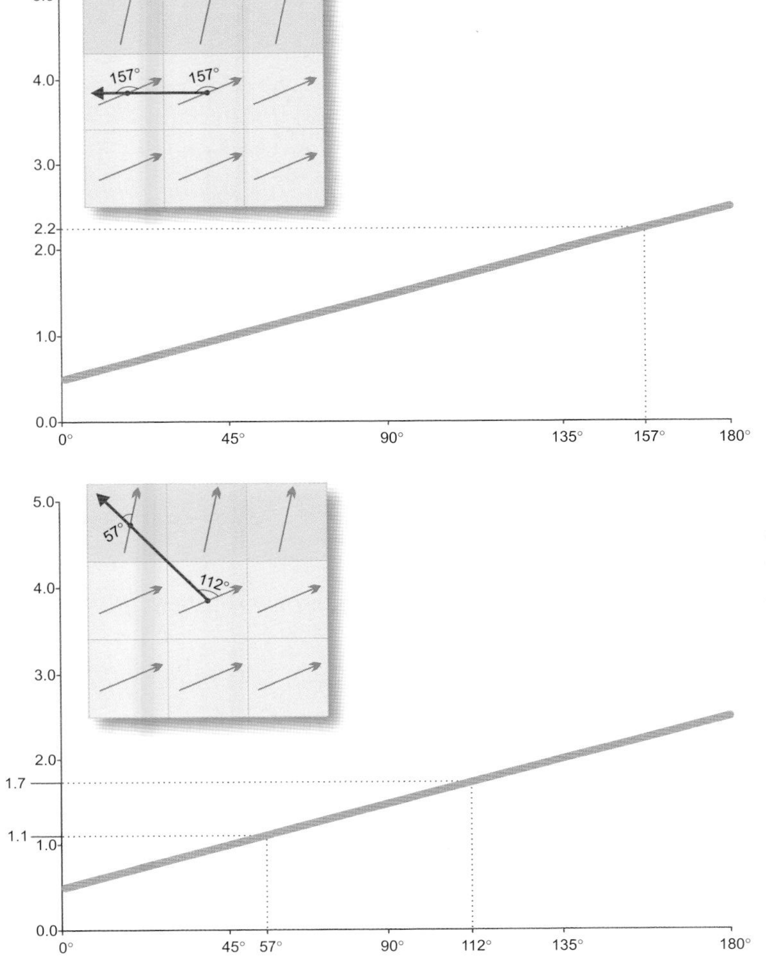

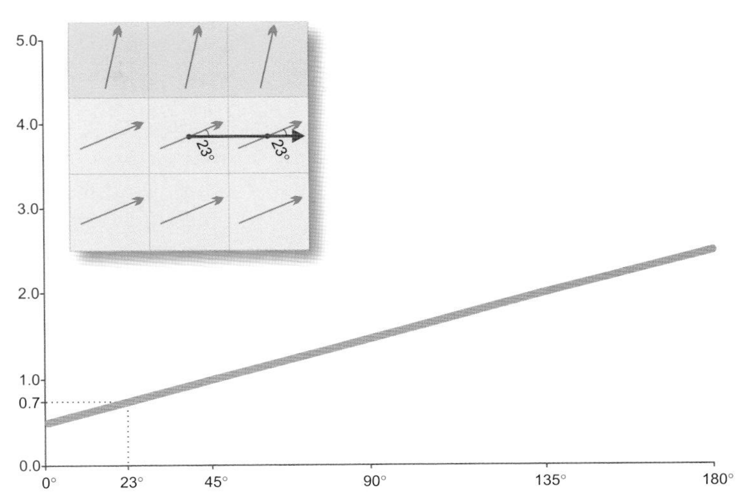

The GIS calculates the angle between the direction of the horizontal force (light gray lines) and the direction of the cell-to-cell movement (dark gray lines). It uses this calculated angle to look up the horizontal factor value. In the example in the upper left, the direction of the horizontal force is 67° from north, and cell-to-cell movement is 90° from north in the east-to-west direction—into the prevailing wind. The calculated angle is 157° (on the x-axis of the chart), so the horizontal factor (on the y-axis of the chart) is 2.2. The measured distance will be multiplied by a factor of 2.2, increasing the cost distance in that direction. In the example in the upper right, the direction of the horizontal force is the same, but the cell-to-cell movement is 90° in the west-to-east direction—generally with the prevailing wind. The calculated angle is 23° and the horizontal factor is 0.7—the cost distance in this direction is less than the measured distance, meaning that the horizontal force makes movement in this direction easier than if wind was not included as a factor. In both these cases, the direction of the horizontal force is the same for the From and To cells (67° from north). If the direction of the horizontal force is different for the two cells, the horizontal factors are averaged. In the example to the left, the horizontal factor for the From cell is 1.7 and for the To cell it's 1.1. So the horizontal factor for movement between the two cells is 1.4.

As with a vertical factor, you can also specify a cut angle beyond which movement is essentially blocked.

To specify the strength of the force, you define the slope of the curve—a flatter slope indicates a weak force (low wind speed, for example), while a steep slope indicates a strong force (high wind speed). With a flatter slope, movement in the direction of the force isn't much easier than movement opposite the force or even than movement assuming no force at all.

In addition to the slope and the cut angle, you specify the horizontal factor value to use for movement that goes in exactly the same direction as the horizontal force (that is, where the difference in the angles is 0°). This factor, known as the "zero factor," defines where the curve intersects the y-axis on the chart and represents the decrease in cost from moving in the same direction as the horizontal force. For a wildfire, for example, moving in the same direction as the horizontal force might decrease the cost by half, so you'd set the zero factor to 0.5. If the prevailing wind direction in a cell is 90°, the cost of moving from west to east across the cell would be half what it would be if there were no wind at all. Of course, unless the direction of the horizontal force is exactly horizontal, vertical, or diagonal, the angle of the cell-to-cell movement will never exactly match it, so the horizontal factor that is actually applied to any given cell will usually be larger than the zero factor.

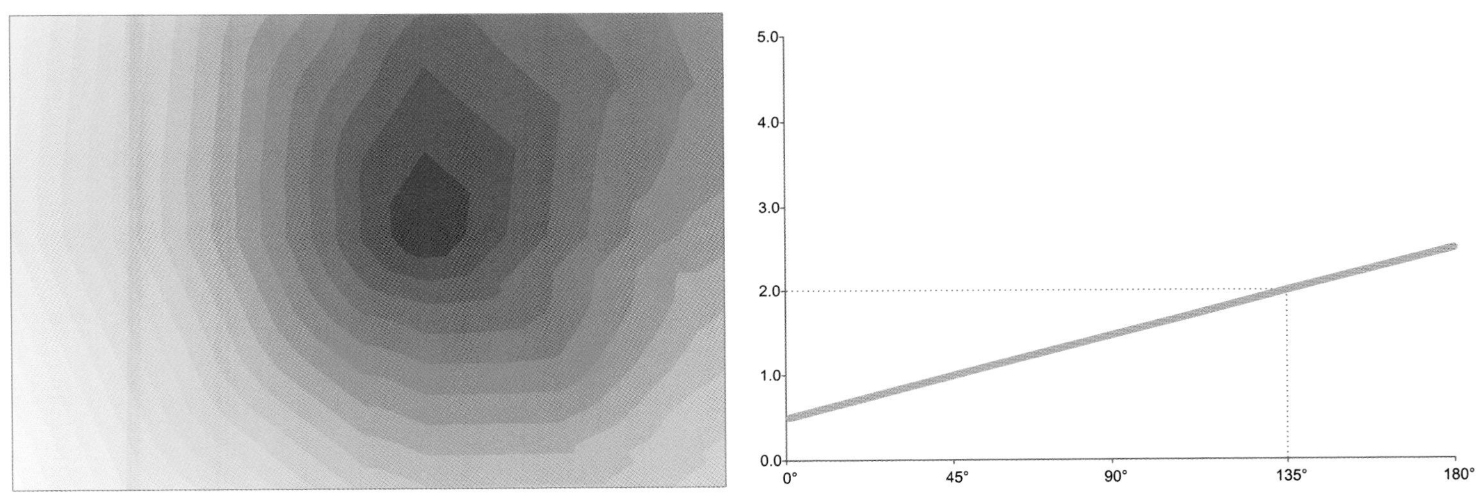

This spread surface was created using only the horizontal factor layer representing prevailing wind direction. In the absence of other influences, the wildfire would tend to burn in the direction of the prevailing wind (north, and then east–northeast). The linear function was used with a zero factor of 0.5 and a slope of 0.01. An angle of 135° is assigned a horizontal factor of 2.0.

With a steeper curve, movement against the force is inhibited more rapidly.

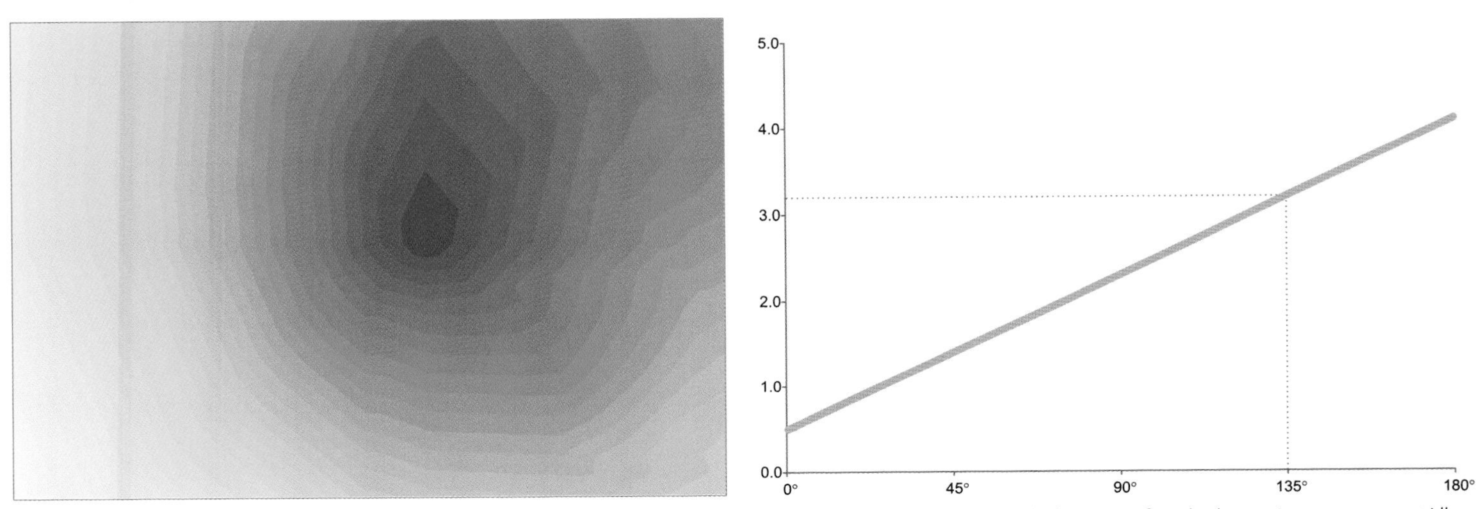

This spread surface was created using the linear function with a zero factor of 0.5 and a slope of 0.02. While the pattern doesn't change significantly, the cost increases more rapidly moving outward from the source. (Each band is a range of cost distance values.) With a slope of 0.02, an angle of 135° is assigned a horizontal factor of 3.2.

You can confine movement to the direction of the horizontal force. You'd do this if, for example, your model assumes that the wind in the study area is consistently strong enough in the prevailing direction to prevent the wildfire from burning into the wind. With the "forward" option, you specify a zero factor for angles up to 45° (usually a value less than 1) and a "side value," which is the horizontal factor assigned to angles between 45° and the cut angle you specify (up to 90°). A slope factor is not used.

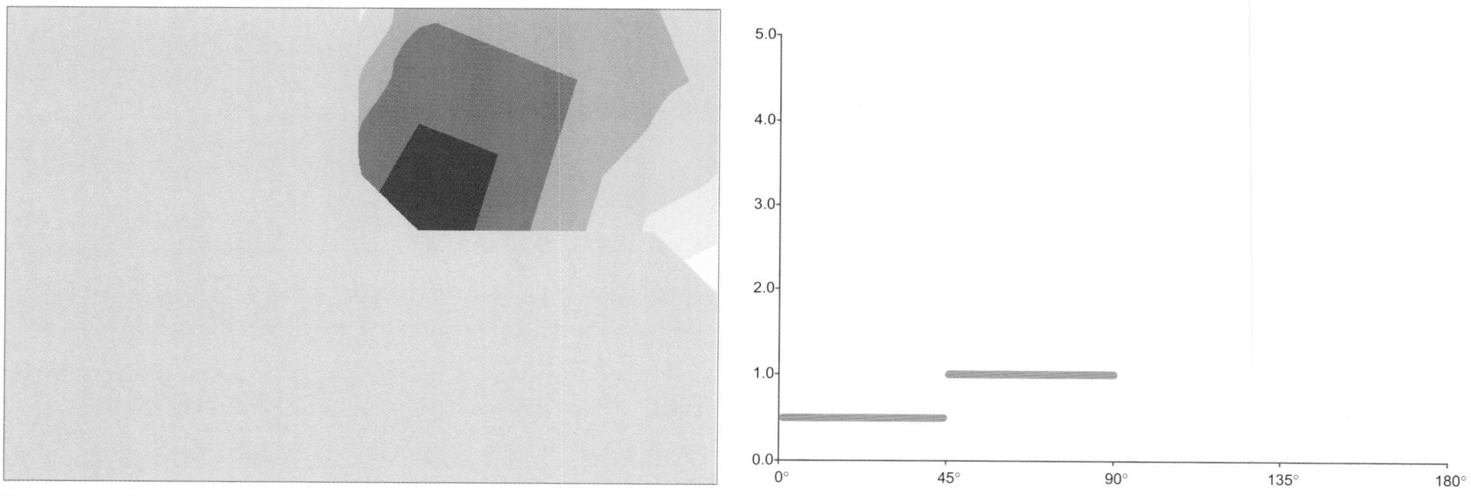

This spread surface was created using only the horizontal factor layer representing prevailing wind direction. The forward option was used with a zero factor of 0.5 and side value of 1. The fire can only move with the prevailing wind—any cells that are upwind of the source are assigned a value of NoData.

By decreasing the zero factor and increasing the side value, you can increase the effect of the force—for example, if the prevailing wind has an extremely high speed.

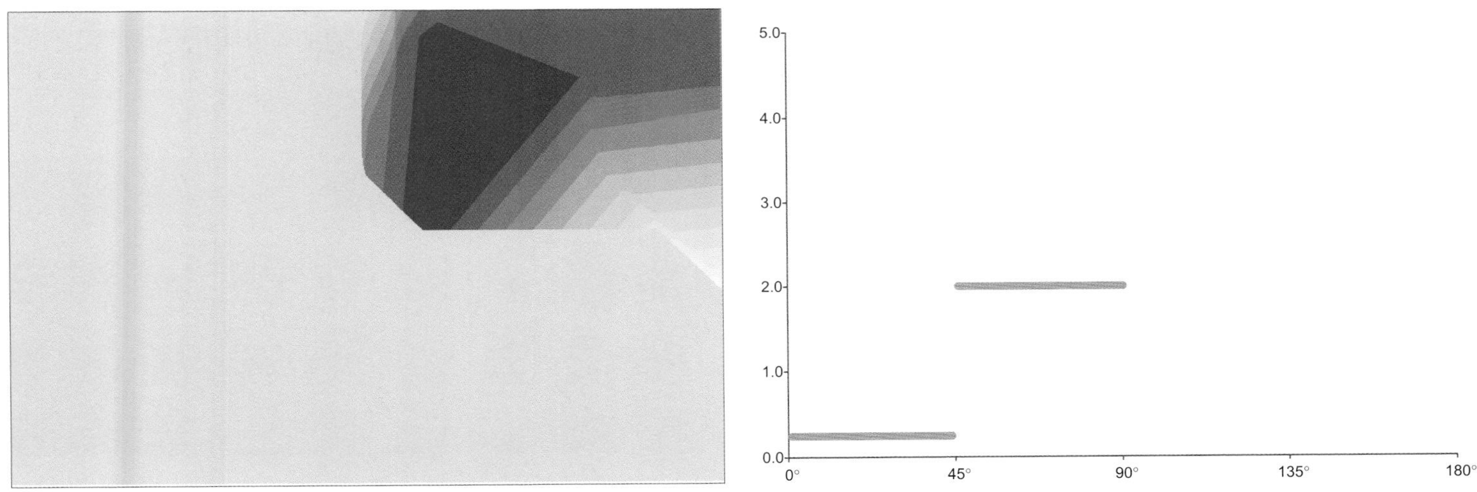

This spread surface was created using only the horizontal factor layer representing prevailing wind direction. The forward option was used with a zero factor of 0.25 and a side value of 2. With these parameters, the directional effect of the wind is increased. (Over the same time period, the fire will go farther in the direction of the prevailing wind.)

By changing the parameters, you can create scenarios based on different wind conditions.

PARAMETER	SPECIFIES
Cut angle	*Angle beyond which movement is blocked*
Slope	*Strength of the horizontal force*
Zero factor	*Horizontal factor value applied to movement in the same direction as the force*
Forward	*Ensures movement is in the same direction as the force only*
Side value	*Used with forward to specify the horizontal factor for angles between 45° and the specified cut angle*

CREATE THE COST DISTANCE SURFACE

Once you've created the cost surface and assembled the surface distance, vertical, and horizontal factor layers, you combine them along with the origin location to create the cost distance surface. (You specify the parameters for the vertical and horizontal factors at the same time.) The GIS creates a surface showing the areas of cumulative cost moving outward from the origin.

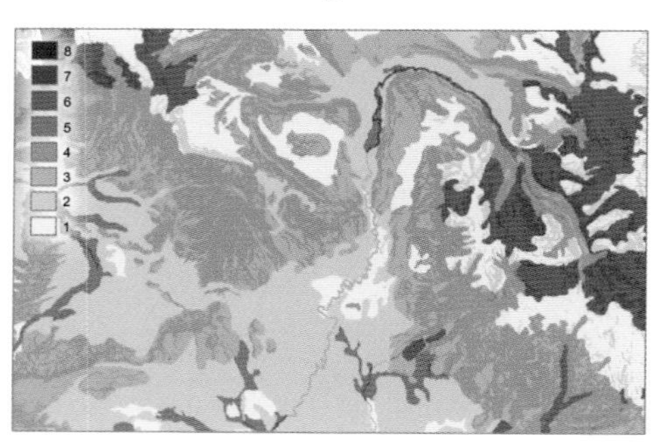

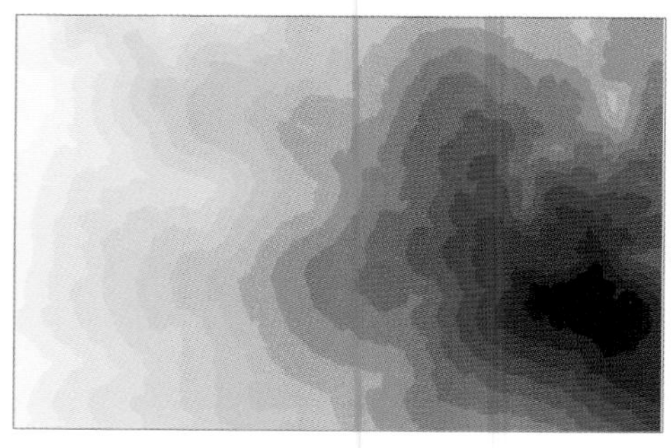

The origin (left) is combined with the cost surface (center) to create the cost distance surface (right). The darker the color, the lower the cost distance value. Cumulative cost increases as distance from the origin increases, but it increases less rapidly in areas where the cost of travel is less.

The various cost factors are essentially weights that the calculated distance between cells is multiplied by. Each factor modifies the original measured distance. If you're using only a cost surface and not including vertical or horizontal factors in your model, the GIS calculates the cost distance values as described below.

The GIS first adds the cost assigned to the From cell to the cost assigned to the To cell and divides by two, averaging the cost. (Travel is considered to be from cell center to center, thus half the travel is across the From cell and half is across the To cell.) This cost is then multiplied by the distance between cell centers. For cells in the horizontal or vertical direction, this distance is the same as the cell size; for cells in the diagonal direction, it is the cell size times 1.41421 (as described earlier).

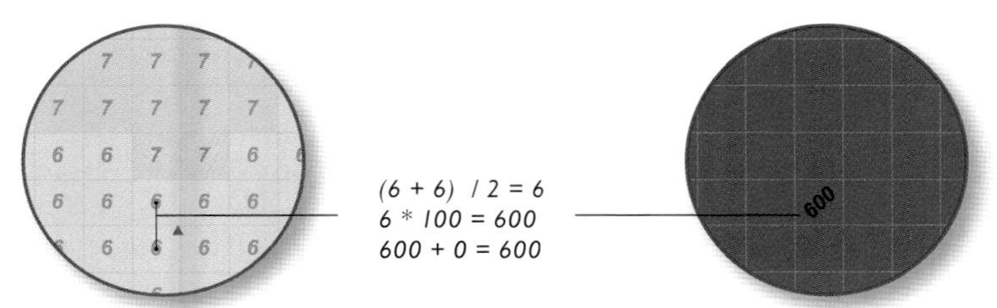

Starting with the origin cell, the GIS calculates the average cost for the From and To cells (the origin cell being the initial From cell). It then multiplies the average cost by the distance between cell centers (100 meters, in this example) to get the cost distance between these two cells. Finally, it adds the newly calculated cost distance to the cost distance previously assigned to the From cell (which is 0 in the case of the origin cell). The To cell becomes a From cell and the process is repeated (below).

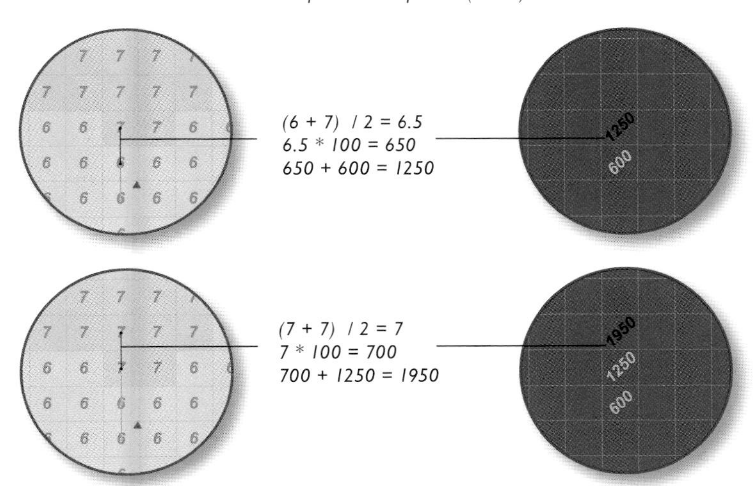

As the GIS searches outward, it calculates the cumulative cost distance to cells farther from the origin.

The GIS assigns the cost distance values moving outward in all directions from the origin cell(s), keeping track of the cells for which it has already calculated the cost distance. There are multiple paths to each cell from the origin. If a newly calculated cost distance is less than the already assigned cost distance value for a cell, the new, lower cost distance value replaces the earlier one.

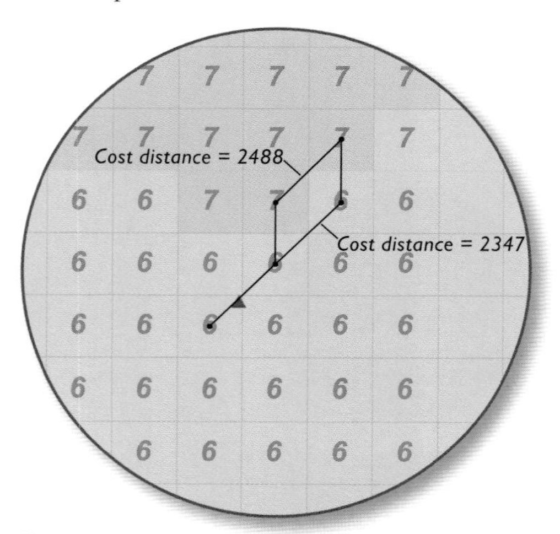

There are many paths to each cell from an origin cell. The lowest cost distance value is assigned.

As noted, the process described above applies to a cost surface with no movement factors. However, many models do, in fact, include vertical and horizontal factors. The GIS calculates the weighted cost distance between each To and From cell using this equation, which includes all the factors:

COST DISTANCE = DISTANCE * ((FROM_COST * FROM_HORIZONTAL_FACTOR) + (TO_COST * TO_HORIZONTAL_FACTOR) / 2) * VERTICAL_FACTOR

For cost and for the horizontal factor, the assumption is that half the movement is across the From cell and half is across the To cell. So in the equation, the horizontal factor for the From cell is multiplied by the cost for the From cell, and the same is done for the To cell. The results are added and divided by two, creating a combined cost-horizontal factor (the term in the outer parentheses in the equation).

On the other hand, the distance value and the vertical factor value are both already calculated from cell center to cell center. Thus they can each be included as a single term in the equation.

To get the cost distance between cells, the distance is multiplied by the combined cost-horizontal factor and the vertical factor for the two cells.

If a cost surface, vertical factor, or horizontal factor is not included in the model, a value of 1 is substituted in the equation for these.

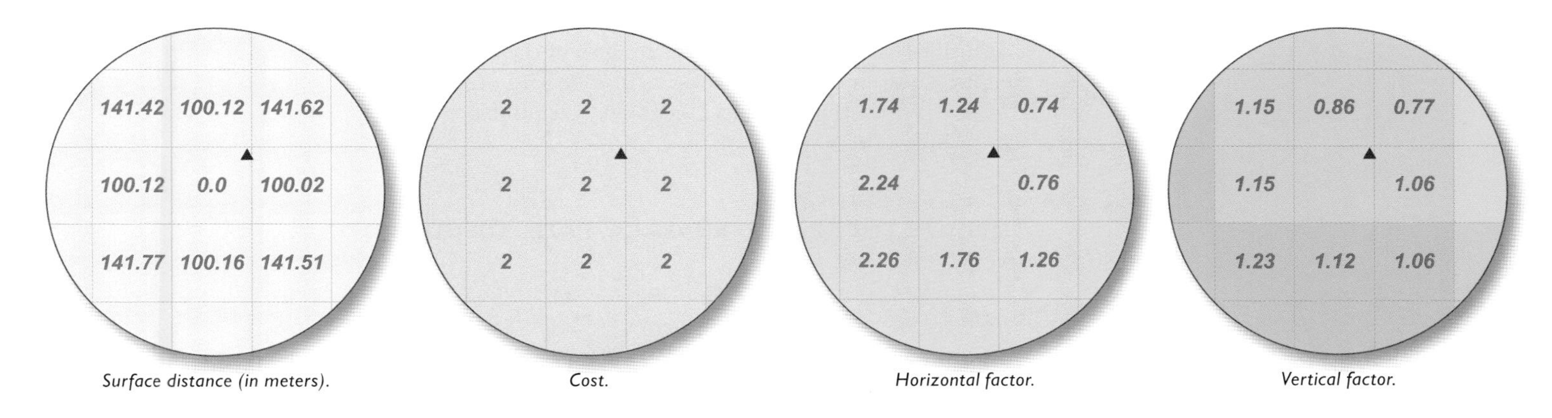

Surface distance (in meters). *Cost.* *Horizontal factor.* *Vertical factor.*

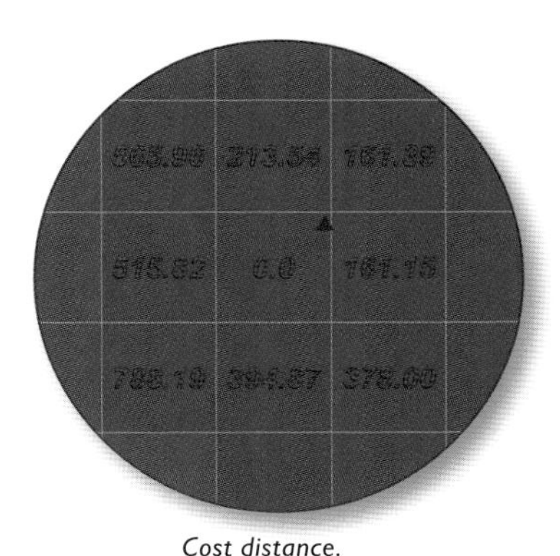

Cost distance.

The cost distance is calculated from the source cell to each surrounding cell. (In this example, the source feature falls in the northeast corner of a particular cell—the coincident cell in each layer becomes the source cell.) Movement to the cells east and northeast of the source entails much lower costs than to the west, southwest, and northwest. This movement comports with the fact that the surface cost is the same for all the cells in that neighborhood (2), while the prevailing wind is to the east–northeast and that direction is also uphill from the source.

In the example above, the cost distance for the cell due north of the source cell is calculated as

COST DISTANCE = 100.12 * (((2 * 1.24) + (2 * 1.24)) / 2) * 0.86

(In this example, the difference of the angle of movement from the angle of the horizontal force happens to be the same for the From and To cells, so the value of 1.24 shown in the diagram is used for both the From and To cells. Similarly, the cost value of 2 is the same for the From and To cells.)

The parameters you specify for the various factors will affect the final cost distance surface. For example, if you increase the strength of the horizontal factor, it will have more influence in the final cost distance surface, and other factors such as the cost surface will have relatively less influence. Balancing the influence of the various input layers is key to creating a valid model.

CREATE THE SPREAD SURFACE, PATH, OR CORRIDOR

Once you've created the cost distance surface, you can create the required output of your model—a spread surface, a point-to-point path, or a corridor.

Creating a spread surface

If the goal of your model is to show the spread of some phenomenon (such as a wildfire) from the origin location, the cost distance surface is, in fact, the end result. It *is* the spread surface, showing the least-cost path in all directions from the origin. Since the phenomenon is moving over time, it will travel farther (for a given period) in directions that have a lower travel cost.

This type of model is useful for planning purposes, rather than for predicting the hour-by-hour movement of a particular event or phenomenon. For example, by creating a model that predicts the movement of the fire, using information such as terrain and wind, you can show which areas the fire is likely to move toward once it begins to burn. These are the areas likely to be in the path of the fire, and thus at high risk should a fire occur at or near an origin location (such as a campground). You could then locate fire lookouts and position firefighters and equipment accordingly or take proactive measures, such as clearing fuel or building firebreaks in susceptible areas.

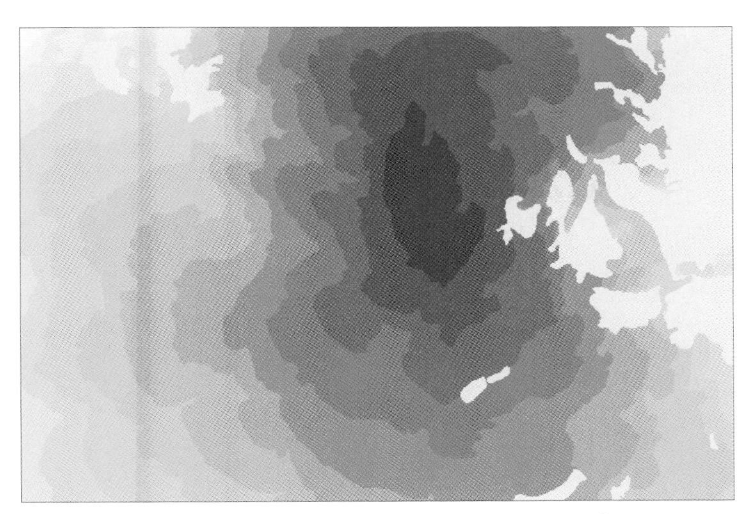

This spread surface for a wildfire was created using land cover as a cost factor, terrain as a vertical factor, and wind as a horizontal factor.

You can model the spread from more than one source at a time by including multiple locations on the source input layer. For example, you could specify several fire sources from lightning strikes or campgrounds and see where the fires would burn. If they converge in some areas and burn away from others, you'd know that the burn areas are particularly at risk. You could then position firefighters and equipment in those areas during hot, dry weather.

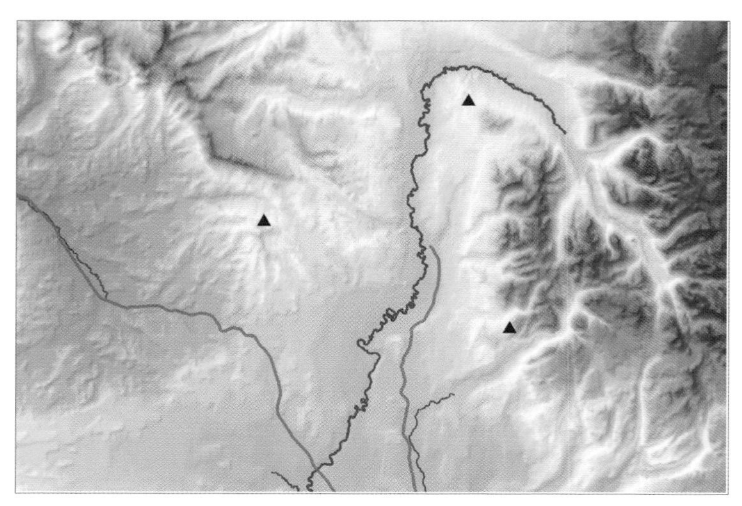

This map shows three potential fire source locations in forested areas (perhaps started by lightning strikes).

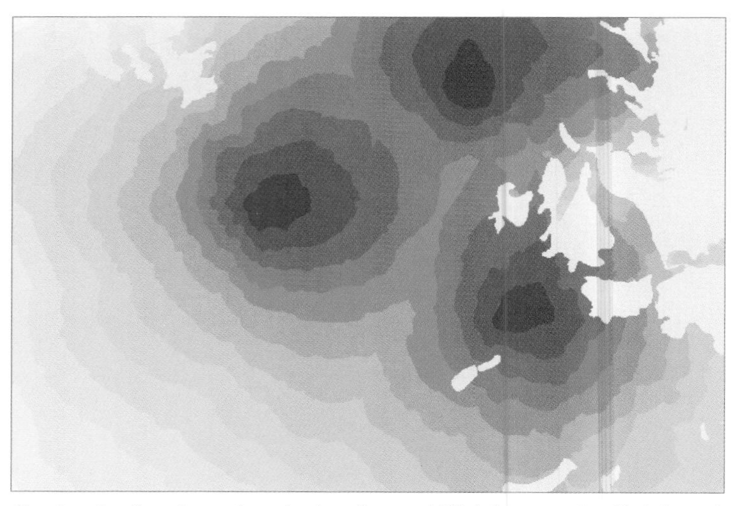

The spread surface shows where the three fires would likely burn over time if wind speed and direction remained constant. A fire igniting in the area of the origin in the middle of the map has the most potential to spread, while a fire igniting near the origin location in the southeast would likely burn itself out as it moved east into rock and snow.

Creating a point-to-point path

If the goal of your model is to create the least-cost path between two locations, you use the cost distance surface along with the origin and a destination location to create the path. As with the origin, the destination is also a point, area, or line feature—it does not have to be of the same type as the origin. So you could create the least-cost path from a town (point feature) to a park (area feature).

Before creating the path, you'll need to create a cost direction layer. (You can do this when you create the cost distance surface or after the fact.) This layer essentially stores information on the least-cost path from each cell to the origin.

The cost direction layer represents the direction of travel from each cell to the neighboring cell having the least cost.

To create the cost direction layer, the GIS uses the cost layer to identify—for each cell—which neighboring cell is the least cost to travel to. It assigns the cells in the cost direction layer a value of 1 through 8 corresponding to the direction of the neighboring least-cost cell. For example, a cell for which the least costly cell to travel to is the one to the left is assigned a value of 5. The GIS uses the cost direction values to trace the path. It knows, for example, when it reaches a cell with a value of 5, the path needs to continue into the cell to the left.

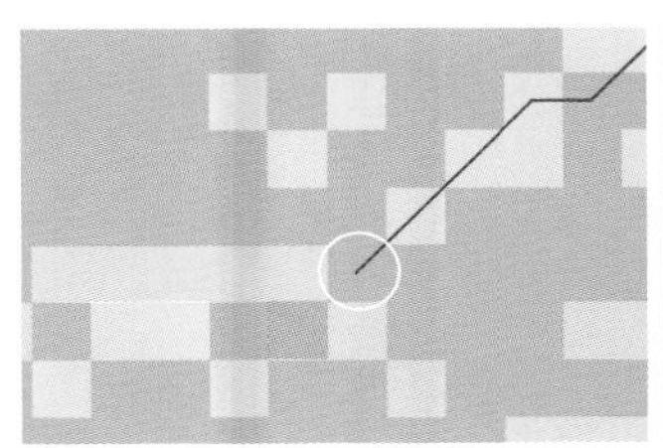

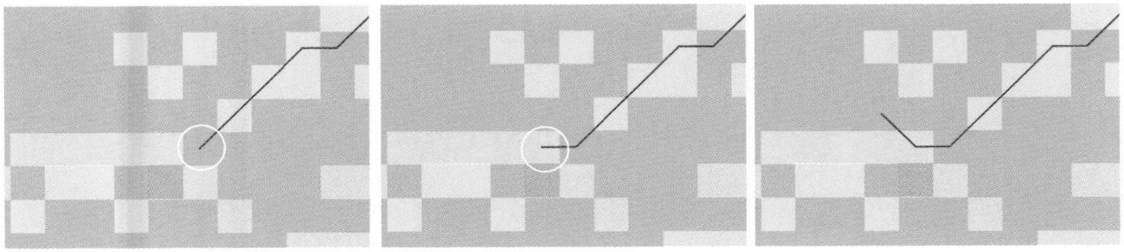

This cell has a value of 5, so the path continues into the cell to the immediate left.

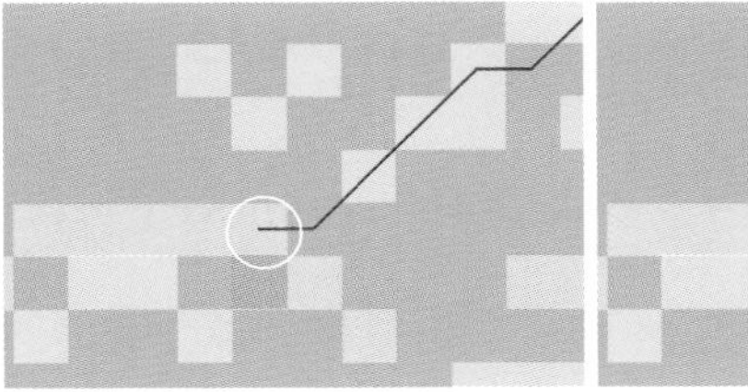

The next cell has a value of 6, so the path continues to the upper left.

6	7	8
5	0	1
4	3	2

The values indicate the direction of the least-cost cell.

How the GIS calculates the least-cost path

The GIS starts at the destination cell and, using the cost direction layer, begins tracing a path back to the origin. Since the GIS knows for each cell which adjoining cell offers the least cost to travel to, it moves to that cell. As it goes, it tags each cell as belonging to the path and adds it to a new layer. It stops when it reaches the first cost direction cell with a value of zero—the origin.

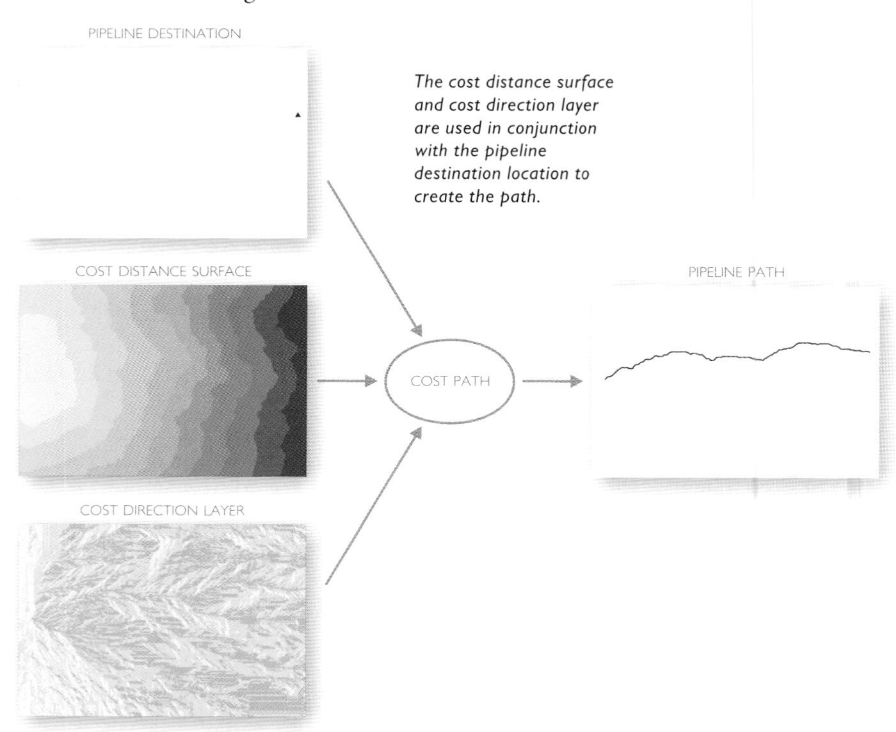

The cost distance surface and cost direction layer are used in conjunction with the pipeline destination location to create the path.

You can then display the path layer to see the least-cost path. The resulting raster path is only one cell wide, so it can be hard to see when displayed with other layers. Converting the path to vector format lets you display the path using a wider line, or using dashes, double lines, or other symbols.

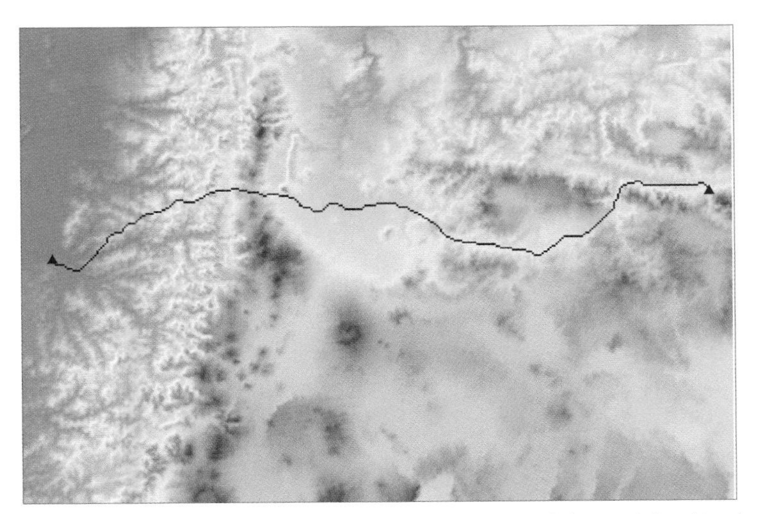
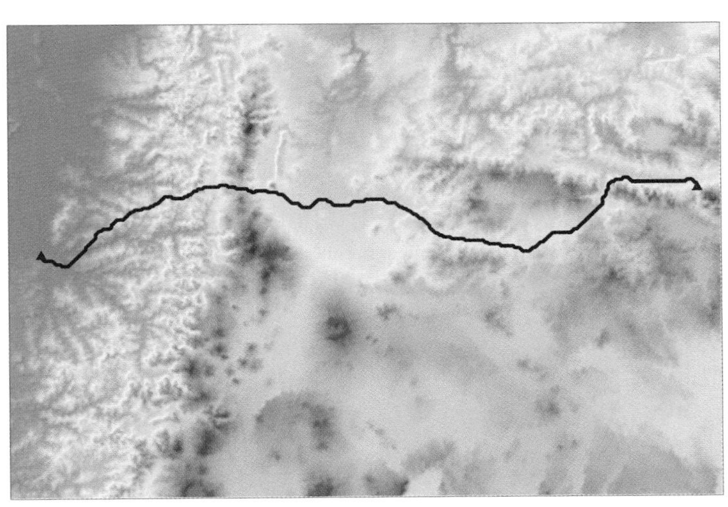

Original raster path (left) and the vector version drawn using a wide line symbol, making the path easier to see.

The GIS adds the total cost of the path—which it gets from the cost distance surface—to the attribute table for the path layer. The cost distance units represent the various criteria that went into your model, combined with actual linear distance.

You can use the total cost to compare alternate paths (that you've created by changing the parameters of your model and rerunning it). Since the units are relative, however, you only know that one path has a higher overall cost than another. You can get a better sense of comparative cost by calculating the percentage difference. For example, if one path has a value of 25,738 and another has a value of 32,641, you know the cost of the second path is about 27 percent more than that of the first.

Calculating a path between several origins and destinations

You can have the GIS create paths between several origins and destinations. For each destination, the GIS finds the least-cost path to the origin that is nearest (in terms of cost). So,

- If there is a single origin (dark green) and several destinations (beige), the GIS will create the least-cost path between each destination and the origin.

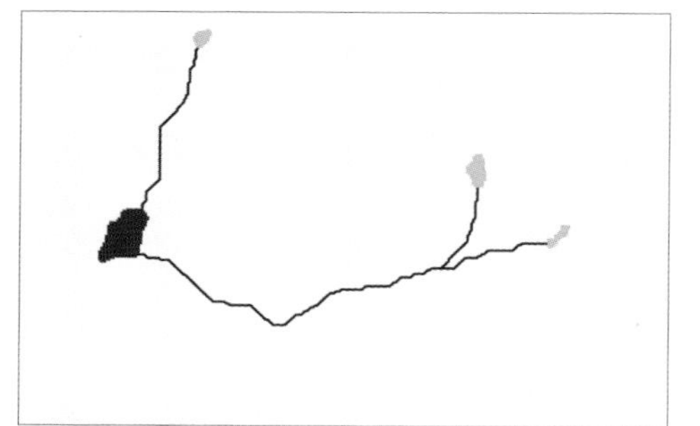

- If there are several origins and one destination, the GIS will create only a single path—the one from the destination to the origin that is the least cost to travel to.

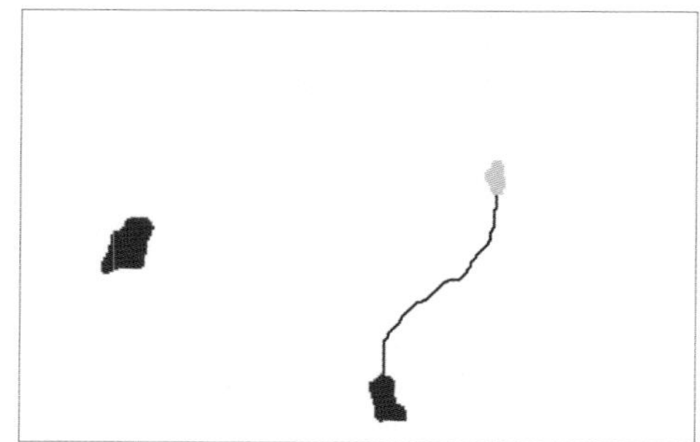

- If there are several origins and destinations, the GIS will create a path between each destination and the origin which is the least cost to travel to (for a given destination).

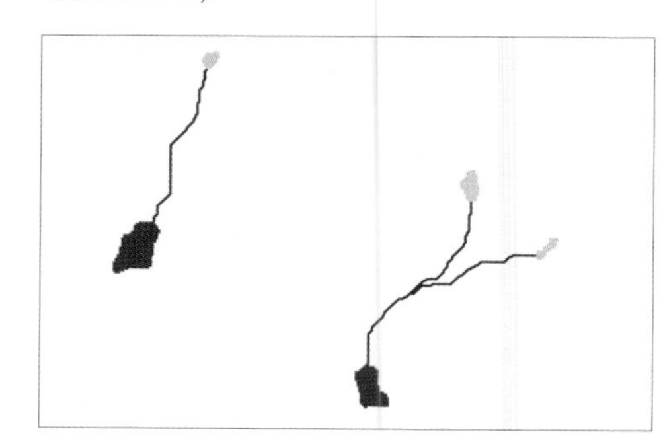

Using barriers

If you include a barrier when you create the cost distance surface, the GIS will detour around the barrier when it creates the path. For example, if you're creating a road corridor, you could designate a buffer area around streams as a barrier, and any possible paths will not pass through the buffers. You'd designate the stream buffer as a barrier by including it and assigning it a value of "Restricted" when you create the cost surface layer.

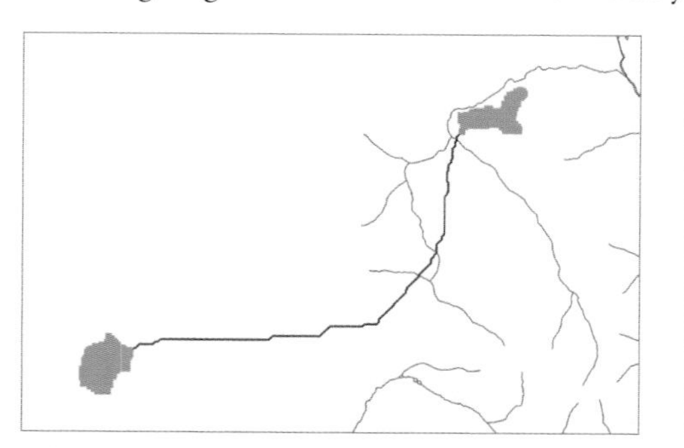

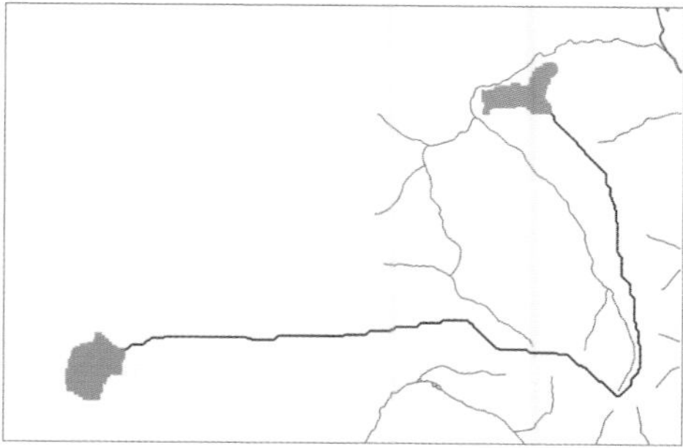

When streams are used as a barrier (right map), the road does not cross them.

Creating a corridor

When you create a point-to-point path, the GIS calculates a single line (one cell wide) between the locations. Alternatively, you can create a corridor showing all cells having a low cost. Corridors are useful if you're generating a path that needs to be wider than the width of a single cell, such as a passage for wildlife moving between protected natural areas.

Corridors are also useful at an early stage in your project. By creating a corridor, you can define the general location of the path and narrow down the exact location at a later stage, taking into account specific engineering or other requirements.

Possible location for a wildlife corridor based on land cover, proximity to water, and distance from roads. Rivers (blue lines) and roads (red lines) are also shown.

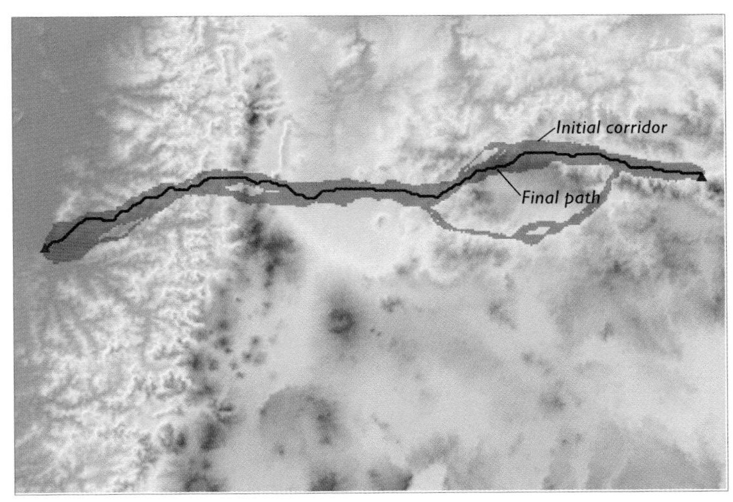

Pipeline corridor and final path.

To create a corridor, you create two cost distance surfaces (as described earlier)—one for each of the two locations you're connecting. The GIS calculates the least cumulative cost for a path that connects the two locations and passes through a cell by adding the two cost distance surfaces. The cost of a cell on the output layer is the total cost to reach it from both locations. You can then select the cells with a cost less than a specified value to delineate a corridor.

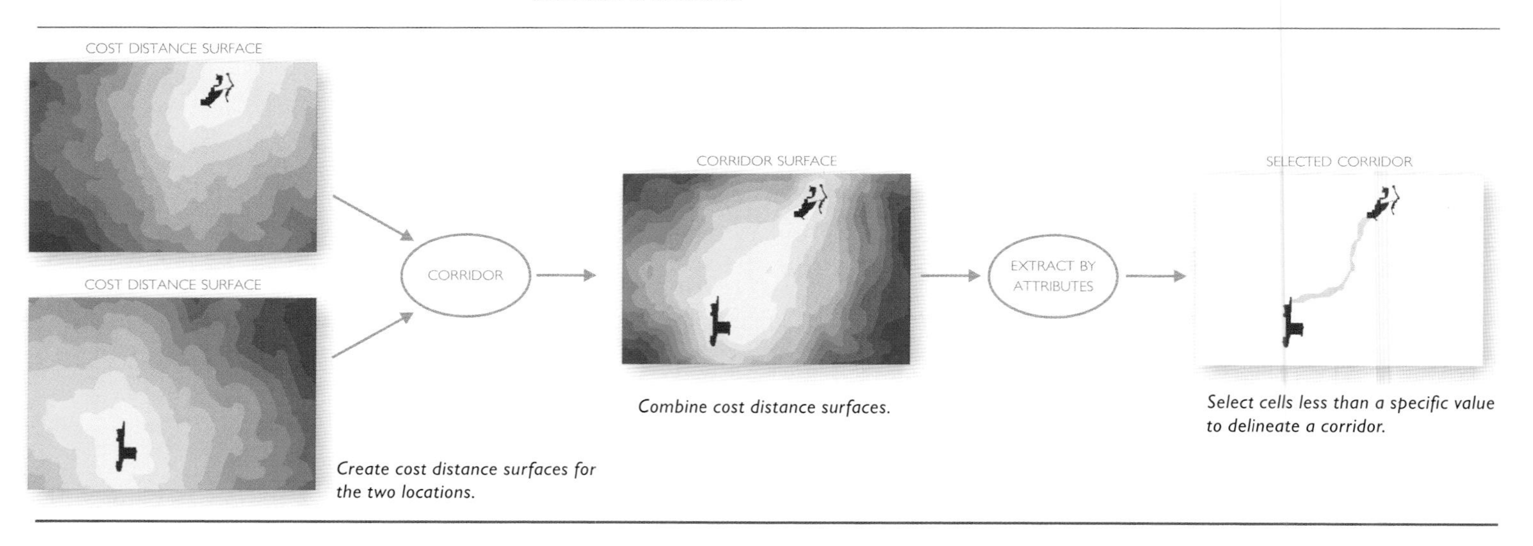

Create cost distance surfaces for the two locations.

Combine cost distance surfaces.

Select cells less than a specific value to delineate a corridor.

If you're exploring alternate paths, you can classify the surface, or use a continuous ramp of colors, to show the least-cost paths. Using a classification scheme such as Jenks' natural breaks can help highlight the least-cost paths, since it emphasizes gaps between groups of values on the continuous surface.

Combined cost distance surface displayed using a continuous color ramp. Lighter areas indicate lower cost.

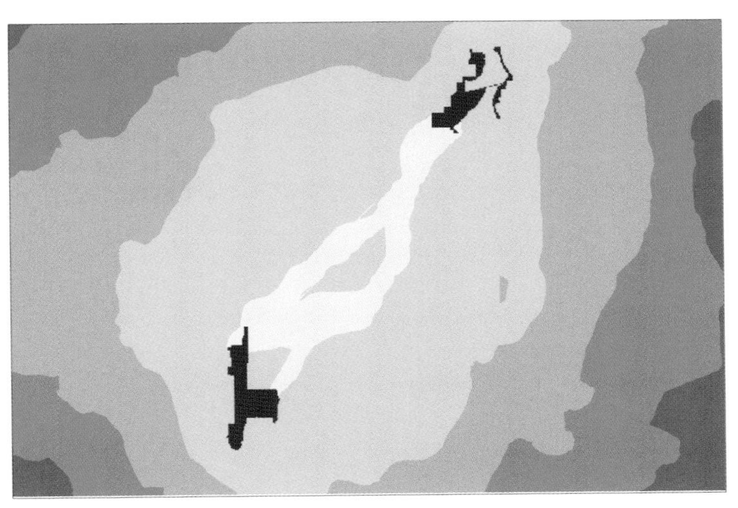

Surface classified using natural breaks.

If you've used monetary values to calculate the cost distance, this approach will let you select all the possible paths costing less than a specific amount. You can also create several alternate paths at one time using this approach. When you select the cells less than a specific cost, the GIS will delineate all possible paths with costs less than that value.

Selecting all cells less than a specific value results in alternate paths.

Creating a path or corridor that connects several locations

If you're creating an overland path that connects several places, you'll need to create paths between each location in succession. For example, if you're creating a wildlife corridor that needs to connect several protected areas, you'd start by creating the corridor between the first two areas. You'd then create the corridor between the next two protected areas, and so on.

First leg of corridor.

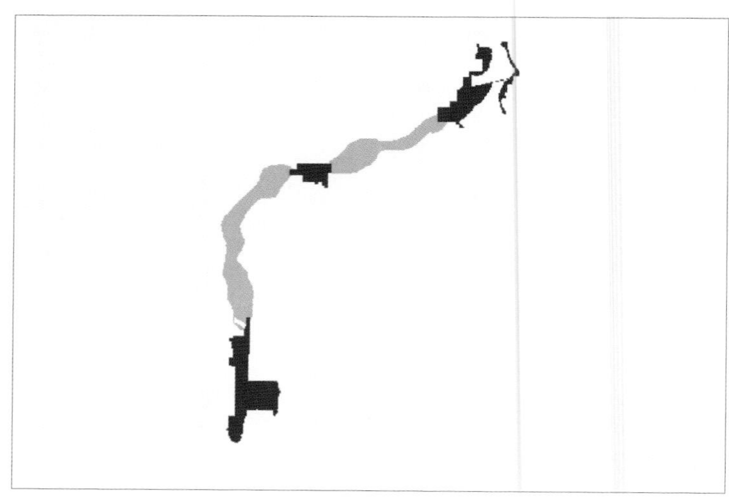

First and second leg of corridor.

Creating alternate paths and corridors

Your analysis may require that you create several alternate paths for a particular purpose. Usually, each alternate favors one set of criteria over another. For example, one path for a new road might give more importance to low construction costs, while an alternate gives more importance to protecting agricultural land and sensitive wildlife habitat.

To create alternate paths, you'll need to re-create the cost distance surface for each path. You change the weights of various cost inputs, or add or subtract inputs, based on the criteria for each alternative.

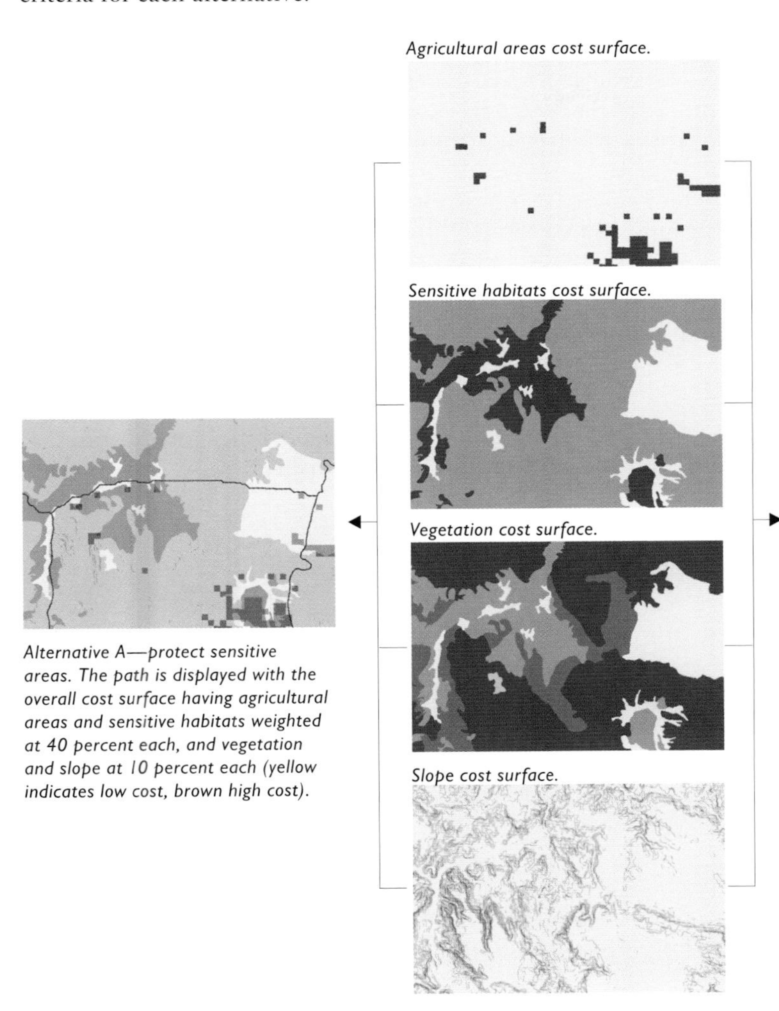

Alternative A—protect sensitive areas. The path is displayed with the overall cost surface having agricultural areas and sensitive habitats weighted at 40 percent each, and vegetation and slope at 10 percent each (yellow indicates low cost, brown high cost).

These maps show two scenarios for a road corridor. The model on the left weights agricultural lands and sensitive habitats higher, giving more importance to the protection of these features. The one on the right weights vegetation and slope higher, giving more importance to lower construction costs.

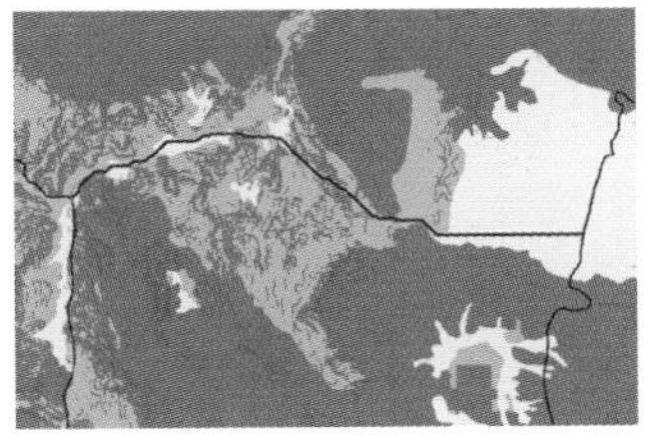

Alternative B—minimize construction cost per mile. The path is displayed with the overall cost surface having vegetation and slope weighted at 40 percent each, and agricultural areas and sensitive habitats at 10 percent each.

EVALUATE THE RESULTS

Whether you're creating a spread surface, point-to-point path, or corridor, you'll want to evaluate the results of your model to make sure it is valid. This mainly rests on your own knowledge of the area and of the subject matter—knowing what you (or other experts) know about the study area, does the map make sense? This evaluation is especially important since most of the factors in the model (cost, the vertical factor, and the horizontal factor) may be hard to quantify in terms of their influence on movement.

A particular pipeline path, for example, may have the lowest cost according to the model, but you may conclude, based on your knowledge of pipeline design and construction, that the alignment of the path and the land it traverses makes construction not feasible. You may need to reassign cost values or weights, or include additional source layers, and rerun the model in order to better capture the real-world conditions. This will also allow you to see which factors have the most influence on the model.

If the path or corridor traverses areas that appear to be high cost, check the values you assigned on the cost layers. You can also display the results of your model with the original input layers to make sure the results reflect your original criteria.

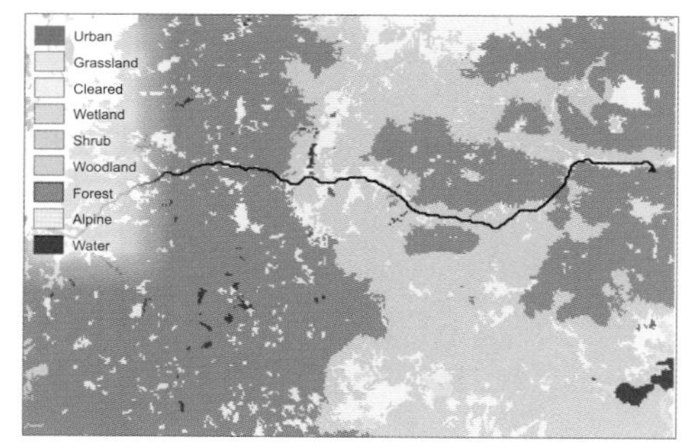

Pipeline path with land cover. The path primarily traverses open land (shrub, grassland, and woodland) while avoiding dense vegetation (such as forest) and water.

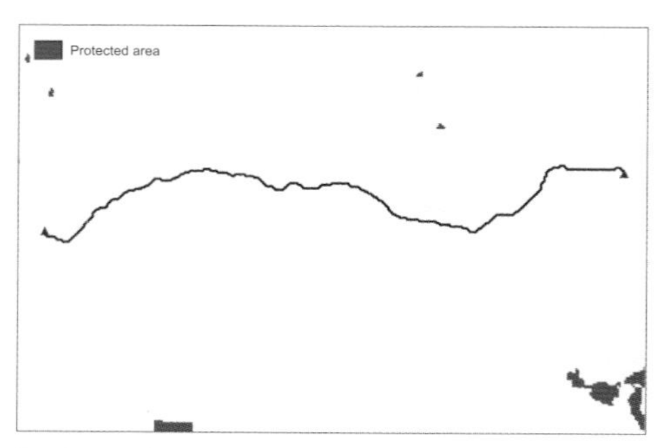

Pipeline path with protected areas. The path avoids crossing any protected areas.

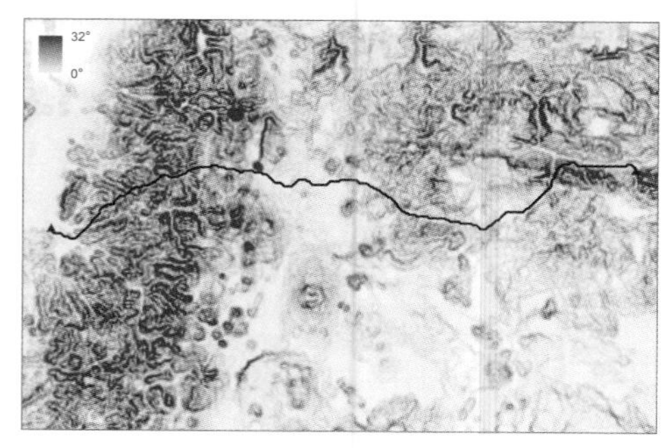

Pipeline path with slope. The path avoids steep slopes (to the extent possible).

You shouldn't be surprised if some individual cells in the least-cost path happen to have a high cost. It may sometimes be necessary to traverse areas that have a high cost on one or more of the individual cost layers.

You'll also want to consider what the spread surface, path, or corridor actually represents. For a wildfire, for example, the map does not necessarily show the actual burn area, but rather shows which areas are closest to the origin and the least cost for a wildfire to burn through, and therefore where the fire is likely to move farthest, fastest.

For a wildlife corridor, the corridor represents the route that offers the least impediment to travel based on the criteria you defined—wildlife may or may not actually travel within the corridor.

DISPLAY AND APPLY THE RESULTS

Showing the spread surface, path, or corridor with other layers of data can help you and others make decisions about areas at risk or about the feasibility of creating the path or corridor. For example, when designating a wildlife corridor, you'd want to know which portions of the corridor pass through public land and which through private land. In addition, displaying layers that were included in the model, such as rivers, can help clarify why the path or corridor is located where it is.

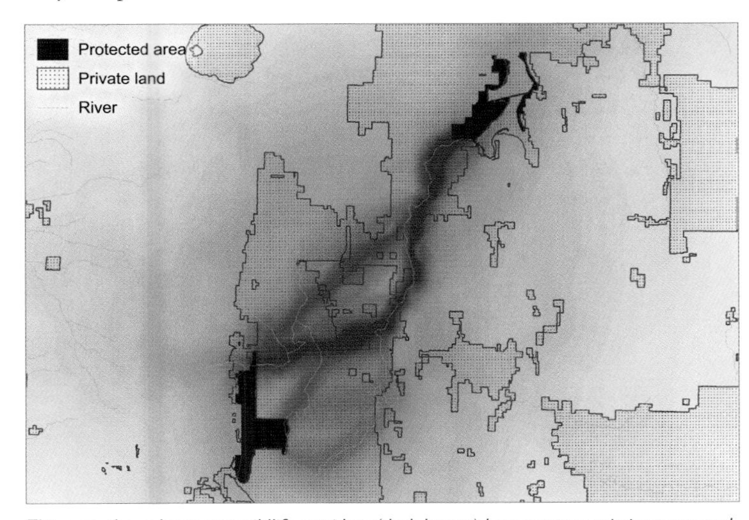

This map shows least-cost wildlife corridors (dark brown) between two existing protected areas (green). One corridor follows a river, while an alternate portion is a more direct route.

You can show alternate scenarios for a path with the cost surfaces that were used to create them. That will allow people to understand the trade-offs between the scenarios. Discussion can focus on the places where the paths differ.

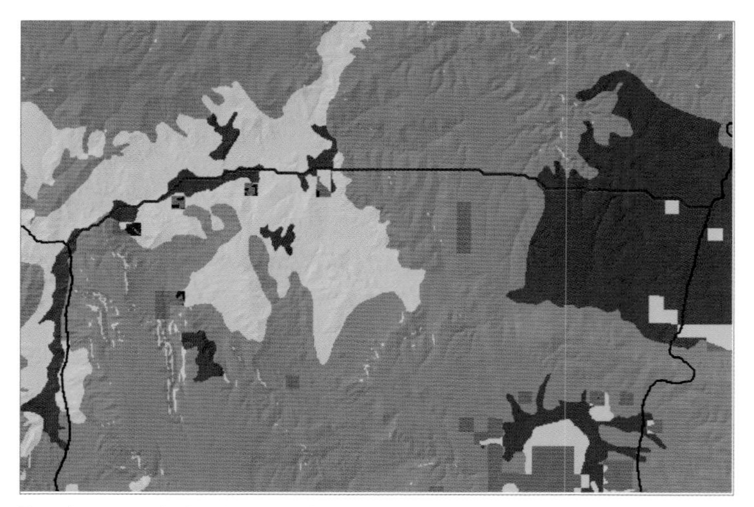

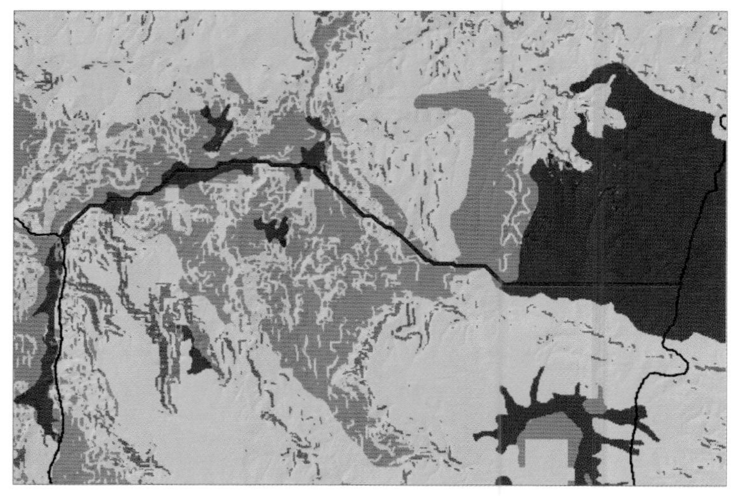

Two alternate paths (thick red line) for a road connecting two highways. Least-cost areas are shown in dark green and highest cost areas in dark red. The alternative on the left gives more importance to protecting environmentally sensitive areas, which are assigned high costs. The path avoids these areas and travels through areas with low environmental costs. The alternative on the right gives more importance to reducing construction costs—areas that are more expensive to build a road over are assigned high costs. The path avoids the ridges near the center of the map and takes a more southerly, flatter route. Displaying the path with the cost surface makes it easier to see the rationale behind the path location. The western portion of the path is basically the same in both scenarios, so discussions can focus on the trade-offs for the eastern portion of the path.

For a spread surface, you can classify the cost distance surface into bands of color or display it with a continuous color ramp to show the gradual change in values.

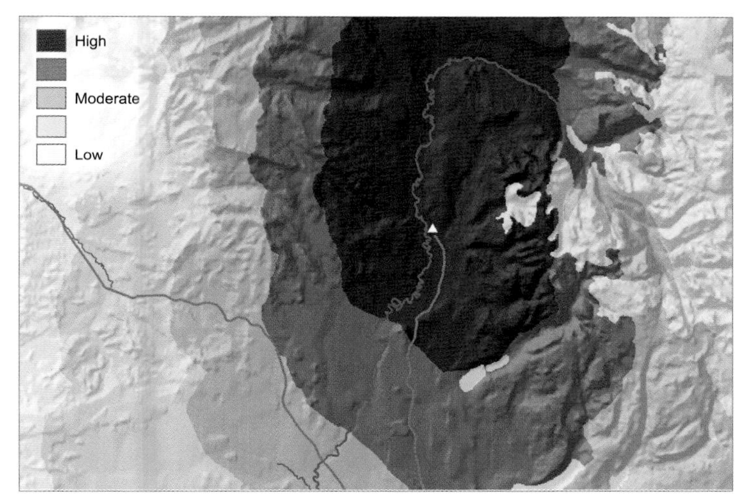

Wildfire spread surface classified into five classes to show areas at high, moderate, and low risk for a fire originating at a campground (white triangle). Major roads and rivers are also shown.

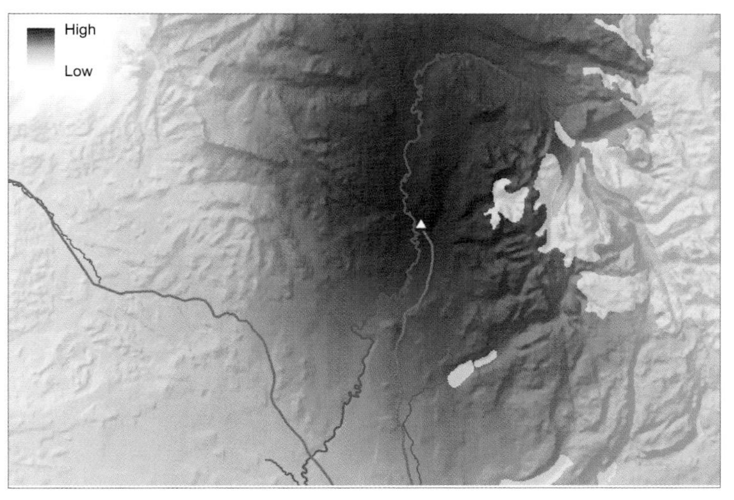

This map symbolizes the same surface with a continuous color ramp. The effect is subtler, without the sharp boundaries between value ranges.

You can also create polygons from the classified surface to show the areas of highest risk with relevant features such as structures, land cover, timber stands, and so on.

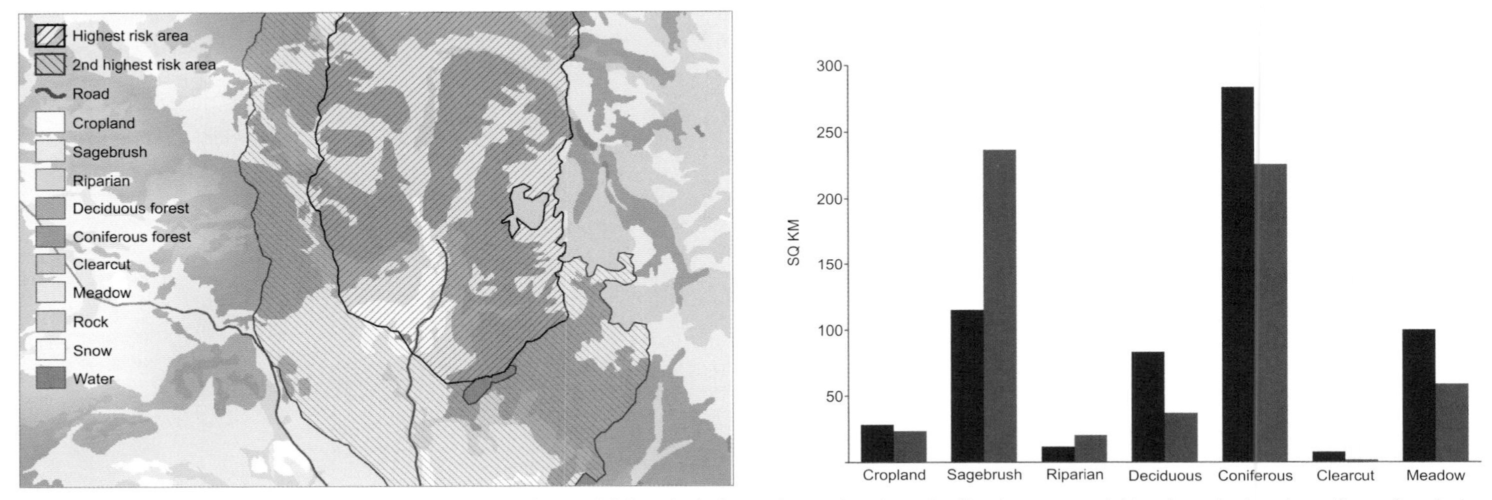

The borders of the two highest risk areas are displayed with land cover. While in the highest risk area there is a mix of land-cover types (although predominantly coniferous forest), the second highest risk area is primarily coniferous forest in the west and southeast, and primarily sagebrush in the south and southwest. This information would help firefighters prepare appropriate strategies for different stages of the fire. The chart shows the amount of each land-cover type in the two risk areas—it's clear that the second highest risk area (orange bars) is mainly sagebrush and coniferous forest.

Information such as data from economic models, along with political or social factors, may come into play in any decisions. The maps created by the GIS can serve as good points of departure for discussions. Showing the layers that went into the model will help people understand why the path or corridor is located where it is or how the spread surface was created.

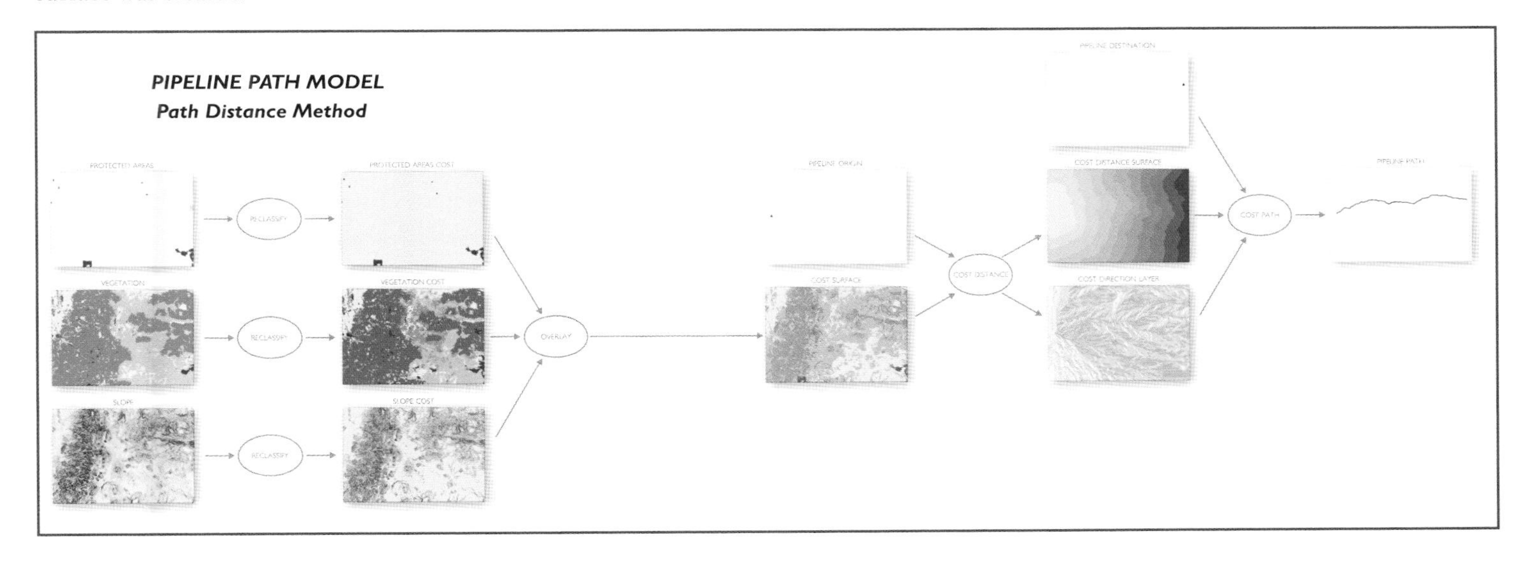

references and further reading

Cormen, Thomas H., Charles E. Leiserson, Ronald L. Rivest, and Clifford Stein. 2009. *Introduction to Algorithms*. Cambridge, MA: MIT Press. A comprehensive textbook on computer algorithms. Includes a discussion of Dijkstra's algorithm for shortest path calculations, among many others.

Dijkstra, E. W. 1959. "A note on two problems in connexion with graphs," Numerische Mathematik 1: 269–271. First publication of Dijkstra's algorithm for calculating the shortest path between two nodes on a network.

Zeiler, Michael. 2010. *Modeling Our World: The Esri Guide to Geodatabase Concepts*. Redlands, CA: Esri Press. Describes the various data models used by ArcGIS. Includes a discussion on the concepts behind the network databases used for transportation applications.

Zhan, Cixiang, Sudhakar Menon, and Peng Gao. 1993. "A directional path distance model for raster distance mapping," 1993 COSIT Conference, 434-443. New York, NY: Springer-Verlag. Presents an approach for weighted path distance mapping using rasters, the method that underlies finding the least-cost overland path.

Modeling flow 5

Model flow to find out where things will accumulate, such as rainfall runoff or sediment loads, or to find the likely paths things will take, such as tracing a toxic material through a stormwater system.

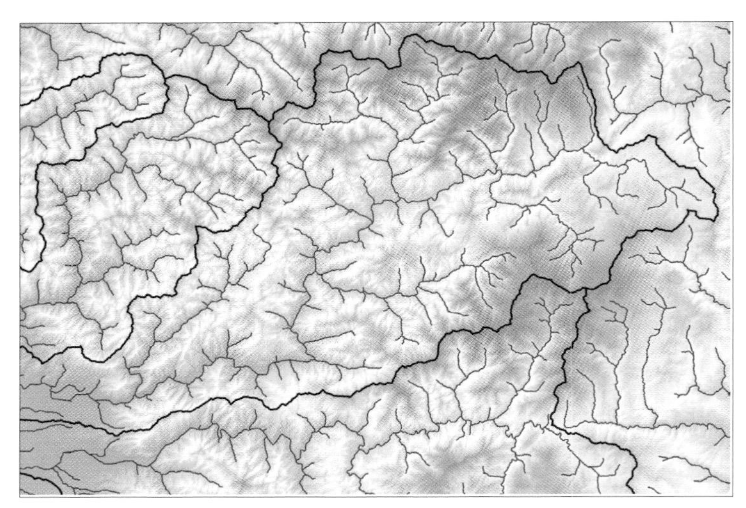

Flow models were used to create the stream channels as well as the hydrologic basin boundaries (shown in red).

A flow model used to identify where motor oil dumped into a stormwater system will go. The oil was dumped into three inlets (blue boxes); the red lines show the pipes the oil will flow through. Street centerlines are also shown, in gray.

Flow is modeled either overland or over a network of features such as water pipes, electrical lines, or a stream system. With overland flow, you model the accumulation of water (or other substance) from across the surface of the study area, converging at a location—such as when modeling how much water will flow through the outlet of a watershed after a storm. Overland flow is modeled using raster data. With flow over a network, you model the flow upstream or downstream from an origin location through the pipes (or other elements of the network). Network flow is modeled over a geometric network.

To define the parameters of your model and choose the appropriate method, you need to first define the problem you're addressing and the information you need from the model. You also need to consider issues such as the behavior of the phenomenon you're modeling, the external influences on the flow, and the time period over which the flow occurs.

DEFINE THE PROBLEM

In addition to the broad goals of your model (where are the stream channels? where will the oil travel through the stormwater system?), you'll want to identify the specific information you need from the model. This will help you define the input and the parameters for the model.

Most flow models address some basic information that is usually at the heart of what you need to know:

- where the flow goes
- what it travels over or what the affected area is
- how far it travels
- how long it takes to get there

In the case of hydrologic flow, you can measure the flow volume. For example, you could model the volume of water reaching the outlet of a watershed during a rainstorm of given duration and intensity. You might also want to measure how much of something the flow carries with it, such as the amount of sediment carried by stormwater runoff.

If you're modeling flow over a network, you can find out what's connected to or affected by the flow. For example, you could model where there is likely to be flooding on streets that empty into a storm drain system if the drains back up. (The core of the model is the storm drain system, but the streets are connected to the drains and could be affected by a major storm if the drainage system is overloaded.)

WHAT INFLUENCES THE FLOW

External factors may be present—beyond the nature of the phenomenon itself—that influence the flow. You should identify these and quantify them to the extent possible, so they can be incorporated in your model.

What the phenomenon travels over, or through, will influence where it flows, to some extent, but even more so, the rate at which it flows. Water will travel more quickly over exposed rock than through a wetland; stormwater will travel more quickly through pipes that are on steep slopes than it does through pipes that are on gentle slopes.

OVER WHAT TIME PERIOD

Flow inherently occurs over some length of elapsed time. Identifying or defining the time period will help you set the bounds of your model and identify the input data you need.

The flow event could be essentially instantaneous, such as when a power transformer goes out, or it could develop over a longer period (hours or days, such as with a rainstorm). In the latter case, your model will likely show the maximum flow over the course of the entire event.

Flow accumulation models are primarily used in hydrologic analysis to determine where water flows and accumulates on a terrain surface. This allows you to define stream channels and hydrologic basins and to measure the amount of rainfall runoff that will accumulate at a given downstream point (such as the outlet of a watershed) as well as how long it will take to travel there. Similarly, you can measure the accumulation of a substance carried by the flow, such as a pollutant or suspended soil particles resulting from erosion. Such analyses are useful in environmental planning, forestry, and wildlife biology, as well as in hydrology.

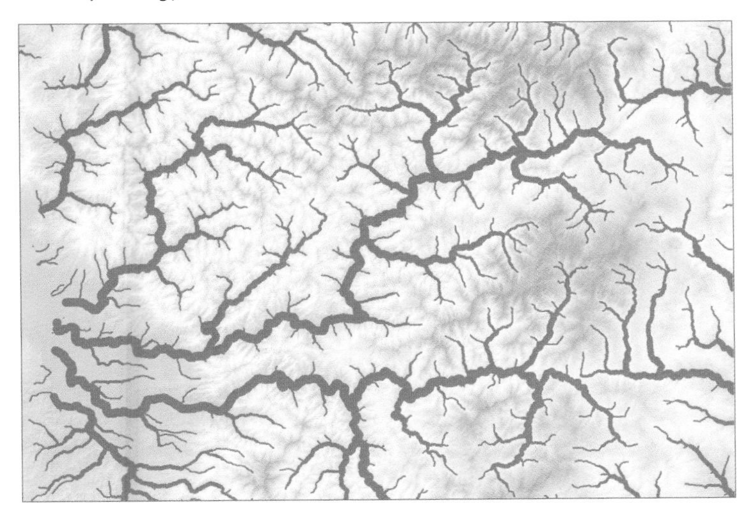

Streamflow layer created using a flow accumulation model and displayed with shaded relief.

The various flow accumulation models all depend on a flow direction surface, which models how water—or another substance—flows across the raster surface, from cell to cell, following the steepest downhill paths. (The flow direction surface is in turn created from an elevation surface.) Once you've created the flow direction surface, you can create a model to obtain the results required for your analysis: delineating drainage channels, delineating drainage basin boundaries, calculating flow volume, or calculating travel time through the drainage system.

The steps for modeling flow accumulation over a surface are:

1. Obtain the elevation surface
2. Create the flow direction surface
3. Create the required output
4. Evaluate the results

OBTAIN THE ELEVATION SURFACE

Modeling flow accumulation requires an elevation surface, such as a digital elevation model (DEM). For most applications, you'll want to obtain the finest resolution DEM available to ensure that the results are accurate. If your study area is at the county or regional level, a DEM with a cell size of 10 meters or 30 meters will suffice. If you're studying a specific watershed, you may want to use even finer-resolution data if it's available.

Modeling flow accumulation starts with an elevation surface.

Preparing the elevation surface

Most publicly available digital elevation models (that is, those available from government agencies such as the USGS), are fine for making maps, for deriving data such as slope and aspect, or for modeling general movement such as a path or corridor. However, to delineate drainage channels or drainage basin boundaries, your elevation surface will likely require additional processing to produce acceptable results. This is because anomalies in the surface can interrupt the downhill flow, which causes errors in the flow direction surface when it's created (as described in the next section). You'll want to identify and fix errors in the surface that can prevent stream channel segments from connecting or that cause basin boundaries to cross stream channels.

Quite often, these errors are in the form of sinks—a cell (or group of cells) surrounded by cells of higher elevation.

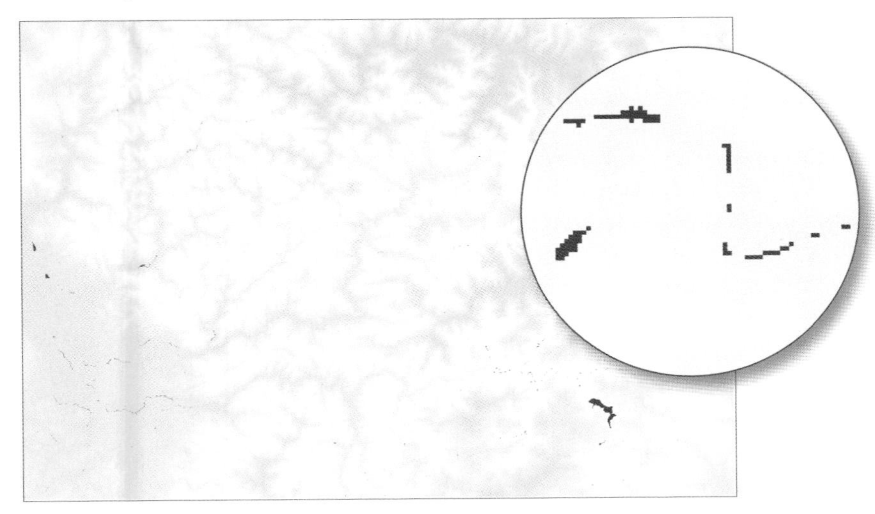

Sink cells (dark brown) with the elevation surface.

If there is a sink in a stream channel, for example, the flow will travel into the cell (or cells) but will not travel out, thus creating a break in the stream.

Sinks can be identified and "filled" using tools in the GIS. The cell value is changed to be equal to the value of the surrounding cell with the lowest elevation value, so water no longer accumulates in the cell, but rather flows across it as it's supposed to.

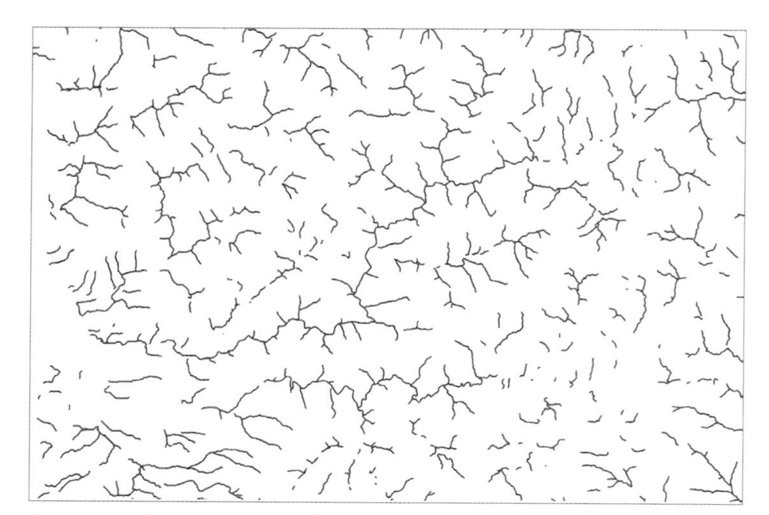

Stream channels created from an elevation surface containing sinks. Stream segments are disconnected where the flow is into a sink.

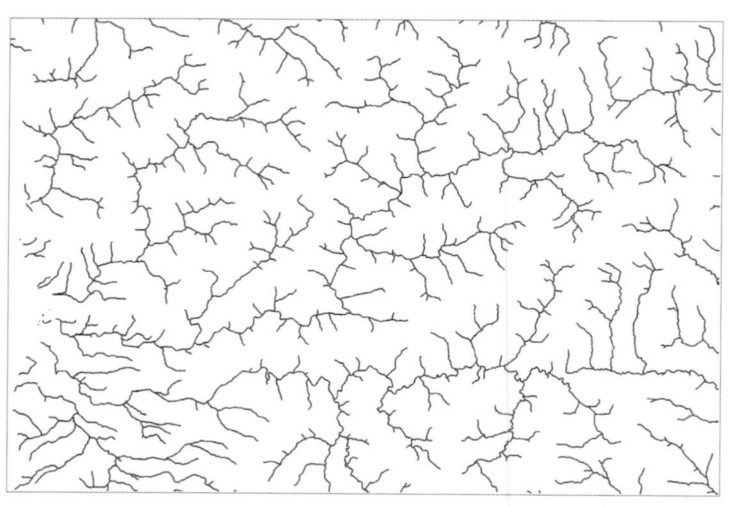

Stream channels created from an elevation surface that has had the sinks filled. Stream segments connect to form a complete network.

Sinks are sometimes the result of sampling errors when the elevation data used to create the DEM was collected or other errors introduced when the DEM was processed. Researchers have found that, generally, sinks with a drop of more than thirty meters from adjacent cells do not occur naturally, except in areas of karst topography or glacial landscapes. If not in a karst or glacial area, these deep sinks can be considered errors (although the only way to be sure is to examine all the sinks in the elevation layer and make a determination). When filling sinks, you can specify that only those with a drop greater than a certain value (such as 30 meters) be filled and that all others remain.

Sinks sometimes also occur in flat or low elevation areas, such as along wide stream channels, in marshes, or along a coastline. If your study area includes any of these, you'll want to make sure the elevation raster uses floating point values rather than integer values (but only if the elevation values are accurate to one or more decimal places). The difference in cell elevations may be less than the round feet or meter values contained in an integer raster. Using a floating point raster will minimize the occurrence of sinks in flat areas since these rasters store values to several decimal places (as many as needed to distinguish elevations between cells that are close in value).

A peak in the surface is a cell surrounded by cells of lower elevation value. These can also alter the flow—by moving a stream channel over by a cell, for example—but are less of an issue than sinks (which pull the flow towards them). Most peaks, of course, are natural features of the terrain. But a single cell which has a much higher elevation value than surrounding cells might be an error in the data. Peaks—as well as sinks, to some extent—can be mitigated by smoothing the surface using filtering tools in the GIS (although some detail may be lost in the process).

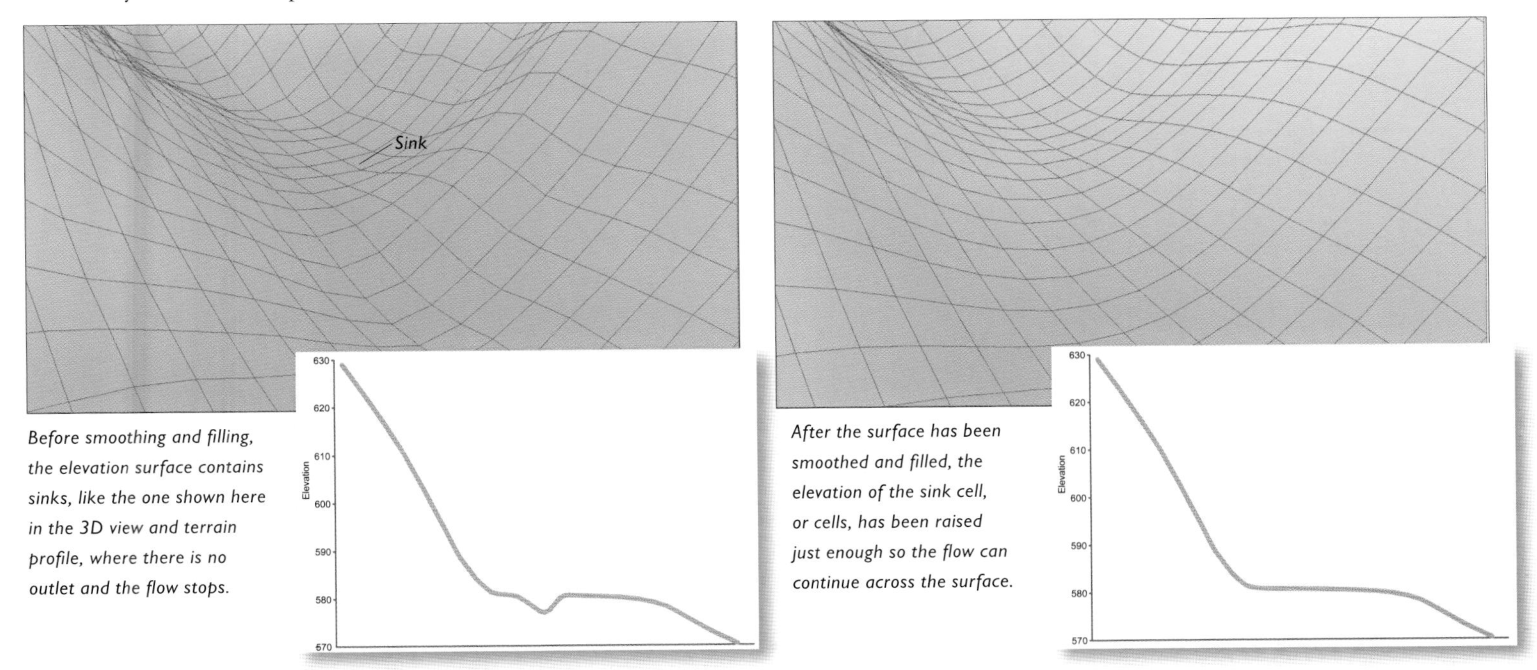

Before smoothing and filling, the elevation surface contains sinks, like the one shown here in the 3D view and terrain profile, where there is no outlet and the flow stops.

After the surface has been smoothed and filled, the elevation of the sink cell, or cells, has been raised just enough so the flow can continue across the surface.

Dealing with waterbodies and other flat areas

If your study area includes ocean or another flat area (such as a large lake) at the edge of your study area, the GIS will attempt to calculate flow accumulation across the flat surface (from the shoreline to the edge of the raster) resulting in, essentially, drainage channels across the ocean surface. (Creating drainage channels from a flow accumulation surface is discussed later in this section.)

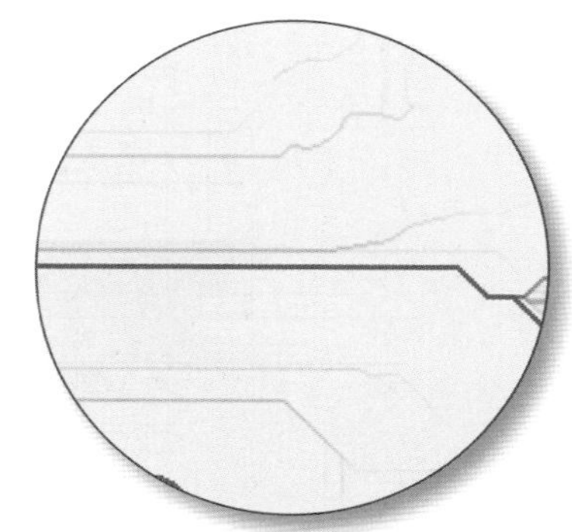

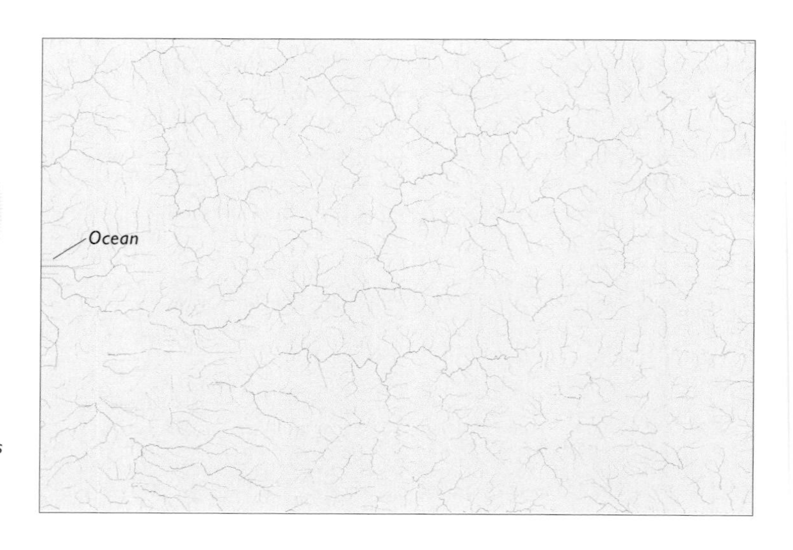

The flow accumulation surface includes an area of ocean at the left edge of the study area. The close-up shows that when the drainage channels (blue lines) reach the shore, they continue across the ocean surface as straight lines.

To avoid this, you'll want to assign a value of NoData on the elevation layer to cells in these flat areas before creating the flow direction surface. For ocean, assign NoData to cells having an elevation of 0. For other flat areas (such as large lakes and dry lake beds) that have an elevation greater than 0, create a raster layer from a polygon layer of the flat features, assign NoData to the raster cells comprising the flat areas, and use this layer as a mask when creating the flow direction surface.

For small lakes and ponds within the study area, flow direction will be calculated across the surface in one direction—from the inlet to the outlet—so no preprocessing is needed.

CREATE THE FLOW DIRECTION SURFACE

The flow direction surface underlies the flow accumulation models. It is created from the corrected elevation surface. The flow direction surface identifies, for each cell—the From cell—the adjacent cell that water would flow to (that is, the adjacent cell that constitutes the steepest downslope)—the To cell. Each cell in the flow direction raster is assigned the direction of the To cell—so if the To cell is directly to the east, the From cell is assigned a value of "east." There are eight possible directions, corresponding to the eight cells surrounding each cell. In the GIS, the directions are actually stored as numeric codes to allow for calculations—east is 1, southeast is 2, south is 4, southwest is 8, and so on, up to a value of 128 for northeast).

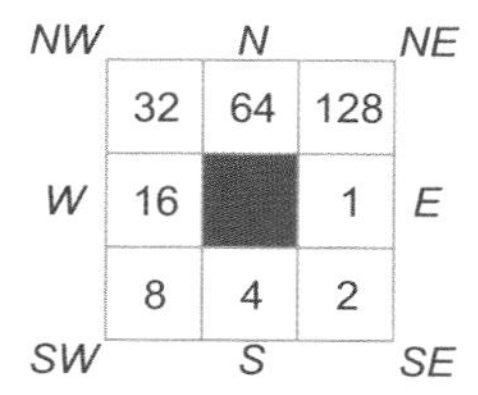

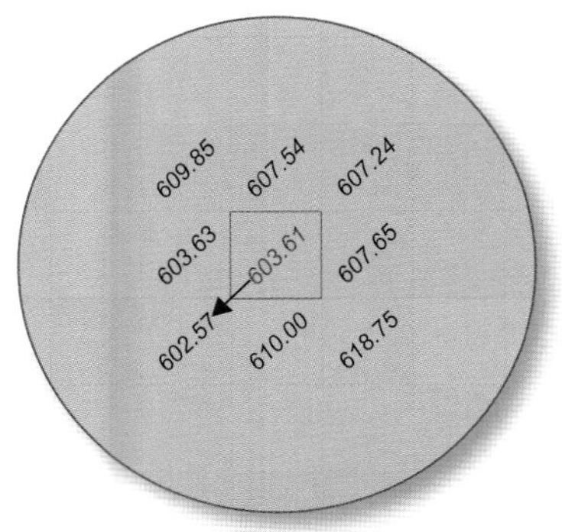

For each cell, the adjacent cell having the steepest downslope is identified—that's the cell where water will flow. The slope is calculated as rise (the difference in elevation) divided by run (the distance between cell centers). In this example, only one adjacent cell is downslope from the From cell (gray square).

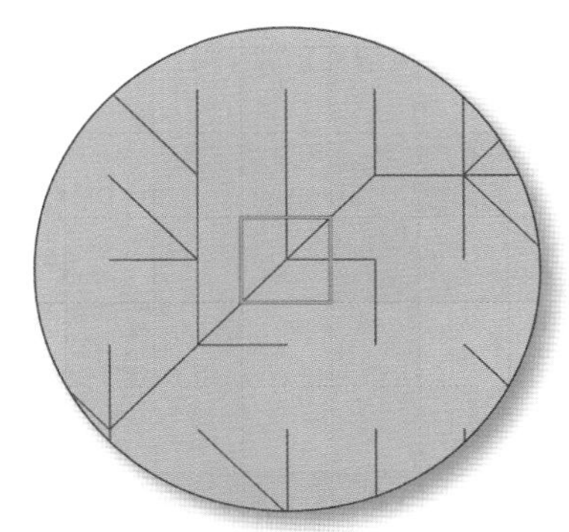

Lines show the direction of flow from each cell to the next. The dendritic pattern of stream flow is typical.

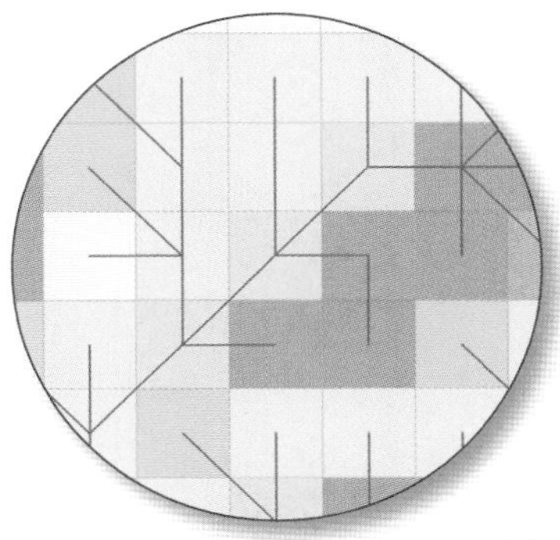

In the flow direction raster, each cell is assigned the direction of the cell it flows to. The cell in the middle flows to the cell to its southwest, so it is assigned a value of "southwest," or 8, color-coded as medium blue in the diagram (and on the map below).

Taken together, the cell values map the flow over the surface, allowing the GIS to calculate flow accumulation and track flow upstream to identify ridgelines (which become drainage area boundaries).

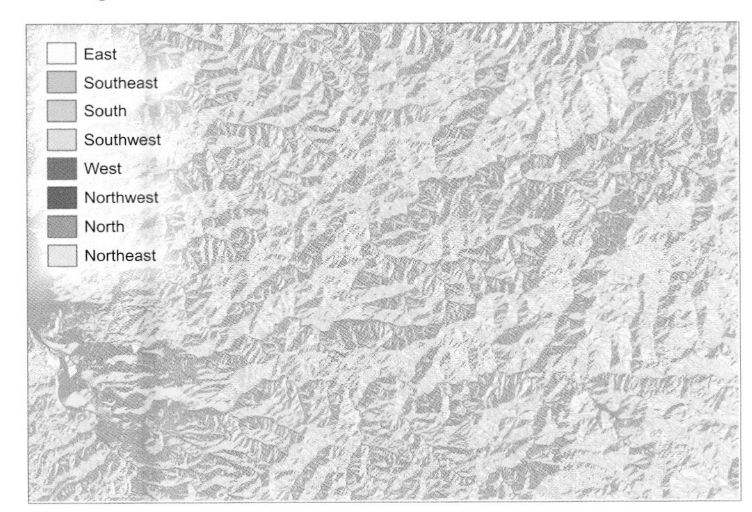

A flow direction surface models the direction of flow from each cell to the surrounding cell with the steepest downslope.

CREATE THE REQUIRED OUTPUT

The flow direction surface is used as input to models that delineate drainage networks, channels, and drainage area boundaries, and that calculate flow volume and travel time through a drainage area. Your analysis may include one or more of these models.

Delineating drainage channels

Using the flow direction surface you can measure the accumulated flow—that is, the number of cells upstream of any point—at any location on the surface. You do this by creating a flow accumulation layer using the flow direction surface as input. The value of each cell in the flow accumulation layer is the number of cells that flow into—or are upstream of—that cell. You can then calculate the total area upstream of the location (by multiplying the number of cells by the area of a cell).

Flow accumulation layer. By assigning each cell a value equal to the number of cells flowing into it (calculated from the flow direction surface), the GIS identifies cells where flow accumulates. As shown in the close-up, values increase for cells downstream.

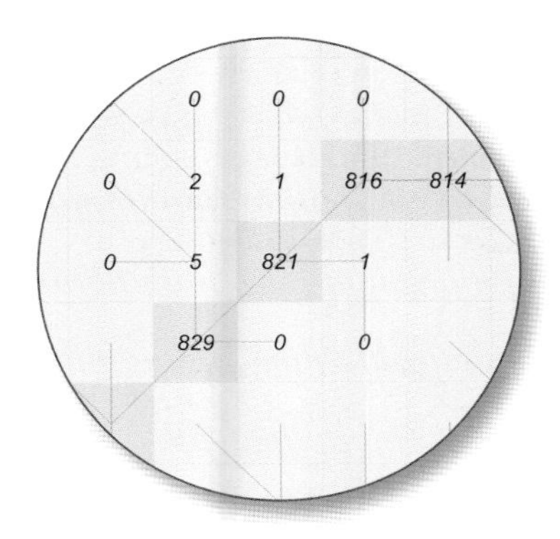

This close-up shows how flow accumulation is calculated. Once the flow from each cell to the next is established, via the flow direction surface, the GIS can calculate how many cells flow into each cell (that is, are upstream of each cell). The lines show the flow from cell to cell. Any cell that has no cells flowing into it is assigned a value of 0. A cell that has other cells flowing into it is assigned a value equal to the sum of the values of adjacent cells flowing into it plus the total number of those adjacent cells. So, for example, the cell with a value of 5 in the diagram has three cells flowing into it: one with a value of 2 (it has two cells flowing into it), and two with a value of 0 (they have no cells flowing into them). The sum of these values is 2. You then add the three cells themselves, for a total value of 5 cells flowing into the cell. Similarly, the cell with a value of 829 has three cells flowing into it: one with a value of 5, one with a value of 821, and one with a value of 0. The sum is 826. You add the three cells flowing into the cell for a total value of 829. You can see that for cells that constitute the drainage network, the flow accumulation value increases rapidly, and is much higher than the values of surrounding cells.

By specifying a threshold or cutoff value for flow accumulation, you can define which cells constitute a channel. All cells greater than this value will be part of the drainage system. For example, if you know that in your study area a stream will form when the drainage upstream is at least 100 hectares in size, you'd calculate the number of cells equal to 100 hectares and use that value as the threshold. In this example, the cell size is 32.823 square feet, so each cell is about 1,077 square feet in area. One hundred hectares equals 10,763,910 square feet; dividing by the area of a cell yields a threshold of around 10,000 cells. Once you've decided on the threshold value you want to use, you can display the layer using two classes (values above and below the threshold value).

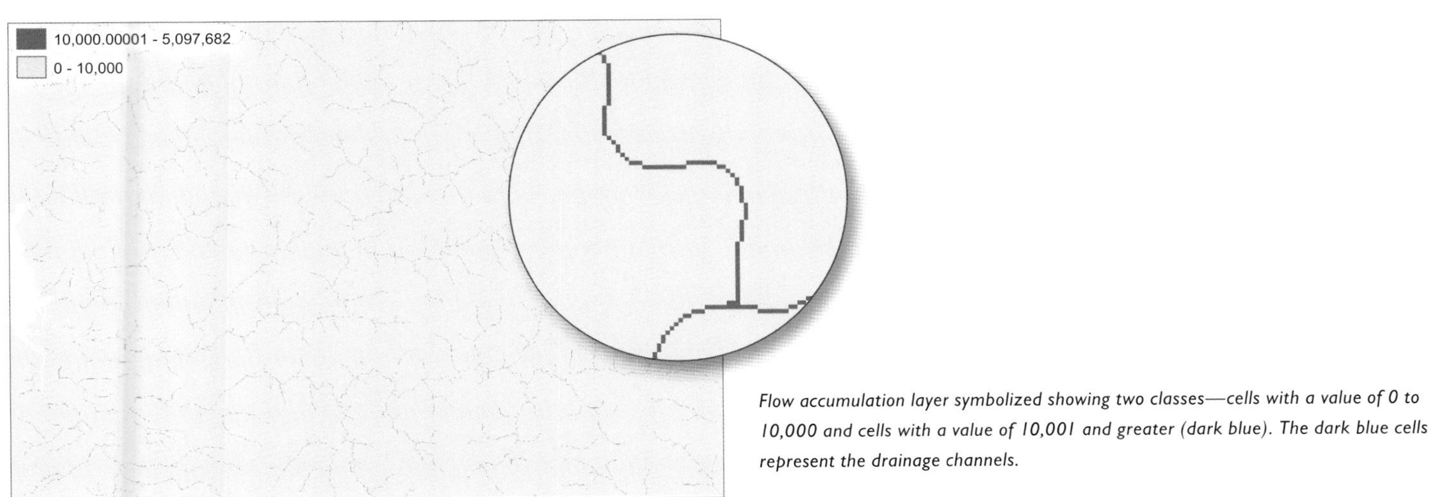

Flow accumulation layer symbolized showing two classes—cells with a value of 0 to 10,000 and cells with a value of 10,001 and greater (dark blue). The dark blue cells represent the drainage channels.

Because the channel is only one cell wide, it may be difficult to see on the map, depending on the size of your study area and the cell resolution. It's often easier to see if you convert the raster streams to vector line features. To do this, first reclassify the raster, assigning a value of NoData to all cells less than the threshold value and another value (such as 1) to cells equal to or greater than the threshold value. Then convert this reclassified raster to vector line features.

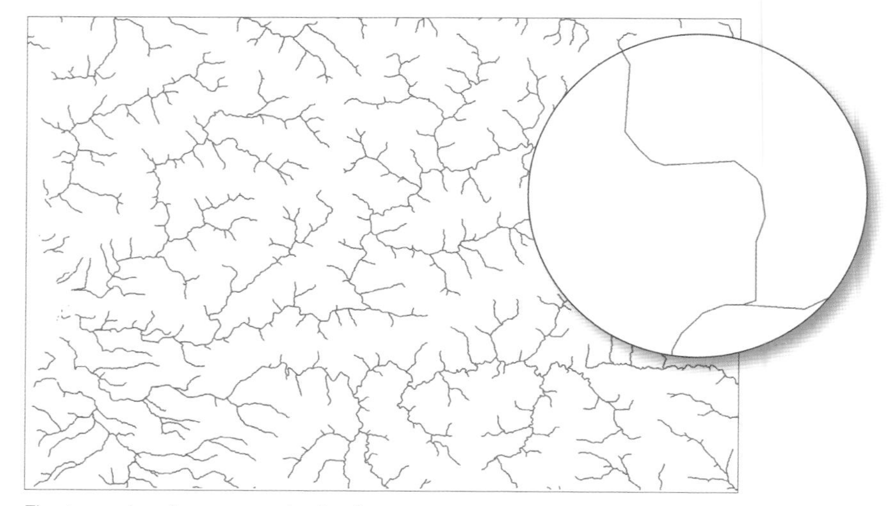

The streams have been converted to line features.

By increasing or decreasing the threshold, you can define the resolution of the drainage system. For example, using a larger threshold value—of, say, 1,000 hectares, or 100,000 cells—will identify only major streams. Using a smaller threshold—say, 10 hectares—would include drainages that contain flowing water only during a rainstorm.

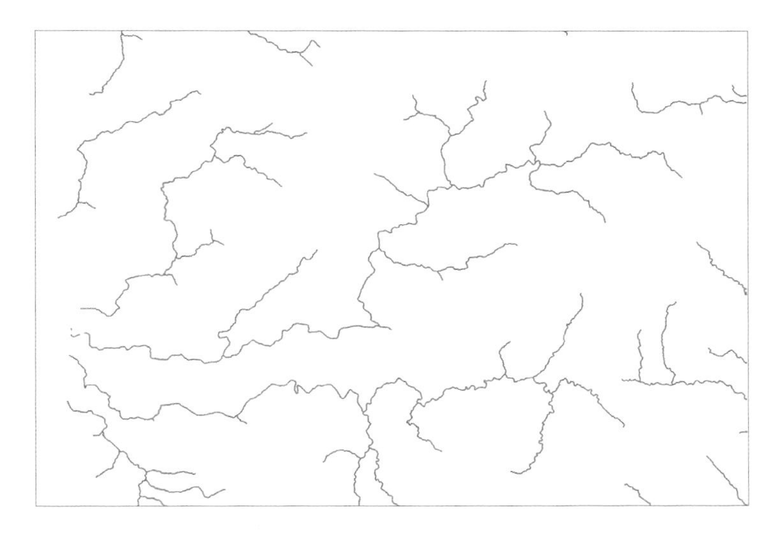

Stream system created using a threshold value of 100,000 cells (about 1,000 hectares in this dataset)—only major streams are included in the system.

You can display the vector streams using line widths or colors that reflect the number of cells flowing into each cell (the accumulation), to give a sense of the relative size of the streams. To do this, first set all cells less than the threshold to NoData. Then reclassify the flow accumulation raster into the number of stream classes you ultimately want to display. Finally, convert the reclassified raster to vector line features.

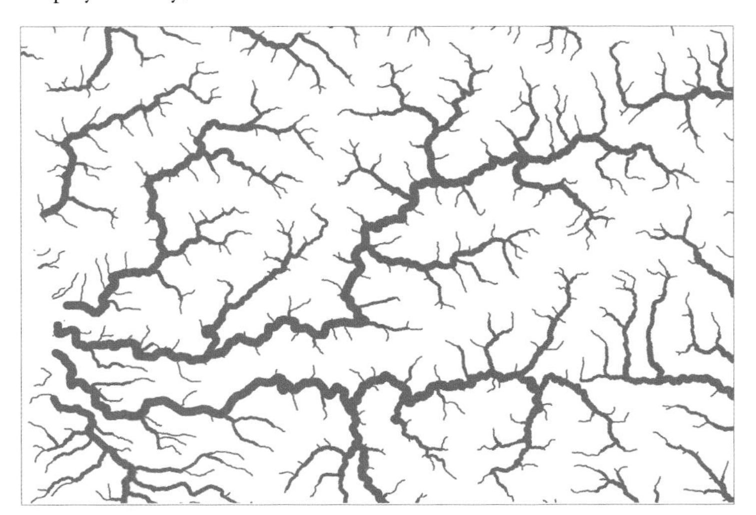

Stream system symbolized by assigning thicker line width to streams with higher flow accumulation values.

Stream system symbolized by assigning darker colors to streams with higher flow accumulation values.

The process for delineating drainage channels.

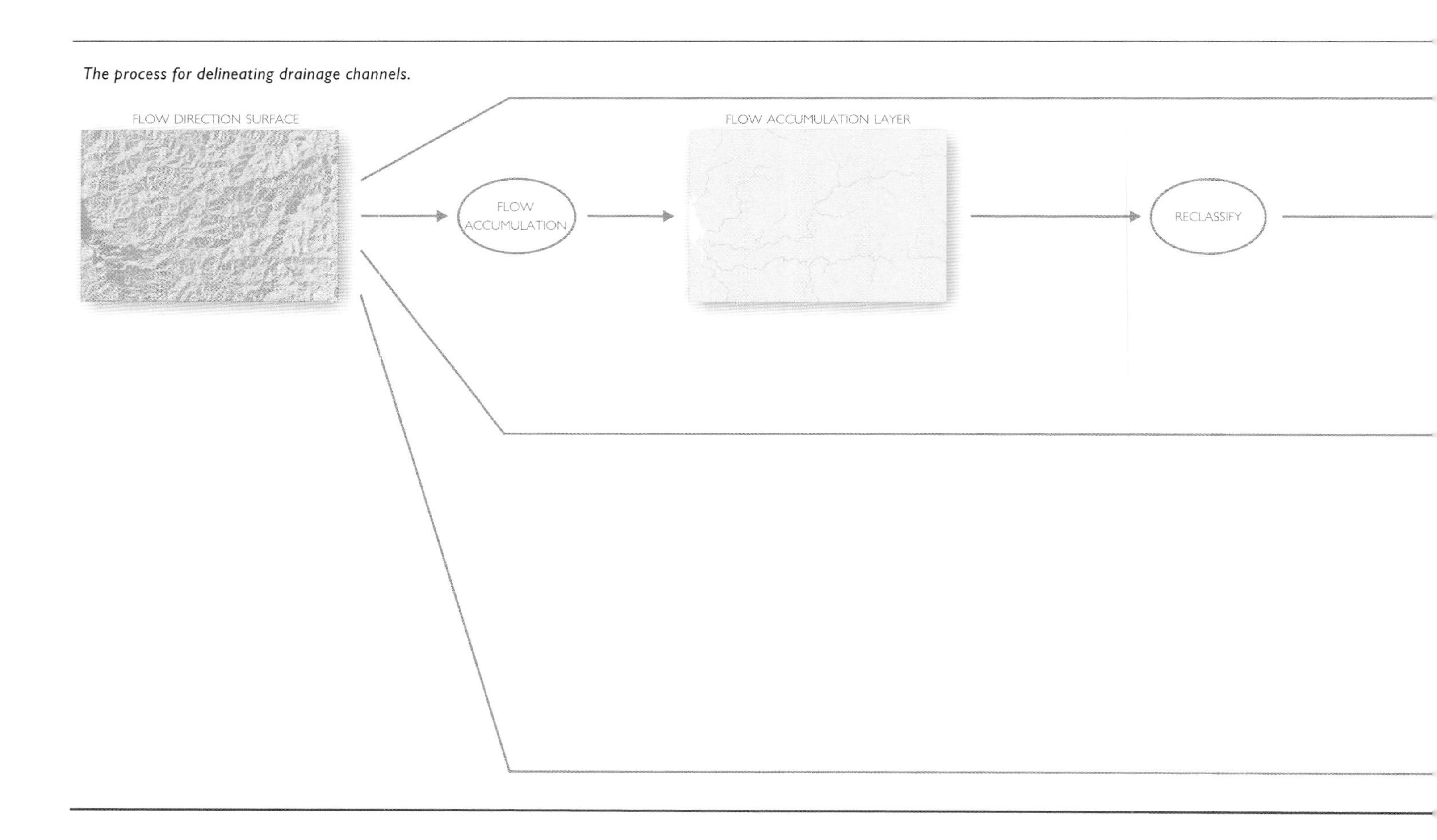

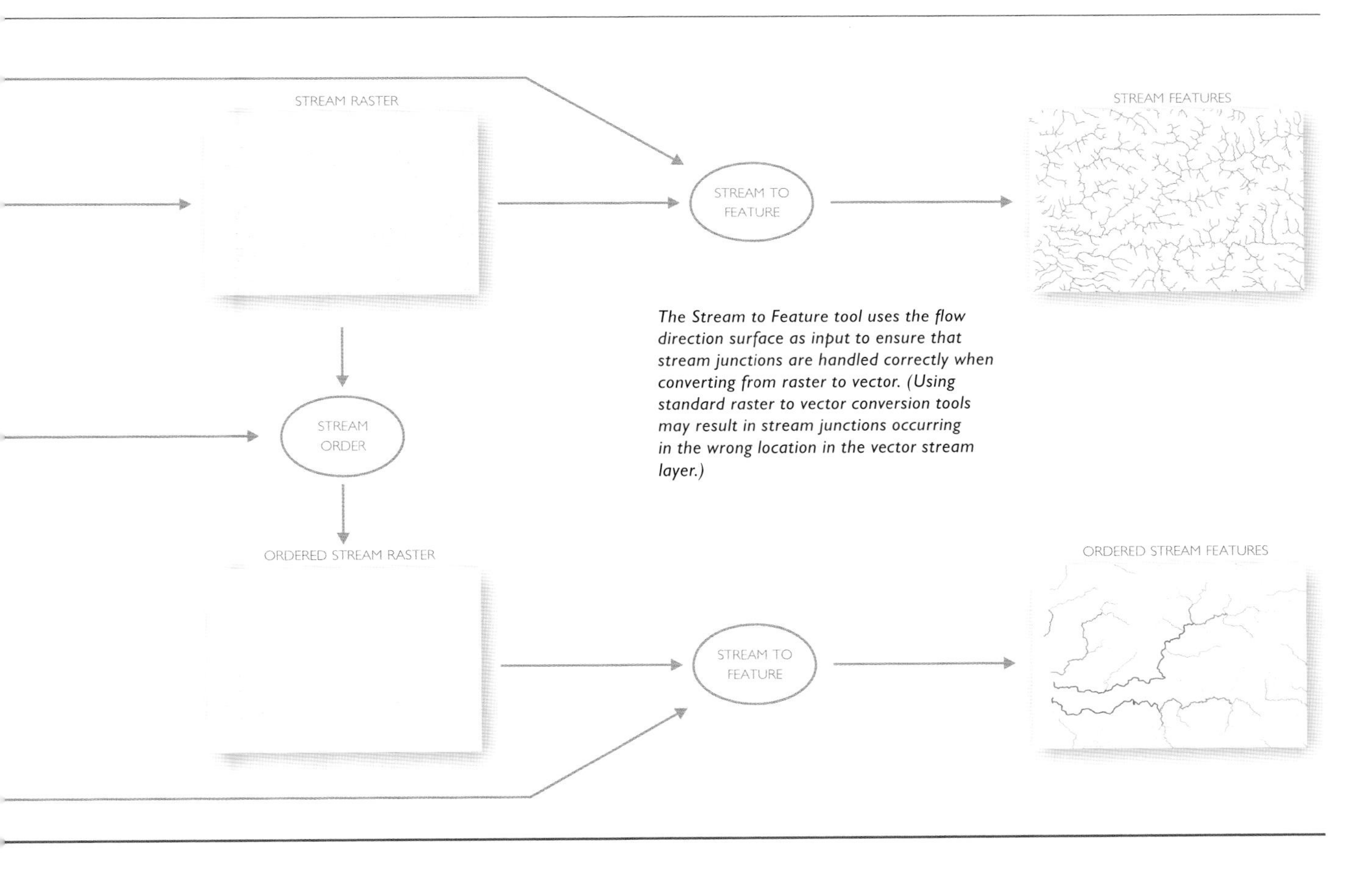

The Stream to Feature tool uses the flow direction surface as input to ensure that stream junctions are handled correctly when converting from raster to vector. (Using standard raster to vector conversion tools may result in stream junctions occurring in the wrong location in the vector stream layer.)

Alternatively, you can assign a stream order value (indicating essentially how many streams flow into each stream) and use these values to symbolize the stream network. You assign stream order to the reclassified stream raster and then convert the ordered streams raster to line features. A common way of assigning stream order is the method developed by geographer Alan Strahler. In the Strahler method, a first-order stream has no other streams flowing into it. A second-order stream is formed when two first-order streams join, a third-order stream is formed when two second-order streams join, and so on.

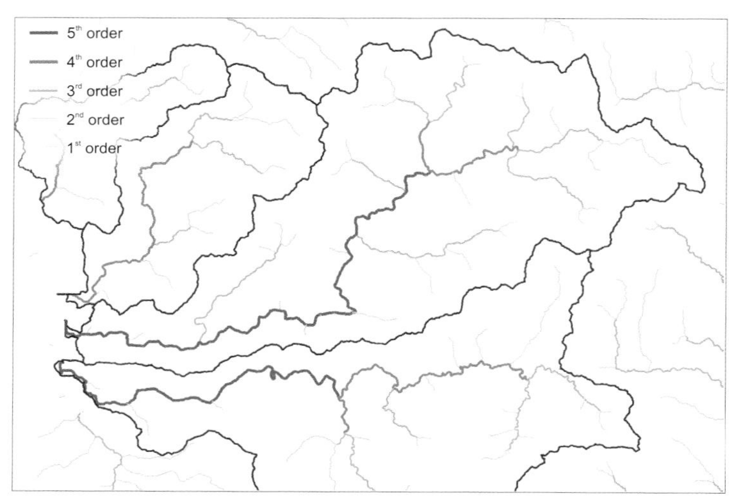

Streams classified by Strahler stream order and symbolized using a combination of line width and color, shown with boundaries for major watersheds in the study area.

Once you've created the output layer, you can display it with other layers, such as shaded relief, to add context to the information in the map.

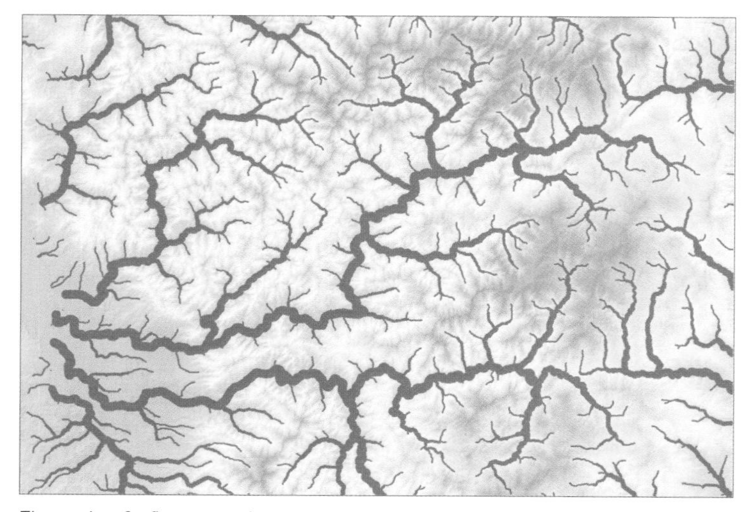

The results of a flow accumulation model were used to create this map showing streams symbolized by amount of flow, along with shaded relief.

You may want to display the results using a perspective view on top of the elevation surface to better visualize the terrain.

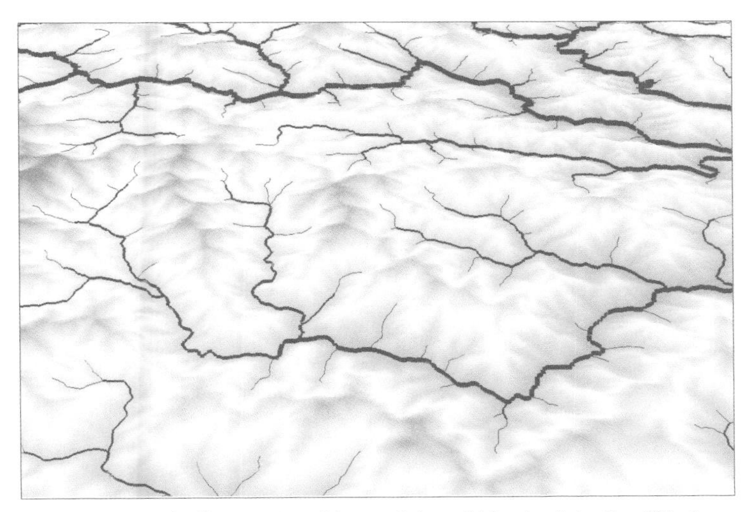

A perspective—or bird's eye—view of the terrain is useful for visualizing flow. This view shows streams symbolized by flow accumulation.

Delineating a drainage area

Once you've delineated the drainage system, you'll likely want to focus on particular drainage areas for additional analysis such as calculating flow volume or the travel time of a pollutant through the system. For example, you may be interested in analyzing flow in the drainages in one portion of your study area.

You use the flow direction surface along with the streams and flow accumulation layers to delineate drainage area boundaries for your study area. The GIS delineates the drainage area by backtracking the flow direction surface from a starting point, known as the "pour point." It identifies all cells upstream of the pour point—to the ridge—thus defining the area draining to the location.

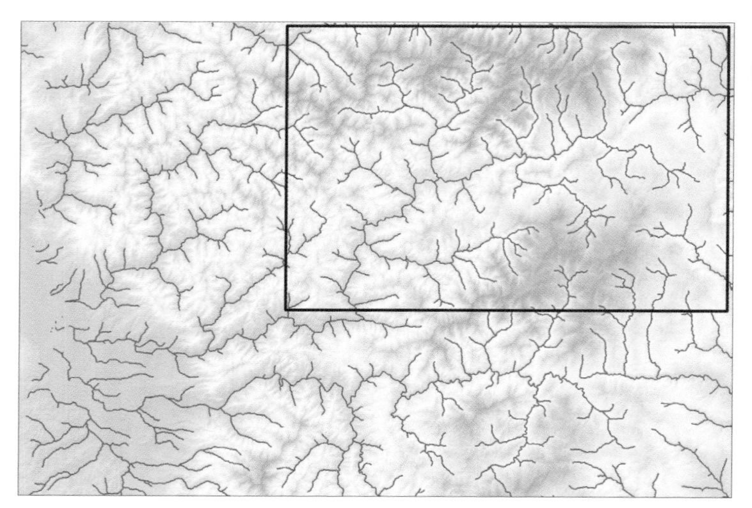

Many flow accumulation analyses focus on a subset of streams such as the one in the northeast corner of the study area (indicated by the box).

You can delineate the area above any location on a channel by using a segment of the channel as the pour point. Or you can delineate the area above an existing point location—such as a stream monitoring gauge or debris check dam—by using the point feature as the pour point.

Delineating the area above a channel segment

To delineate the drainage area above a particular channel segment, assign a unique ID to each segment (or link in the drainage system), identify the segment you want to delineate the drainage area for, and create a new raster layer containing only that segment or segments. Finally, delineate the drainage area upstream of the segment, using the segment as the pour point.

Using the channel segment is a reliable way to make sure that the pour point is at the outlet of the drainage area you want to delineate (since it is on the stream channel defined by the flow accumulation layer), and hence all the area upstream will be included in the drainage area.

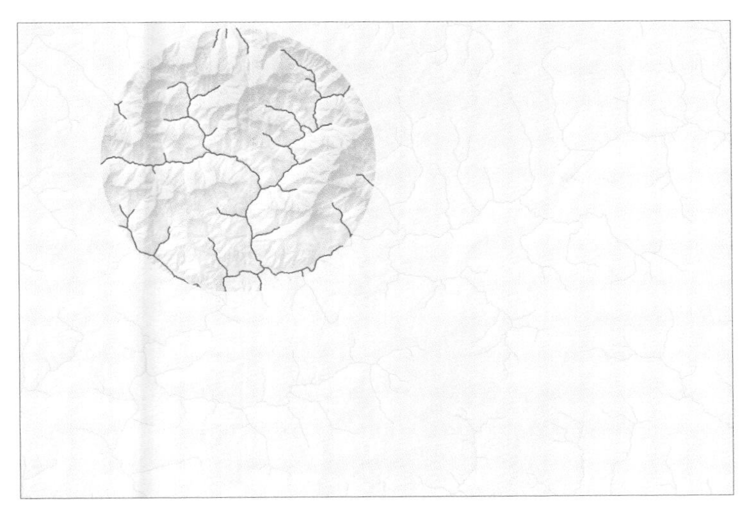

The circle highlights the specific stream system for which the drainage area will be delineated.

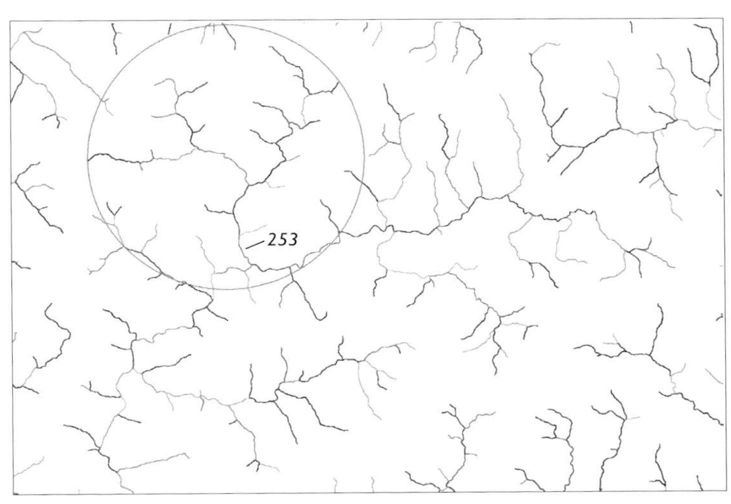

Each stream segment is assigned a unique ID. Streams are shown here color-coded by the segment ID. The most downstream segment of the stream system has an ID of 253.

You create a new layer that contains only the required segment—all other cells are assigned a value of NoData.

Using the input cells from the segment raster along with the flow direction surface, the GIS delineates the drainage area for the stream system.

You can delineate several drainage areas at one time by selecting more than one segment and using those segments to create the pour point layer.

In this example, two adjacent stream segments (left) are used to delineate adjacent watersheds (right).

The process for delineating a drainage area from a channel segment.

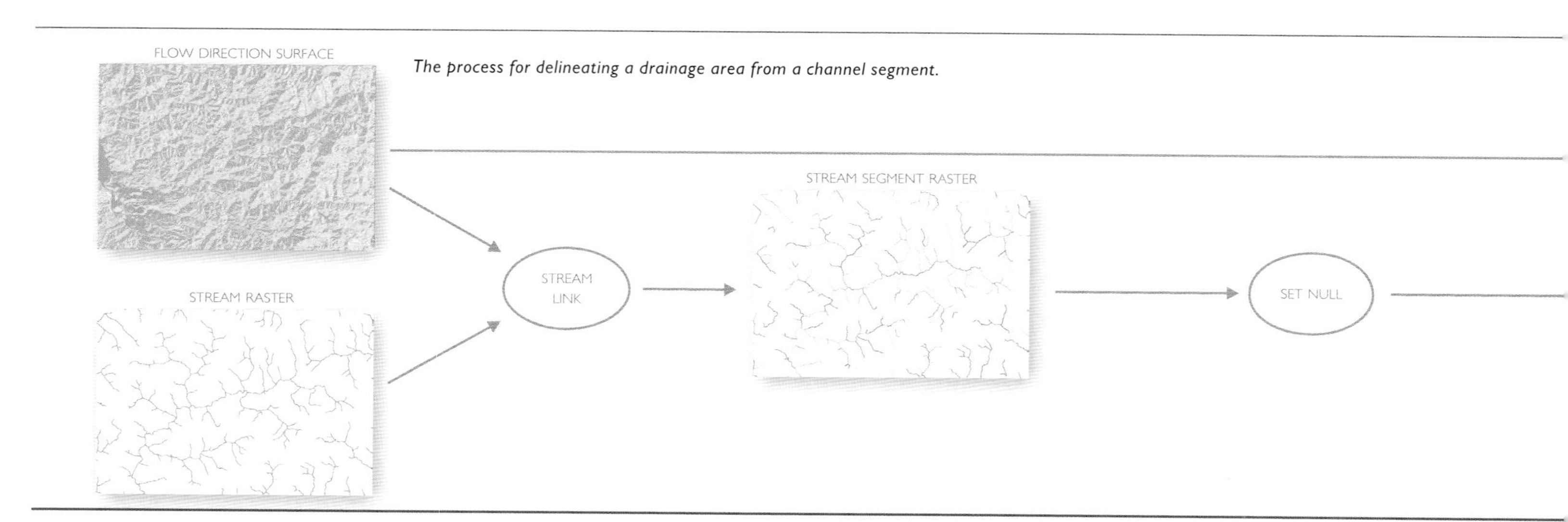

If one of the segments is downstream of another, the drainage area boundary will stop at the boundary of the upstream area. The result will be the drainage areas for the lower and upper reaches of the stream.

Selected stream segments shown with the full stream system. The segment in red is downstream of the two adjacent segments.

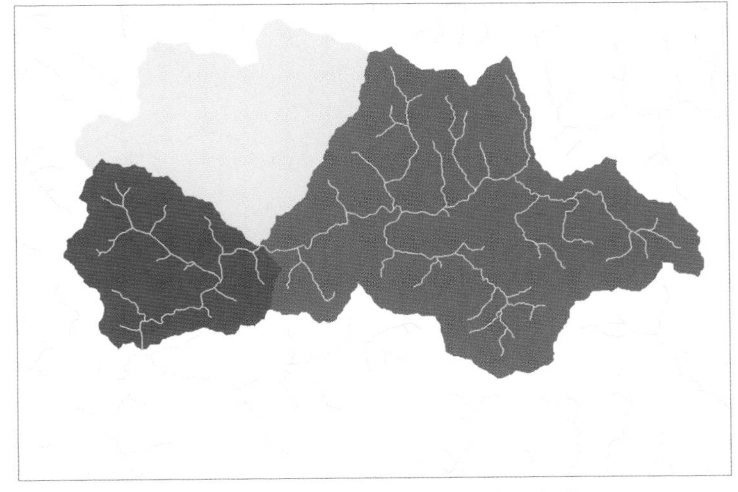

When the drainage areas are delineated, the area created from the downstream segment abuts the upstream drainage areas. (If the drainage area for the downstream segment alone were delineated, it would encompass all three drainage areas.)

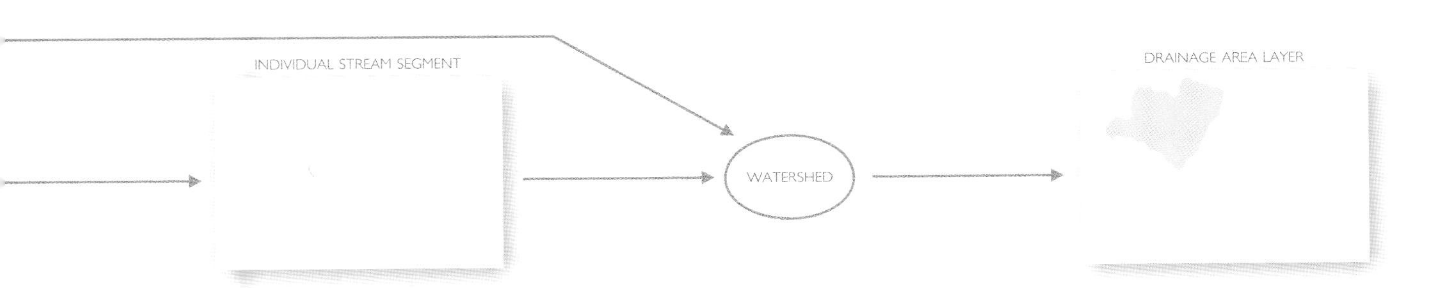

Delineating the area above a point location

To delineate the drainage area above a stream gauge, check dam, or other existing point location, you first need to ensure that the point is actually located on a drainage channel—otherwise, the drainage area will not be delineated properly.

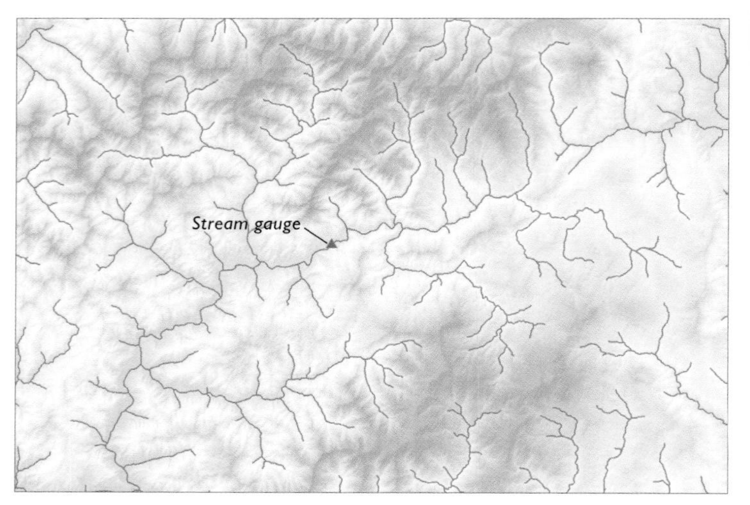

The specified pour point (triangle)—representing a gauging station. At this scale, the gauge appears to be located on the stream channel.

In the field, the gauge (or dam) is actually located at or on the channel. In the GIS, however, that may not be the case. If the gauge was digitized from a paper map, it may not be stored in the GIS at its exact x,y coordinates. If the x,y coordinates were entered into the GIS to create the point location, they may not have been correct. (For example, if they were obtained using a low resolution GPS, the coordinates may be inaccurate.) Also, the elevation surface from which the drainage channels are derived is created from sampling locations in the field so is not an exact replication of the ground surface. (This is especially true if the elevation surface has a large cell size.) Any or all of these may lead to the pour point feature not being located exactly on a drainage channel.

To make sure the pour point is located on a channel, you snap it to the flow accumulation layer from which the channels were derived. When you snap the pour point feature to the flow accumulation layer, you specify a search distance. A new raster layer is created containing a single cell that is coincident to the cell with the highest flow accumulation value within the search distance.

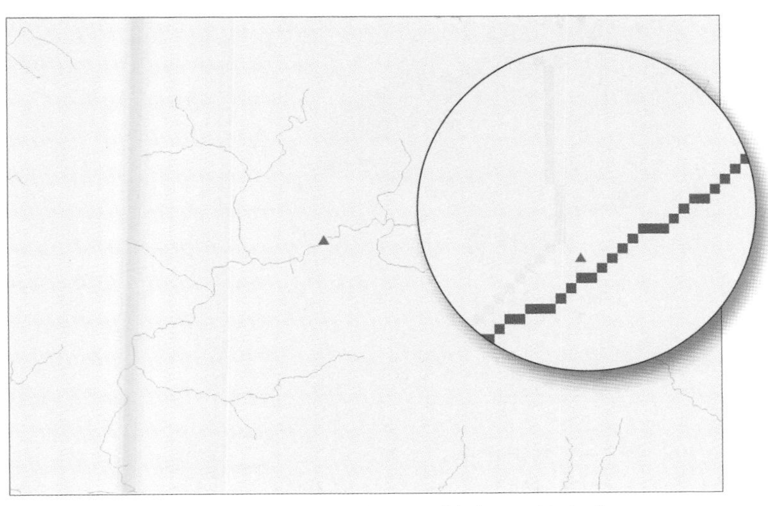

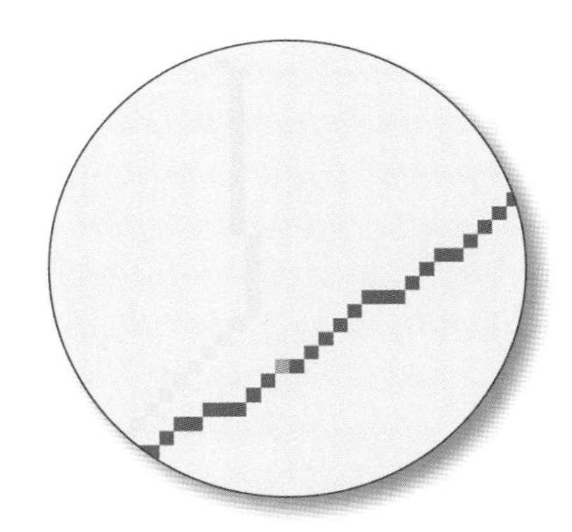

The pour point feature (a stream gauge, in this example) shown with the flow accumulation layer (darker cells indicate higher accumulation). While on the map the gauge appears to be located on the flow accumulation drainage channel, the close-up view shows that it is, in fact, offset from the channel. After snapping (close-up view on right), the pour point is represented by a cell—shown in orange—that is coincident with a cell on the flow accumulation drainage channel.

The GIS uses the resulting raster containing the pour point cell to delineate the drainage area.

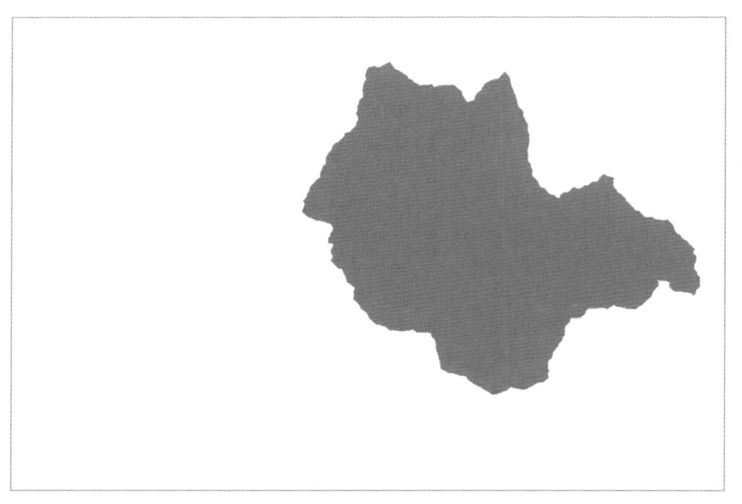

The GIS creates a layer identifying the cells that drain to the pour point cell, delineating the drainage area. All other cells are assigned a value of NoData.

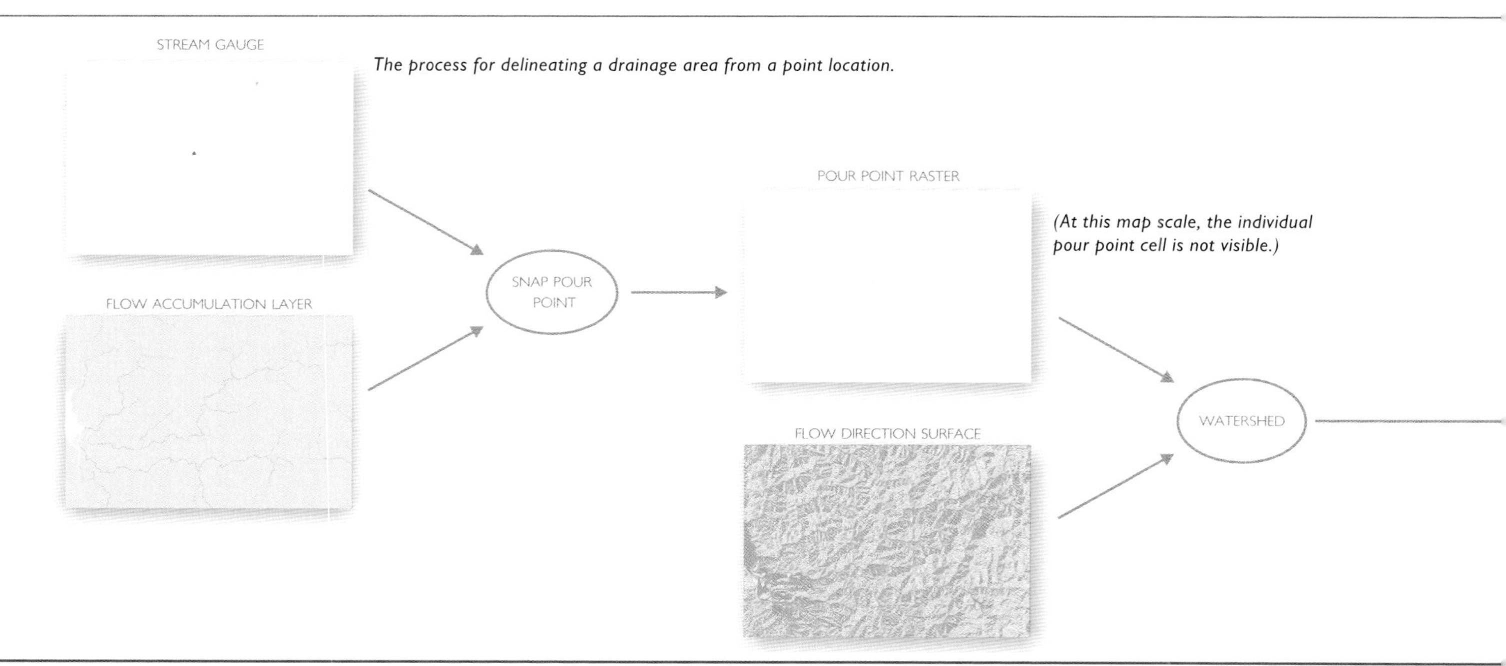

The process for delineating a drainage area from a point location.

If the original pour point feature happened to be located exactly on top of a cell in a drainage channel, you'd get the same result without snapping. But if the feature is offset by even a single cell, the drainage area will not be delineated correctly, since it will consist of only those cells draining to the cell on the flow accumulation layer corresponding to the pour point. Quite often the result will be a very small drainage area.

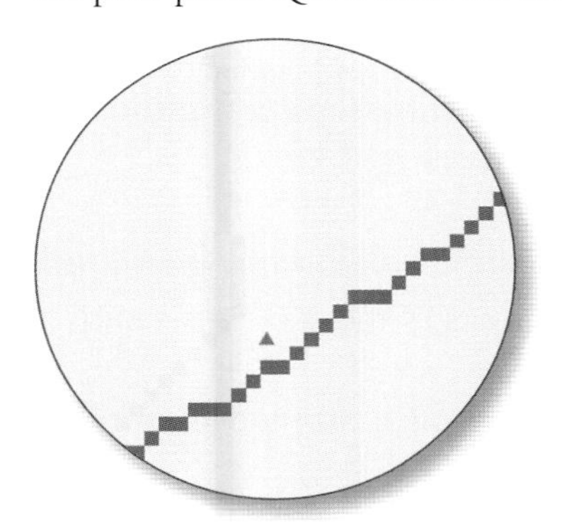

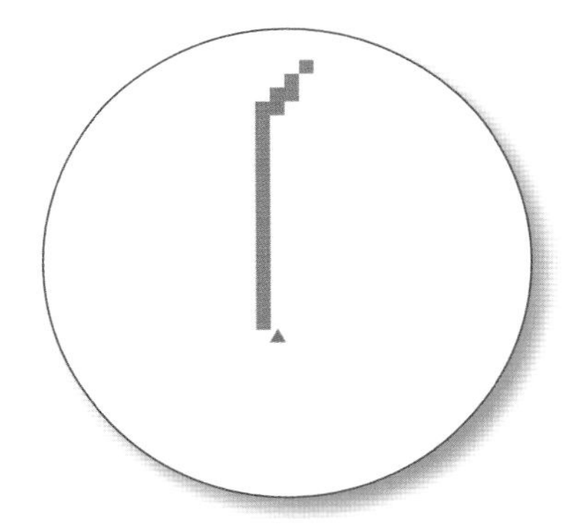

Here again is the close-up view of the pour point feature (stream gauge) on the left, showing the offset from the flow accumulation drainage channel. If this location is used as the pour point—without snapping—the delineated drainage area consists of only the few cells upstream of the location (close-up view on the right), rather than the entire watershed.

As with stream segments, you can create several drainage areas at one time by using an input layer that contains more than one point feature.

Summarizing information about the drainage area

Once you've delineated the drainage area, you can calculate its areal extent by multiplying the number of raster cells in the drainage area by the area of a cell (the resolution of the cell, squared). The GIS stores the number of cells for each area in a field (called Count in ArcGIS). For example, if the cell size is 32.823 feet, then the area of a cell is 1077.349 square feet, and you'd calculate the areal extent for each drainage area as

AREA = COUNT * 1077.349

You can then calculate the area in units such as hectares or square kilometers (or simply include a conversion factor when you calculate the area).

AREA IN ACRES = AREA / 43560

or

AREA = (COUNT * 1077.349) / 43560

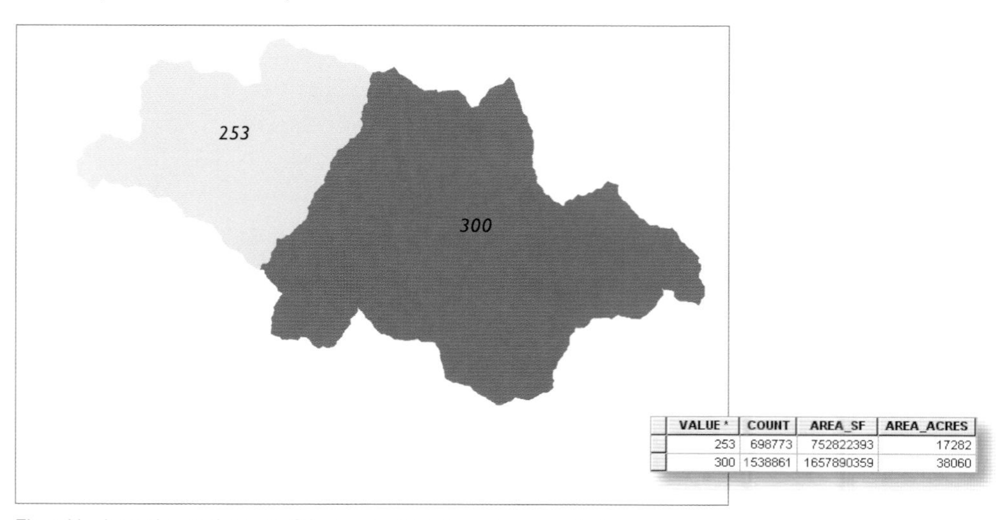

VALUE *	COUNT	AREA_SF	AREA_ACRES
253	698773	752822393	17282
300	1538861	1657890359	38060

The table shows the areal extent of the two drainage areas, in square feet and in acres. The Value field contains the unique ID of each drainage area.

ArcGIS includes a tool—Zonal Geometry—that will automatically calculate the area of each drainage area (in map units), along with other statistics such as the length of the perimeter and the location of the centroid. It also measures the shape and orientation of the area. You can convert the area in map units (usually square feet or meters) to units such as acres, hectares, or square kilometers—as described above—so there are fewer digits, making the table easier to read and interpret.

VALUE	AREA	PERIMETER	THICKNESS	XCENTROID	YCENTROID	MAJORAXIS	MINORAXIS	ORIENTATION
253	752813890	176652.36	11171.61	7421030.5	734634.13	18218.543	13152.979	16.365065
300	1657871600	276039.81	16848.906	7453329.5	719893.75	26814.393	19680.361	167.34042

The Zonal Geometry tool in ArcGIS calculates the areal extent of each drainage area (in map units) along with other measures.

In addition to calculating the characteristics of a drainage area, you can summarize characteristics of the landscape within the drainage area. For example, you may want to know the mean, minimum, and maximum slope within the drainage area. (This information might help you, for example, determine which recently burned watersheds are at higher risk of debris flow during a storm.)

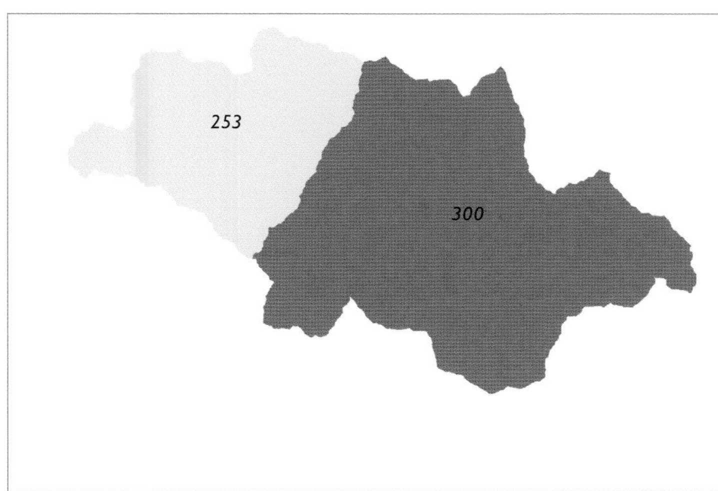

Drainage areas.

Slope.

In ArcGIS, you can use the Zonal Statistics tool to summarize data with continuous values—such as slope, elevation, or precipitation—and store the results as a table. Again, each drainage area is a zone, and the values of another raster (such as slope) are summarized for each zone.

VALUE	COUNT	AREA	MIN	MAX	MEAN	STD
253	698773	752813890	0	59	30	9.3
300	1538861	1657871600	0	62	19	11.3

The Zonal Statistics tool in ArcGIS calculates summary statistics—including minimum, maximum, and mean values—for each drainage area (zone) that overlays another raster layer (slope, in this example). While the minimum and maximum slope for both drainage areas is about the same, the mean slope for the drainage area to the northwest (with an ID of 253) is much larger.

You can also summarize categorical data for each drainage area. For example, you can calculate the amount of each type of land cover in each drainage area. (This might be useful for habitat conservation, using watersheds as planning units.)

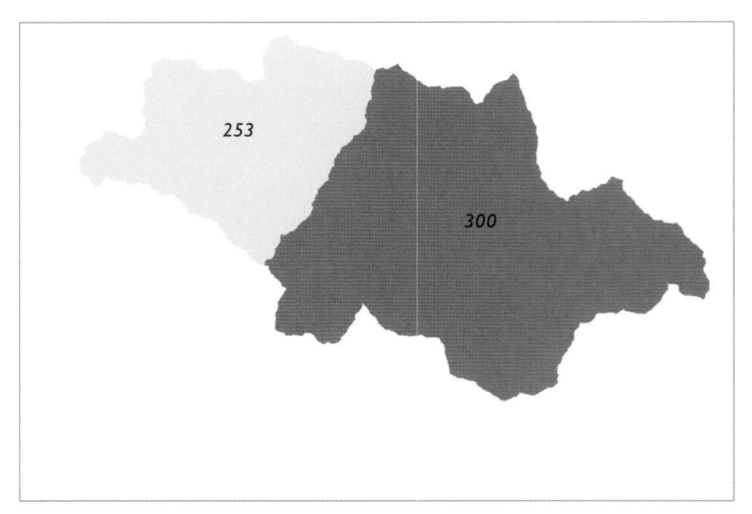

Drainage areas.

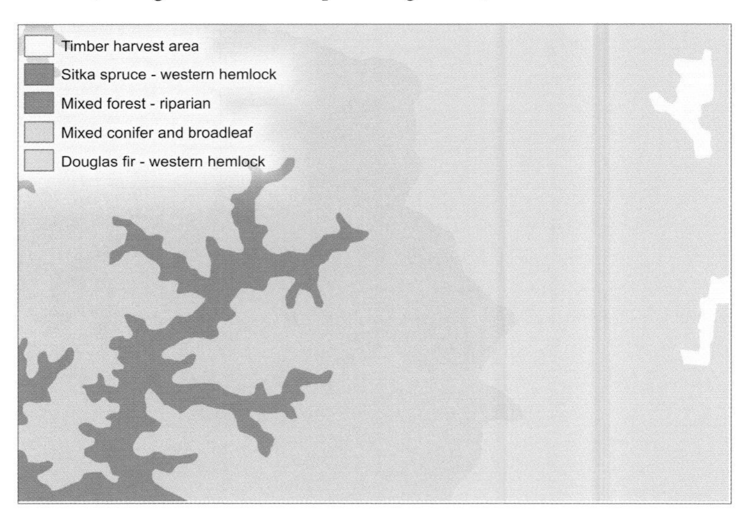

Land cover.

In ArcGIS, the Tabulate Area tool will calculate this for you, using the raster layer of drainage areas and a raster or polygon layer of categorical data.

VALUE	Timber harvest area	Mixed forest - riparian	Mixed conifer and broadleaf	Douglas fir - western hemlock
253	0	52518015	700295869	0
300	82955	62800118	690970441	904018107

Tabulate Area calculates the number of square feet of each land-cover type in each drainage area.

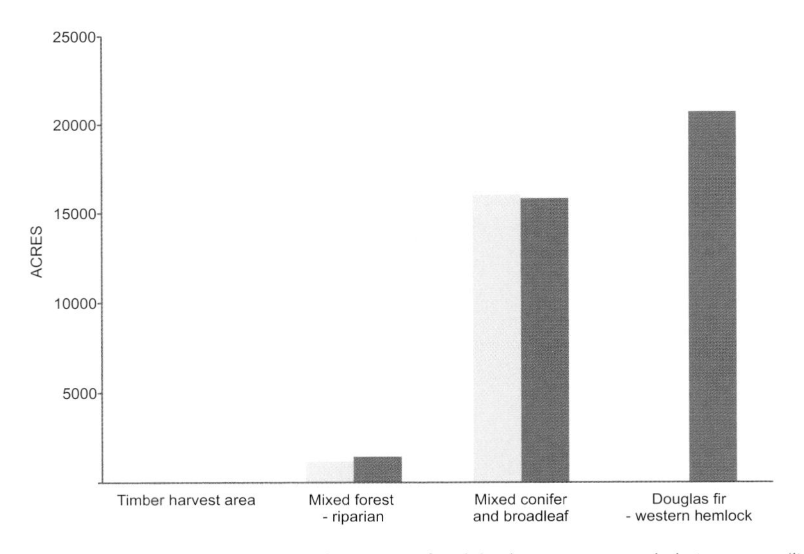

A bar chart can be used to compare the amount of each land-cover type in each drainage area (light green is area 253 and dark green is area 300). In the chart, square feet have been converted to acres.

Once you've calculated the area of each type, you can calculate percentages by dividing by the area of the drainage area and multiplying by 100.

VALUE	AREA	Timber harvest area	%	Mixed forest - riparian	%	Mixed conifer and broadleaf	%	Douglas fir - western hemlock	%
253	752813890	0	0.00	52518015	6.98	700295869	93.02	0	0.00
300	1657871600	82955	0.01	62800118	3.79	690970441	41.68	904018107	54.53

By adding the Area field from the Zonal Geometry table to the table created by Tabulate area (using a table join), you can calculate the percentage of each drainage area covered by each land-cover type. Pie charts (below) can be used to show the percentage of each type in each drainage area.

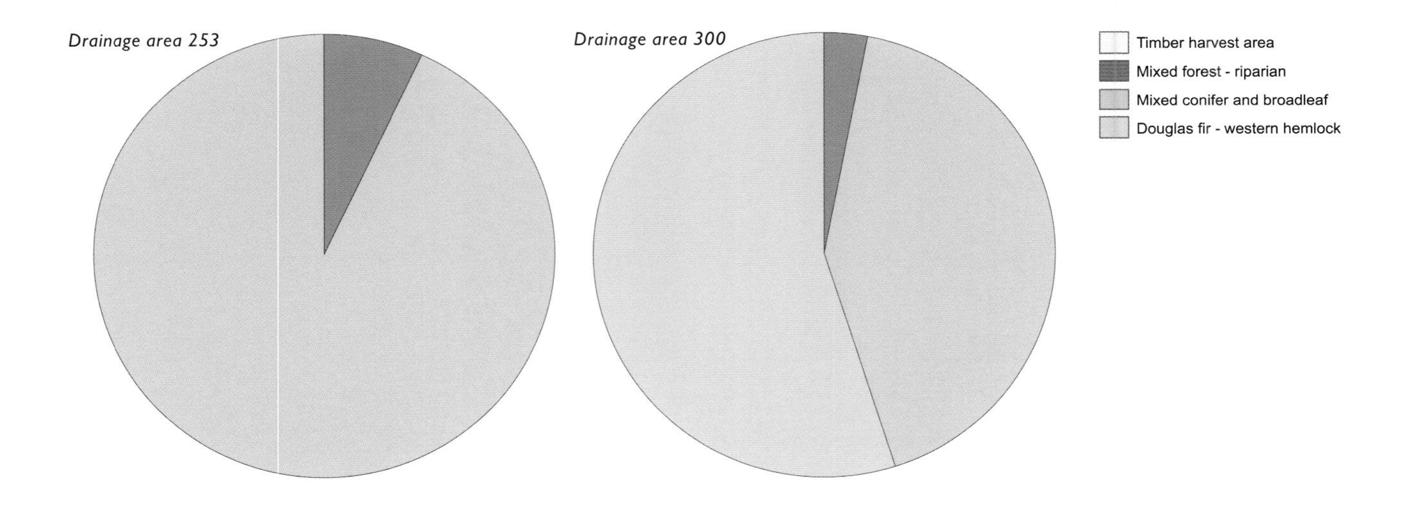

Alternatively, you can convert the raster drainage areas to polygon features and use polygon overlay with the categorical data (which must also be polygon data).

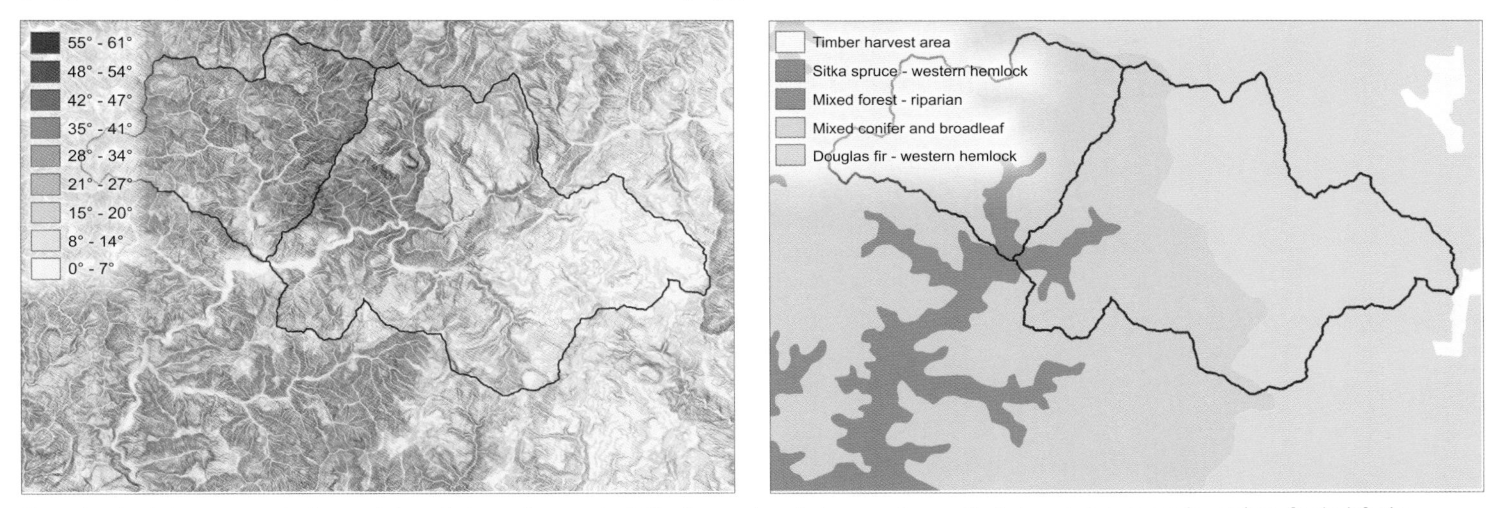

Converting the drainage area raster layer to polygon features allows you to display the area boundaries, as well as use the features in polygon overlay analysis. On the left, the boundaries with slope; on the right, the boundaries with land cover.

Converting the drainage areas to polygons also allows you to display the drainage area boundaries on top of other data layers, such as slope or vegetation.

Calculating flow volume

To calculate the volume of a substance that accumulates in the drainage area, you specify a weight layer when creating the flow accumulation layer. Each cell's cumulative value will be multiplied by the weight before being added to the value of the downstream cell.

You can, for example, calculate the amount of water that will accumulate at each location—including the outlet—over the course of a year or during a particular storm. To do this, you'd use a layer of what's known as "effective rainfall" as the weight layer.

You can think of effective rainfall as the amount of rain that falls at a given location, assuming it all runs off and flows to the adjacent cell. The absorption into the soil is already calculated in (as you'll see). One commonly used method for calculating effective rainfall is described by hydrologists Ven Chow, Larry Mays, and David Maidment in their book *Applied Hydrology* (see "References and further reading" at the end of the chapter), as well as other hydrology texts. Basically, you start with layers of rainfall, land cover, and soils. You categorize land-cover types by how impervious they are and soil types by how well they absorb water. The soil categories—of which there are four—are known as hydrologic soil groups, with group A soils being the most absorptive and group D the least. You then assign a runoff coefficient (known as a runoff curve number) to each combination of land-cover category and hydrologic soil group. Finally, the curve number is plugged into an equation that modifies the raw rainfall values.

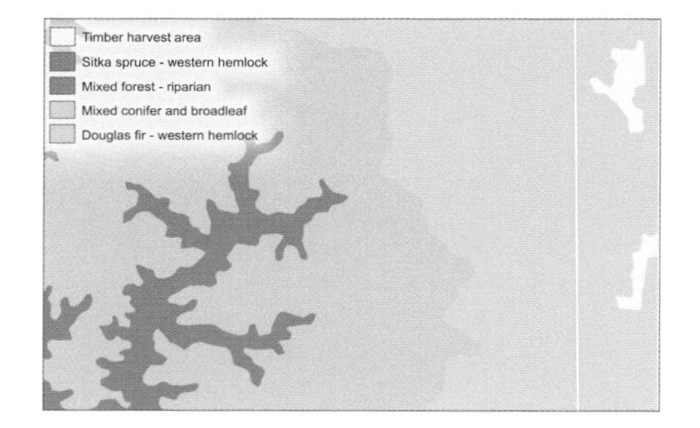

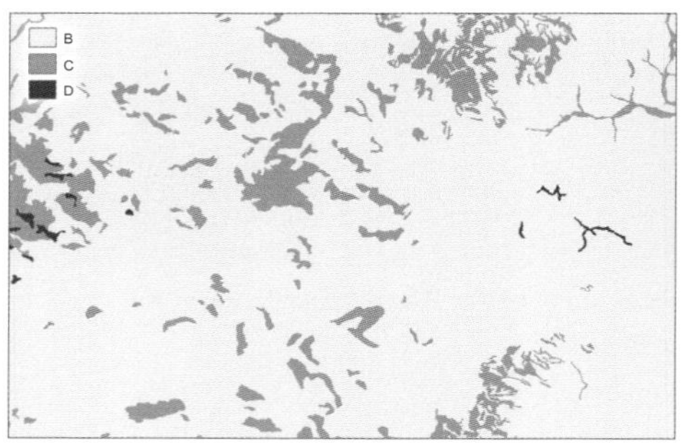

Land cover, left map, and soils classified by hydrologic soil group, center. (Only three hydrologic groups are present in this map.) The layers can be combined to find a runoff curve number, which is then used to modify average annual precipitation (right map). The result is a layer of effective rainfall (map at top of next page).

Typically, volume is calculated for particular drainage areas that have already been delineated (as described in the previous section). One way to do this is to use the drainage area layer as a mask layer. The weighted flow accumulation will be calculated only for the area covered by the drainage areas—the output weighted flow accumulation layer will have NoData values for the area outside the drainage areas.

Raster layer of average annual effective rainfall (in inches).

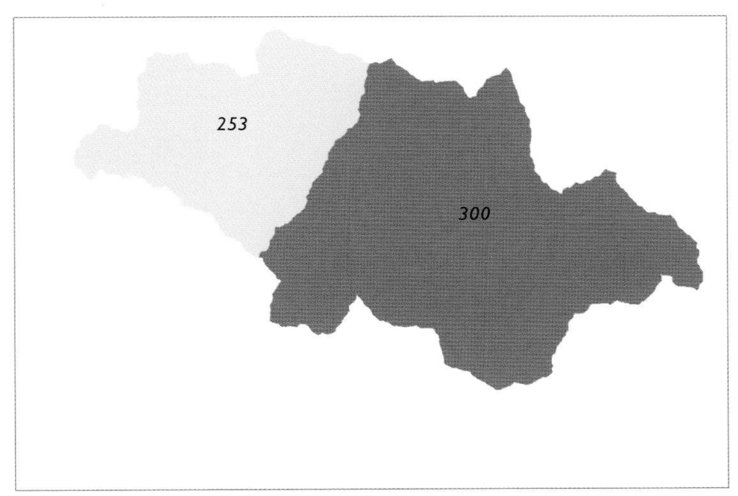

Drainage areas.

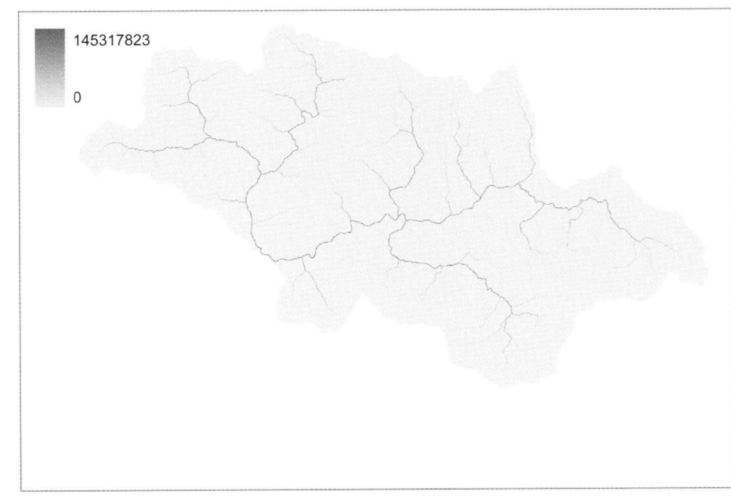

Weighted flow accumulation created using average annual effective rainfall as a weight layer.

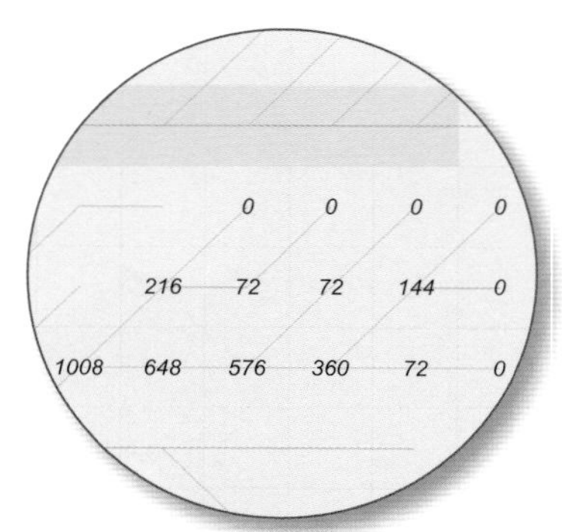

Calculating weighted flow accumulation. The runoff (or effective rainfall) from each cell in this area is 72 inches annually. The numbers show the amount of runoff entering each cell from the cells that flow into it. The cell with a value of 144, for example, receives the 72 inches of runoff from the cell to its northeast and the 72 inches from the cell to its east, for a total of 144. The cell with a value of 360 receives the 144 inches of runoff that flow into the cell to its northeast and the 72 inches that flow into the cell to its east, plus the 72 inches that fall directly into each of those two adjacent cells, for a total of 360 inches.

The values of the resulting weighted flow accumulation raster show the amount of rainfall flowing into each cell. To calculate the volume of water, you need to multiply this value by the area of a cell.

You calculate a conversion factor to convert the flow accumulation values to the desired volume units. For example, if the cell size is in feet and the rainfall is in inches, you'd convert inches to feet, and multiply the value by the cell resolution squared (to get area) to create the conversion factor. In this example, the cell resolution is 32.823 feet.

CONVERSION FACTOR = (32.823 * 32.823) * (1 / 12)

= 1077.349 * 0.083

= 89.420

You then create a new raster by multiplying the weighted flow accumulation raster by the conversion factor. The value of each cell in the flow accumulation raster is converted to volume in the desired units (such as cubic feet).

If the drainage area you're working with is large, converting to units that are used for large volumes, such as acre-feet, will make the values on the raster easier to read and interpret. You can include an additional conversion factor to convert to these units. For example, the factor to convert cubic feet to acre feet is 0.00002296. So to convert the weighted flow accumulation raster values to acre-feet, you'd use a conversion factor of 0.002:

89.420 * 0.00002296 = 0.002

The result will be the total volume of rainfall—in acre-feet—reaching the outlet (or any other cell) during the course of a year, on average. Again, the assumption is that all the rain that falls becomes runoff. While effective rainfall accounts for rainfall that's absorbed into the soil, some of the rainfall will also evaporate. In practice, the calculated volume is multiplied by a coefficient of less than 1 to account for evaporation.

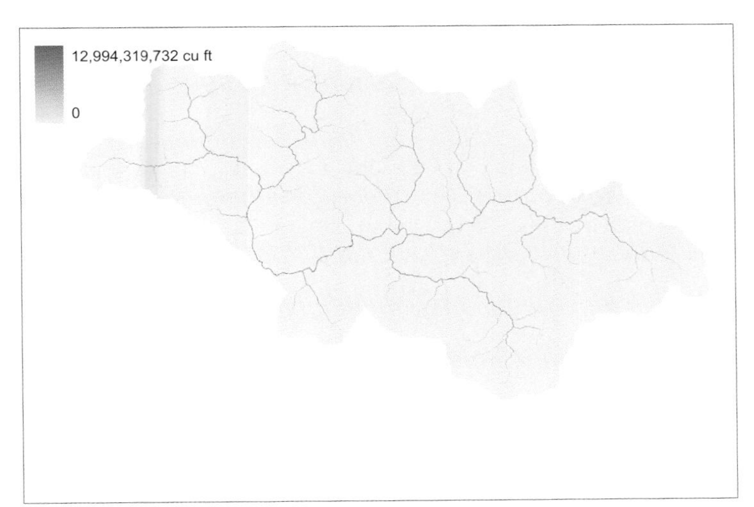

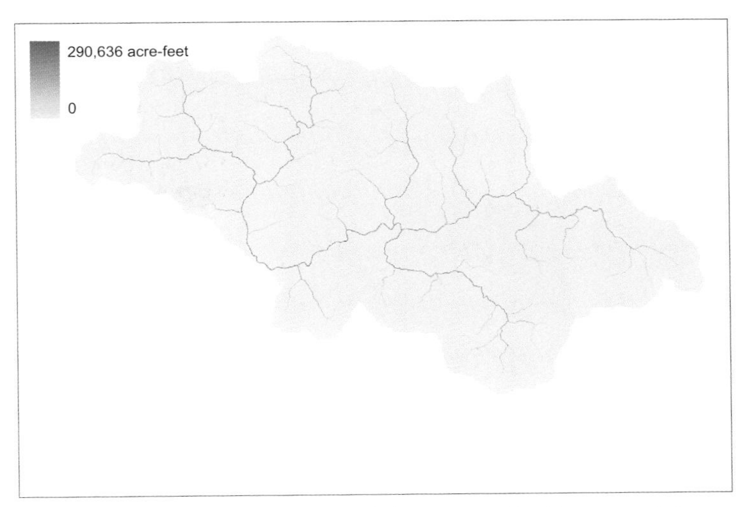

The weighted flow accumulation layer, converted to cubic feet (left) and acre-feet (right). In both cases, the flow pattern is the same as with the original weighted flow accumulation layer—only the units of measurement are different.

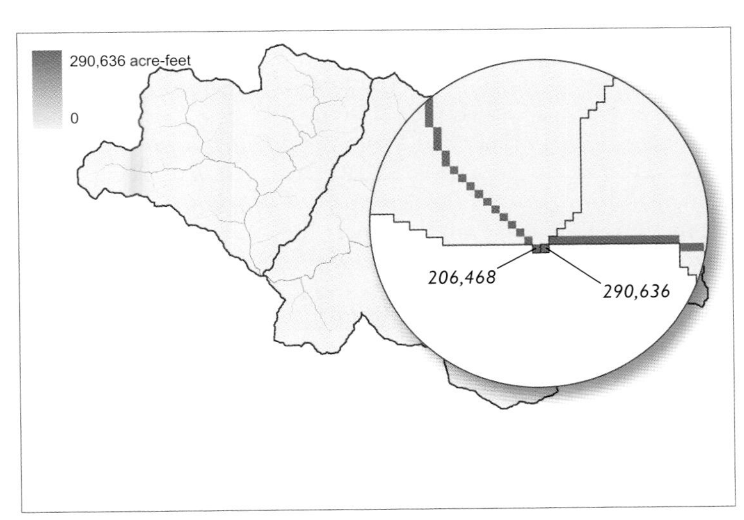

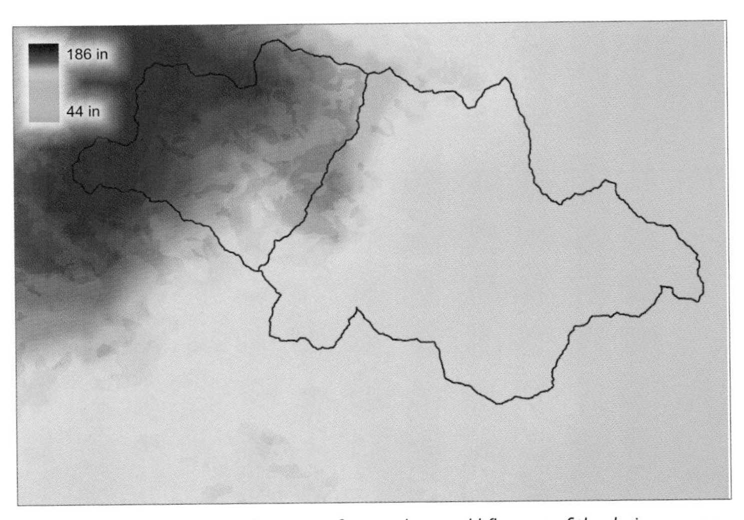

If you zoom to the outlet of the drainage area (left map and inset), you can determine the value of the outlet cell—the total amount of water that would flow out of the drainage area in an average year (again, based on effective rainfall). In this example, you can compare the two drainage areas. The smaller area is a little less than half the size of the bigger one, but it has two-thirds as much runoff. This comports with the fact that there is much higher average annual rainfall over the extent of the smaller drainage area (right map).

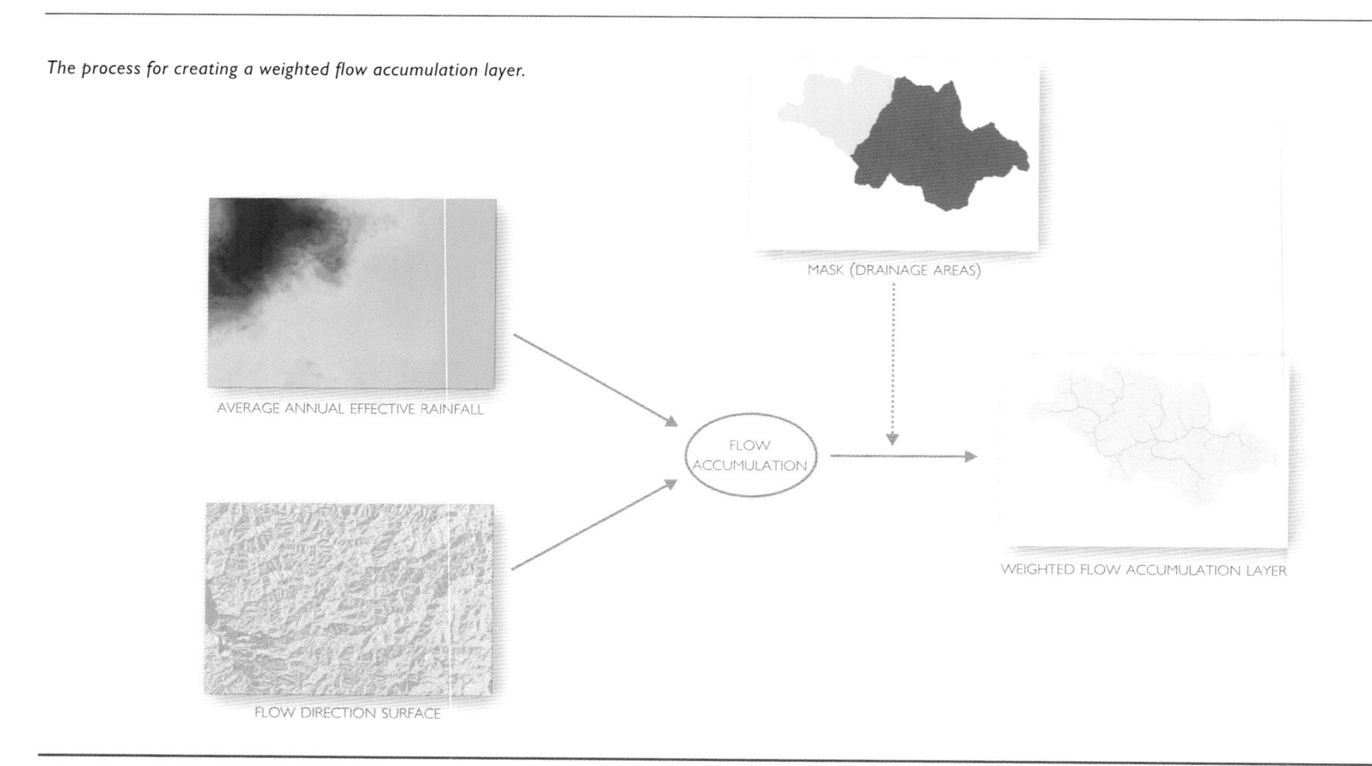

The process for creating a weighted flow accumulation layer.

Similarly, if you know the amount of sediment produced per unit area over a period of time (a year or a rainfall season), you can create a weight raster by first calculating the amount of sediment per cell and then assigning these values to the flow accumulation raster. The results will show the amount of sediment that will accumulate at each location over the time period. You'd use a similar approach to calculate the accumulation of a pollutant such as nitrogen or phosphorous.

Calculating travel time through a drainage area

You can calculate the time it takes for water from a rainstorm to reach the outlet of a drainage area from the farthest point in the area. This is known as the time of concentration. The time of concentration tells you how long it will take for the peak flow to reach the outlet.

To calculate time of concentration, the GIS calculates the longest flow path for the drainage area—that is, the path a drop of water falling at the head of the drainage area (such as the highest point on a ridge) would travel to reach the outlet. It also calculates the time it takes for water to cross each cell and creates a raster layer of these values, known as a travel time layer. The GIS then traces through the cells along the longest flow path, accumulating the time to traverse each cell. The sum of cell values is the time of concentration.

ArcHydro, an application for hydrologic and hydraulic modeling that is used with ArcGIS, includes a tool for calculating time of concentration—it essentially automates the process in a model. You specify the drainage areas for which you want to calculate time of concentration and the amount of precipitation falling in the drainage area during the storm.

You also specify layers representing factors that determine the rate of flow of water across the surface. These include the flow direction, slope (the velocity increases when the water travels over steep slopes), the roughness of the surface (velocity increases when the channel bed is smooth), and the amount of paved surface in the drainage area (such as roads or built areas, which will have 100 percent runoff). The more accurate and detailed the layers, the better the results of your model.

The tool creates the travel time layer and calculates the longest flow path. It then calculates the time of concentration for each drainage area.

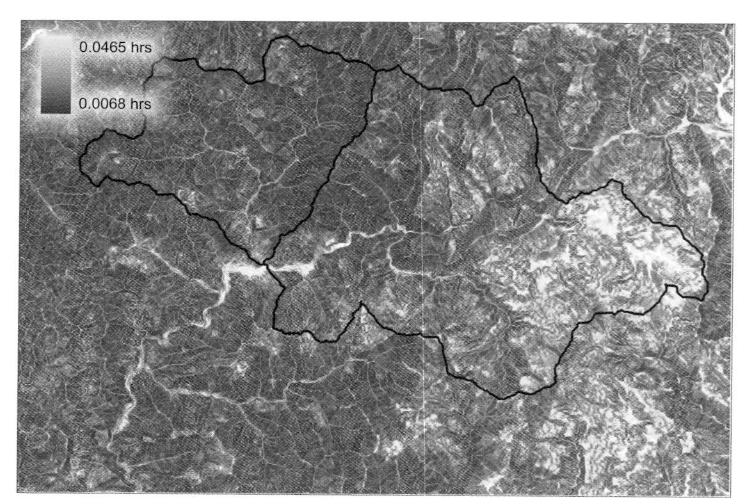

A travel time layer, showing the time (in hours) required for rainfall runoff to flow across each cell. Cells in channels and other areas with flatter slopes have higher values (light colors) since it takes longer for water to traverse those cells. The boundaries for two drainage areas for which time of concentration is to be calculated are also shown.

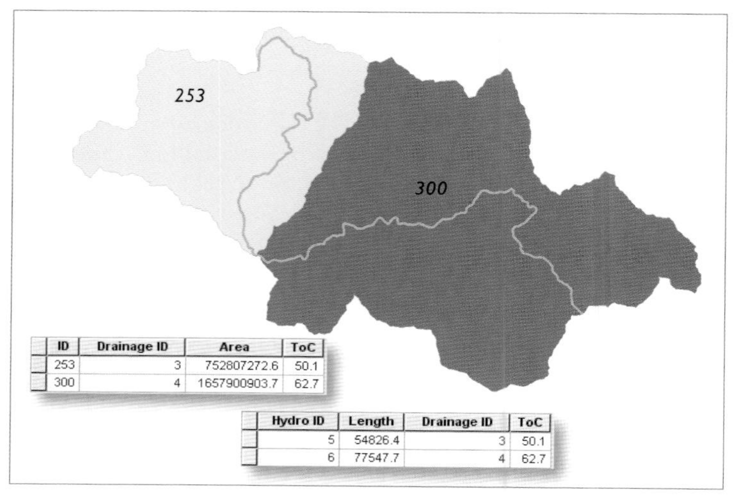

ID	Drainage ID	Area	ToC
253	3	752807272.6	50.1
300	4	1657900903.7	62.7

Hydro ID	Length	Drainage ID	ToC
5	54826.4	3	50.1
6	77547.7	4	62.7

The map shows the two drainage areas along with the longest flow path for each (yellow lines). The drainage areas attribute table is on top. The time of concentration (ToC) for the smaller area is about two days (50.1 hours); the time of concentration for the larger area is about twelve hours longer. The attribute table for the longest flow path line layer (bottom) includes the length of each path, in feet—about 10 miles for the smaller drainage area and 14.5 miles for the larger one.

EVALUATE THE RESULTS

You'll want to closely examine the results of your model to make sure they make sense. The elevation surface may still contain anomalies—even after processing—that will have altered the results. While the computer takes the surface values literally, and calculates the flow accordingly, your eye may detect patterns on the surface that reveal anomalies or errors. You'll want to make sure that basin boundaries follow ridges and that they don't overlap each other or cross stream channels. You'll want to make sure stream channels don't run uphill, that they don't cross each other, and that they completely connect. Even with an elevation surface that has been corrected to the extent possible, you may find that channels don't connect, and you will have to edit the result to make the channels completely connect in a stream network.

Tracing flow over a network allows you to model the movement of material from an origin location through (or across) a set of connected features, such as pipes or streams. For example, you'd model the flow of motor oil dumped into stormwater inlets to find the downstream portions of a stormwater network that are affected.

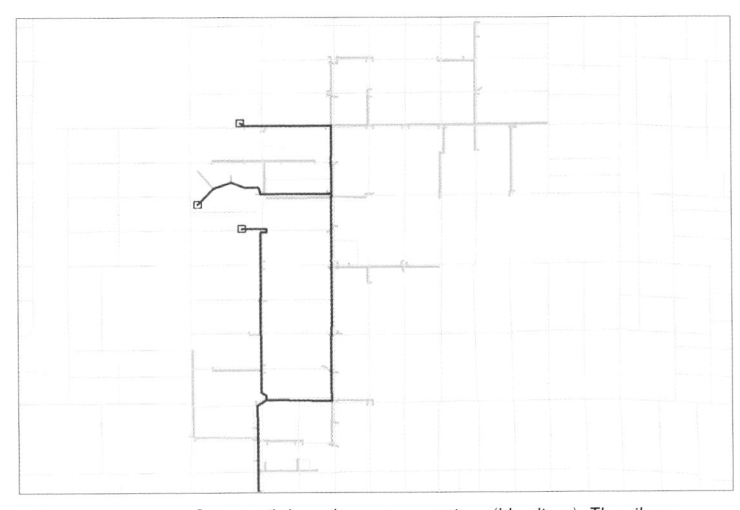

A downstream trace of motor oil through stormwater pipes (blue lines). The oil was dumped into three inlets (blue boxes); the red lines show the pipes the oil will flow through. Street centerlines are also shown, in gray.

More so than with modeling flow over a raster surface, with network flow much of the modeling occurs within the structure of the data (rather than as a consequence of combining various data layers to create a result). Specifically, a network flow model requires a geometric network. Geometric networks store information about how the elements of the network (such as stormwater mains, inlets, and manholes) connect to each other. The direction of flow through the network is also stored with the network elements. Once the connectivity and flow direction are established, modeling the flow is simply a matter of tracing along the connected network edges from the origin of the flow—either with the flow direction (downstream) or against the flow direction (upstream).

Because it is based on flow direction, network tracing is mainly applicable for networks where the flow is downhill, from high elevation to lower elevation, and is in one overall direction, such as with stormwater or stream networks where the flow converges at an outlet.

(This is in contrast to networks where material is pumped through the system under pressure and in which the material can travel several different routes to get to the same location. In a water network, for example, if water pressure drops on one street due to heavy use, flow comes from another direction to keep the pressure constant. Because of the multiple routes the material may travel, flow direction for portions of the network can't be determined, so doing an upstream or downstream trace does not produce useful results.)

The process for modeling flow over a network is:

1. Specify the geometric network
2. Set the flow direction
3. Perform the trace
4. Evaluate the results
5. Display and apply the results

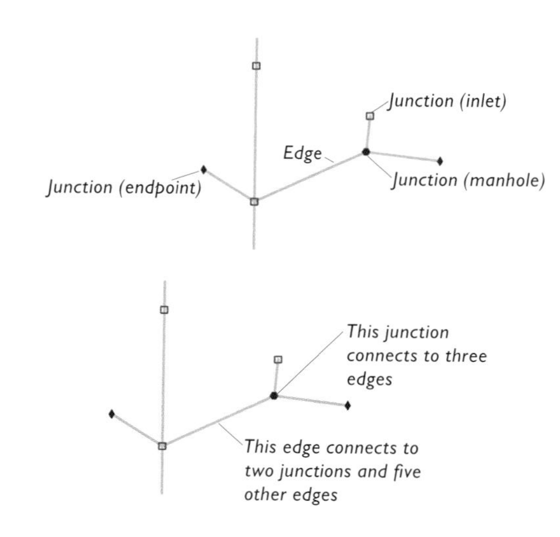

SPECIFY THE GEOMETRIC NETWORK

A geometric network is used to model the behavior of a material, such as water, energy, or data, as it moves through or over pipes or wires. In order to model behavior, the elements of the network—and how they relate to each other—have to be defined in specific ways. For example, a stormwater network would include stormwater mains (lines), inlets (points), and manholes (points). Geometric networks ensure that all lines (termed "edges") in the network connect (at "junctions"). Point features on the network—the inlets and manholes—are treated as junctions. Junctions are automatically created where edges (stormwater mains) connect—for example, where two mains of different diameters are joined. They are also automatically created at the endpoints of edges, if there isn't already a junction representing a point feature at that location. The network stores the connectivity of the features, so you know which mains connect to which and which main an inlet or manhole is located along. That allows the GIS to trace the flow of material from one edge to the connected ones.

A geometric network for your application may already exist in your GIS database. If not, you can create one from a set of existing line and point features (the mains, inlets, and manholes). ArcGIS, for example, includes tools to do this.

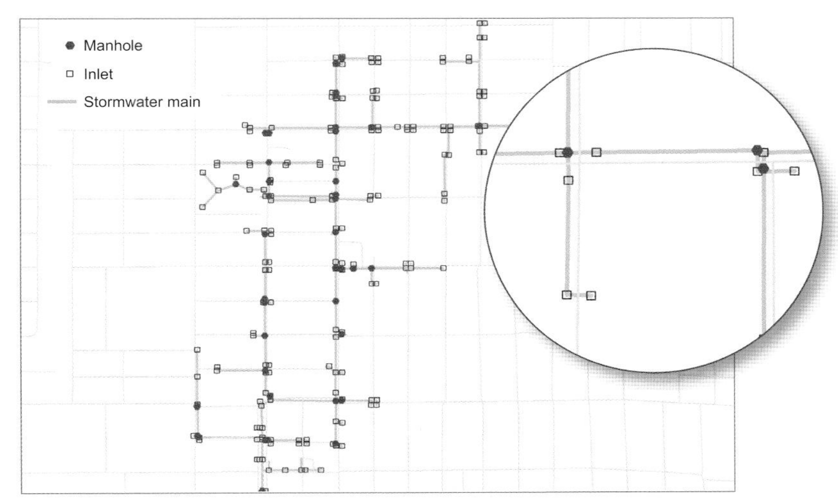

A portion of a stormwater network showing mains, inlets, and manholes. Street centerlines (gray) are also shown. In the close-up, you can see that mains parallel the street centerlines, inlets are located along mains (generally on either side of the street centerline, along the curbs), and manholes are often at the intersections of mains. The inlets, manholes, and mains are managed in the GIS as a set of connected features.

SET THE FLOW DIRECTION

In addition to the connectivity of the features, a geometric network stores the flow direction of each edge in the network. The flow direction is defined by a "from" endpoint and a "to" endpoint for each edge. When you do a trace, the flow moves along the edge either with the flow direction (for a downstream trace) or against the flow direction (for an upstream trace).

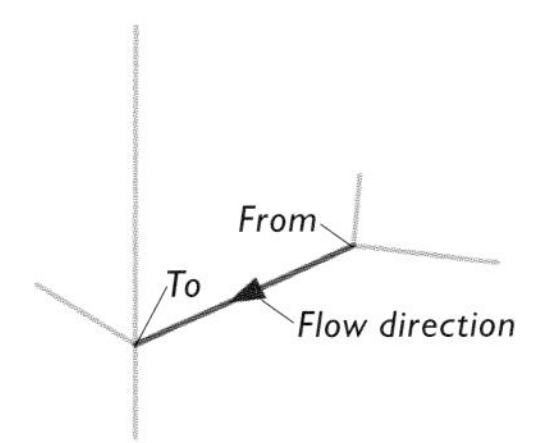

For a network that flows in one overall direction (such as a stormwater or stream network), you need to ensure that the flow direction of each edge is in the downhill direction, or toward the outlet. Otherwise, the trace will not give valid results or not complete correctly. For example, if you're doing a downstream trace over a stormwater network and the flow direction of one of the edges in the path of the trace incorrectly points uphill, instead of downhill, the trace will stop when it gets to that edge.

There are two common ways to set the flow direction for a geometric network. One is to use the digitized direction of the network edges. The other is to include sinks and sources in your network.

Using digitized direction to set flow direction

In many cases, line features that were intended to be used for network flow analysis will have been digitized with flow direction in mind. In other cases, the line features may not have been created for use in flow analysis, so the from and to points may have been assigned arbitrarily by the person who created the original data (by entering geographic coordinates from a survey or GPS or by manually digitizing from a map).

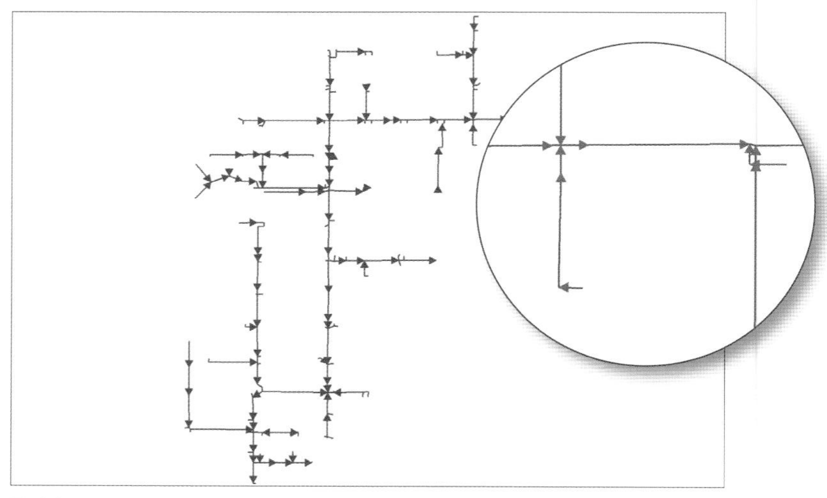

Each line segment in the stormwater mains layer is drawn with an arrowhead indicating the From and To ends of the segment. In this example, the digitized direction of some of the horizontal stormwater mains is in the opposite direction to the downhill flow of the stormwater network.

The GIS has tools for displaying the digitized direction of each edge and for flipping the from and to points. Once the digitized direction of the lines is correct, you set the flow direction for the network edges using tools in the GIS.

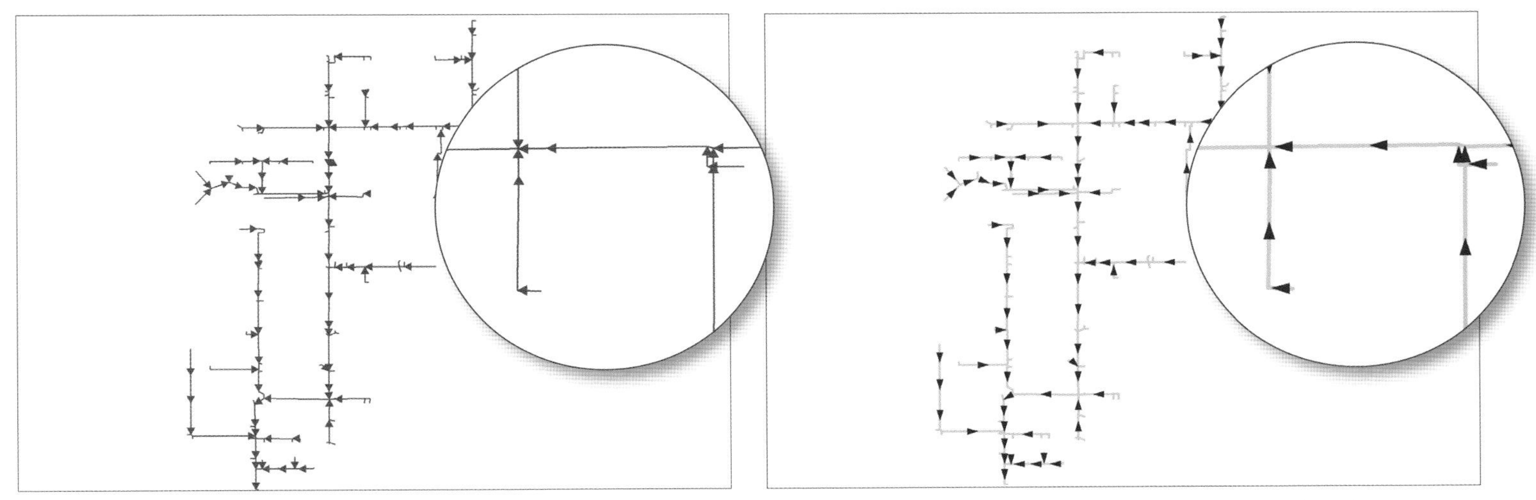

Once the lines have been flipped by editing them in the GIS and reversing the From and To ends, the digitized direction matches the flow of the stormwater network and can be used to set the flow direction of the network.

The map displays the flow direction of the network, set using the digitized direction of each stormwater main segment.

Using sinks and sources to set flow direction

Another way to set flow direction is to include a point feature as a "sink." (The sink is stored as a junction in the network.) In a geometric network, a sink is a location that the network flow converges on, such as a water treatment plant in a stormwater or sewer network. After specifying a sink, you can set the flow direction for the network, and the flow direction for each edge will flow towards the sink. You might use a sink to set the flow direction if you have a large dataset in which line direction has been assigned arbitrarily, so editing individual lines and flipping their direction would be a major task.

In networks where the material is forced through the network, you can include point features that serve as "sources" that push the flow outward from those locations. Sources might be pumps in a water network or transformers in an electric network. As with sinks, sources can be used to set flow direction (the flow direction for each edge is away from the source).

Networks that include sinks and/or sources often have multiple possible routes—from the source to the sink or other destination—via loops or lines that connect at more than one place. This is to ensure that, for example, water flow continues unabated if there is a drop in pressure in one area, electricity continues to flow if there is an outage at one transformer, or sewage doesn't back up if a main is blocked.

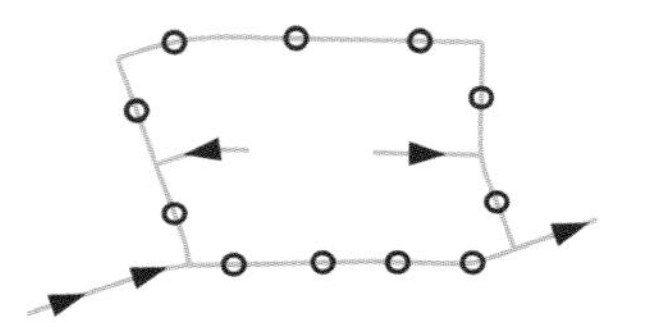

Indeterminate flow (indicated by the circles) occurs when, for example, flow could travel along either of two or more edges, as is the case with this loop in a sewer network.

Because of the circular flow on portions of these networks, it isn't possible to establish flow direction for some edges. When flow direction can't be determined for an edge—because the flow could be in either direction—the flow direction is said to be indeterminate. When a trace reaches a network edge that has indeterminate flow, the trace stops. Thus, these types of networks are not conducive to modeling flow. You can, however, still do limited traces upstream or downstream from edges having indeterminate flow.

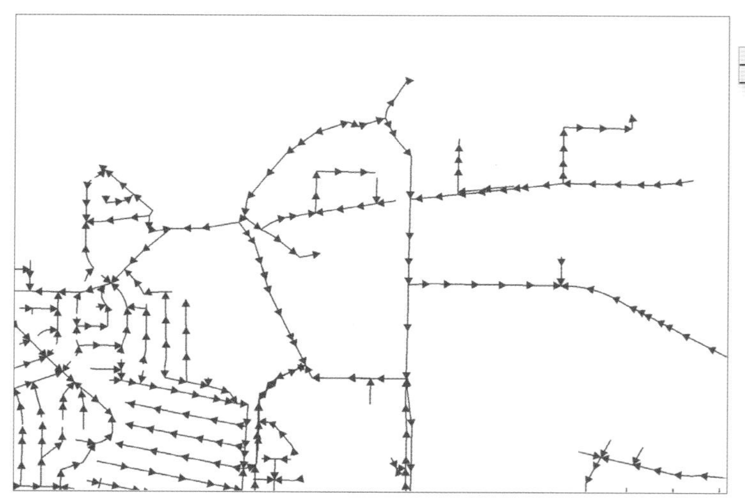

This map shows a portion of a sewer mains layer. You can see that the segments were not digitized in a particular direction.

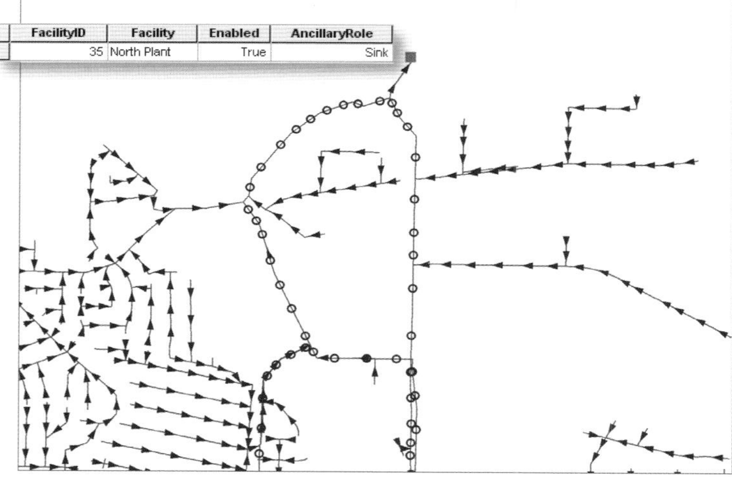

A network created from the sewer mains and a treatment plant (blue square at top of map). The treatment plant serves as a sink in the network (the value in the AncillaryRole field in its attribute table identifies it as such). Flow direction is established so that all mains flow toward the sink. Where network edges interconnect or create loops, flow can't be established (as indicated by the blue circles). This is known as indeterminate flow.

PERFORM THE TRACE

Once you've established flow direction for the network edges, you can model flow through the network. You do this by performing a trace.

To run the trace, you first specify the origin point (or points), and then run either a downstream or upstream trace function in the GIS. The trace identifies the connected edges in the specified direction (downstream or upstream) from the origin. It also identifies all the junctions connected to those edges.

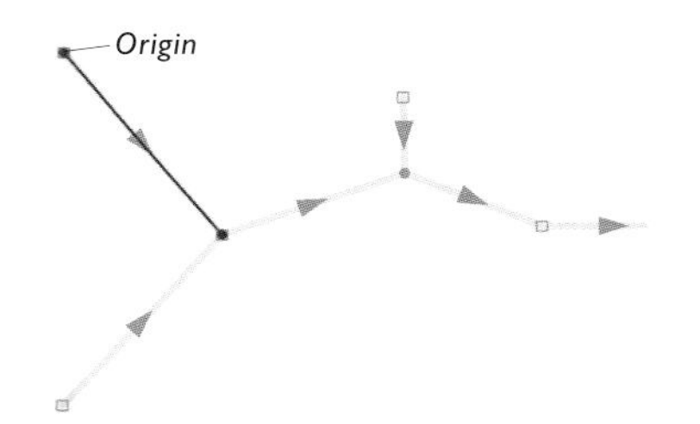

In a downstream trace, the GIS starts at the junction representing the origin location and traces along the connected edge in the direction of the flow. The junction at the other end of the edge becomes one of the selected features associated with the trace.

Selected

Not selected

The GIS next checks the edges connected to the newly selected junction. If the trace from the selected junction to the junction at the other end of an edge is against the flow direction, that edge is not included in the trace selection. If the trace is with the flow direction, the edge is selected and included in the trace.

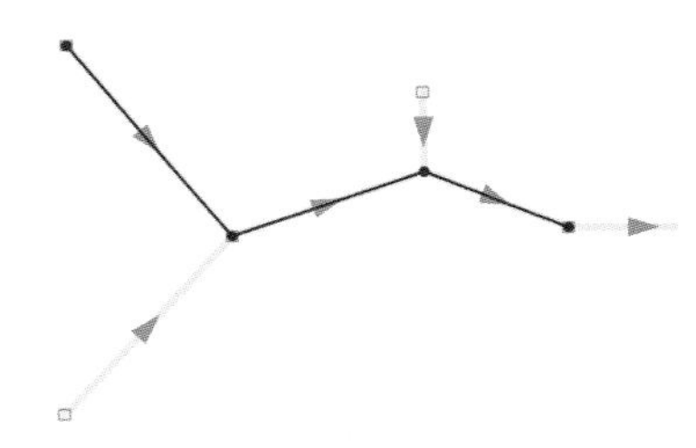

The GIS continues the process, tracing along the edges with the flow direction (downstream).

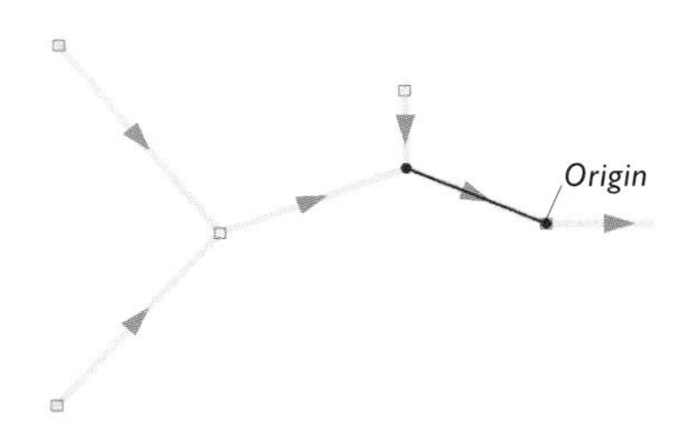

In an upstream trace, the GIS starts at the junction representing the origin location and traces along the connected edge opposite to the direction of the flow.

At each junction, the GIS checks the flow direction of each connected edge. It selects any edge where the trace is against the flow direction.

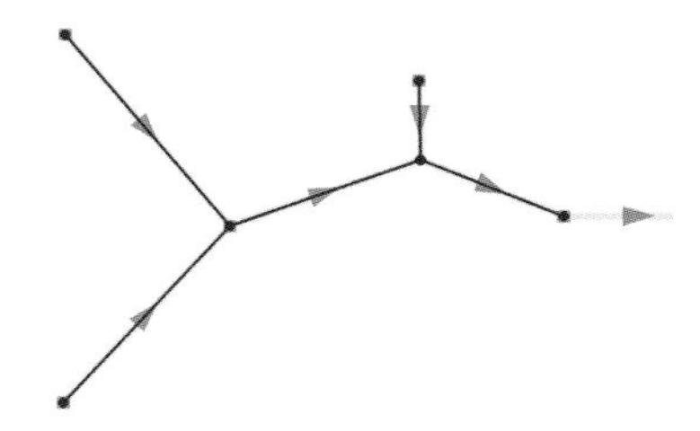

The GIS continues the process, tracing along the edges against the flow direction (upstream) until it reaches all the junctions with no edges connected to them (end point junctions).

For example, if an inspection of a stormwater drainage system reveals that there is motor oil at a particular manhole, you might first trace downstream from the manhole to identify where the oil is traveling. You specify the manhole as the origin and then do a downstream trace. The trace automatically identifies the manholes and inlets along those mains. You can display the selected mains and manholes on a map and in a table. Field crews can check those manholes to see how far the oil has traveled and determine the extent of the spill.

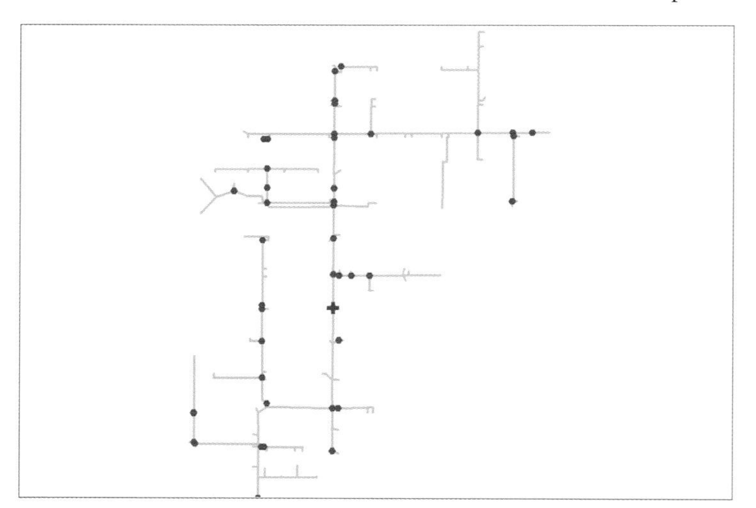

The red cross indicates the location of the manhole that serves as the origin of the trace. (In this map, inlets are not displayed.)

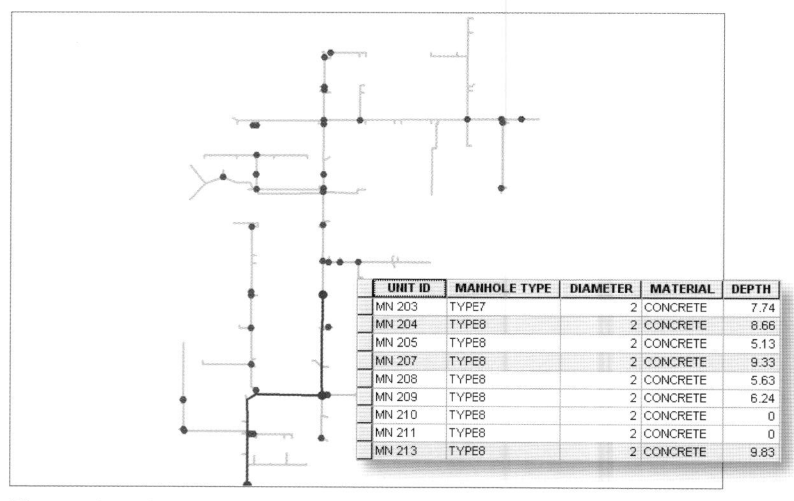

UNIT ID	MANHOLE TYPE	DIAMETER	MATERIAL	DEPTH
MN 203	TYPE7	2	CONCRETE	7.74
MN 204	TYPE8	2	CONCRETE	8.66
MN 205	TYPE8	2	CONCRETE	5.13
MN 207	TYPE8	2	CONCRETE	9.33
MN 208	TYPE8	2	CONCRETE	5.63
MN 209	TYPE8	2	CONCRETE	6.24
MN 210	TYPE8	2	CONCRETE	0
MN 211	TYPE8	2	CONCRETE	0
MN 213	TYPE8	2	CONCRETE	9.83

This map shows the result of a downstream trace—the edges and manholes highlighted in red are downstream from the origin, determined using the flow direction of the network edges. The table shows the selected manholes (highlighted in gray).

Next, you would trace upstream from where the oil was discovered to try to locate the source of the spill. The trace will identify all the mains upstream, along with manholes along those mains. You could send field crews to those manholes to look for evidence of the spill.

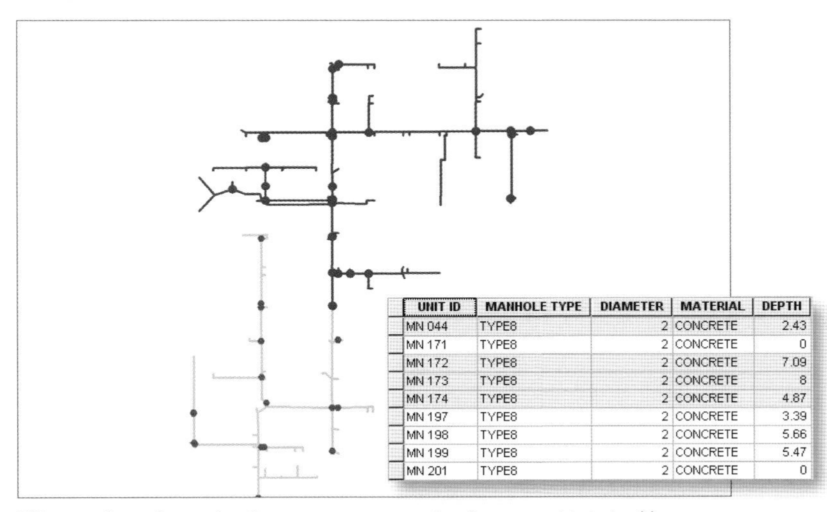

UNIT ID	MANHOLE TYPE	DIAMETER	MATERIAL	DEPTH
MN 044	TYPE8	2	CONCRETE	2.43
MN 171	TYPE8	2	CONCRETE	0
MN 172	TYPE8	2	CONCRETE	7.09
MN 173	TYPE8	2	CONCRETE	8
MN 174	TYPE8	2	CONCRETE	4.87
MN 197	TYPE8	2	CONCRETE	3.39
MN 198	TYPE8	2	CONCRETE	5.66
MN 199	TYPE8	2	CONCRETE	5.47
MN 201	TYPE8	2	CONCRETE	0

This map shows the results of an upstream trace, using the same origin point (the specified manhole). The result identifies all the stormwater mains upstream of the origin and the manholes located on those mains (highlighted in gray in the table).

Setting barriers

You can specify that certain features in the network are impassable or that the flow can't move through portions of the network. One way to do this is by interactively setting temporary barriers on edges or junctions—you simply select the edge or junction on the map, and specify that it is a barrier. When the trace reaches a barrier, it stops. In the motor oil spill example, you might set barriers at manhole locations where reports coming back from the field indicate that no oil was found. The upstream trace will be limited to mains downstream of these manholes.

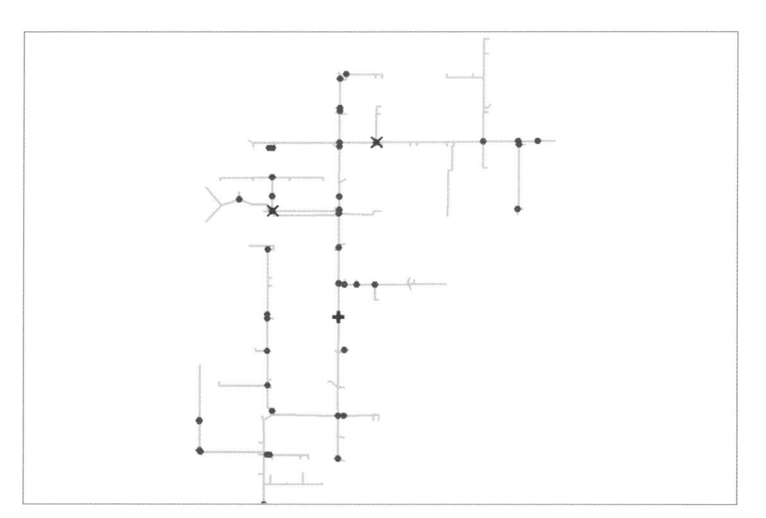

The red cross indicates the location of the manhole that serves as the origin of the trace, while the two red Xs are the locations of barriers to upstream flow. (In this example, they represent manholes where evidence of oil was not found, so the source could not be upstream of these locations.)

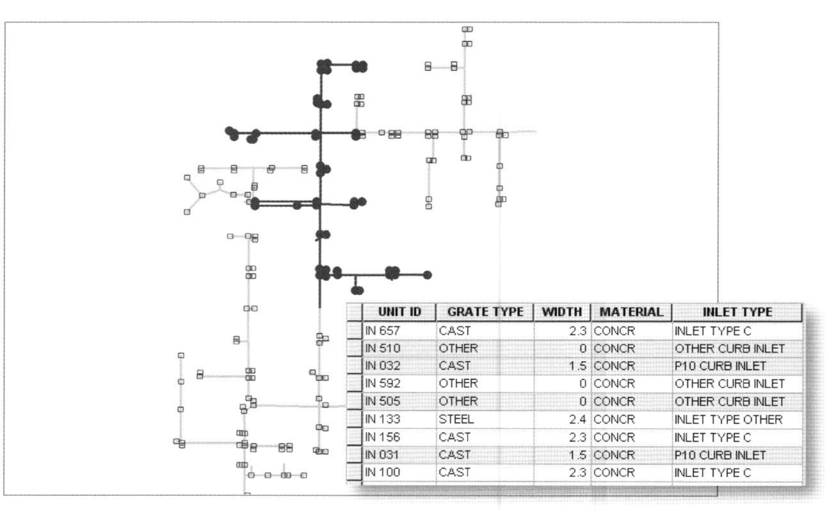

UNIT ID	GRATE TYPE	WIDTH	MATERIAL	INLET TYPE
IN 657	CAST	2.3	CONCR	INLET TYPE C
IN 510	OTHER	0	CONCR	OTHER CURB INLET
IN 032	CAST	1.5	CONCR	P10 CURB INLET
IN 592	OTHER	0	CONCR	OTHER CURB INLET
IN 505	OTHER	0	CONCR	OTHER CURB INLET
IN 133	STEEL	2.4	CONCR	INLET TYPE OTHER
IN 156	CAST	2.3	CONCR	INLET TYPE C
IN 031	CAST	1.5	CONCR	P10 CURB INLET
IN 100	CAST	2.3	CONCR	INLET TYPE C

Setting the two manhole locations as barriers stops the upstream trace at these points. The result identifies the stormwater mains upstream of the origin, but downstream of the barriers, along with the manholes located on those mains (highlighted in gray in the table). Hence, the source of the oil is along one of the identified mains. The selected manholes can be checked to further narrow the source of the oil. (Inlets are also shown on this map, for reference.)

An alternative to setting temporary barriers is to set a junction or edge as impassable in the network feature's attribute table—the barrier is then stored with the network. The junction or edge will act as a barrier for all trace operations until you set it back to its original state. The attribute table for each geometric network layer includes a field named "Enabled." When the field value is set to True for a feature, the flow can continue through the junction or along the edge. When the value is set to False, the flow is blocked for that feature. You might, for example, set the Enabled value to False for a stormwater main that is undergoing repair for the duration of the time it is out of service.

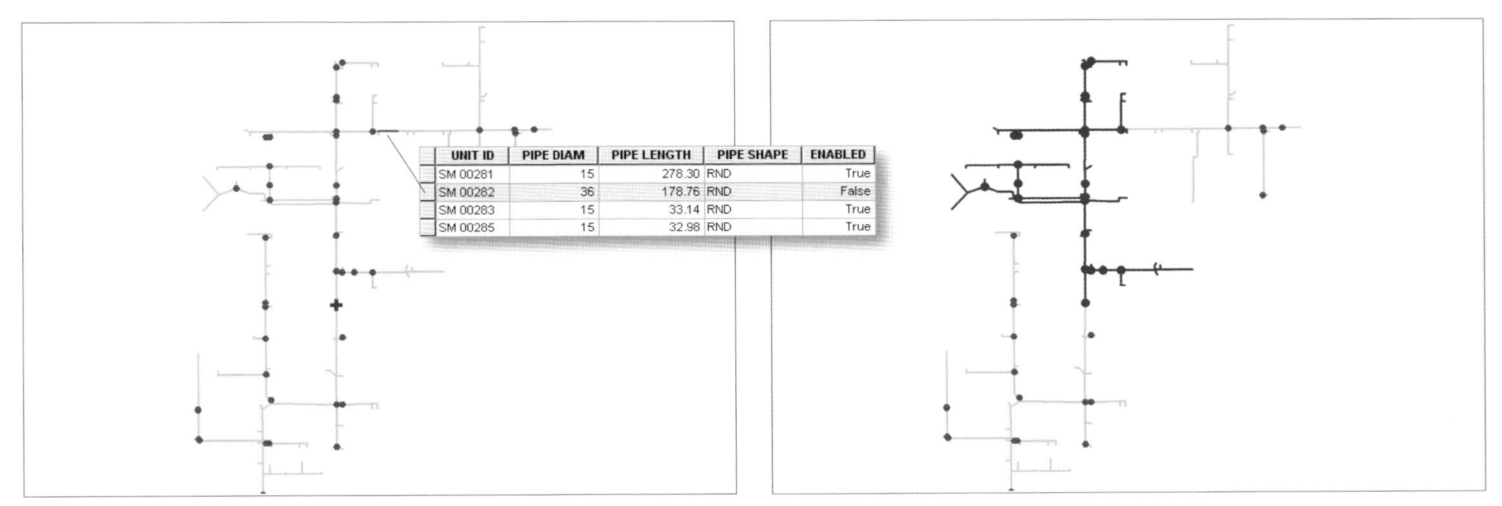

UNIT ID	PIPE DIAM	PIPE LENGTH	PIPE SHAPE	ENABLED
SM 00281	15	278.30	RND	True
SM 00282	36	178.76	RND	False
SM 00283	15	33.14	RND	True
SM 00285	15	32.98	RND	True

You can set a barrier by selecting a network edge (or junction) and setting the Enabled value to False. (The red cross indicates the origin location.)

The results of the upstream trace with the blocked stormwater main included. The trace stops when it reaches the blocked main—the source of the oil could not be upstream of the main.

Using several origin locations

You can use more than one origin location when performing a trace. The flow (upstream or downstream) will be traced from all the origin locations at one time. For example, an industrial area where solvents have been dumped previously may have several stormwater inlets. After specifying these inlets as origin locations, you can trace downstream along the stormwater network to identify the mains and manholes that need to be monitored.

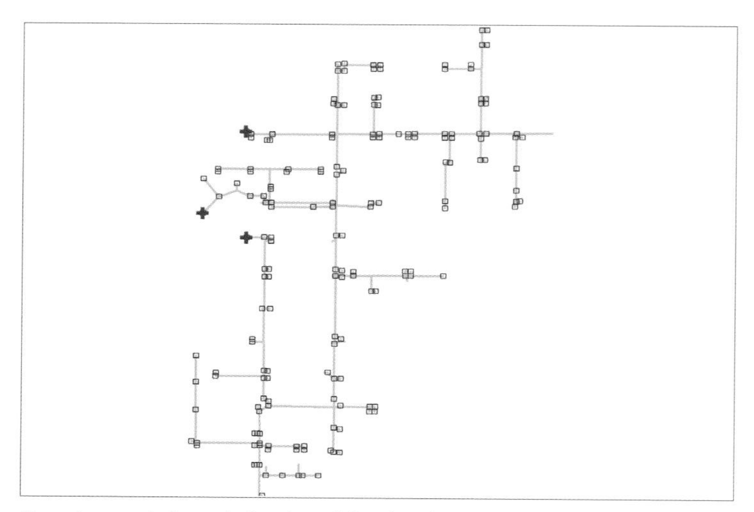

The red crosses indicate the locations of the inlets that serve as the origin of the trace—in this example, stormwater inlets in an industrial area.

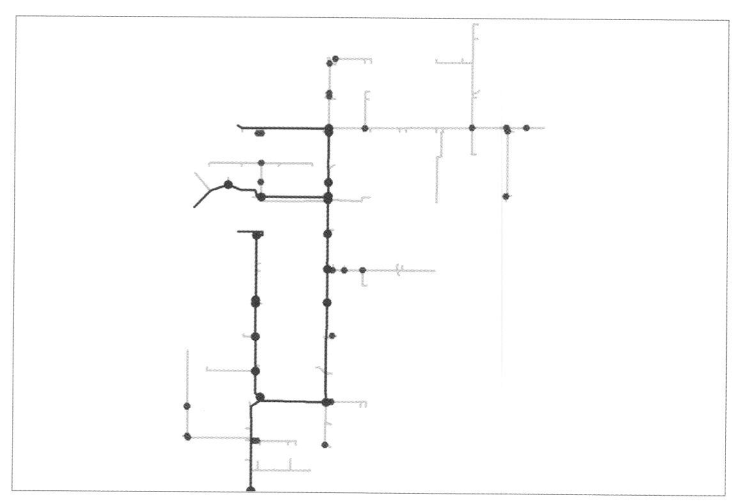

The result identifies the stormwater mains and manholes downstream of all three origin locations.

Calculating upstream flow accumulation

You can calculate the accumulation of a characteristic associated with the network features when you trace upstream. You do this by specifying an attribute of the edges and/or junctions as a weight value when you create the geometric network. You then include this weight value when you do an upstream trace. In a simple example, you might specify the length of stormwater mains as the weight value. When you do the upstream trace, the GIS reports back the total length of the mains selected through the trace—the length values are accumulated as you trace upstream from the origin location. (If you do an upstream accumulation trace without a specified attribute, you get the number of selected features—in this example, the number of stormwater mains selected by the trace.)

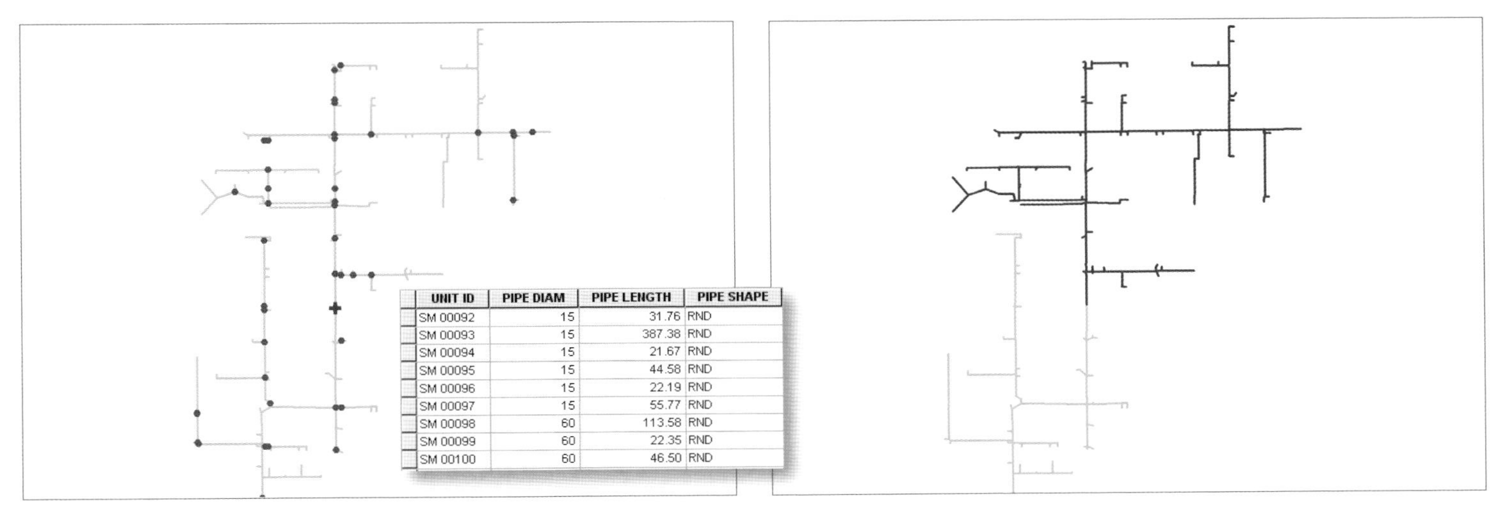

UNIT ID	PIPE DIAM	PIPE LENGTH	PIPE SHAPE
SM 00092	15	31.76	RND
SM 00093	15	387.38	RND
SM 00094	15	21.67	RND
SM 00095	15	44.58	RND
SM 00096	15	22.19	RND
SM 00097	15	55.77	RND
SM 00098	60	113.58	RND
SM 00099	60	22.35	RND
SM 00100	60	46.50	RND

The red cross indicates the location of the manhole that serves as the origin of the trace. The Pipe Length attribute in the table was specified as the weight value when the geometric network was created and is also specified as the weight value to use for the upstream accumulation.

The results of the upstream accumulation trace identify the stormwater mains upstream of the origin location (in red). The GIS also reports the accumulated weight values along the selected network edges—in this example, the total length of mains is 15,984.65 feet.

Similarly, if you knew, or could estimate, the amount of water entering each inlet during a rainstorm of given duration and intensity, you could assign these values as weights to the inlets. You could then calculate the amount of water flowing through the stormwater network at any given location or the total amount reaching the outlet location.

EVALUATE THE RESULTS

Once you've performed the trace, check to make sure the results are valid. Look for any places where the trace ends abruptly. This may indicate a break in the network, an edge having the wrong flow direction, or an edge with indeterminate flow. It may also indicate a junction or edge along the flow path that has been set as a barrier. These may or may not be errors, but should be checked and verified.

Look also for flow that travels in a direction opposite to what you would have expected (upstream in a downstream trace). This may also indicate that edges were assigned the wrong flow direction. Again, you'll want to check the flow direction of the network edges and correct any errors.

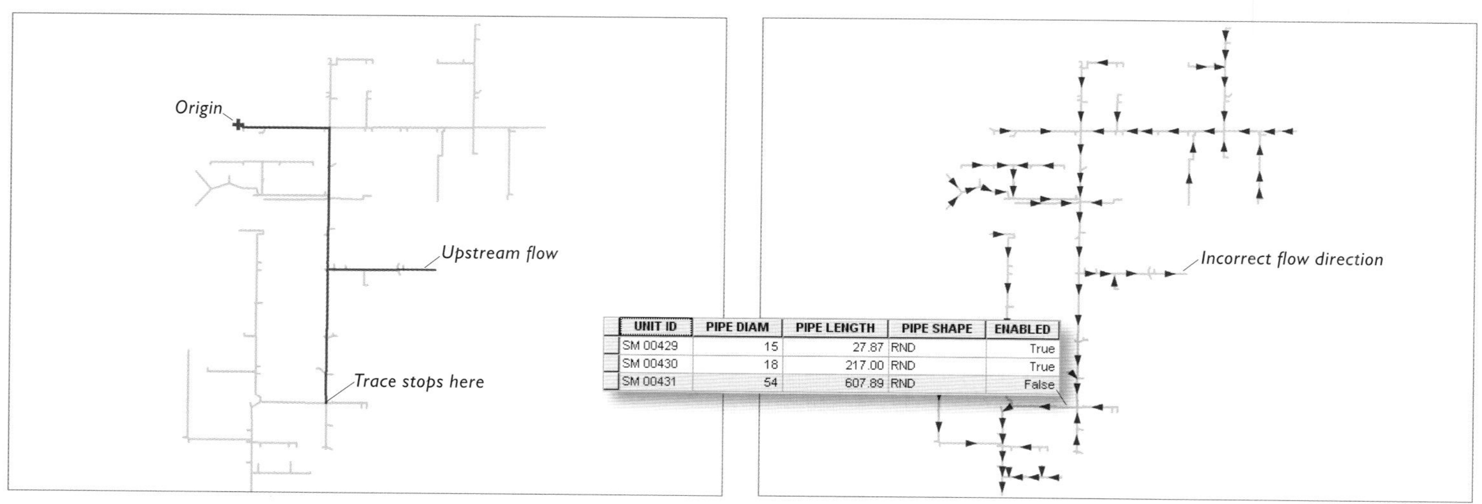

UNIT ID	PIPE DIAM	PIPE LENGTH	PIPE SHAPE	ENABLED
SM 00429	15	27.87	RND	True
SM 00430	18	217.00	RND	True
SM 00431	54	607.89	RND	False

This map shows the results of a downstream trace from the origin location (a stormwater inlet). There are two possible problems with the results: the trace splits and appears to head upstream along the mains in the center of the map; and the trace stops at an intersection of mains for no apparent reason.

When the flow direction for the stormwater mains is displayed, it confirms that the flow direction is indeed incorrect for the mains in the center where the trace goes upstream. The flow direction for these mains needs to be reversed. Where the flow stops near the bottom of the map, the flow direction is correct. However, the main where the trace should have continued is blocked—the Enabled value was set to False. The main may, in fact, be impassable or this may be an error that should be corrected by setting the Enabled value to True.

DISPLAY AND APPLY THE RESULTS

Once you've verified the results of your flow analysis, you'll want to display the results for your intended audience. You can highlight on your map the traced edges and any selected features along them, as well as the origin point(s). You may also want to display reference information such as streets to provide context.

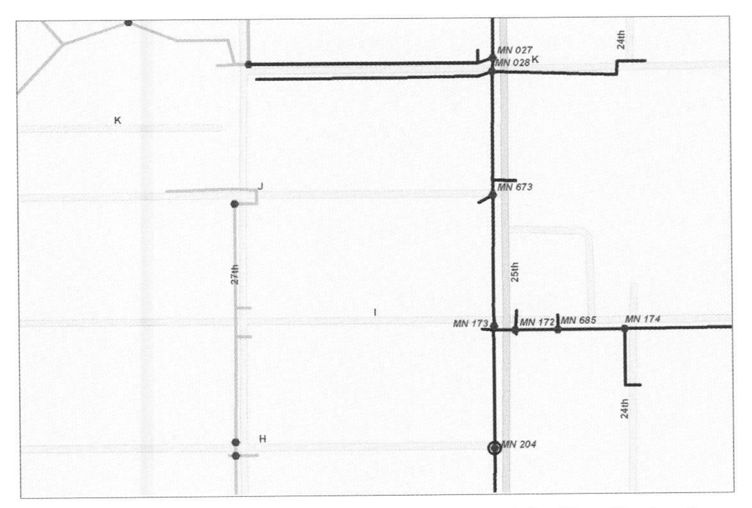

This close-up shows a portion of the stormwater mains potentially affected by the oil spill (in red), both upstream and downstream of the origin (the circled manhole where the spill was first detected). It also shows the manholes along the mains, each labeled with its ID. Unaffected mains (in blue) and streets are shown for reference. The map would be useful for field crews tracking the extent of the spill.

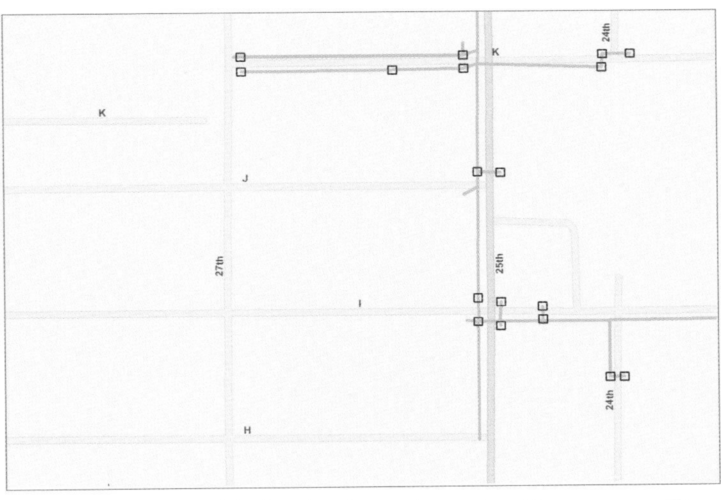

This map shows the inlets (blue boxes) and streets in a portion of the source area. The map would be useful for inspectors, as well as the public, in identifying the source of the spill.

You can also use the selected features to map or select related features. For example, you might map the inlets in the source area with land use to find where industrial users are located. Or you could select industrial or commercial parcels within 300 feet of an inlet and map just those parcels. This would help focus the search for the source of the contaminant in a stormwater network.

Chow, Ven Te, David R. Maidment, and Larry W. Mays. 1988. *Applied Hydrology*. New York, NY: McGraw-Hill. A classic hydrology text. Includes a section on calculating effective rainfall using land cover and hydrologic soil group categories. Runoff curve numbers and associated equations are presented and described.

Maidment, David R., ed. *ArcHydro: GIS for Water Resources*. 2002. Redlands, CA: Esri Press. Includes chapters on modeling drainage area networks and boundaries, as well as using GIS for time series analysis and linking GIS to external hydrologic models. The book also presents the ArcHydro data model for building and storing a hydrologic geodatabase.

Zeiler, Michael. *Modeling Our World: The ESRI Guide to Geodatabase Concepts*. 2010. Redlands, CA: Esri Press. Describes in depth the various data models used by ArcGIS. Includes an extensive discussion on the concepts behind the geometric networks used for utilities applications.

Modeling interaction 6

You can use GIS to model the interaction between facilities that provide a service (such as stores, libraries, or distribution centers) and locations that have demand for those services. These models are often used to find which areas are currently served by a facility (and which are not served by any facility), to find the location for a new facility that will best meet the demand, or to predict the amount of demand at each facility (such as the number of shoppers at a store or visitors to a regional park).

Census blocks (represented by dots) have been allocated to fire stations (red squares) so that each block is assigned to the closest station, within four minutes travel time. Blocks are color-coded by the station they're assigned to. Census blocks in gray are more than four minutes from a station.

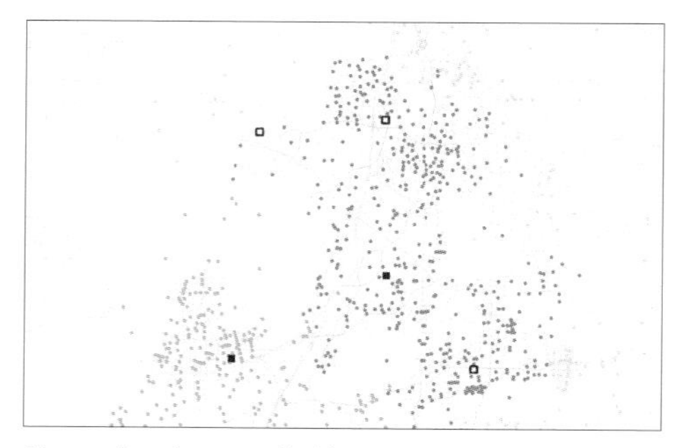

The two best locations for libraries (solid blue squares) have been chosen from five potential locations (hollow squares are unchosen locations). Census blocks are color-coded by the library they're closest to.

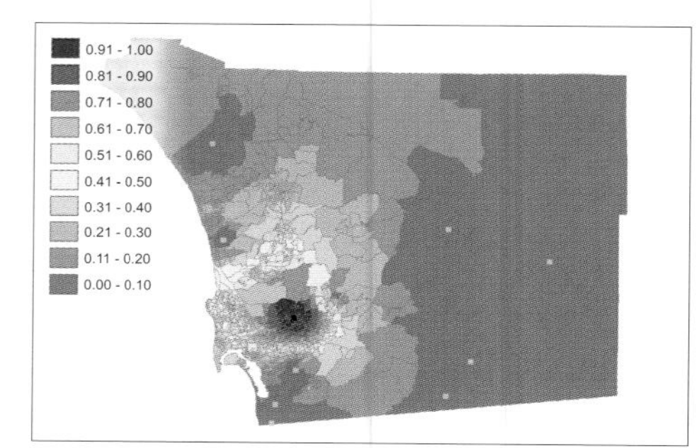

Probability that hikers will travel from each census block group to a particular regional park (black square). Locations of other parks are also shown (gray squares).

One set of models allocates demand from the locations to their closest facility, within specified parameters, such as the maximum distance a location can be from a center.

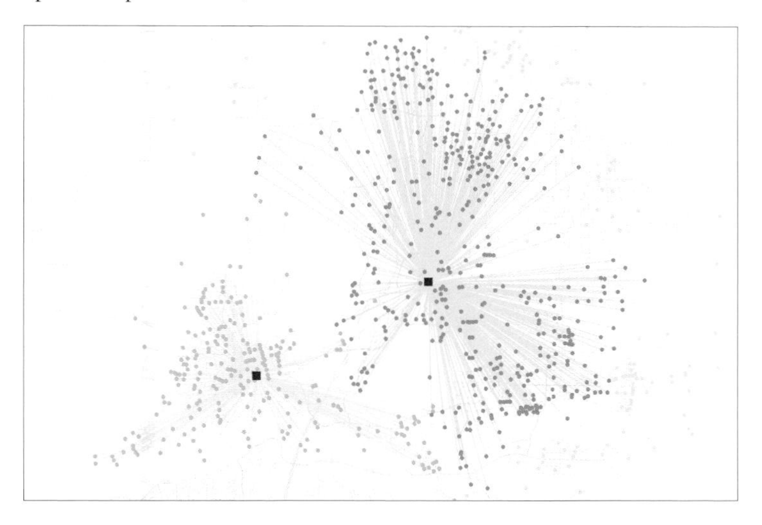

This map shows a solution using two library locations (solid squares). Census blocks are color-coded by which library they have been allocated to, and color-coded lines are drawn to create a spider diagram.

Another set of models attempts to determine which facility, or facilities, people at each location will travel to, given a choice of facilities (each more or less attractive than the other) and the cost of getting to each one. They essentially model people's preferences for one facility over another (sometimes referred to as "spatial choice").

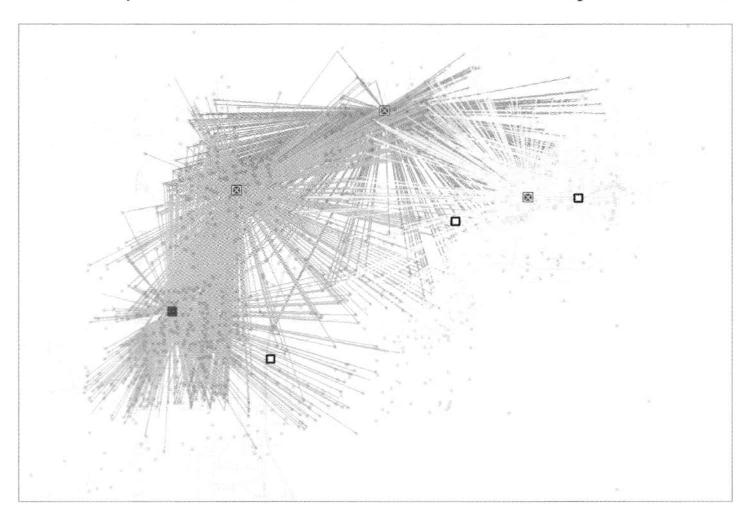

In this model, the location for a new pet store (solid square) has been chosen from a set of potential locations. Candidate locations that weren't chosen are shown as hollow boxes. The locations of existing competing stores are also shown (boxes with Xs). The lines show at which store people in the various census blocks (orange dots) are likely to shop. The square footage of each store was used as the attraction value (the assumption being that the larger the store, the more products it carries and the more attractive it is to shoppers).

To define the parameters of your model and choose the appropriate method, you first need to define the problem you're addressing and the information you need from the model. You also need to consider the characteristics of the facilities and demand points, as well as factors that influence the interaction.

DEFINE THE PROBLEM

To determine which type of model to build, you need to assess the problem you're trying to solve or the question you're asking. If you're simply allocating demand at a location to the nearest facility, you'll build a location-allocation model, discussed in the section "Allocating demand to facilities." If you're modeling the preferences of people given a choice of several facilities to travel to, you'll build a spatial interaction model, discussed in the section "Modeling travel to facilities."

You'll also want to determine what specific information you need from the model. You can assign locations to existing facilities, create market or service areas, choose locations for new facilities, figure out how many facilities you need and where they should be, or calculate the total demand (visitation or usage) for each facility.

THE FACILITIES AND DEMAND POINTS

Interaction is based on the idea of a service being provided by a facility and the demand for that service—from surrounding locations—being allocated to one or more facilities. To ensure the results of your model are valid, you need to quantify the demand as accurately as possible. The demand points can be individual locations, such as houses or businesses, with each point having a value of one (in the case of houses) or with each point having a quantity associated with the location (the number of employees at each business). Or, the demand points can represent areas, such as census blocks or counties. In this case, the demand is an aggregate value, such as the total population in each census block. In some cases, the demand value will represent a specific population at each demand point, such as the number of people over 65.

Facilities can have associated values that are used in creating the interaction model. In some models, for example, you can specify a supply for each facility, and demand is allocated to the facilities so that supply is used up. When there is more demand than supply, there is excess demand. Conversely, when there is less demand than supply, there is excess supply, and demand may be allocated to facilities from demand points that are farther away.

If there is competition between facilities, each facility can be assigned a measure of attraction as perceived by people who might travel to the facilities. The attraction value can be a single characteristic representing how likely people are to go to the facility, such as the square footage of a store (larger stores are likely to draw people from a wider area). Or, it can be based on a combination of characteristics. For a theme park, attraction could include the number of rides at the park, whether there are other theme parks nearby, whether it's near an airport, whether it's near a lake, and so on.

INFLUENCES ON THE INTERACTION

In addition to the characteristics of the facilities and demand points, the results of the model are determined by the movement or travel between the facilities and demand point locations. A number of factors will influence travel between the demand points and the facilities. These include:

- The cost of travel, whether distance, time, or money.

- Whether travel is from the demand points to the facility (as with people traveling to libraries) or from the facility to the demand points (as with a delivery truck traveling from a distribution center to grocery stores).

- Whether travel is calculated in a straight line or over a street or other transportation network. This is sometimes a matter of scale—at the local level, it may be appropriate to model interaction over streets, whereas at the regional or continental scale (where distances between facilities and demand points are longer) modeling straight-line travel may suffice. Similarly, you'd likely want to calculate travel over a network in dense urban areas but might use straight-line distance in rural areas where facilities and demand points are spread out and roads are few. In addition, some implementations of the various interaction models only calculate travel over a transportation network, while others allow you to calculate travel using a network or straight-line distance.

- How rapidly the likelihood of travel to the facility drops off as distance from the facility increases, and whether there is a distance or time beyond which people are unlikely to travel to the facility.

You'll want to consider how these factors apply to your model. How they are implemented for a specific model is discussed in the applicable section later in the chapter.

One type of interaction model allocates demand from surrounding areas to a facility or set of facilities. The facilities might be public facilities that offer a service, such as fire stations, schools, or libraries, or they might be commercial facilities, such as distribution centers for grocery stores or for parcel delivery. The method optionally allows you to choose the set of locations of potential facilities that most efficiently meets all the demand. Hence it is often called location-allocation modeling.

You use location-allocation modeling to create service areas for facilities that provide a service to surrounding areas so that all demand is met. For example, you can assign census blocks to fire stations so that blocks are assigned to the closest station. This will ensure that, in case of a fire, a house will be reached in the shortest time possible given the existing fire station locations.

Census blocks (represented by dots) have been allocated to fire stations (red squares) so that each block is assigned to the closest station. Blocks are color-coded by the station they're assigned to.

You also use location-allocation modeling to analyze access to services that are used by people as needed, such as health clinics or senior day-care centers. This will tell you which areas have limited or no access to the services and, thus, where services may need to be expanded.

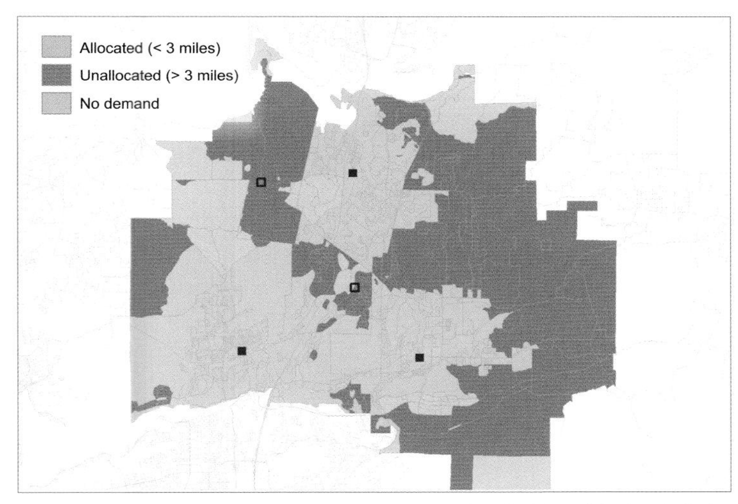

This map shows which census blocks are within three miles of a library (blue) and which are more than three miles from a library (dark tan). Existing libraries are shown as solid blue boxes and two potential locations for a new library are shown as hollow boxes.

If you're building new facilities, use location-allocation modeling to find the optimum location and number of facilities from a set of potential locations. You can include existing facilities in the analysis. While this is somewhat similar to suitability analysis, in location-allocation modeling, the best location is defined specifically as the location (or set of locations) that minimizes the overall distance, travel time, or cost from all the demand points to the facilities. In addition, with suitability analysis you generally identify the best location from all possible locations (all the parcels or census tracts, for example), whereas in location-allocation modeling you choose the best subset of locations from a set of potential locations already identified. (In fact, you can use suitability analysis to identify the potential locations for use in the location-allocation model—see chapter 2, "Finding suitable locations" and chapter 3, "Rating suitable locations.")

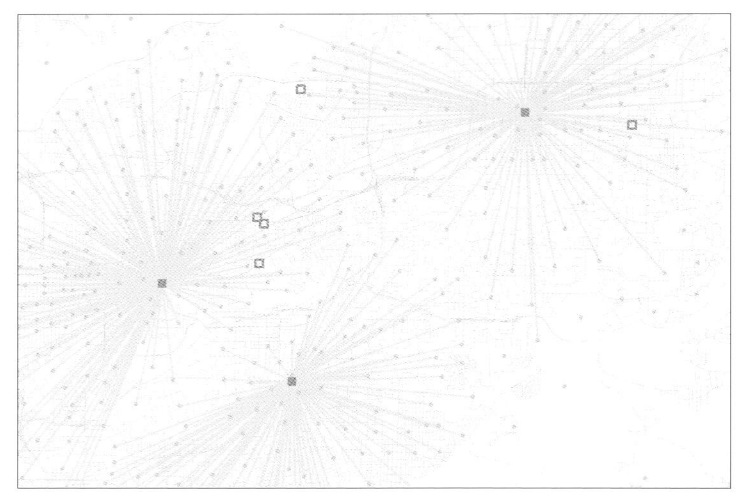

The map shows the three locations chosen for a senior day-care center (solid boxes), out of eight potential locations, so that the most seniors within five miles of a center are covered. The dots represent census blocks, and the lines show which center each allocated block is assigned to.

The process for allocating demand to centers is:

1. Define the solution you need
2. Set up the network
3. Specify the facilities
4. Specify the demand points
5. Run the model
6. Evaluate the results
7. Display and apply the results

DEFINE THE SOLUTION YOU NEED

You'll first want to define the type of solution you need. This will help you decide which attributes you need for the network dataset (network datasets are discussed in the next section, "Set up the network"), as well as for the facilities and the demand points. It will also help you determine the parameters to use for the model.

Are you allocating demand to existing facilities or locating new facilities?
If you are allocating demand to existing facilities to create a service area or to analyze access to services, you will use a dataset of existing facilities. These are specified as "required" facilities when you add them to the model. In addition, if you are allocating based on the supply available at the facilities, you will need to specify the attribute that stores the supply values.

If you're choosing the best locations to build new facilities, you'll specify a dataset of candidate facilities. You can also specify the locations of existing (required) facilities that must be included in the solution.

How many facilities do you need?
When running the model, you specify how many facilities need to be included in the solution. For allocation solutions, where all the facilities are required, you specify that all the existing facilities be included in that solution. That way, demand will be allocated among the existing facilities.

For location solutions, if you already know the number of new facilities needed to meet the demand, you specify that that number be included in the solution. Alternatively, you can have the software tell you the number of facilities needed to meet the demand. This is useful if you're building facilities in an area where there are none. You'd also use this approach if you want to know whether existing facilities are more than meeting the demand and whether some can be closed—you let the GIS choose the locations of the minimum number of facilities needed.

Do you want to minimize transportation costs?
For most applications, demand for each demand point is allocated to the nearest facility (in terms of distance, time, or cost, as measured along the network). For some applications, you may want to choose facilities such that the overall transportation cost from all demand points to all facilities is minimized. You'd do this, for example, to minimize the cost of moving goods from distribution centers to stores or to minimize the overall transportation costs for busing students to schools.

Are there constraints on how far demand points are from facilities?
By default, demand will be allocated to facilities until all demand has been allocated. You can use the network impedance to specify a cutoff distance, time, or cost—any demand points beyond the cutoff value will not be allocated. You'd specify a cutoff impedance that uses distance, for example, if you want to ensure that students are within three miles of a school. You'd specify a maximum impedance that uses travel time if you want to ensure that households are within a six-minute response time of a fire station. (Any students or households not allocated indicate an unmet need and possibly the need for additional schools or fire stations.)

If demand drops off as distance from a facility increases, you can specify the rate at which this occurs. For example, as the distance or travel time to reach a senior center increases, fewer people may be likely to use the center. You include this constraint by specifying an impedance transformation with a parameter. The options for this are discussed in the section "Maximize demand that diminishes with distance" later in this chapter.

The impedance transformation also controls how much weight is given to the travel required when demand is allocated. This, in turn, impacts which set of facilities represents the optimum solution for allocating demand. By default, demand is allocated to the nearest facility. By setting an impedance transformation that, in essence, increases impedance from the original travel values, a different facility—one that is closer to the demand point having high transportation costs—may be chosen. The higher the impedance, the more likely it is that the chosen facility will be closer to the demand point(s) having higher transportation costs.

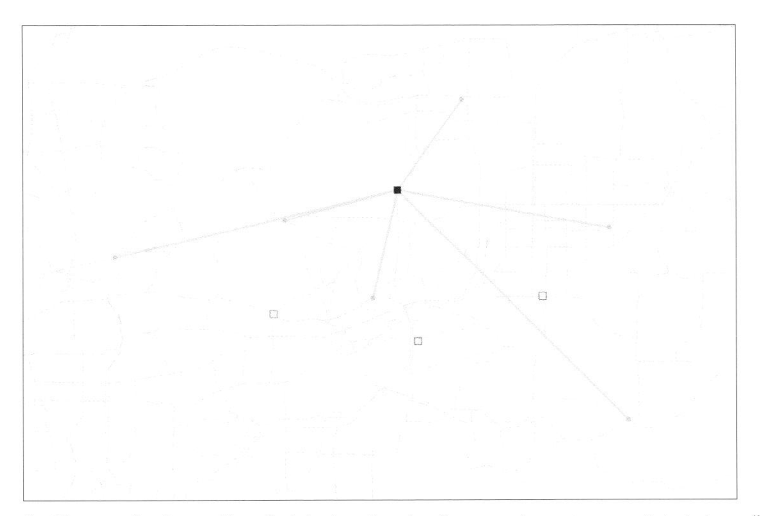

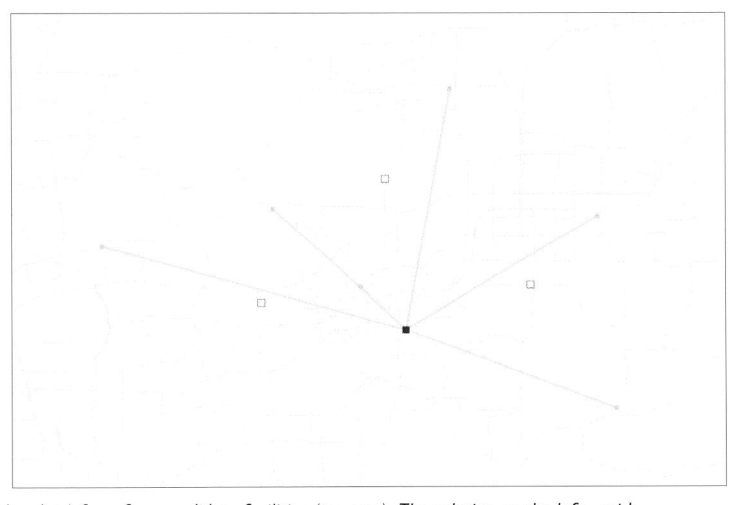

In this example, the goal is to find the best location for a warehouse to serve six pet stores (blue dots) from four candidate facilities (squares). The solution on the left—with the warehouse located at the location at the top (solid blue square)—was created using no impedance transformation, while the one on the right was. Without the impedance transformation, the total distance between the stores and the warehouse is 151,551 feet, and the farthest a store is from the warehouse is 44,041 feet. (All distances are calculated over the highway network, even though the interaction lines are shown as straight lines.) Using an impedance transformation, the total distance is greater (164,154 feet), but the farthest a store is from the warehouse is 39,380 feet. The location is chosen that minimizes the distance any given demand point is from a facility.

Is travel from the facilities to the demand points or vice versa?

The network dataset, as you'll see in the next section, can include one-way streets, restricted turns, and barriers to travel. You can also specify whether—and where—U-turns are allowed. To accurately calculate travel paths and costs, you specify whether travel is from the facilities to the demand points (from fire stations to houses, for example) or from demand points to facilities (as with students to schools). The calculated impedance will account for direction of travel on one-way streets, whether you can turn from one street onto another, whether intersections are "no left turn," and so on.

In this example, when one-way streets are included in the network, the distance traveling from the facility (dark green square) to the demand point (dark green dot) is much shorter (left diagram) than from the demand point to the facility (right diagram).

SET UP THE NETWORK

Location-allocation models require a transportation network dataset. (Network datasets are described in chapter 1, "Introducing GIS modeling" and chapter 4, "Modeling paths." See also the "References and further reading" section at the end of this chapter.) Usually this is a network of streets, although other transportation networks can also be used (walking or bicycle paths, subways, rail and light-rail lines, and so on), as can multimodal networks (those that combine any of the above types of transportation routes).

Network dataset of streets. Edges (streets) connect at junctions (intersections—gray circles). Junctions also occur at ends of edges and may occur along an edge.

You'll want to make sure each edge in the network has an attribute representing impedance. Usually this is related to travel along the network—the distance of each edge, or the travel time or the monetary cost required to traverse it. Impedance is used to limit the allocation of demand to the facility (for example, allocating students to schools within three miles or allocating households to fire stations within a six-minute travel time).

Each edge in the network has an associated length attribute that is calculated when the network is built. Length is used as the impedance, if no other attribute is specified.

NAME	LENGTH	SPEED	ONE WAY
Bailen	335.5	25	
Balboa	323.6	25	
Balboa	296.2	25	
Balboa	201.7	25	
Bear Valley	397.3	35	TF
Bear Valley	360.6	35	FT
Bear Valley	69.2	35	FT
Bear Valley	52.3	35	TF

Attributes for network edges (streets, in this example). Length can be used as an impedance value or used (along with the Speed attribute) to calculate travel time. One-way streets can be included in the model—they are indicated here as allowing travel only in the To-From (TF) or From-To (FT) direction.

Many applications use travel time as the impedance (for example, the response time for fire, police, and emergency medical vehicles). As discussed in chapter 4, "Modeling paths," you can include the measured travel time for each edge or you can calculate it from the travel speed and length of each edge. If travel time is especially important, as with fire response, you'll also want to set up impedances for turns at street intersections (as described in chapter 4).

You can also use a monetary cost as an impedance value. This is stored as an amount for traversing each edge. You'd use monetary cost, for example, when using location-allocation to minimize the total expense of making deliveries from distribution centers to grocery stores in a region (using the delivery cost per mile to calculate impedance). Similarly, you'd use monetary cost to locate schools so as to minimize the total amount required to bus students to schools daily within a school district (using the cost per student per mile to calculate the impedance).

As with other models that use networks, such as shortest path or flow models, you can specify barriers to travel by making edges or junctions impassable. You can also specify one-way streets and allow for various U-turn options (at all junctions, at intersections and dead-ends, at dead-ends only, or not at all). These options are all stored as attributes in the network dataset. The more your network dataset reflects actual travel conditions, the more accurate the results of your model will be.

SPECIFY THE FACILITIES

Facilities are represented as point features in the GIS. Often, existing facilities (such as fire stations, schools, or health clinics) are already stored in GIS databases as point features. If you are building a model to choose the best location for new facilities, you will create the facilities as point features. (You can add the potential locations interactively or load a layer of point features representing the potential locations.)

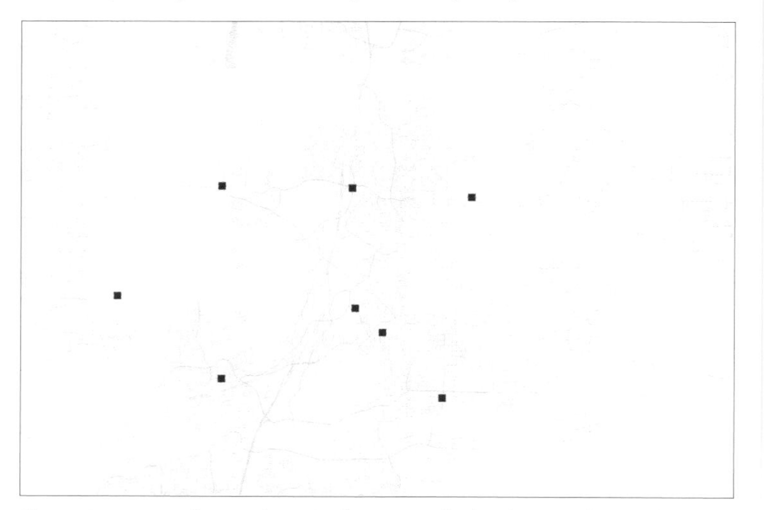

Fire stations are usually stored as point features, so the data layer can be used for facility locations as is.

Facilities that cover a larger area, such as parks, are sometimes stored as polygons showing the boundaries of the features. You'll need to convert these facilities to point features (a point is created at the centroid or other location within the polygon).

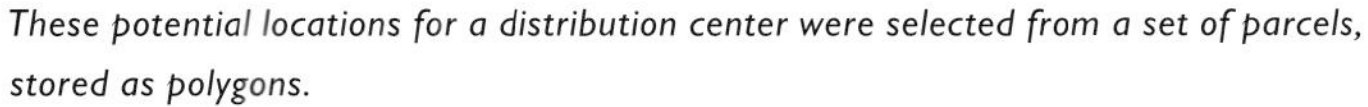

These potential locations for a distribution center were selected from a set of parcels, stored as polygons.

The selected parcel polygons have been converted to point features which can then be used as facilities in the location-allocation model.

When you load the facilities into the model, you specify whether they are "required" or "candidate" locations. If you are allocating demand to existing facilities (such as students to existing schools or households to existing fire stations), specify that the facilities are required and specify that the number of facilities chosen be equal to the number of existing facilities.

If the facilities are yet to be built (such as elementary schools or bus stops in a new planned community), specify that the facilities are "candidates." The candidates may come from your own research or models or from one of the suitability models discussed in chapters 2 and 3—you'd use these models to find suitable locations, then use location-allocation to find the number of facilities to build, and at which specific locations. You'd also specify the facilities as candidates if you're modeling which existing facilities may need to be closed or removed (such as if you're reducing the number of shipping drop-off boxes in a city).

You can include both required and candidate facilities in your solution. For example, there may be several existing fire stations, and you need to build a new one at a location chosen from a set of candidate locations. The existing fire stations would be specified as "required" and the potential locations specified as "candidates." The solution will allocate households to existing stations plus the new location so that the combination of existing stations and the new one best meets demand.

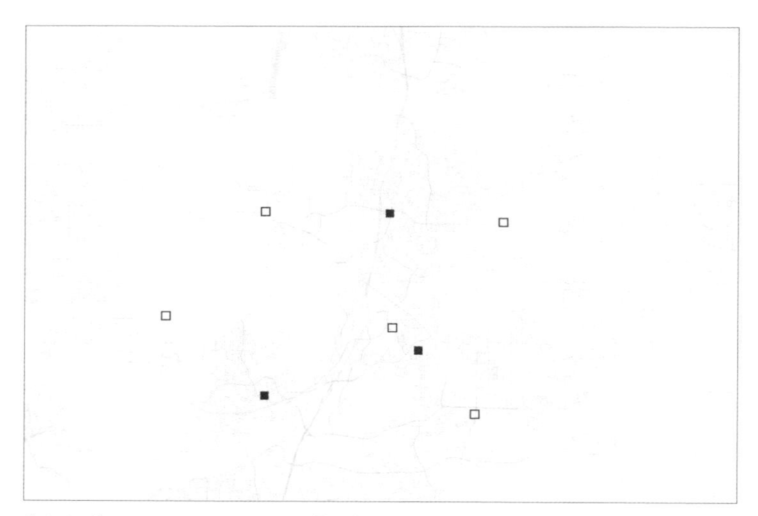

Existing fire stations are represented by the solid squares and candidate locations for a new fire station by the hollow squares. By specifying the existing stations as "required" facilities, you ensure that they will be included in the solution, along with one of the candidate locations.

Once the model has been run, the selected candidate facilities are marked as "chosen." Any required facilities are included in the solution. If, for example, you want to end up with three facilities from a set of seven candidates plus one required facility, you'd specify that you want to select a total of three facilities. The result will show the required facility and the demand allocated to it, along with two additional facilities chosen from the candidates.

If you're allocating demand to existing facilities based on the supply at each facility, the facilities need to have an attribute that holds the supply value. The value may be constant for all facilities (a fire station may serve 4,000 households) or it may vary for each facility (large high schools may have more available slots for freshmen than small high schools do).

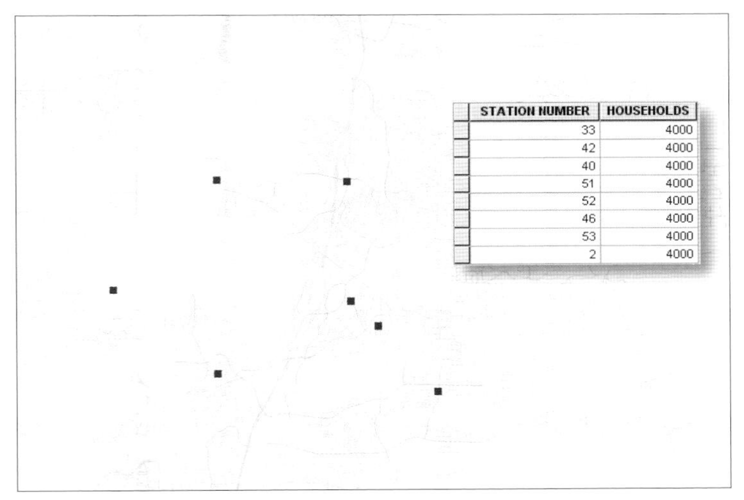

STATION NUMBER	HOUSEHOLDS
33	4000
42	4000
40	4000
51	4000
52	4000
46	4000
53	4000
2	4000

Fire stations with number of households served as the supply field.

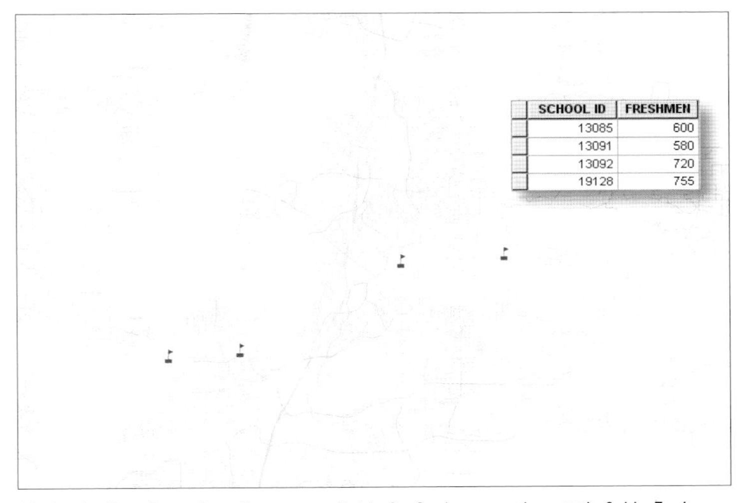

SCHOOL ID	FRESHMEN
13085	600
13091	580
13092	720
19128	755

High schools with number of spaces available for freshmen as the supply field. Each school has a different supply value.

SPECIFY THE DEMAND POINTS

The demand points are the locations served by the facility. They are also—as you might suspect—represented as point features in the GIS.

The demand points can represent the locations of individual features, such as the street addresses of houses, students, customers, or businesses.

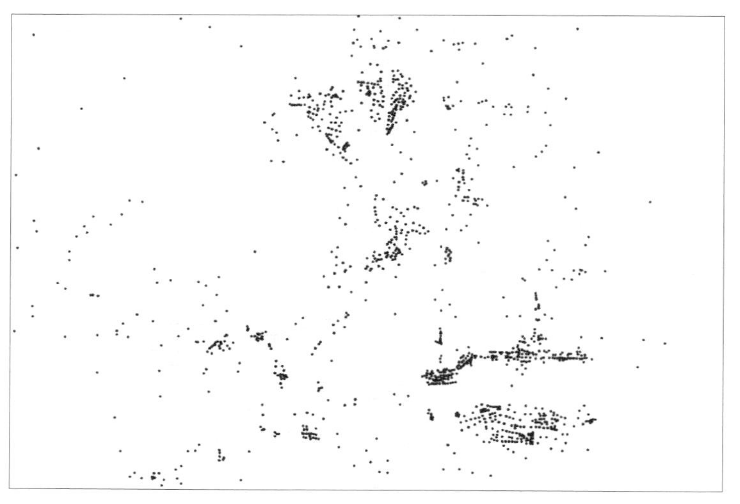

Locations of businesses, with highways. The locations represent demand points for individual features (businesses) in a location-allocation model to find the best places to put transit stops for express buses.

Demand points can also represent lines, such as the streets that are within a certain drive-time of fire stations. In this case, you convert street segments to point locations—generally, a point is created at, or near, the midpoint of each line segment. The points representing the segments are then used as the demand points.

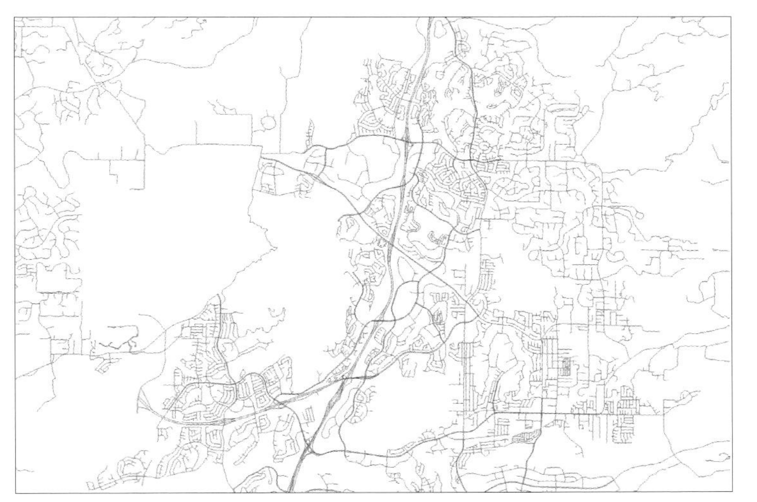

The map on the left shows streets as line features. The map on the right shows street line segments converted to point features. The points are used as demand points in a location-allocation model when finding the best location for a new fire station.

Demand points can also represent data aggregated by area, such as the number of households or the number of high school freshmen in each census block. In this case, a point inside the area (or polygon) becomes the origin location. Often this is the centroid of the polygon. You'd use aggregated data if you don't have individual point locations for the features (the actual location of each household stored as a point feature) or if protecting privacy is a concern. Much demographic data, for example, is available only as aggregated data by census areas (blocks, block groups, or census tracts).

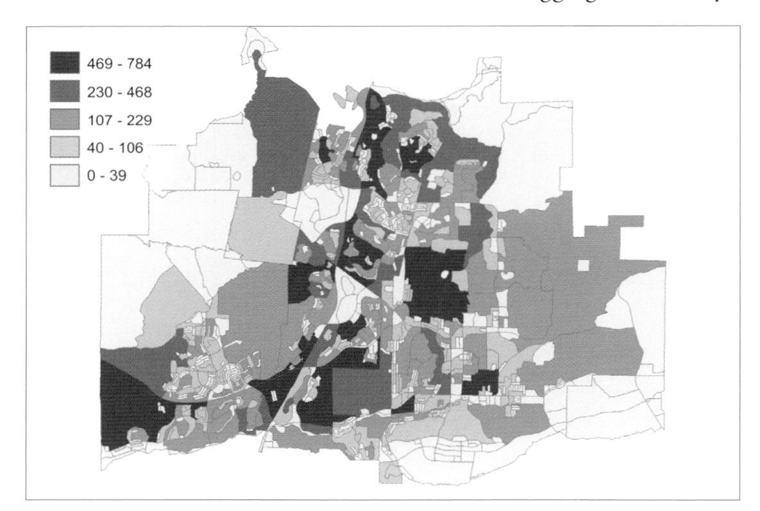

Number of households in each census block.

Census blocks represented by point locations.

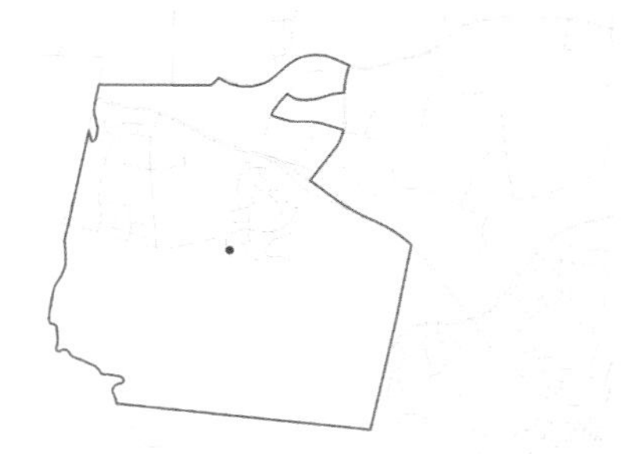

The census block centroid is in the center of the block boundary, but most of the households are likely along the streets in the upper portion of the census block, as much as a mile away from the centroid.

However, for aggregated data, the point may not be an accurate representation of the location of most of the individual locations in the area, especially if the area is relatively large. Plus, it may not be in the center of the area. Since travel is calculated along the network edges (such as streets), where the point is located could determine whether the demand associated with the area gets allocated and which facility it gets allocated to.

When you add the facilities and demand points to the model, they are assigned to the nearest edge (line) or junction in the network.

The distance of the demand point along the edge, from the From-point of the edge, is calculated and stored (each edge has a direction and hence a From-point and a To-point). This partial length is added to the lengths of all the edges between the facility and the demand point to determine the distance between the two.

To calculate the impedance values—such as time or money—the percentage distance is also calculated. For example, if the edge is 504 feet long, and the demand point is 327.6 feet from the From-point of the edge, the demand point is 65 percent of the way along the edge. This percentage is used to calculate the impedance to or from the demand point.

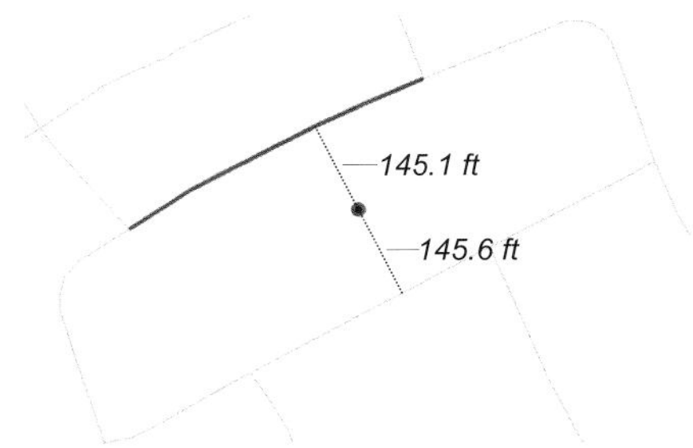

The demand point is closer to the upper street (dark green) so its location is assigned to that network edge.

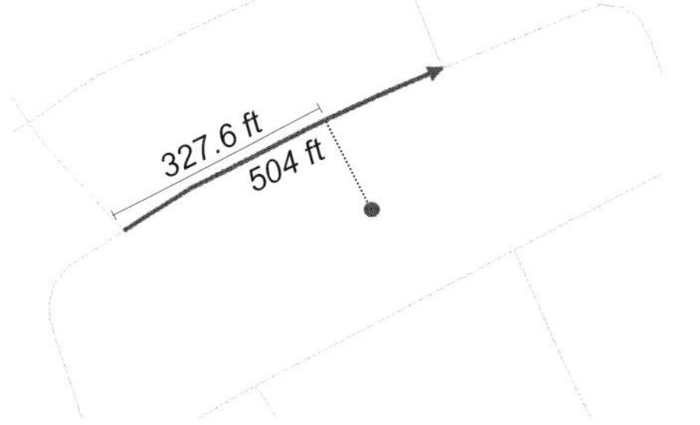

Once the demand point is located on the network, its position along the network edge is calculated—in this example, the point is about 65 percent of the way along the edge.

When you convert polygon or line features to demand points, where the point is located will determine which edge, or junction, the point is assigned to and how far along the edge it is. That can affect the calculated impedance and, in turn, the results of the location-allocation solution. This is generally only a problem for demand points that are near the cutoff distance or are equidistant from two or more facilities.

The distance between the facility (green square) and the demand point is measured between their locations on the network edge. Where the demand point is located will affect the distance and thus the results of the model. In the left diagram, the demand point is located on the upper street (as shown in the previous diagrams) and the distance from the facility is 1,923.7 feet. If the demand point were shifted slightly and located on the opposite street (diagram on the right), the distance would be 2,082.9 feet. With a cutoff distance of 2,000 feet, the demand point would be allocated in the first case, but not in the second.

For some applications this issue is not critical—if you're allocating library patrons (aggregated to census block centroids) to libraries, whether a census block is assigned to one library or another may not matter to the results of your analysis; if the block is an equal distance from two libraries, the patrons may in reality use either or both libraries. On the other hand, if you're allocating students to schools that have a supply limit, whether the students in a particular census block are assigned to one school versus another takes on more significance. For these applications you'll want to examine the demand points that are in locations where the placement of the point will affect the results and make sure they are in the location that makes the most sense. For example, if the point represents a block that is in a residential neighborhood but is adjacent to a major highway, it's likely most of the students actually live on the residential street rather than the highway, so make sure the point is located on the residential street (you can do this by editing the layer of demand points and moving points as necessary).

Demand points have an associated demand value, or weight, stored in the layer's attribute table. For point locations representing a single demand point, such as a customer or student, the demand value is one. In some cases the demand value is a quantity associated with the demand point. For example, in a model to allocate workers to bus stops, each point represents an individual business and the demand value is the number of employees at each. For demand points representing aggregated data, the demand value always represents a quantity associated with each location, such as the number of households in each census block or the population of each census tract.

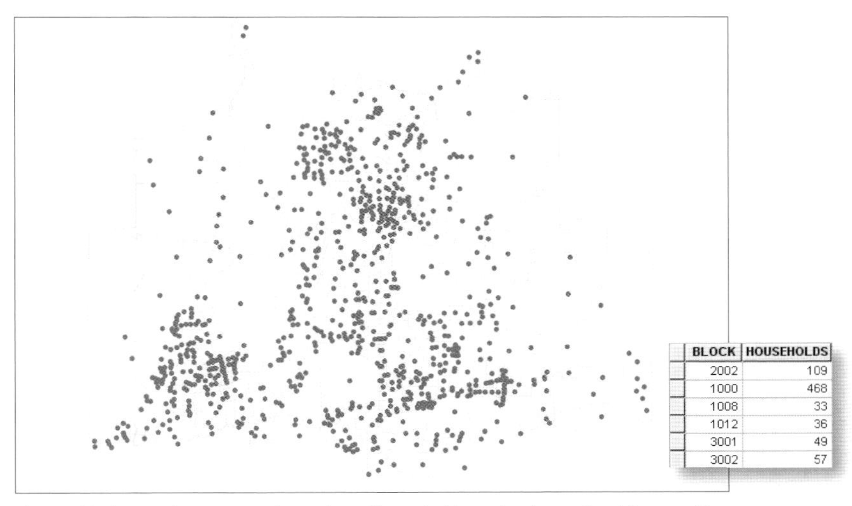

BLOCK	HOUSEHOLDS
2002	109
1000	468
1008	33
1012	36
3001	49
3002	57

Census block point locations, with number of households as the demand weight stored in the attribute table.

By default, when you specify an impedance cutoff or impedance transformation in the model, the parameters you specify apply to all the demand points. However, you can assign some or all the demand points individual cutoff or transformation values. You do this by assigning the appropriate values in the demand point layer's attribute table. (If a demand point has a value for one of these parameters in the attribute table, that value is used, otherwise the value used is the one you specify when you run the model.) This lets you handle special cases such as requiring shorter response times for fire trucks in denser, urban areas, or seniors being willing to travel longer distances to an adult day-care center in rural areas versus suburban areas.

RUN THE MODEL

Once you've set up the network and specified the facilities and demand points, you can run the model. When running the model, you specify the type of location-allocation problem you're solving—the model is optimized to provide a solution to that particular problem. Regardless of the problem type, the same underlying method is used to reach a solution—demand from a given location (demand point) is allocated to the nearest facility. Solutions vary in how the interplay between distance and demand is handled. And, in fact, depending on the data and the parameters you use, different solutions may produce the same results. However, you specify the problem type that's the most appropriate for your application—whether you want to maximize allocated demand, match demand to supply at a facility, maximize demand that diminishes as distance increases, or allocate demand so that overall transportation costs are minimized.

With each of the solutions, the results will show you the set of facilities chosen, which demand points are allocated to which facilities, the total demand allocated to each facility, and the total weighted impedance for each facility (such as the total distance or total time for all students to travel to their assigned school). If you're choosing locations from among candidates, you'll also see which candidates provide the best solution (in terms of allocating demand in the most efficient way).

Maximize demand allocated

One solution maximizes the amount of demand allocated, with all the demand from each demand point allocated to the single facility closest to it. You can specify a distance or time cutoff beyond which demand is not allocated. This is referred to generally as the MaxCover problem type.

If you know how many facilities you need in the solution, you specify this number (these can be either existing facilities, new facilities, or a combination). In ArcGIS, this is known specifically as the Maximize Coverage problem.

If, on the other hand, you want to find out how many facilities are needed to meet the available demand, you have the model determine this and choose the facilities that meet the most demand. In ArcGIS, this is known as the Minimize Facilities problem (still considered a type of MaxCover problem).

Maximize coverage

If you want to allocate demand to existing facilities, specify all the facilities as being required, then specify that the number of facilities to choose is equal to the number of existing facilities. For example, if you want to create service areas for eight existing fire stations—using streets as demand points—you'd specify the existing stations as required facilities, then specify that you need eight facilities in the solution. The streets will be allocated to their closest station. The results show how much demand is assigned to each station (the number of streets assigned to each).

The goal of this solution is to assign streets to existing fire stations so that each street is assigned to its nearest station (over the street network), and all streets are assigned. The maps shows the eight required facilities, the demand points representing street segments, and the street network dataset.

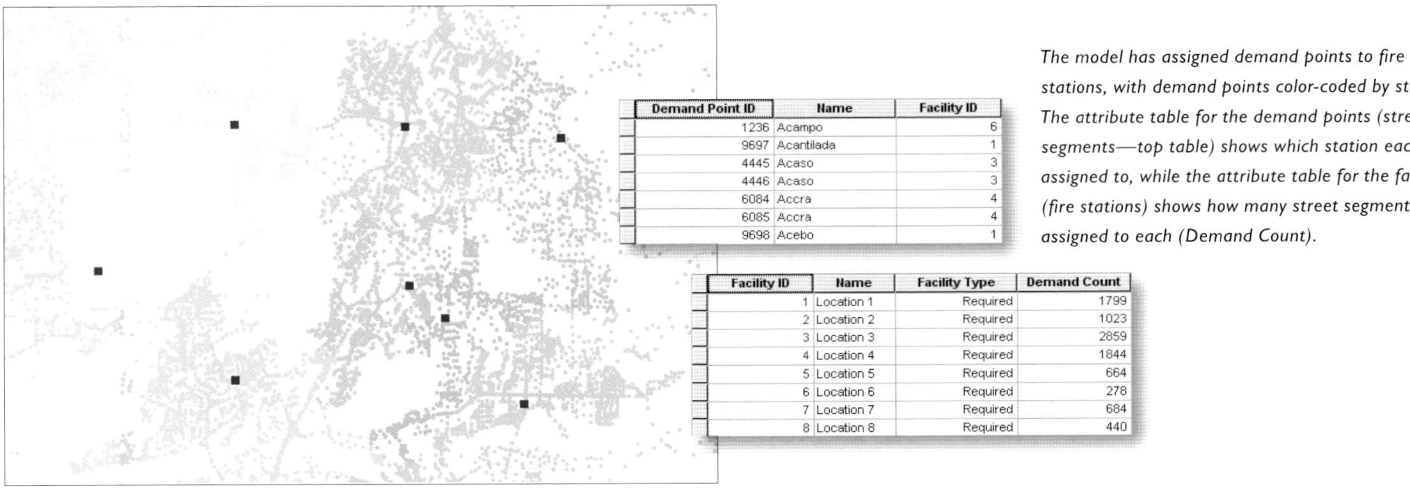

Demand Point ID	Name	Facility ID
1236	Acampo	6
9697	Acantilada	1
4445	Acaso	3
4446	Acaso	3
6084	Accra	4
6085	Accra	4
9698	Acebo	1

Facility ID	Name	Facility Type	Demand Count
1	Location 1	Required	1799
2	Location 2	Required	1023
3	Location 3	Required	2859
4	Location 4	Required	1844
5	Location 5	Required	664
6	Location 6	Required	278
7	Location 7	Required	684
8	Location 8	Required	440

The model has assigned demand points to fire stations, with demand points color-coded by station. The attribute table for the demand points (street segments—top table) shows which station each is assigned to, while the attribute table for the facilities (fire stations) shows how many street segments are assigned to each (Demand Count).

You can specify a maximum time or distance cutoff—only demand points within the time or distance of a facility will be allocated to the facility. For example, you can specify that streets must be within a four-minute travel time of a fire station. Demand points that are beyond the distance or time of any of the facilities will not be allocated (this will show you where demand is not being met).

The model has assigned demand points (street segments) to facilities (fire stations) within a four-minute travel time. Demand points are color-coded by station—points that are more than four minutes from a fire station are unassigned (gray dots). The attribute table for the facilities (fire stations) shows how many street segments are assigned to each (Demand Count).

Facility ID	Name	Facility Type	Demand Count
1	Location 1	Required	1491
2	Location 2	Required	708
3	Location 3	Required	1371
4	Location 4	Required	979
5	Location 5	Required	333
6	Location 6	Required	95
7	Location 7	Required	600
8	Location 8	Required	124

If the goal of your model is to find where to add facilities, specify the existing facilities as required and any potential locations as candidate facilities. You'd do this, for example, if there are two existing libraries, you're building a new one, and there are three potential locations. There are a total of five facilities: two required and three candidates. Specify that the solution include three facilities. The results will include the two existing libraries, plus the new location that—in conjunction with the existing libraries—best allocates the demand.

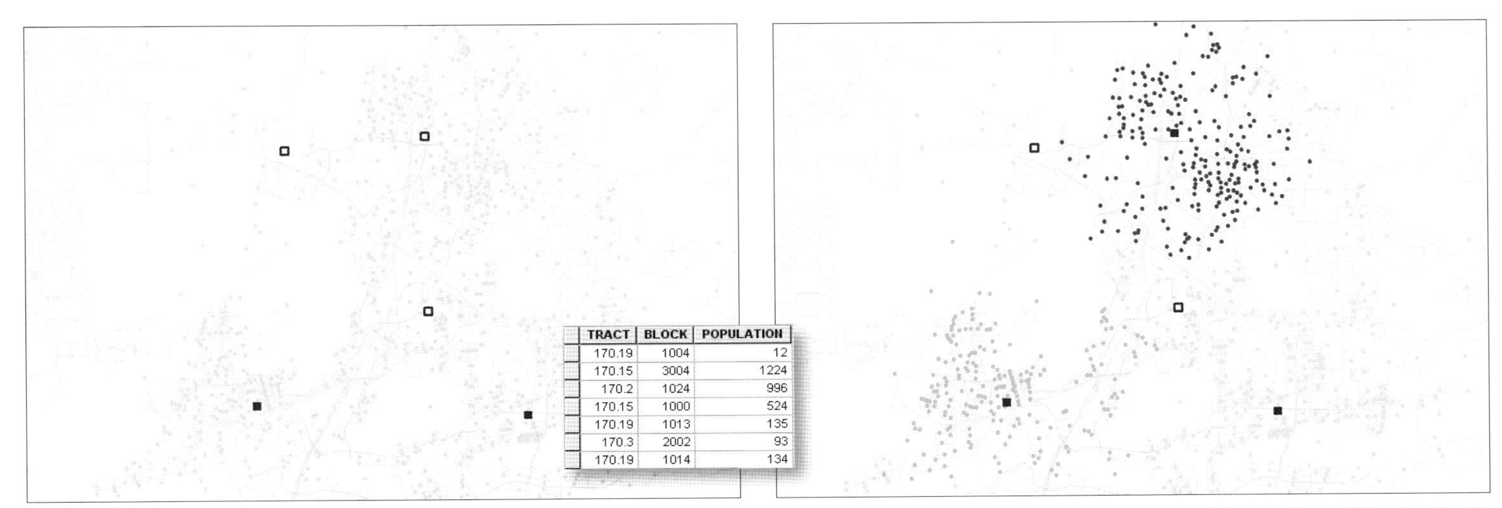

TRACT	BLOCK	POPULATION
170.19	1004	12
170.15	3004	1224
170.2	1024	996
170.15	1000	524
170.19	1013	135
170.3	2002	93
170.19	1014	134

In this model, there are two existing libraries (solid blue squares) and three candidate locations to build one new library (hollow squares). The demand points (blue dots) represent census block centroids, with a weight value of the population of each block (as shown in the attribute table). The goal is to find the location for the new library that will ensure the most people are within three miles of a library, so a cutoff of 15,840 feet is used.

The model solution keeps the two existing libraries and chooses the candidate location (in the upper center) that will make sure the most people are within three miles of a library. The demand points are color-coded by the library they're closest to, while the light blue dots are more than three miles from a library and are unassigned.

If you're locating new facilities in an area where none exist and you know how many facilities are needed, specify the potential locations as candidate facilities. You'd do this, for example, if you need to build two libraries for a new planned community—based on the predicted population—and you have five potential locations. Specify the potential locations as candidates—the results will show which three locations best meet the demand, and thus where the libraries should be built.

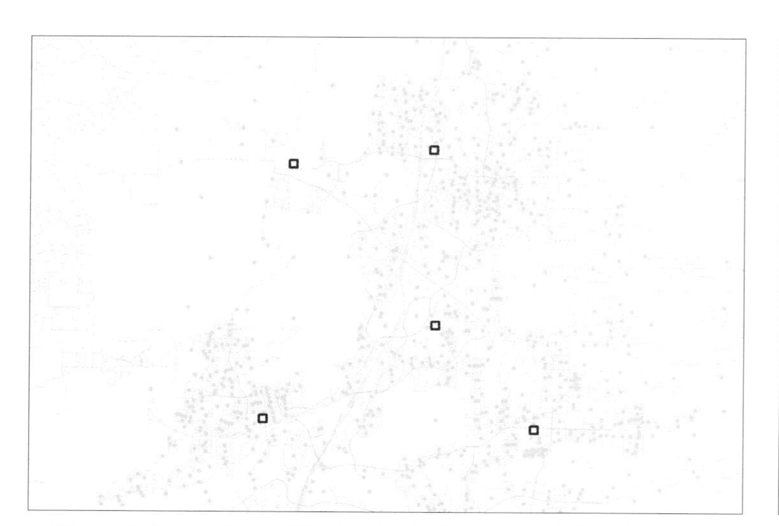

In this model, there are five candidate locations (hollow squares—left map) for two new libraries. The demand points (blue dots) represent census block centroids, with the weight value being the population of each block. The goal is to find the locations for the new libraries that will ensure the most people are within five miles of a library, so a cutoff value of 26,400 feet is used. In the solution (right map) the locations in the lower left and center were chosen (solid blue squares) and the census blocks assigned to the closest library. Blocks that are more than five miles from a library are unassigned (light blue dots).

Minimize facilities

Rather than specifying how many facilities you need the solution to include, you can have the GIS calculate the minimum number of facilities necessary to cover the demand. You specify the potential locations as candidate facilities. You also specify a distance or time cutoff (without the cutoff, one facility would meet all the demand). The GIS then chooses the subset of candidate facilities that meets the demand.

You'd use this approach if, for example, you're building facilities where none exist—such as fire stations or libraries in a new planned community—and you need to find out the number of facilities required to cover the demand.

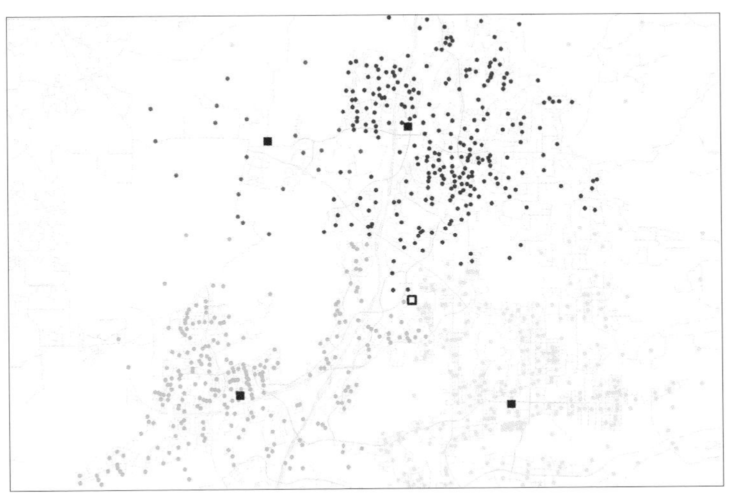

In this model, there are five candidate locations for libraries (blue squares, above). The demand points (blue dots) represent census block centroids, with a weight value of the population of each block. The goal is to find the minimum number of libraries—and their locations—that will ensure the most people are within five miles of a library. A cutoff value of 26,400 feet is used. In the solution (right map), the GIS chose four locations (solid blue squares). Census blocks are assigned to the closest library. Blocks that are more than five miles from a library are unassigned (light blue dots)—in this solution only a few blocks are unassigned.

This approach is also useful if you have facilities that are nonpermanent or easily moved—such as school bus stops (which may only be a designated corner) or shipping drop-off boxes (which can easily be relocated)—and you want to determine the most efficient number and location of the facilities. Even though the facilities are existing, you'd specify them all as candidates (rather than required) before running the model.

In this model, the facilities are potential school bus stops, indicated by the gray squares representing each street segment. The demand points are census block centroids (blue dots), with the number of elementary school students as the weight value. The goal is to have the GIS choose the minimum number of stops, and their locations, so that no students have to walk more than a quarter mile to a stop—a cutoff value of 1,320 feet is used.

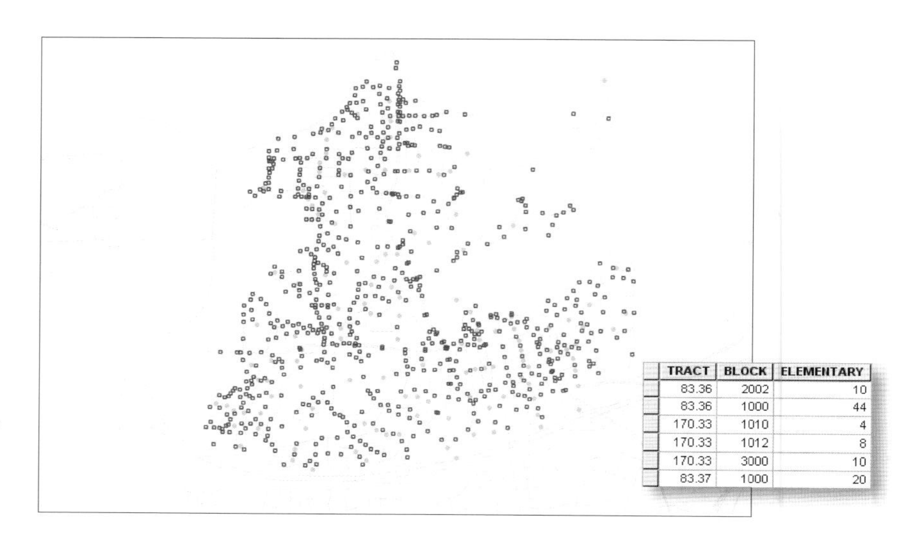

TRACT	BLOCK	ELEMENTARY
83.36	2002	10
83.36	1000	44
170.33	1010	4
170.33	1012	8
170.33	3000	10
83.37	1000	20

In the solution, the GIS has chosen 35 stops. In the map, a line is drawn between each stop and the blocks assigned to it. (Even though the lines are straight, the walking distance for students is calculated over the street network.) The attribute table for the facilities shows—for each stop—the name of the street it is on, the number of blocks assigned to it, and the total number of students in those blocks—that is, the number of students that would use the stop. The solution has assigned a large number of students to each stop—you may need to rerun the model and specify a larger number of stops (which would likely entail adding more bus routes).

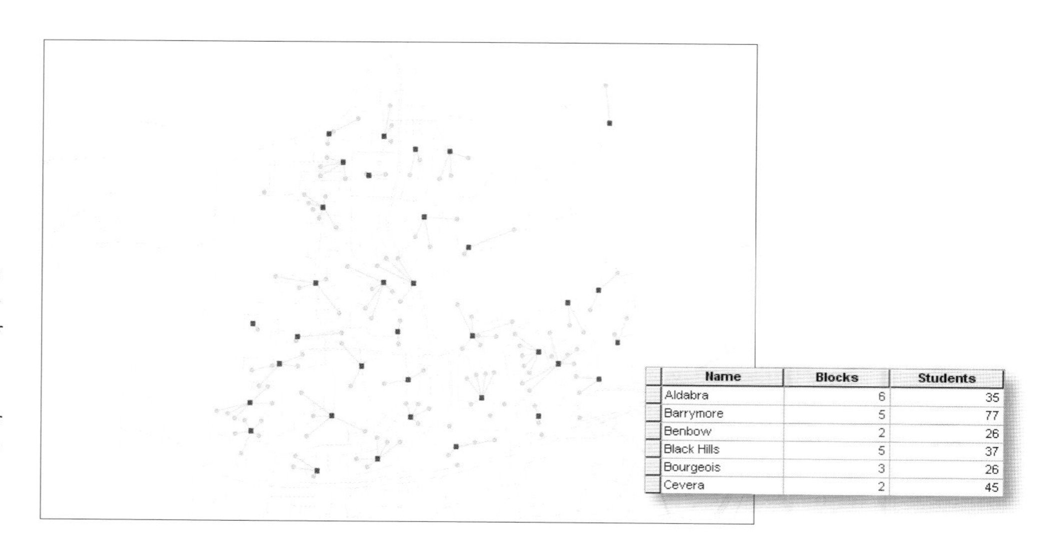

Name	Blocks	Students
Aldabra	6	35
Barrymore	5	77
Benbow	2	26
Black Hills	5	37
Bourgeois	3	26
Cevera	2	45

Another use for this approach is to find out which, if any, facilities need to be closed due to a drop in demand. Suppose sales are down at your chain of coffee shops and you need to close some shops. You'd identify the shops having the lowest revenues and specify these as candidate facilities. You'd also specify the cutoff (the maximum time or distance a customer can be from a shop). The results will show you which shops can be closed without reducing your customers' access to a shop. (Presumably, sales would subsequently increase at the remaining, reduced number of shops.)

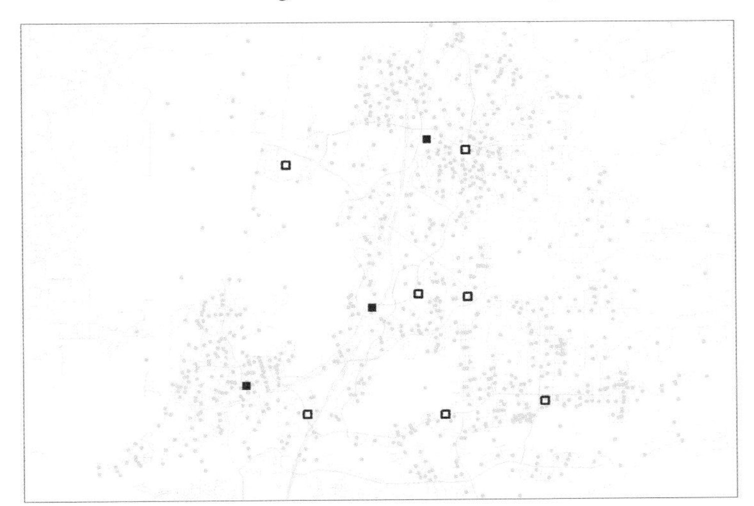

In this model, the facilities are existing locations for a chain of coffee shops. Three of the shops are performing well and will remain open (solid squares). The rest are candidates for closing (hollow squares). Even though they are existing, they are specified as candidates in the model so the GIS can consider whether or not the location is necessary. The demand points are census block centroids (blue dots), with the total population as the weight value. The goal is to have the GIS choose which shops, if any, can be closed while still ensuring that people have to drive no more than five minutes to a shop.

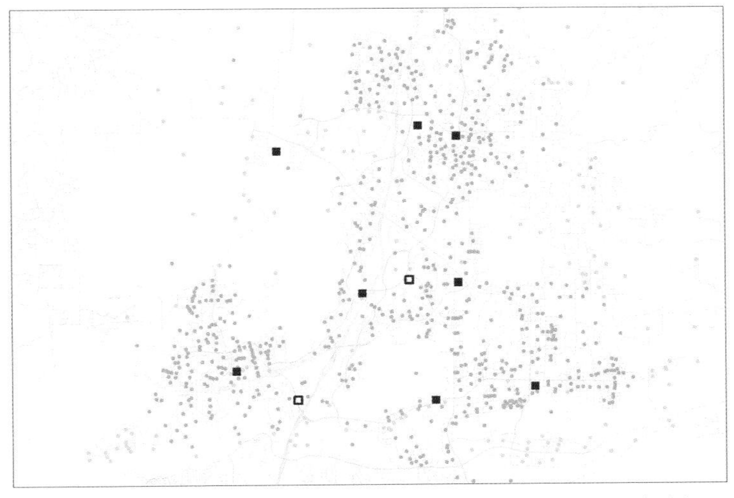

In the solution, the model calculates that closing two shops (the hollow squares) while keeping eight shops (the three that are doing well plus five others, shown as solid squares) would not reduce the number of people within a five-minute drive of a shop. Allocated blocks (those within five minutes of a shop) are shown as tan dots. Some blocks have no population, and hence no demand, and are unallocated.

Match demand to available supply at a facility

This problem type assigns demand to facilities based on the supply at the facility and the amount of demand from the surrounding area. Demand is allocated to the closest facility until the available supply is met. In ArcGIS, this is known as the Optimum Allocation problem. An example of matching demand to supply is allocating households to fire stations, assuming that each fire station can cover about 8,000 households (in this case, the supply value is the same for each facility). Another example is assigning students taking a college entrance exam to test centers, based on the number of available spaces at each center (in which case, supply varies from facility to facility).

To run the model, you specify all the facilities as being required so that demand is allocated to existing facilities. You also specify the attribute of the facilities that contains the supply values. The result shows which demand points have had all their demand allocated to a facility and which have had some or none allocated to a facility (generally, these are the demand points that are farthest from any facility). The result also shows which facilities have had all their supply fulfilled and which, if any, have excess supply. If there is excess supply at any of the facilities, it usually means all the demand at the demand points has been allocated. Conversely, if there is excess demand, all the supply at the facilities will have been fulfilled, leaving some demand points with unallocated demand.

In this scenario, students (represented by census block centroids shown as blue dots) are allocated to college entrance exam test centers, based on the supply of test slots at each center (as shown in the table). For this first test date of the year, relatively few students are taking the test and all students can be allocated. In fact, one center has extra supply (orange square). The partial table of census blocks shows that there is no extra demand.

Name	Demand	Extra Demand
Location 17	1	0
Location 18	0	0
Location 19	3	0
Location 20	1	0
Location 21	0	0
Location 22	0	0

Facility ID	Supply	Extra Supply
1	90	58
2	50	0
3	70	0
4	70	0

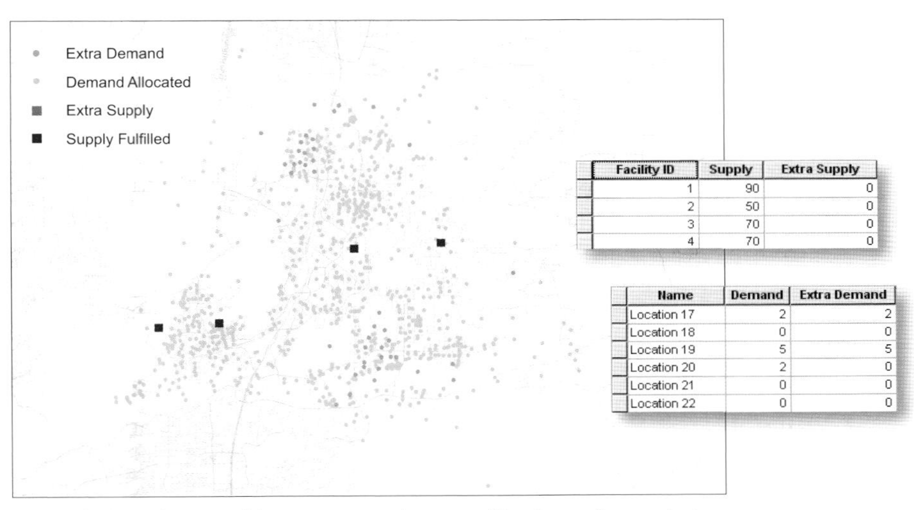

Facility ID	Supply	Extra Supply
1	90	0
2	50	0
3	70	0
4	70	0

Name	Demand	Extra Demand
Location 17	2	2
Location 18	0	0
Location 19	5	5
Location 20	2	0
Location 21	0	0
Location 22	0	0

For the third test date test of the year, more students are taking the test (procrastinators plus students retaking the test). All supply at the test centers is fulfilled (blue squares) as shown in the table, and there is extra demand for some census blocks (orange dots). The partial table of census blocks shows that demand has increased and, for some blocks, there is extra demand.

As with the other problem types, you can refine the allocation by specifying a maximum impedance. You'd specify a maximum impedance that uses distance if you want to ensure that students are within three miles of a test center, along streets.

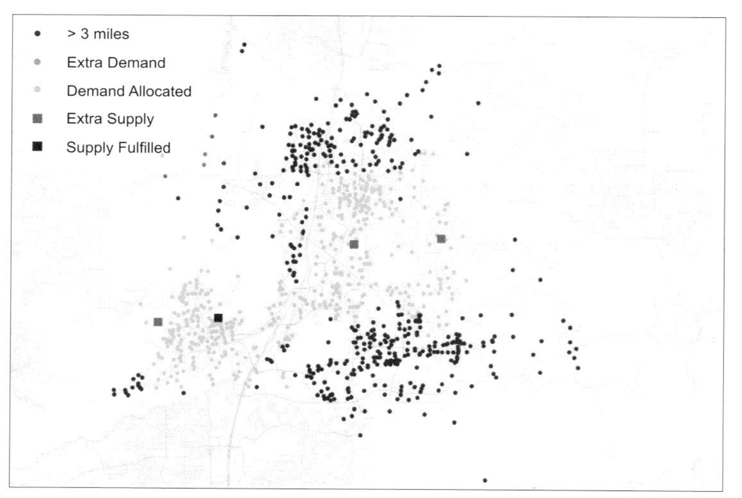

When a distance limit of three miles is applied (no student should be more than three miles from a test center), only students near centers are allocated and students in more distant census blocks (red dots) are not. All but one center have excess demand and can accommodate more students. The solution would be to increase the allowed travel distance or add test centers in areas not currently served.

You'd specify a maximum impedance that uses travel time if, for example, you want to ensure that households are within a six-minute response time of a fire station, given a supply of 8,000 households per station.

Students or households will not be assigned to the facility if the maximum impedance is exceeded, even if demand has not been completely allocated. (Any students or households not allocated indicate an unmet need—you'd want to increase supply at the existing facilities, or else add test centers or build new fire stations.)

Maximize demand that diminishes with distance

A third type of location-allocation problem you can model is when the number of people likely to use a facility decreases as distance (or travel time) to the facility increases. (This problem is known in ArcGIS as Maximize Attendance.) Often, stores and other retail facilities experience this kind of demand drop-off (sometimes termed distance decay or distance friction)—people are less likely to visit a store the farther away it is. However, modeling interaction for retail facilities often includes additional factors where the choice of the customer is involved, such as the presence of competitors and the attractiveness or pull of each store (such as how large it is, what it's near, and so on). These types of market share models are discussed later in this chapter in the section "Modeling travel to facilities."

There are cases, though, where facilities have no competitors and each facility essentially offers the same service as any other (so one is not likely to draw more people than another) but where people still have choice about whether to use the facility or not—if it's close enough they may use it, but if it's too far away they may not. One example is public bus stops—people who live close to a bus stop are more likely to use the bus; as the distance from the stop increases, fewer people are likely to use the bus and will opt to drive instead. Another example is senior day-care centers—people near a center are more likely to use it, and to use the closest one, but as the distance increases and it takes more time and effort to get there, fewer people will make the trip. If it's too far, they won't bother to go at all.

Since this model calculates how many people (or what proportion of available demand) will go to a facility, you likely wouldn't use it to allocate demand to existing facilities where the demand from all locations has to be allocated (as with fire stations or schools). Rather, you would use it to locate new facilities (such as bus stops or day-care centers) to cover as many potential users of the service as possible, given the distance decay phenomenon.

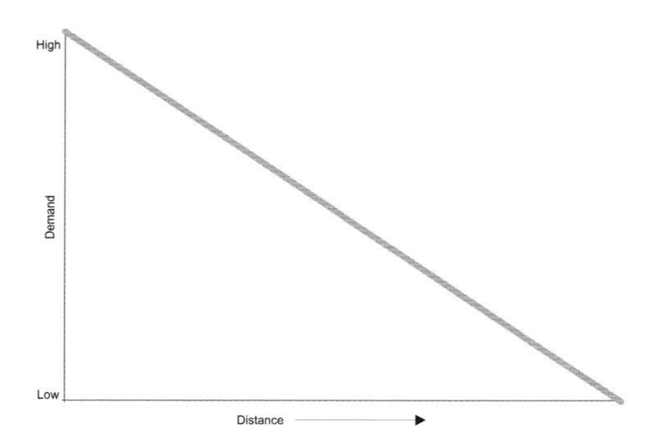

In a linear transformation, demand decreases at a constant rate as distance increases.

As with the other problem types, demand is allocated to the nearest facility, within a cutoff distance. To specify the rate at which demand decreases (that is, the likelihood of someone visiting a facility decreases) as distance increases, you specify an impedance transformation. The impedance transformation describes the relationship between the distance from the facility and the percentage decrease in demand. It is often represented as a curve on a graph. The simplest transformation is a linear one—demand decreases at a constant rate as distance increases.

Essentially, for each demand point, the GIS looks up a percentage on the curve based on the distance from the facility, multiplies the demand at the location by the percentage, and allocates that transformed demand to the facility. The transformed demand is the number of people that could be expected to use the facility from that location. So, for example, if there are 18 seniors in a census block, and the proportion, based on the distance from the facility, is 50 percent, nine seniors would be expected to travel to the day-care center from that block—a demand of nine is allocated to the center. The GIS calculates the transformed demand for each demand point to obtain the total allocated demand for each facility.

You can specify an impedance transformation with a cutoff. For example, you may know that people will not travel more than five miles to a senior day-care center, so you'd specify five miles as the cutoff.

Number of potential senior center attendees per census block group.

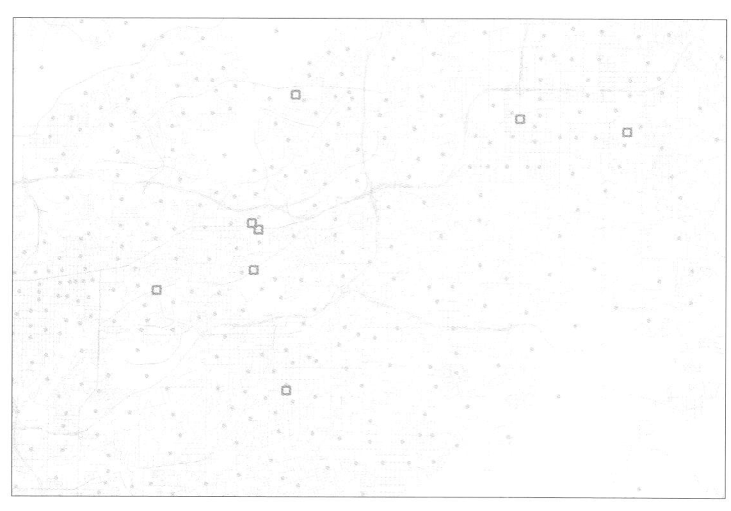

Potential senior center locations (squares) with block group centroids (dots) as demand points.

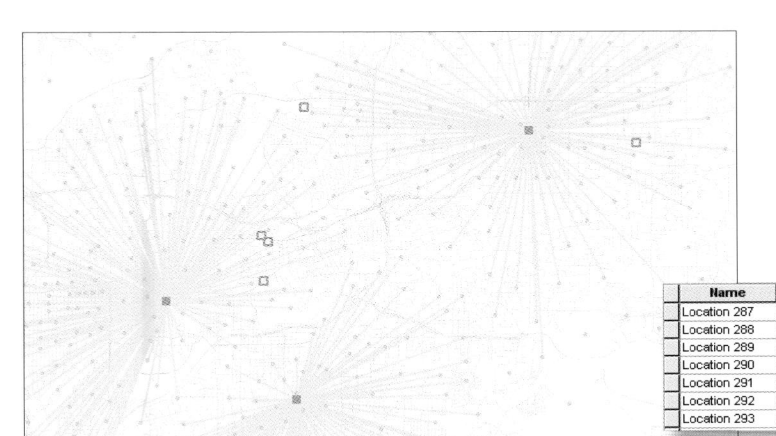

Name	Potential Attendance	Facility ID	Allocated Attendance
Location 287	8	8	8
Location 288	5	8	5
Location 289	22	8	22
Location 290	13	8	13
Location 291	22	8	22
Location 292	14	8	14
Location 293	10	8	10

Facility ID	Status	Allocated Block Groups	Total Attendance
1	Candidate	0	0
2	Candidate	0	0
3	Candidate	0	0
4	Chosen	162	2797
5	Candidate	0	0
6	Candidate	0	0
7	Chosen	105	2238
8	Chosen	102	1579

Choosing three facilities without diminishing demand, using a cutoff distance of five miles (26,400 feet). All demand at each demand point is allocated (demand point attribute table). In the facilities attribute table, you can see the total demand allocated to each facility.

Choosing three facilities while accounting for diminishing demand with increasing distance, using a linear transformation with a cutoff distance of five miles (26,400 feet). A portion of the demand at each demand point is allocated (demand point attribute table). In the facilities attribute table, you can see the total demand allocated to each facility—since only a portion of the demand at each demand point is allocated, the total demand allocated for each facility is less than in the solution above.

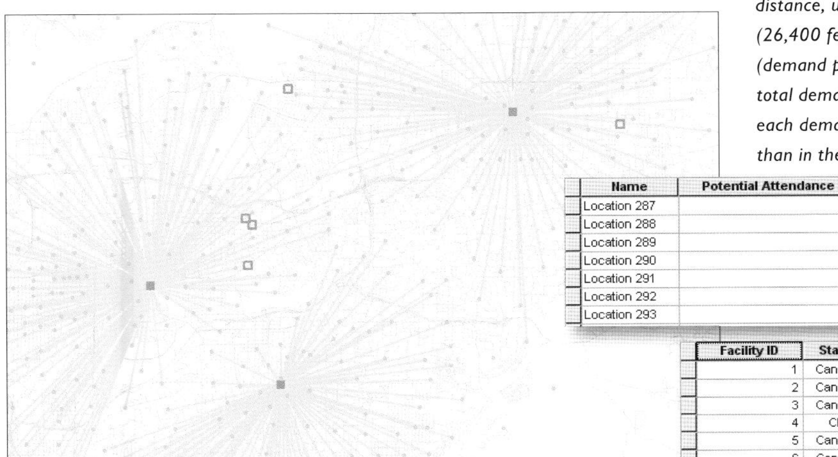

Name	Potential Attendance	Facility ID	Allocated Attendance
Location 287	8	8	7.8
Location 288	5	8	4.3
Location 289	22	8	18.4
Location 290	13	8	10.2
Location 291	22	8	20.2
Location 292	14	8	12.4
Location 293	10	8	8.2

Facility ID	Status	Allocated Block Groups	Total Attendance
1	Candidate	0	0.0
2	Candidate	0	0.0
3	Candidate	0	0.0
4	Chosen	162	1149.9
5	Candidate	0	0.0
6	Candidate	0	0.0
7	Chosen	105	970.2
8	Chosen	102	741.2

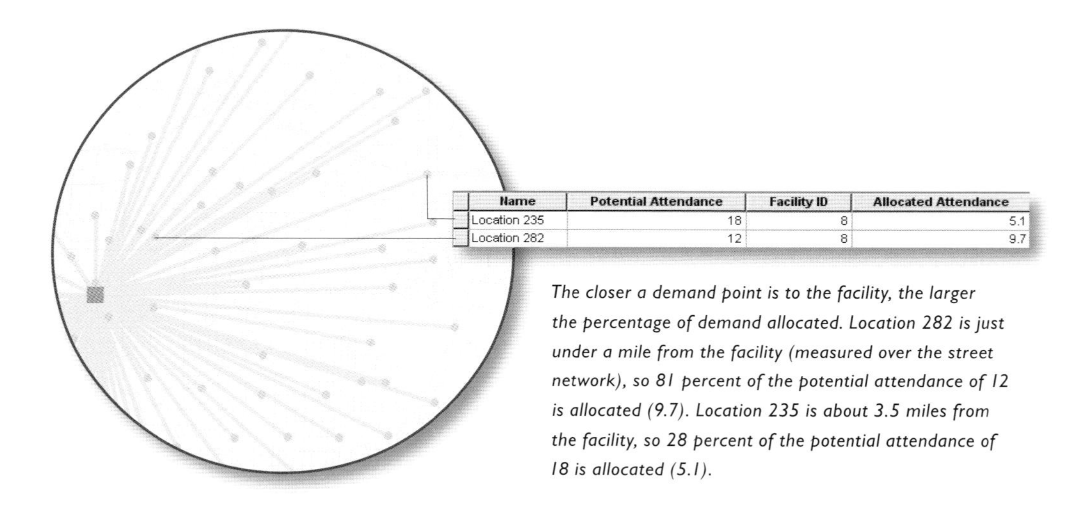

Name	Potential Attendance	Facility ID	Allocated Attendance
Location 235	18	8	5.1
Location 282	12	8	9.7

The closer a demand point is to the facility, the larger the percentage of demand allocated. Location 282 is just under a mile from the facility (measured over the street network), so 81 percent of the potential attendance of 12 is allocated (9.7). Location 235 is about 3.5 miles from the facility, so 28 percent of the potential attendance of 18 is allocated (5.1).

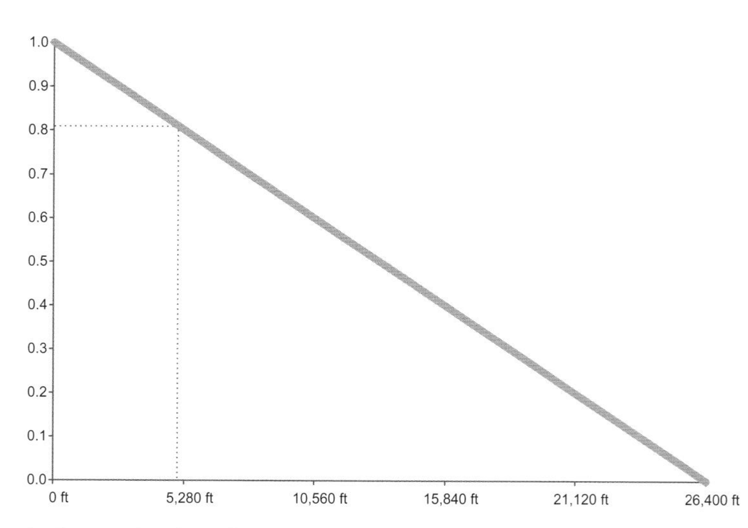

In this example, a linear function, or graph, is used to determine the percentage of demand allocated (y-axis) given the distance of the demand point from the facility (x-axis). For Location 282, which is 4,995 feet from the facility, the percentage is 81 (0.81).

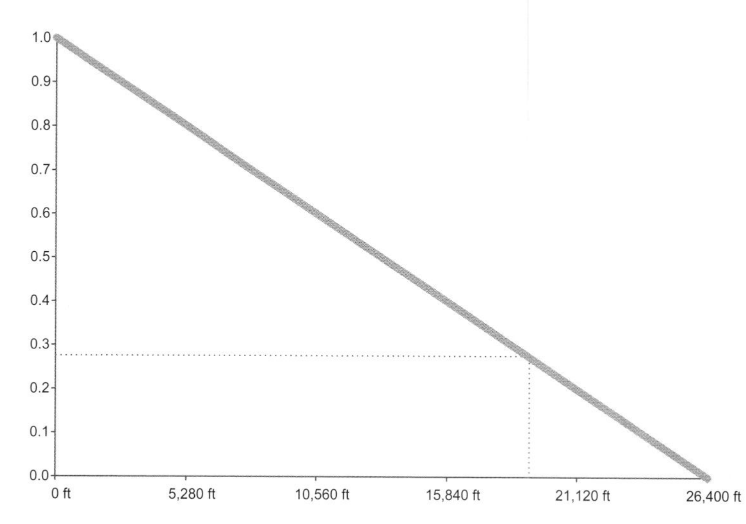

For Location 235, which is 18,946 feet from the facility, the percentage is 28 (0.28).

In many applications, demand doesn't, in fact, drop off at a constant rate as distance increases, but rather drops off at an increasing rate as distance increases. This is true, for example, for many retail applications. A power or exponential function is used to capture this phenomenon. (These functions are discussed in the next section, "Modeling travel to facilities.") In practice, it may be challenging to replicate exactly the behavior of people traveling to a facility, but data on where trips originated—often collected by stores or through visitor surveys—can help you select the function that most closely represents this behavior.

Minimize overall transportation cost

A fourth location-allocation problem type is one that minimizes the overall transportation costs. That is, you locate facilities to move the most goods or people using the least total distance, time, or money. This is known as the Minimize Impedance problem. It's also referred to as the *p*-Median model.

This problem often applies when travel between the facility and all demand points occurs on a regular basis. You'd specify this problem, for example, if you want to locate grocery distribution centers so the cost of delivering to all the stores in the region is minimized.

The model attempts to minimize the total distance that people have to travel to a facility, so is often used for locating public facilities such as libraries or health care centers. (Other problem types—such as those that maximize overall demand allocated, for example—may result in some people having to travel much farther to one facility than other people do to another facility.)

Minimizing overall transportation cost is not useful for locating facilities when there's only potential or occasional travel between demand points and facilities. For example, a fire station doesn't respond to all the buildings surrounding it on a daily basis and may never respond to most buildings. So trying to locate fire stations based on minimizing transportation costs between the stations and the buildings doesn't make sense. The goal for fire stations is to locate them to make sure they can cover the most area within a given response time.

The model attempts to allocate demand and locate facilities by minimizing total weighted impedance. For each facility, the allocated demand is multiplied by the total impedance to the facility (from all the allocated demand points). These values are then summed for all facilities—the solution chooses facilities so that this summed value is minimized.

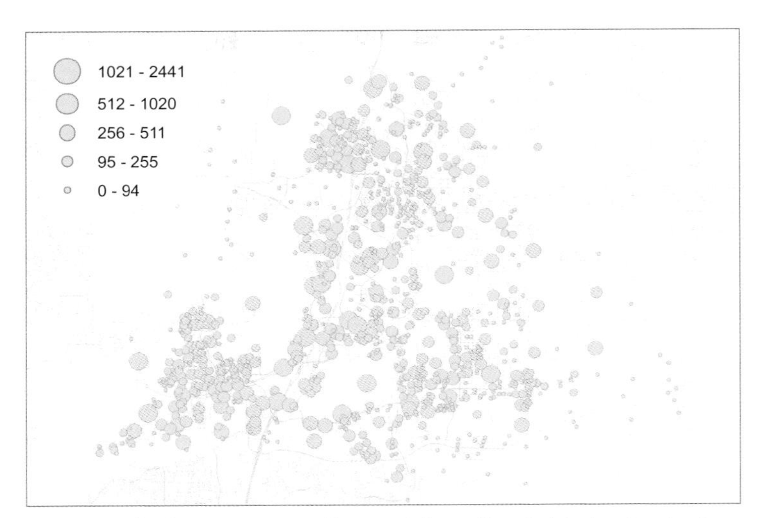

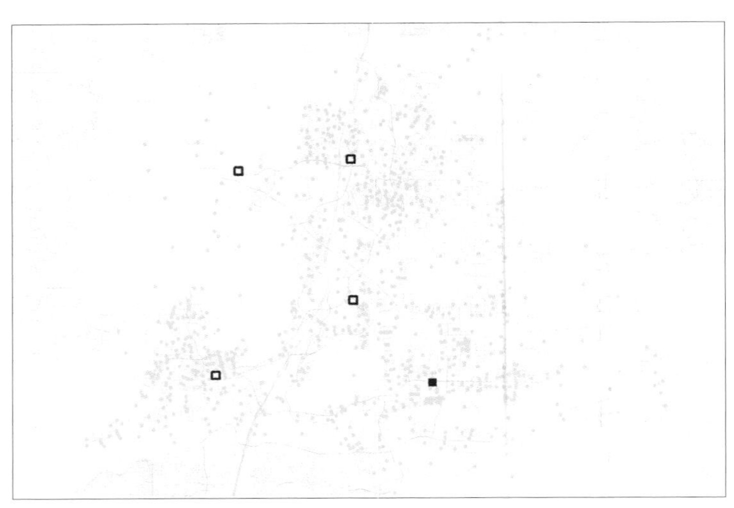

In this example, there is one existing library and the goal is to find the best location for two new libraries—from four candidate locations—so that the total travel time to a library is roughly equal from all neighborhoods in the community. The map on the left shows census block centroids symbolized by the population of each block. The map on the right shows the centroids (blue dots), the existing library (solid square), and the four candidate locations (hollow squares).

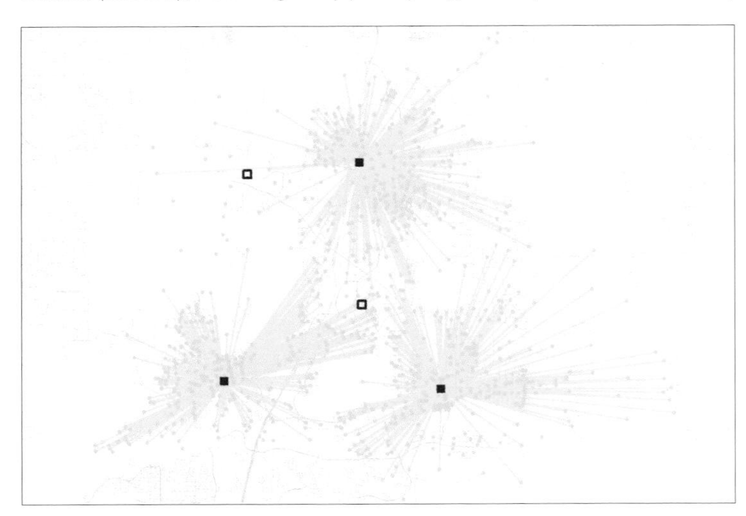

The solution chooses the two additional locations that minimize the overall travel time from all census blocks to their nearest library. No impedance cutoff is used so that all demand points are considered when choosing locations (the unallocated blocks have no population, and thus no demand).

Comparison of problem types

GOAL	ARCGIS PROBLEM TYPE	BEST USE	EXAMPLE APPLICATION	NOTES
Maximize demand allocated	*Maximize Coverage/ Minimize Facilities*	*Allocating demand or locating facilities when all demand must be allocated or all areas serviced*	*Allocating streets to fire stations; locating health clinics or emergency evacuation areas*	*With Maximize Coverage, you specify the number of facilities to choose; with Minimize Facilities, the GIS determines the optimum number of facilities*
Match demand to available supply at facility	*Optimum Allocation*	*Allocating demand to facilities that have limited supply*	*Allocating students to schools; allocating stores to warehouses*	*Allocation only; does not choose the best locations*
Maximize demand that diminishes with distance	*Maximize Attendance*	*Locating facilities when facilities all offer similar services*	*Finding best location for senior day-care centers, supermarkets, or gas stations*	*Uses linear impedance transformation*
Minimize overall transportation cost	*Minimize Impedance*	*Locating or allocating facilities so that overall transportation cost is minimized or to reduce overall distance traveled from demand points to facilities*	*Allocating stores to distribution centers; choosing locations for libraries*	*Usually used with impedance of time or monetary cost*

How the GIS allocates demand and chooses facilities

No matter which type of interaction problem you're modeling, the GIS uses the same core tool—an origin-destination cost matrix. The matrix is a table which lists the total impedance (distance, time, or money) for the shortest path along the network between each facility (origin) and each demand point (destination).

Consider an application in which you want to choose two locations for a warehouse, out of four potential locations, to serve six stores. The goal is to minimize travel distance from the warehouses to the stores, and thus minimize transportation costs.

ORIGIN	DESTINATION	DISTANCE
A	1	22,535
A	2	14,557
A	3	10,780
A	4	36,449
A	5	39,768
A	6	42,479
B	1	34,811
B	2	16,817
B	3	12,619
B	4	15,949
B	5	27,312
B	6	44,041
. . .	. . .	. . .

While the interaction between each origin and each destination is represented as a straight line on the map (light green), the distance that is stored in the origin-destination cost matrix is calculated over the street network. For example, the network distance between origin A and destination 1 is 22,535 feet (dark green line).

Based on the number of candidate facilities and the number you're choosing, the GIS creates a list of all possible combinations of facilities. In this example, there are six possible solutions:

- Choose facility A and facility B
- Choose facility A and facility C
- Choose facility A and facility D
- Choose facility B and facility C
- Choose facility B and facility D
- Choose facility C and facility D

Using the origin-destination cost matrix, for each potential solution the GIS assigns the demand points to the facilities. Each demand point is assigned to the facility it's closest to.

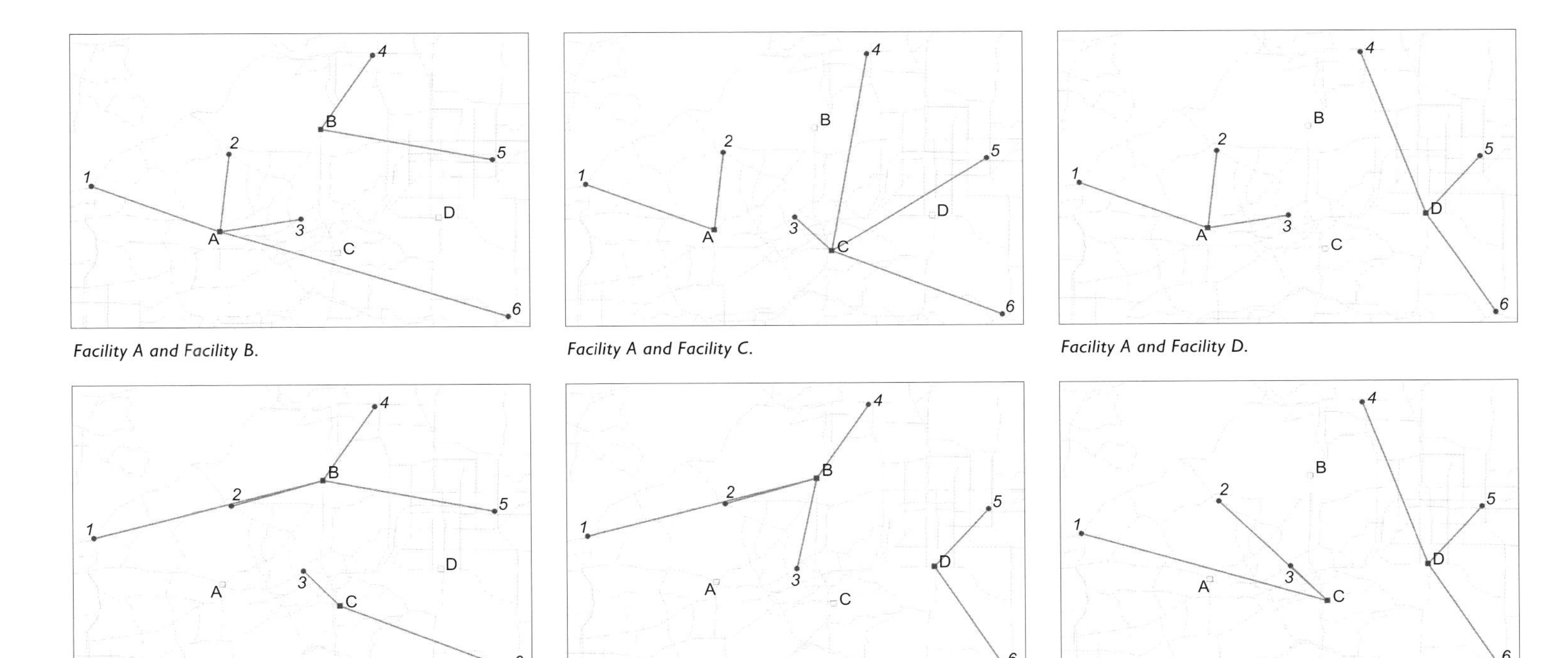

Facility A and Facility B.

Facility A and Facility C.

Facility A and Facility D.

Facility B and Facility C.

Facility B and Facility D.

Facility C and Facility D.

For example, in the first potential solution above—choosing facilities A and B—the GIS uses the origin-destination cost matrix to determine that location 1 is closer to facility A (22,535 feet) than to facility B (34,811 feet). So it assigns location 1 to facility A. Locations 2, 3, and 6 are also closer to facility A than to facility B. Locations 4 and 5 are closer to facility B. The GIS makes all these assignments of demand points to facilities for each potential solution, as shown in the diagrams above.

It then sums the distance (or other impedance) from all demand points to their respective facilities.

SOLUTION	TOTAL DISTANCE
A and B	*133,612*
A and C	*139,320*
A and D	*111,309*
B and C	*134,341*
B and D	*115,184*
C and D	*134,135*

Based on the solution you need, the GIS then chooses the optimum solution. It considers both impedance and demand. Depending on the solution, it favors one over the other—but even if it tries to minimize impedance, for example, it will still secondarily try to maximize demand allocated. Similarly, if it's trying to maximize demand allocated, it will do so by minimizing impedance (by assigning demand points to the closest facility rather than just any facility). In this example, the goal is the solution that has the lowest total distance, which is the one that chooses facility A and facility D.

If there are only a few facilities, there are few combinations and the GIS can quickly search through the table to find the optimum solution. But as the number of facilities increases, the number of combinations—and potential solutions—grows rapidly. For example, if you have forty candidate intersections and you need to choose twelve for bus stops, the number of potential combinations is 5,586,853,480.

Searching through all the possible solutions to find the guaranteed optimum one would take too long. So the GIS uses an intelligent search method to home in on the combination of facilities that is most likely the optimum solution. Essentially, it starts with a subset of combinations and finds the optimum solution out of these. It then chooses another subset and searches these to see if there is a better solution. In the process, it discards the least good solutions. Once it no longer finds better solutions in subsequent searches, it settles on the best solution it has found. (This approach is known as a "heuristic"—the vehicle routing problem described in chapter 4, "Modeling paths," uses a similar approach).

If you're simply allocating demand to a set of facilities and not choosing locations from a set of candidate facilities, the solution is simpler—all that's needed is the origin–destination cost matrix. Essentially there is one solution—the one that uses all facilities. The GIS assigns each demand point to its closest facility, within the cutoff distance, and the results show the demand allocated to each facility.

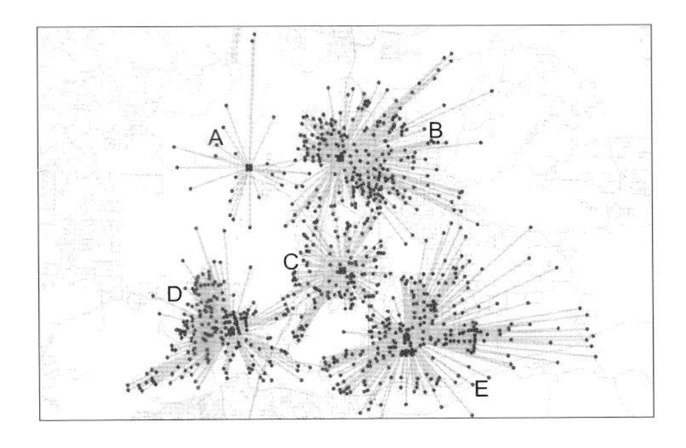

FACILITY	DEMAND	TOTAL DISTANCE
A	25	323,773
B	245	2,756,286
C	130	1,098,815
D	212	1,750,827
E	278	2,932,909

This example shows five existing facilities with demand allocated among them. The GIS creates an origin-destination cost matrix and uses it to assign each demand point to its closest facility. It then can sum the demand assigned to each facility and the total distance from the demand points to their assigned facility. If there is a weight associated with the demand points (such as population for census block centroids), the GIS also sums the weighted demand and the total weighted distance.

EVALUATE THE RESULTS

The results of the analysis are contained in the attribute tables for the facilities and the demand points, as well as in a table containing the interaction between each demand point and the facility it is allocated to.

The table for the facilities lists whether the facility was required to be included in the solution, whether it was chosen to be included, or whether it wasn't chosen (in which case it is still listed as a "candidate" location). It also lists the total number of demand points assigned to required and chosen facilities (for example, the number of census block groups assigned to each library), as well as the total demand at those demand points (for example, the total population assigned to the library).

The table also lists the total impedance (distance or time, usually) between the facility and all the demand points assigned to it, as well as the total weighted impedance (the total impedance between the facility and a demand point multiplied by the weight for that demand point, summed for all facility–demand point pairs). This information is useful for some applications but not others. For example, you may want to know the total impedance (travel time) between a distribution center and all the stores it serves. On the other hand, knowing the total impedance (distance) between census blocks and the library they're assigned to may not be useful information.

Facility ID	Facility Type	Demand Count	Demand Weight	Total Length	Total Weighted Length
1	Required	213	34419	1788585.93	288166228.95
2	Candidate	0	0	0	0.00
3	Chosen	214	41127	2044386.87	424600251.90
4	Required	205	47854	1665898.29	424421068.19
5	Candidate	0	0	0	0.00

The attribute table for facilities (libraries, in this example) shows which ones were required as part of the solution and which were chosen by the model. It also shows the number of demand points (census blocks) and total demand (population) assigned to each facility.

The table for the demand points lists the weight for each and indicates which facility the demand point was assigned to and how much of the weight was allocated. Demand points with no demand (a weight of 0) are not allocated—the facility value is listed as <Null>. Similarly, demand points that have demand but are beyond the cutoff impedance are not allocated.

To identify the demand points that have demand but are beyond the impedance cutoff (as opposed to those that have no demand), select demand points that have demand greater than zero, but no assigned facility (the facility ID value is null). By summing these selected demand points, you can determine the total demand that was not allocated.

Name	Weight	Facility ID	Allocated Weight
Location 5	139	3	139
Location 6	638	3	638
Location 7	0	<Null>	<Null>
Location 8	12	<Null>	<Null>
Location 9	1224	3	1224
Location 10	996	<Null>	<Null>
Location 11	524	3	524

The attribute table for demand points (census blocks, in this example) shows the demand at each (the population of the census block, contained in the weight value), the facility the demand point was assigned to (if any), and the demand that was assigned. In this partial table, you can see that blocks 5, 6, 9, and 11 were all assigned to facility 3, and that all their demand was assigned. Census block 7 has no demand (no population) and was not assigned, while blocks 8 and 10 have demand but were not assigned to a facility because they were beyond the cutoff impedance.

When you run the model, a set of line features is generated representing the interaction between each demand point and the facility it's assigned to. The attribute table for these line features contains information about the interaction, including the demand assigned (the weight value), the impedance between the demand point and the facility, and the weighted impedance between the demand point and the facility (the impedance multiplied by the weight). This would represent, for example, the total distance traveled if everyone in a census block made one trip to the nearest library.

Name	Weight	Total Weighted Length	Total Length
Location 5 - Facility 3	139	743726.17	5350.55
Location 6 - Facility 3	638	8487951.42	13304.00
Location 9 - Facility 3	1224	11281350.57	9216.79
Location 11 - Facility 3	524	6514150.90	12431.59

The attribute table for interaction lines shows the properties of the interaction between each demand point that was assigned to a facility and that facility. This partial table includes the four assigned demand points from the demand point table. (The demand points that were not assigned do not have any interaction or a corresponding line feature and are not included.) Selecting all the records for a particular facility and summing the weight, total impedance, and total weighted impedance gives the same values that are found in the facilities attribute table.

You can use the information in the tables to display the results on a map (discussed further in the next section). Displaying the results on a map can help you spot possible problems with your model. To better allocate demand, you may need to increase the cutoff impedance or choose more facilities. (Of course, these changes will have implications for the services provided—increasing the cutoff impedance means some people will have to travel farther to a facility and choosing more facilities means increased costs for building the additional facilities.)

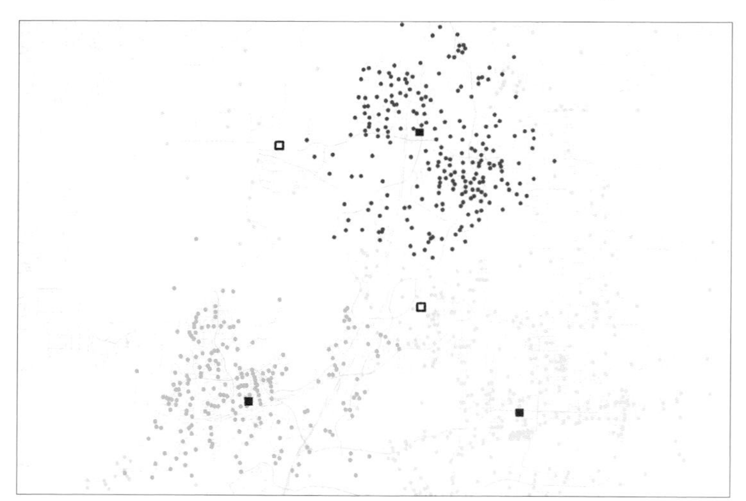

In this solution, population in census blocks was assigned to three library locations. Two existing libraries were included as required facilities (solid blue squares in lower left and right) and one additional library was chosen from among three candidates (solid blue square in upper center; candidates not chosen are hollow squares). Census block centroids are color-coded by the facility they were assigned to. Census blocks with no population or beyond the specified three-mile cutoff were not assigned and are shown in light blue.

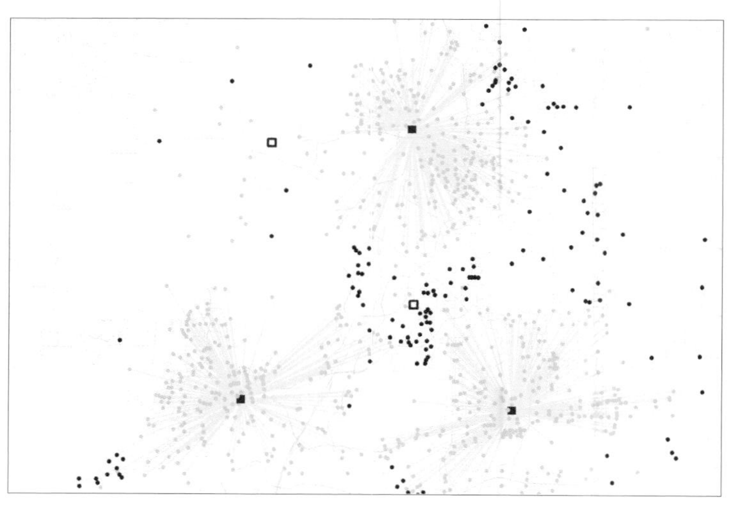

In this map, the same solution is symbolized using lines to show which library each census block was assigned to. Census blocks with no population are shown in light tan, while those with population but beyond the three-mile cutoff are shown in brown. The clusters of brown dots indicate areas that would be likely locations for an additional library. Alternatively, the three-mile cutoff distance could be increased to allocate more blocks.

The results are highly dependent on the demand at the demand points and the impedance values associated with the network edges. If the results are suspect (for example, most demand points are allocated to one facility with few points allocated to the others), you will want to check these values to identify and correct any errors.

The quality of the network dataset you're using can also affect the results. For example, the network may have gaps where edges don't connect, one-way streets are misidentified, turns are incorrectly identified as allowed (or not), streets are missing, or streets and intersections are temporarily blocked. If the results are not what you might expect, check these network properties.

A quick way to check whether a demand point is allocated correctly is to create a route between the demand point and the facility (as described in chapter 4, "Modeling paths"). If the route cannot be created, it may be that there is a gap or barrier in the network. If the route is created, but the distance (or other impedance) seems incorrect, it may be that the impedance values for the edges are not correct or that the connectivity for the network is incomplete. For example, it may be that edges are allowed to connect only at endpoints rather than at any location along their length. (Edges sometimes connect where there is a T-intersection.)

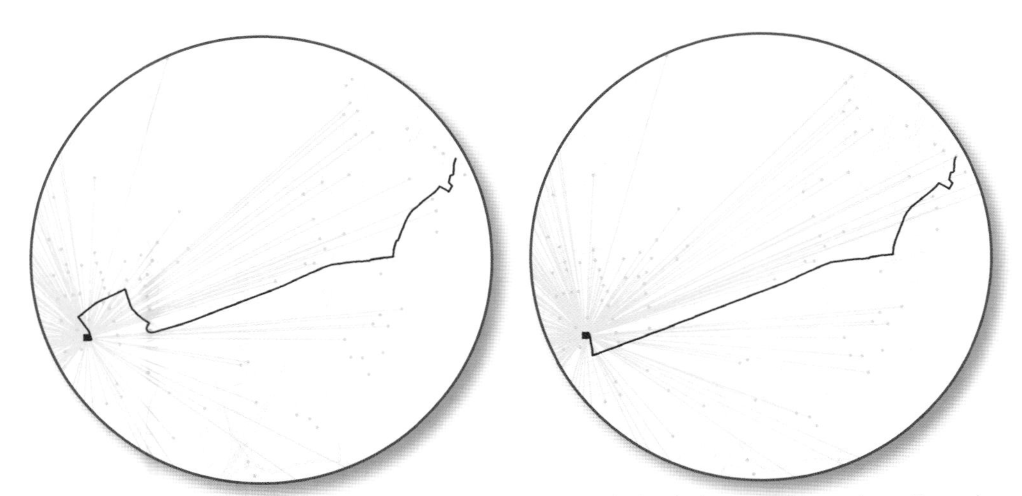

In these close-ups, the interaction lines between demand points and the facility they're assigned to are shown. Demand points with no line are unassigned. The left close-up shows the calculated route (red line) over the network from an unassigned demand point to the facility over a street network with limited connectivity (edges connect only at endpoints). The distance of the route is 17,066 feet, putting the demand point beyond the three-mile cutoff (15,840 feet). The route on the right was created using a network with full connectivity. The distance from the same demand point to the facility is 15,218 feet, placing the demand point within the three-mile cutoff and allowing it to be assigned to the facility.

DISPLAY AND APPLY THE RESULTS

As you've seen, you can display the results of a location-allocation model in a variety of ways. By selecting features using fields in the attribute tables for the facilities, demand points, and interaction lines, and by using attribute values to symbolize the features, you can present a range of information from the results of the model.

If you're mainly interested in which areas are served by the chosen facilities—and which aren't—you can categorize the demand points by those that are allocated within the cutoff impedance, those that have demand but are beyond the cutoff impedance, and those that have no demand. If the demand points are centroids that represent polygons (such as census blocks or tracts), you can use the point features in each category to select the corresponding polygon features. You can then display the original polygon features using the same categories. This may give you a better view of which parts of the study area are served by the chosen facilities.

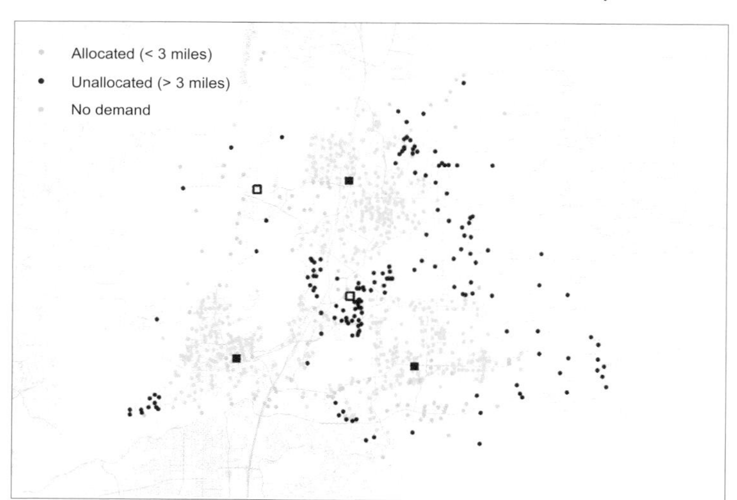

This map shows the three chosen library locations (solid squares) and the two unchosen candidates (hollow squares). It also shows which census blocks (represented by centroids) were allocated, which were beyond the three-mile cutoff distance, and which are unpopulated (and thus have no demand).

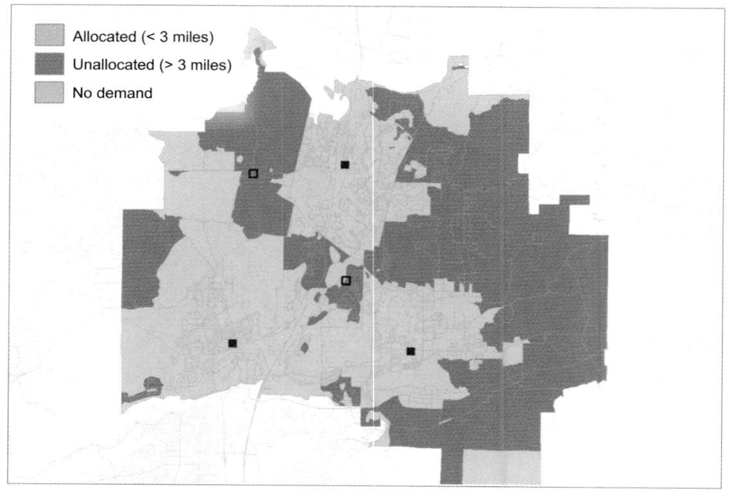

This map shows the same solution as the map to the left, but with color-coded census block polygons rather than centroids. The map gives a more general view of which areas are within three miles of a library and which aren't. (Recall that distances are calculated over the street network rather than in a straight line.)

If you're mainly interested in which demand points are allocated to which facilities, you can color code the demand points by the facility they're assigned to. One way to do this is to draw a line from each demand point to the facility it's been allocated to. (Even though the lines are straight, the allocation is based on travel along the network.) These maps are sometimes called spider diagrams. They can give a visual representation of comparative demand for each center, as well as show which centers origins are allocated to. To make it more obvious, you can color-code both the demand points and the lines.

Again, if the demand points represent polygon centroids, you can use the allocated points to select the corresponding polygons and color-code the polygons by the facility they're assigned to. This will define the service area for each facility.

This map shows a solution using two library locations (solid squares). Census block centroids are color-coded by which library they have been allocated to, and color-coded lines drawn to create a spider diagram. The lines show the relative distances and directions that demand is drawn from.

This map is created from the same solution but shows census block polygons color-coded by which library they're allocated to. The map could serve as a basis for creating service districts.

If you're interested in how much demand is allocated to each center, you can display this visually in several ways. One is to use a spider diagram with graduated symbols for the lines to show the relative weight of each allocated demand point (the greater the demand, the wider the line). The overall effect shows where—and from which directions—most of the demand originates for each chosen facility. Alternatively, you can map the facilities themselves using graduated symbols to show the relative demand at each.

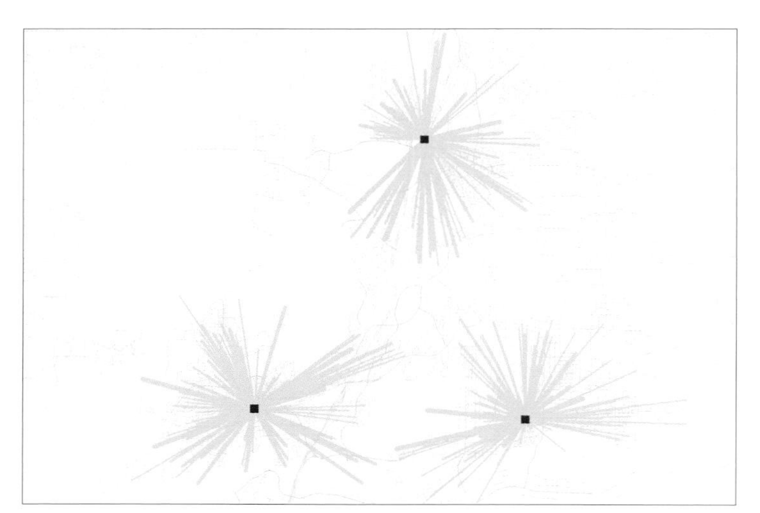

This map shows the solution using three library locations along with allocated census block centroids. The thickness of the lines indicates the relative demand (population) of each census block, showing where most of the demand for each library originates.

This map shows the same solution, but this time the size of the squares (representing the library locations) indicates the demand allocated to each library.

By changing the parameters of the model—the number of facilities to choose, which facilities are required, the cutoff impedance—you can fairly easily create results for different scenarios.

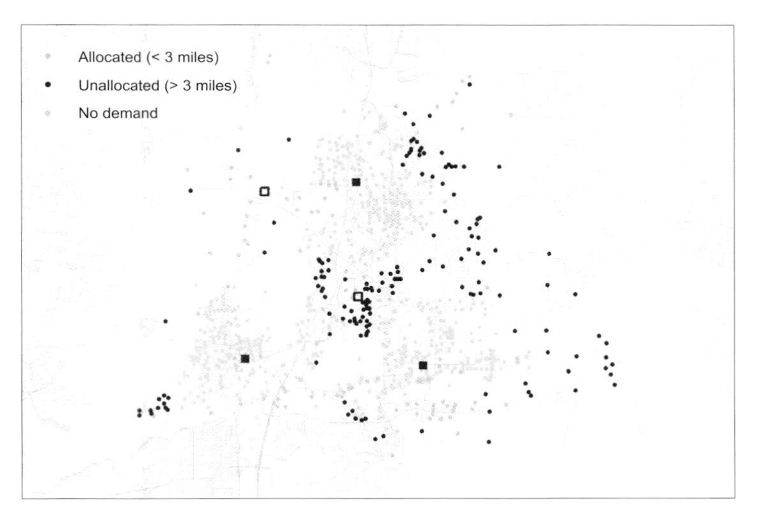

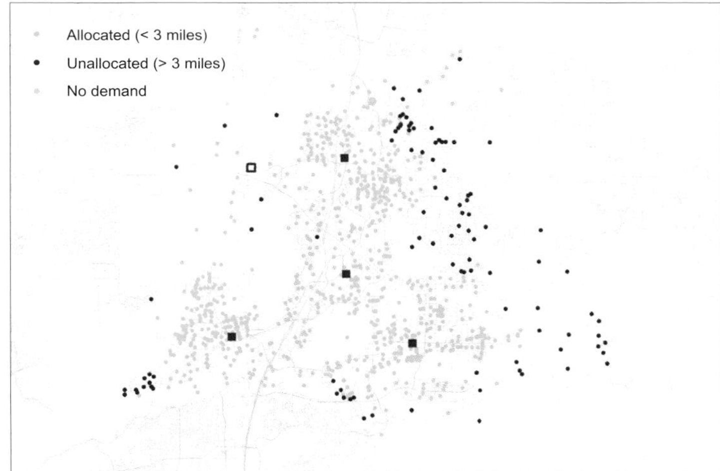

These three maps show different scenarios for choosing library locations and allocating demand to those locations. The map in the upper left shows the three locations (solid squares) that would serve the most people within three miles of a library. The map in the upper right shows how many more census blocks would be served if a fourth location was included. The map in the lower right shows the coverage with only three libraries but assuming that five miles is an acceptable distance to travel to a library.

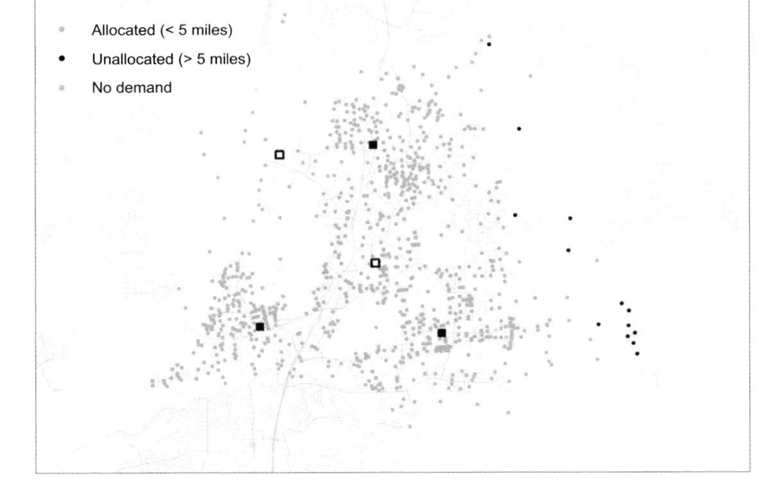

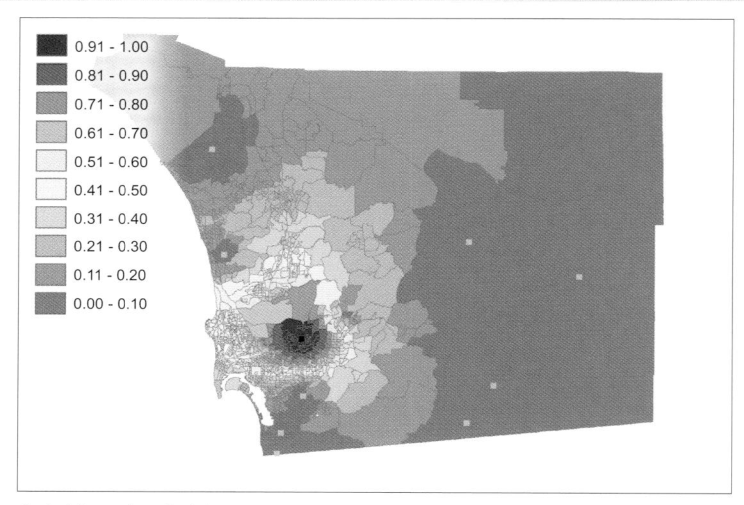

Probability values for hikers traveling from census block groups to a large park with a high attraction value (black square). Locations of other parks (gray squares) are also shown. The probability is low in block groups that are near another park.

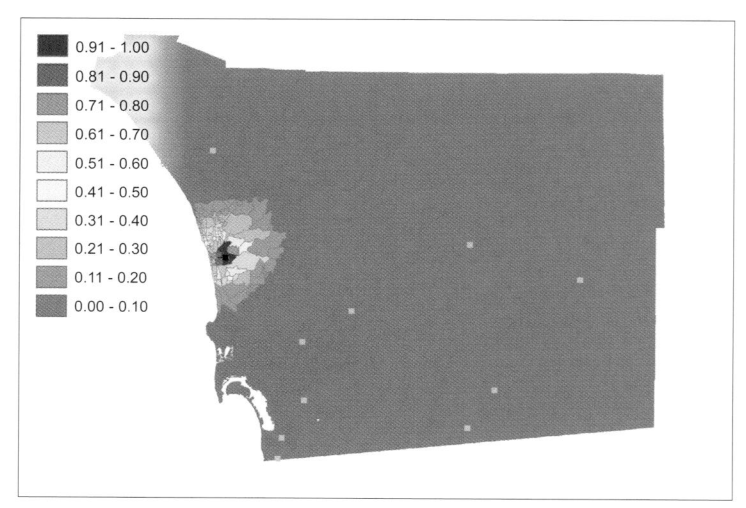

Probability values for hikers traveling to a park with a low attraction value (black square). The probability is high only in block groups very close to the park.

Model travel to facilities to find out how many people are likely to travel to each facility. The method uses the attraction of each facility, the location of competing facilities, and the cost of travel (distance, time, or monetary cost) to or from the facilities to calculate what proportion of the demand at each location a facility is likely to capture. The total demand for each facility can then also be calculated. Unlike simply allocating demand to the nearest facility—as discussed in the previous section—the model attempts to consider people's choices or preferences when they are deciding which facility to travel to. The assumption is that a person confronted with a choice of facilities to visit will make their choice based not only on the distance to each facility but also on how they perceive the attractiveness of each. When modeling travel to facilities, demand can be (and usually is) split between facilities—that is, some demand for a demand point may be allocated to one facility and some to another, or even to several others. (This is also unlike allocating demand, where all demand at a location is allocated to the single nearest facility.)

You'd model travel to facilities to find out, for example, which regional parks people are likely to travel to from various locations in a county and how many people are likely to travel to each park. If the potential visitors are hikers, this would be dependent on the number of hikers in each census block group (the demand), the attraction weight of each park (a relative value taking into account the size of the park, the variety of habitats, the presence or absence of a lake, and so on), and the distance between each block group and each park. All parks being equal, hikers would likely travel to the nearest one; a park that was more attractive, however, would draw additional people even though it is farther away.

These models are referred to generally as spatial interaction models. (They are also known as "gravity" models. While the gravity analogy is not precise, the attraction associated with a facility is similar to gravitational pull in that the greater the attraction value for a facility, the more people it will draw. In addition—as with gravity—the strength of the attraction decreases with distance.)

Spatial interaction models are used to predict customer visits or sales at a store or other facility, to find underserved areas in which to locate a new facility, or to create alternate scenarios such as analyzing the impact of adding or closing a store or modifying the attraction value of existing facilities (for example, by building more trails at a park or increasing the size of the park).

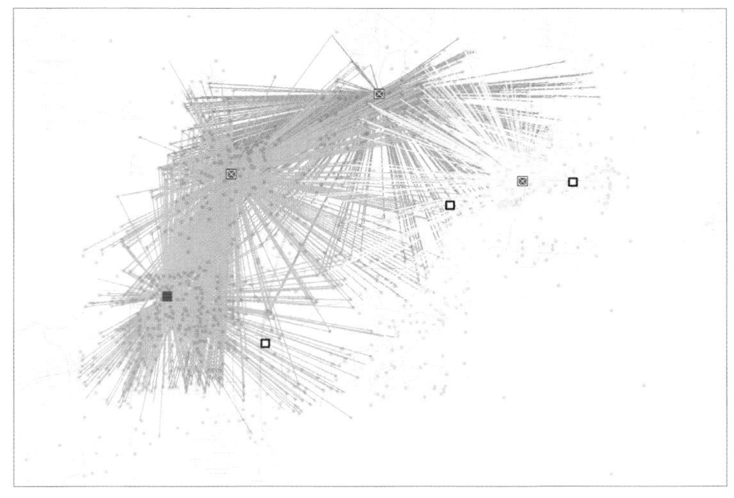

In this spatial interaction model, the location for a new pet store (solid square) has been chosen from a set of potential locations. The locations of existing competing stores are shown (boxes with Xs) as well as candidate locations that weren't chosen (hollow boxes). The lines show the store where people in the various census blocks (orange dots) are likely to shop. The square footage of each store was used as the attraction value (the assumption being that larger stores carry more products and are, therefore, more attractive to shoppers).

The process for modeling attraction and competition is:

1. Specify the facilities and attraction values
2. Specify the demand points and associated demand
3. Run the model
4. Evaluate the results
5. Display and apply the results

SPECIFY THE FACILITIES AND ATTRACTION VALUES

As with location-allocation models, the facilities are point locations, usually representing the location of a building (the location of a store or library, for example).

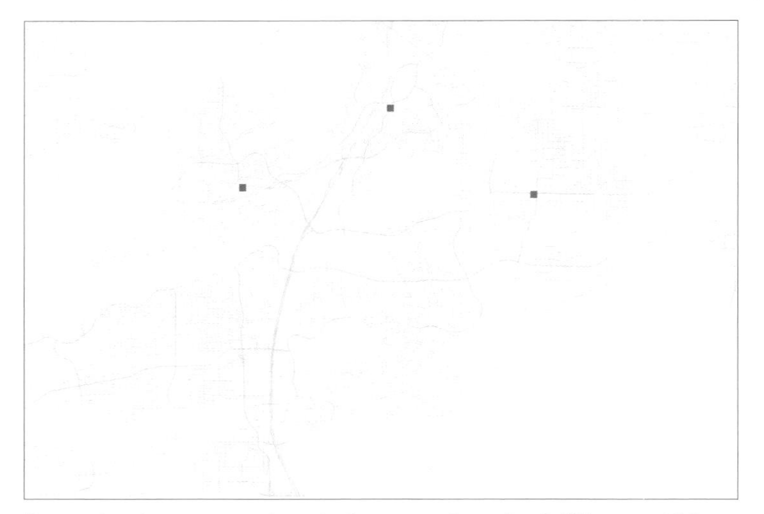

Pet store locations represented as point features can be used as facilities to model the stores at which people are likely to shop.

Facilities can also be points representing the location of a polygon feature such as a county park (usually the centroid of the polygon is used.) The facility layer should include all the available choices—the locations of the facilities you're modeling interaction for as well as the locations of any competitors.

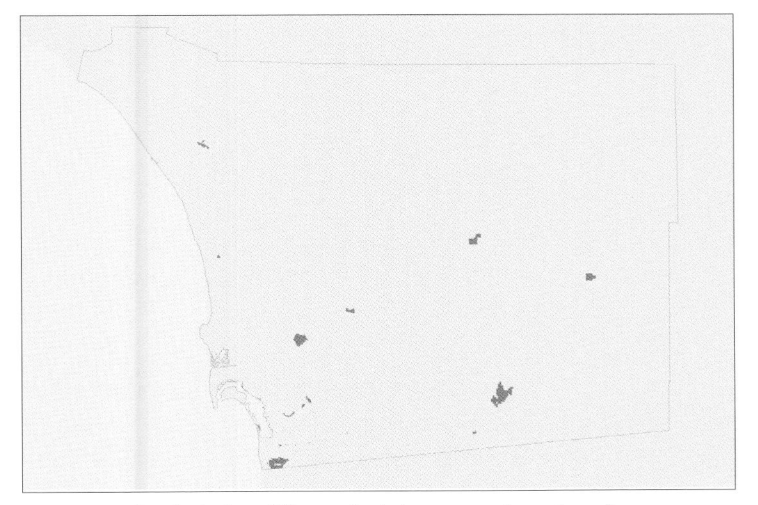

Major regional parks (at least 100 acres in size) represented as polygon features.

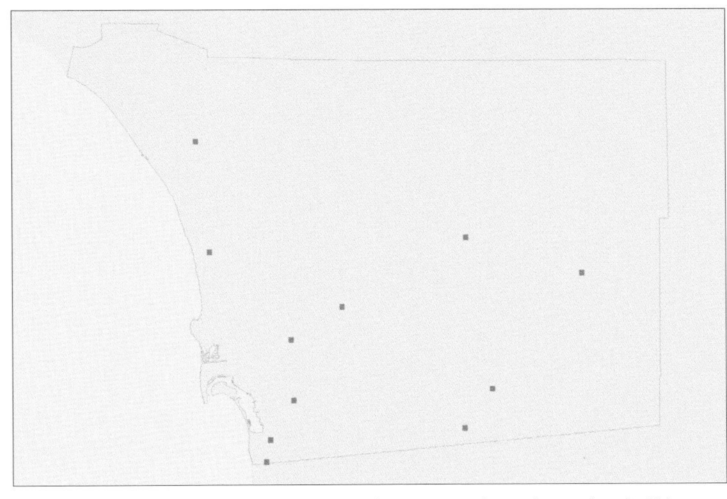

The parks represented as point features—polygon centroids can be used as facilities to model which parks people will visit.

Each facility has an attraction value associated with it. The greater the value, the more likely the facility is to attract people. The attraction value is often a quantity associated with the facility. For example, if you're modeling the potential usage of regional parks by hikers, you might use the size of the park in acres as the attraction value (since larger parks will have more trails and more variety of landscapes). If the population is campers, you might use the number of campsites at each park as the attraction value. For potential shoppers at pet stores, you might use the square footage of each store or the number of different brands each store carries.

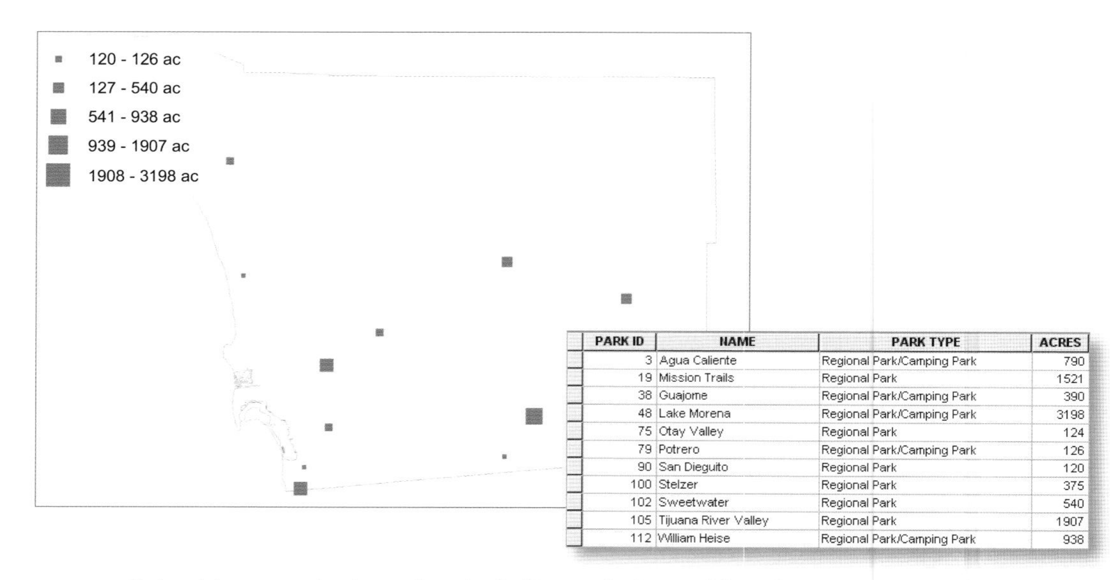

PARK ID	NAME	PARK TYPE	ACRES
3	Agua Caliente	Regional Park/Camping Park	790
19	Mission Trails	Regional Park	1521
38	Guajome	Regional Park/Camping Park	390
48	Lake Morena	Regional Park/Camping Park	3198
75	Otay Valley	Regional Park	124
79	Potrero	Regional Park/Camping Park	126
90	San Dieguito	Regional Park	120
100	Stelzer	Regional Park	375
102	Sweetwater	Regional Park	540
105	Tijuana River Valley	Regional Park	1907
112	William Heise	Regional Park/Camping Park	938

Each park has an associated attraction value. In this example, the size of the park, in acres, is used as an indicator of how attractive the park is to the population of hikers in the county.

While attraction values are usually quantities, you can also use qualitative characteristics as attraction values—such as the relative sense of wilderness at various parks or the perceived popularity of restaurants. To use these characteristics as attraction values, you assign them values on a scale that you define (such as 1 to 5 or 1 to 9, with the highest number representing the highest level of attraction). A ratio scale is used so that a facility having an attraction value of, say, 4 is twice as attractive as a facility having a value of 2. That's because, in the model, an attraction value of 4 will generate exactly twice as much interaction from a given demand point as an attraction value of 2. In practice, it can be difficult to accurately assign attraction values to qualitative characteristics, which is why quantities are often used as surrogates for actual perceived attractiveness. If you do use qualitative characteristics as attraction values, the values you assign should be based on research such as user surveys or consumer preference studies.

You can create an overall attraction value by combining several characteristics of the facility. For example, if you're modeling usage for a regional park, the target audience might be the population at large, and the attraction value might be a combination of the most important characteristics reflecting various uses (the size of the park, the number of campsites, whether there is a lake).

To calculate overall attraction using several attributes, use an approach similar to the suitability model techniques discussed in chapter 3, "Rating suitable locations." (The difference is that you calculate attraction values using fields in the attribute table for the facilities layer, rather than combining layers representing the individual attributes, as you do in suitability analysis.) Since each characteristic is likely measured in different units (park size is in acres while the number of campsites is simply a count), you assign values to the various individual attributes of the facility on a common scale, such as 1 to 9. Again, a ratio scale is used so that a value of 9, for example, represents three times the attraction of a value of 3. In addition, the levels of attraction should be consistent between the characteristics—so an attraction value of 6 for park size represents a level of attraction equal to an attraction value of 6 for number of campsites.

You then add the scaled values and divide by the number of values in the scale so that you end up with values on the original scale. For example, if you want to calculate an overall attraction value for parks, you might combine the size of the park (in acres), the number of campsites, and whether or not there's a water feature. Using a scale of 1 to 9, with 9 indicating more attractive and 1 less attractive, you'd divide the acres values into classes and assign a value of 9 for that attribute to parks in the class of largest parks, 8 to the next smallest parks, and so on. You'd do the same for the number of campsites (with parks having no campsites assigned a value of 1). For water features, you'd assign a value of 9 to parks with a water feature and a value of 1 to parks without. You'd then add the three attraction values and divide the result by three (the number of input attraction values) to assign an overall attraction value to each park on the same scale as the input values (1 to 9).

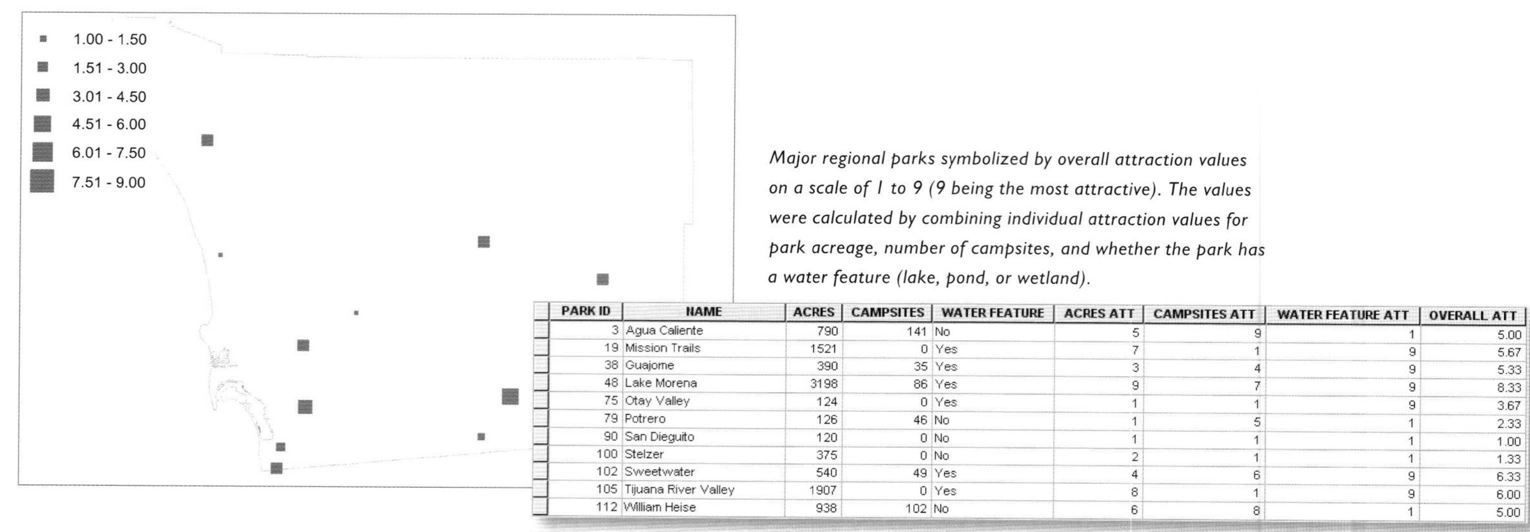

Major regional parks symbolized by overall attraction values on a scale of 1 to 9 (9 being the most attractive). The values were calculated by combining individual attraction values for park acreage, number of campsites, and whether the park has a water feature (lake, pond, or wetland).

PARK ID	NAME	ACRES	CAMPSITES	WATER FEATURE	ACRES ATT	CAMPSITES ATT	WATER FEATURE ATT	OVERALL ATT
3	Agua Caliente	790	141	No	5	9	1	5.00
19	Mission Trails	1521	0	Yes	7	1	9	5.67
38	Guajome	390	35	Yes	3	4	9	5.33
48	Lake Morena	3198	86	Yes	9	7	9	8.33
75	Otay Valley	124	0	Yes	1	1	9	3.67
79	Potrero	126	46	No	1	5	1	2.33
90	San Dieguito	120	0	No	1	1	1	1.00
100	Stelzer	375	0	No	2	1	1	1.33
102	Sweetwater	540	49	Yes	4	6	9	6.33
105	Tijuana River Valley	1907	0	Yes	8	1	9	6.00
112	William Heise	938	102	No	6	8	1	5.00

The Acres, Campsites, and Water Feature fields were part of the original attribute table. The Acres Att, Campsites Att, and Water Feature Att fields were added to hold attraction values on a scale of 1 to 9. The values in the attraction fields were calculated by reclassifying the values in the original fields. Parks with high values in the Acres field have high values on the scale in the Acres Att field, and so on.

It's possible—even likely—that some attributes of the facility are more important than others in terms of attraction value. As with suitability modeling, you can create a weighted overall attraction value. To do this, you'd determine what percentage of the total each attribute represents. For example, if you're modeling park usage for the summer months, you might decide that the presence or absence of a water feature accounts for 60 percent of the attraction, the size of the park accounts for 30 percent, and the number of campsites for 10 percent. You'd multiply the assigned attraction values by the corresponding percentage before summing the values for each park.

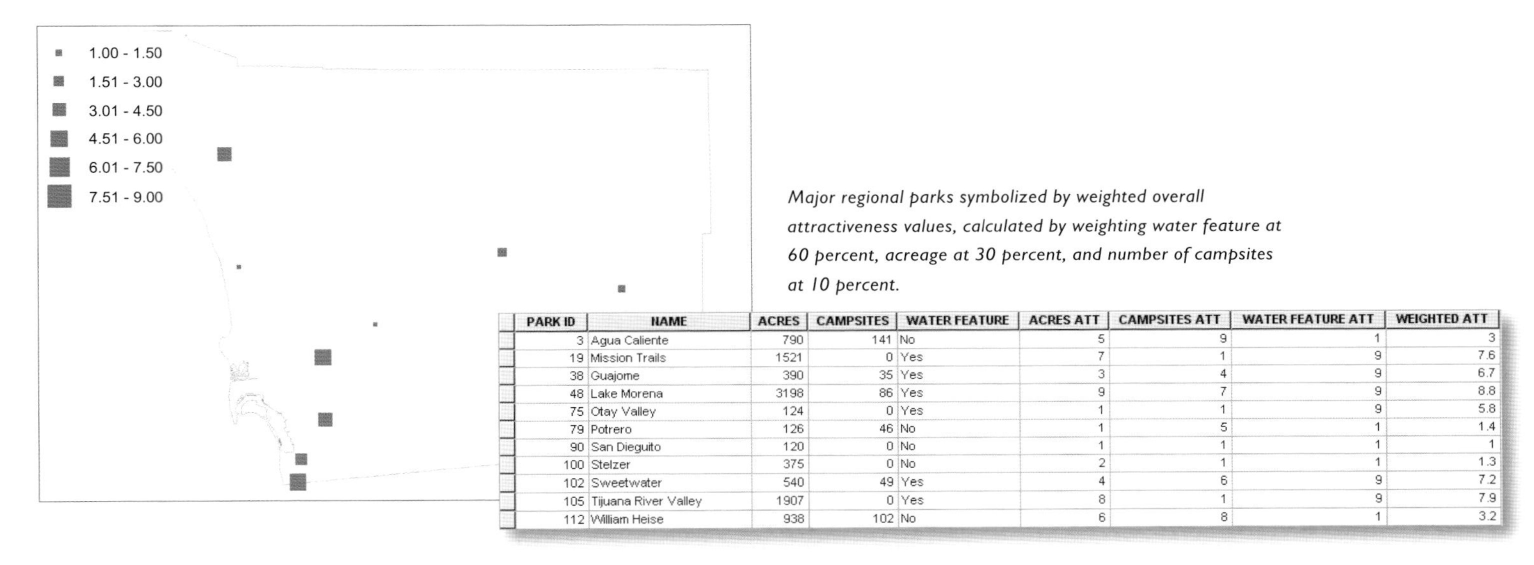

PARK ID	NAME	ACRES	CAMPSITES	WATER FEATURE	ACRES ATT	CAMPSITES ATT	WATER FEATURE ATT	WEIGHTED ATT
3	Agua Caliente	790	141	No	5	9	1	3
19	Mission Trails	1521	0	Yes	7	1	9	7.6
38	Guajome	390	35	Yes	3	4	9	6.7
48	Lake Morena	3198	86	Yes	9	7	9	8.8
75	Otay Valley	124	0	Yes	1	1	9	5.8
79	Potrero	126	46	No	1	5	1	1.4
90	San Dieguito	120	0	No	1	1	1	1
100	Stelzer	375	0	No	2	1	1	1.3
102	Sweetwater	540	49	Yes	4	6	9	7.2
105	Tijuana River Valley	1907	0	Yes	8	1	9	7.9
112	William Heise	938	102	No	6	8	1	3.2

Major regional parks symbolized by weighted overall attractiveness values, calculated by weighting water feature at 60 percent, acreage at 30 percent, and number of campsites at 10 percent.

In practice, the weights, as well as the attraction values themselves, are obtained from data on actual usage at facilities. This could be visitor surveys for parks, market research, or customer data collected by stores. In fact, some implementations of spatial interaction models, such as one in ArcGIS Business Analyst, allow you to feed such survey data directly into your model. Ordinary Least Squares (OLS) regression is used by the model to determine which characteristics of the facilities are the most important and a coefficient is assigned to each characteristic. The coefficients are used as weights in the model to assign an overall attraction value to each facility.

SPECIFY THE DEMAND POINTS AND ASSOCIATED DEMAND

As with location-allocation models, demand can be associated with individual point locations (the addresses of pet owners) or can be aggregated data represented by the centroid of a polygon (the number of hikers in each census block group).

When using aggregated population data, in some cases you might use the total population of the block, tract, or county. In many cases, though, you'll use a subset of the population corresponding to the facilities you're modeling.

For example, if you're modeling the usage of regional parks by hikers, with demand represented by block groups, you'd use the number of hikers in each block group as the demand value. If you knew the actual number of hikers in each block group, you'd use this data—but that's unlikely, so instead you'd use a subset of population data. You might use the total number of young and middle-aged adults in each census block group, since these are the most likely age groups to be regular hikers. While not all of the people in this age range will be hikers, you might assume that the same percentage of people of this age in each block group are. Since the demand values are relative, you can create the model using this data.

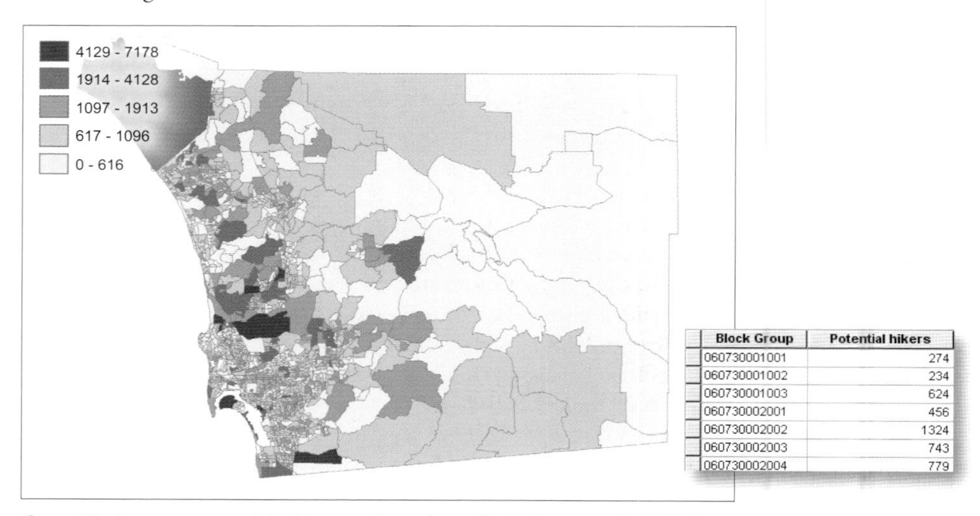

Block Group	Potential hikers
060730001001	274
060730001002	234
060730001003	624
060730002001	456
060730002002	1324
060730002003	743
060730002004	779

Census block groups are used, in this example, as demand point locations. Each block group has a demand value, the potential population of hikers in the block group.

You can also use a quantity other than population as the demand value. For example, if you're modeling travel to pet stores, with demand represented by census tracts, you might use consumer data such as the amount spent annually on pet supplies in each tract as the demand value or, alternatively, an index of the propensity of residents in each tract to spend money on pet supplies (such index data is compiled by marketing research firms).

RUN THE MODEL

Once you've specified the facilities and their attraction values, and the demand points and their demand, you can run the model to see how many people would likely travel to each facility.

The GIS estimates the likelihood that people at each demand point would travel to each facility using the attraction value of the facility, the demand value at the demand point, the distance from the demand point to the facility, and the locations of other facilities.

The specific information you get from the results of the model depends on the implementation of the spatial interaction model you're using. ArcGIS, for example, includes several tools that allow you to model attraction and competition. These are all based on what's known as the Huff Model (described in the next section).

The Huff Model tool (available as a download from the ArcGIS Resource Center) creates a new layer of demand points. (A version of the Huff model is also available in ArcGIS Business Analyst). The attribute table for the new layer contains, for each point, the total demand for the point along with the probability of travel from that demand point to each facility. If, for example, the facilities are major regional parks, the demand points are census block groups, and the demand associated with block groups is the number of hikers in each, the probability value tells you what percentage of the total population of hikers in a block group will go to each available park.

(Of course, there may be other parks that compete for hikers and campers—such as nature preserves and state parks in the same county and in neighboring counties and states. But for the purposes of this example, the study area is limited to a single county and the facilities to major county parks.)

You can see that for the first block group in the table, for example, almost half (48 percent) are likely to go to Mission Trails Park, while none are likely to go to Potrero Park. The probability values for a given demand point sum to 1.

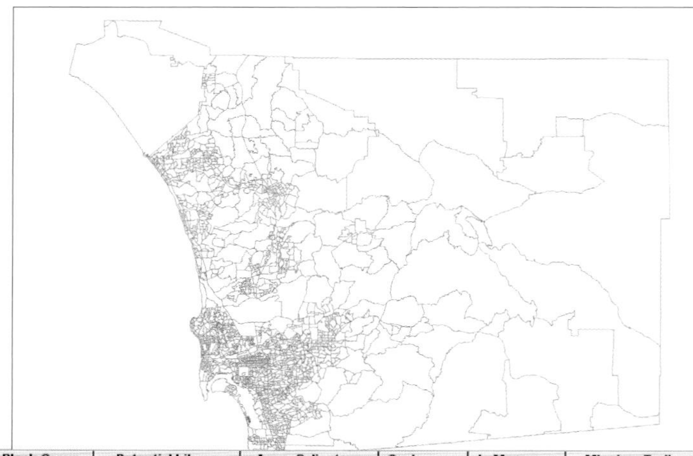

The result of the Huff Model tool is a layer of demand points (block groups, in this example). Each feature has associated with it (as fields in the attribute table) the probability of the target population for that feature traveling to each of the facilities. So in the parks example, each census block group has a field for each park with the probability of hikers traveling from the block group to that park (the fields suffixed with "p").

Block Group	Potential hikers	Agua_Caliente_p	Guajome_p	L_Morena_p	Mission_Trails_p	Otay_V_p	Potrero_p	San_Dieguito_p	Stelzer_p	Sweetwater_p	Tijuana_R_V_p	William_Heise_p
060730001001	274	0.01	0.01	0.06	0.48	0.02	0.00	0.01	0.03	0.13	0.22	0.02
060730001002	234	0.01	0.01	0.06	0.47	0.02	0.00	0.01	0.03	0.13	0.23	0.02
060730001003	624	0.01	0.01	0.06	0.46	0.02	0.00	0.01	0.03	0.13	0.24	0.02
060730002001	456	0.01	0.01	0.06	0.48	0.02	0.00	0.01	0.03	0.14	0.22	0.02
060730002002	1324	0.01	0.01	0.06	0.46	0.02	0.00	0.01	0.03	0.14	0.23	0.02
060730002003	743	0.01	0.01	0.06	0.45	0.02	0.00	0.01	0.03	0.14	0.24	0.02
060730002004	779	0.01	0.01	0.06	0.47	0.02	0.00	0.01	0.03	0.14	0.23	0.02

By multiplying the probability by the available demand at each demand point, the number of hikers that are likely to travel to each park from each block group can be calculated. And summing the demand allocated to a park from each block group will give you the total demand for that park.

Block Group	Potential hikers	Mission_Trails_p	Mission_Trails_demand	San_Dieguito_p	San_Dieguito_demand
060730001001	274	0.48	130	0.01	3
060730001002	234	0.47	109	0.01	3
060730001003	624	0.46	286	0.01	8
060730002001	456	0.48	218	0.01	5
060730002002	1324	0.46	616	0.01	14
060730002003	743	0.45	336	0.01	8
060730002004	779	0.47	364	0.01	9

The table shows the probabilities and allocated demand for two parks. By summing the demand field for a park, you can obtain the total demand for the park. (Only a portion of the table is shown here.)

Naturally, for these numbers to be valid, the initial population must be accurate. If you're estimating the number of hikers in each block group, then the total demand at each park will also be a very coarse estimate.

Another implementation of the Huff Model is the Market Share problem type, an option in the ArcGIS Network Analyst Location-Allocation tool. It creates the same output as the other location-allocation options described in the previous section ("Allocating demand to facilities"). In this case, since the demand from a single demand point is allocated to more than one facility, the attribute table for the interaction lines contains multiple records for each demand point—one for each facility demand was allocated to. The file contains only the allocated demand and not the probability value.

Facility ID	Facility Type	Square Feet	Allocated Demand
1	Competitor	15200	37306
3	Required	3200	77459
4	Competitor	14600	47463

Name	Demand	Allocated Demand
Location 9	11	0.00
Location 10	116	91.82
Location 11	183	107.85
Location 12	435	238.38
Location 13	25	0.00

Demand Point ID	Facility ID	Allocated Demand
9	4	11.00
10	1	24.18
10	3	91.82
11	1	75.15
11	3	107.85
12	1	196.62
12	3	238.38
13	1	0.21
13	4	24.79

The map and tables show the output from the Market Share model. The map shows the allocated demand for each pet store—the target store (solid square) and two competitors (boxes with Xs). The top table lists the three facilities (pet stores)—both the target store (required) and the competitors—and the demand (population) allocated to each (the attraction value is square feet). The middle table shows a portion of the list of demand points (census blocks, in this example) with the total demand (population) in each and the demand allocated to the target store (any demand allocated to competitors is not included). The bottom table shows a partial list of the interaction between each demand point and each facility, including competitors. As you can see, demand can be split between two or more facilities. So for demand point 11, for example, the total demand of 183 (shown in the middle table) is split between the target store (facility 3) and one of the competitors (facility 1). Only the demand value of 107.85 allocated to the target store is included in the list of demand points (middle table).

Whichever spatial interaction model you use—whether you're calculating the probability values or the allocated demand from each demand point to a particular facility—the result shows the pattern of usage for the facility (that is, where people are likely to travel from to visit that facility).

As you might expect, a facility with a low attraction value would draw most people from the immediate area and few people from far away.

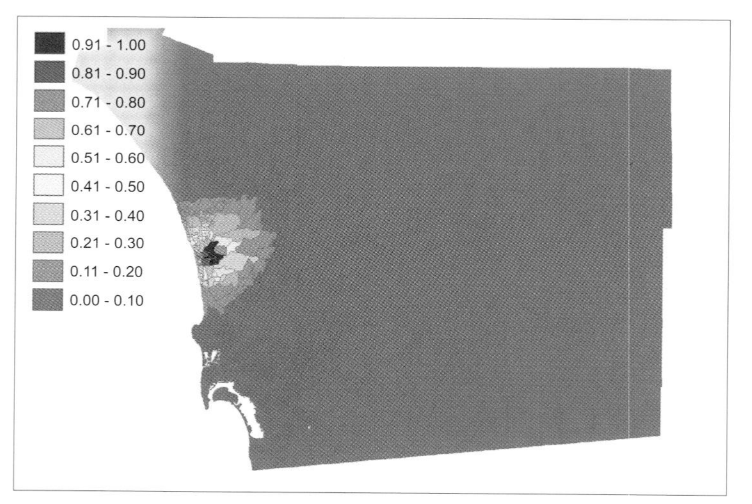

Probability values for San Dieguito Park, which has a relatively low attraction value of 120 acres. (In this example, acreage is the only attraction attribute). Only block groups very near the park have high probability values (red and orange)—that is, only hikers living near the pork are likely to use it; hikers from other parts of the county (block groups with low probability value, in blue) are unlikely to travel to the park.

A facility with a larger attraction value would draw people from a wider area. However, travel decreases from areas that are near another facility. If the competing facility has a relatively high attraction value, more surrounding demand points will have low probability for the target facility. If the competing facility has a low attraction value, fewer surrounding demand points will have low probability values.

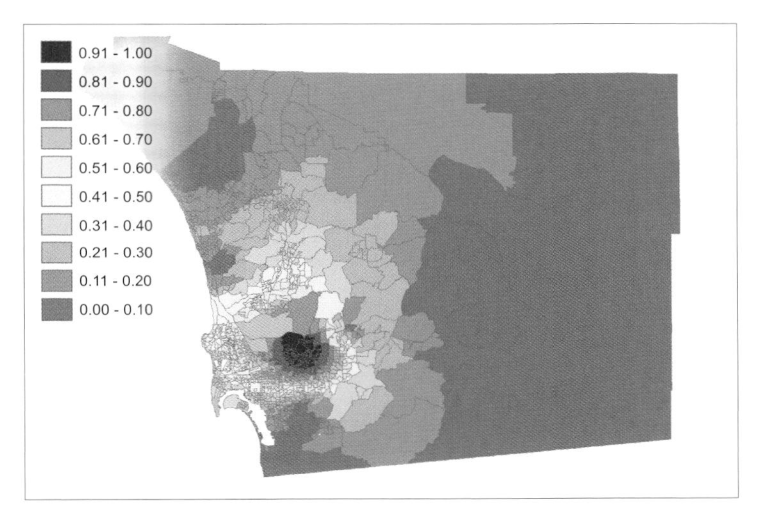

Probability values for Mission Trails Park, which has a relatively high attraction value (1,521 acres). More block groups farther from the park have high probability values, meaning that more hikers living farther from the park (but no too far) are likely to use it.

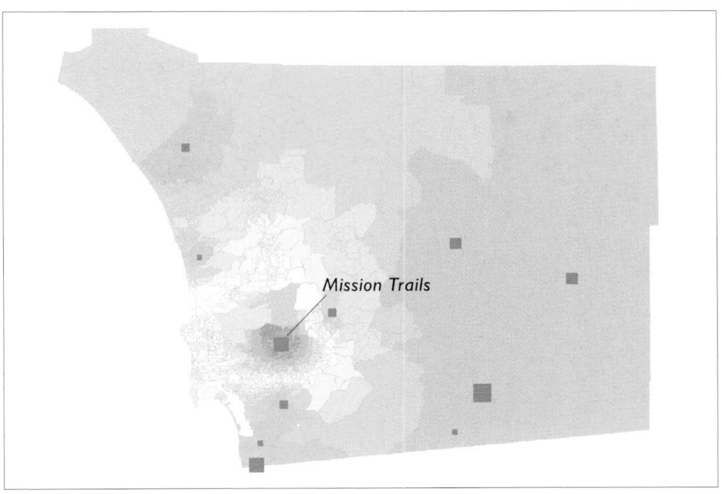

Block groups that are near other parks have low probability values (blue). Hikers living near these parks are unlikely to travel to Mission Trails when they can hike near home—especially if the nearby park has a high attraction value (larger green squares).

How the GIS calculates the probability of travel

Spatial interaction models take various forms but, as mentioned earlier, a commonly used one is the Huff Model, developed by economist David Huff for modeling store sales.

Essentially, the Huff Model calculates the probability that people will travel from a location (a demand point) to a facility, given the attraction value of the facility and the distance from the location to the facility. If two facilities are an equal distance from the location, people will travel to the one with the higher attraction value. But people may travel to a less attractive facility if it is closer—it depends on how much less attractive and how much closer. The probability of travel is calculated as a ratio of a measure of the interaction between the demand point and the particular facility, divided by the sum of interaction measures between the demand point and all available facilities.

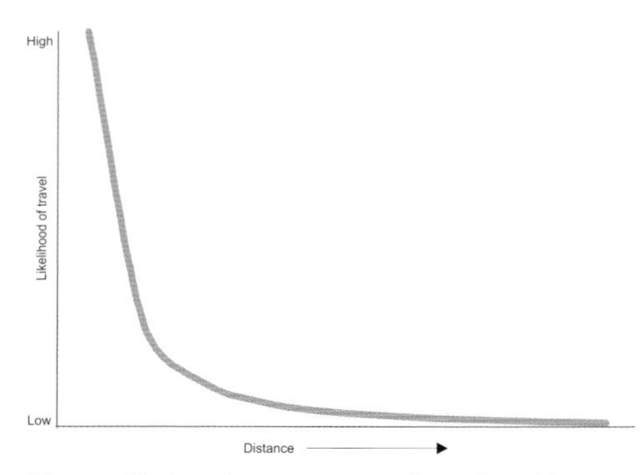

Distance friction using a power transformation with a parameter of 2.

The measure of interaction the model uses is the attraction value of the facility multiplied by the distance between the demand point and the facility.

It's generally true that as distance from the facility increases, the likelihood that people will travel there drops off at an accelerating rate. This phenomenon is called distance friction and it has to be incorporated into the interaction calculation. To do this, a distance friction transformation is used. (In ArcGIS Network Analyst, it is the impedance transformation discussed earlier in the section "Allocating demand to facilities".)

Along with the distance friction transformation, you specify a parameter. The larger the parameter, the more rapidly the travel drops off. Essentially, with a larger distance friction parameter, the *distance* exerts more influence on the decision to travel to the facility; with a smaller distance friction parameter, the *attraction value* exerts more influence on the decision.

The negative of the parameter is used in the calculations, so that with a larger parameter, the transformed distance value is smaller and the interaction value is smaller. When the distance value itself is large (that is, the facility is far from the demand point), it becomes much smaller after the distance friction parameter is applied, so the interaction value for that demand point is a smaller proportion of the summed interaction values. The probability of travel from that demand point to that facility is calculated as being low.

FACILITY	WEIGHT	DISTANCE	INTERACTION VALUE
A	3,200	11,761	0.0000231
B	15,200	28,196	0.0000191
C	14,600	40,643	0.0000088

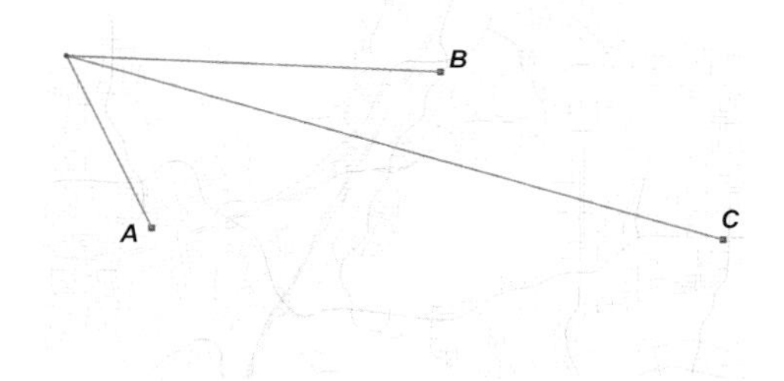

FACILITY	INTERACTION VALUE	PROBABILITY	ALLOCATED DEMAND
A	0.0000231	0.45	68.40
B	0.0000191	0.38	57.76
C	0.0000088	0.17	25.84
Sum	0.0000510	1	152

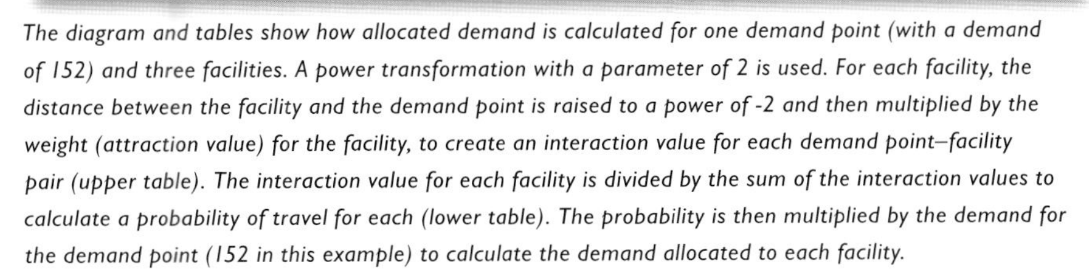

The diagram and tables show how allocated demand is calculated for one demand point (with a demand of 152) and three facilities. A power transformation with a parameter of 2 is used. For each facility, the distance between the facility and the demand point is raised to a power of -2 and then multiplied by the weight (attraction value) for the facility, to create an interaction value for each demand point–facility pair (upper table). The interaction value for each facility is divided by the sum of the interaction values to calculate a probability of travel for each (lower table). The probability is then multiplied by the demand for the demand point (152 in this example) to calculate the demand allocated to each facility.

Often, either a power or exponential transformation is used. In suburban and rural areas both people (demand points) and facilities are more spread out and driving long distances to reach a facility is common. In these cases, a power transformation may allow you to adequately model distance friction. The same is true if your study area covers large areas such as a county, region, or state, and there are long distances between demand points and facilities.

In general, an exponential distance friction transformation allows you to model distance friction over short distances. It is often used for spatial interaction models in densely populated urban areas where there are many facilities and people are less likely to travel very far to reach a facility. This is especially true since travel itself may be difficult due to congested streets and crosswalks, traffic lights, construction, and so on.

With both the power and exponential transformations, you specify a distance friction parameter. The value you use for the parameter is somewhat dependent on the uniqueness of a facility and the number of available facilities. For example, the distance you'd travel to a gas station drops off pretty rapidly—there are many of them and they all pretty much have the same attraction value. Hence, you're unlikely to travel very far to one. For modeling travel to gas stations, you'd use a larger distance friction parameter. On the other hand, people are willing to travel longer distances to facilities that are unique or that are fewer in number (say a five-star restaurant or an amusement park). The distance drop-off is much more gradual, so you'd use a smaller distance friction parameter.

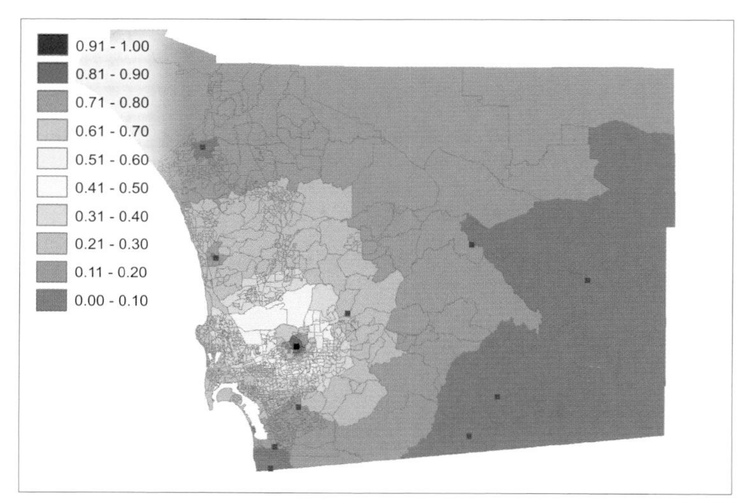

Probability values for hikers traveling to Mission Trails Park (black square), using a power function with a distance friction factor of 1. The assumption is that people are likely to travel some distance to hike at a park. For most of the county, there is at least a 10 percent to 30 percent likelihood of people traveling to Mission Trails (only along the eastern edge of the county is the likelihood less than 10 percent). The likelihood increases somewhat for people living in areas near Mission Trails. However, because—using this distance friction factor—people are likely to travel farther to a park, people living near Mission Trails are also more likely to travel to another park, so the probability values for block groups near Mission Trails Park are not large.

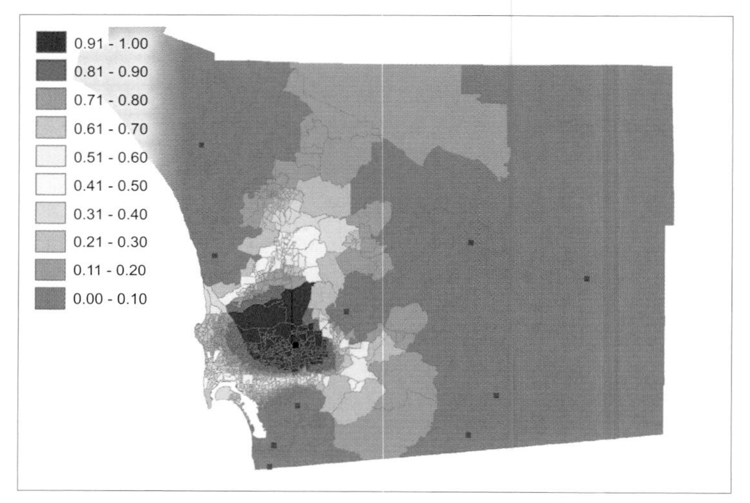

Probability values for hikers traveling to Mission Trails Park, using a power function with a distance friction factor of 4. The assumption is that people are less likely to travel very far to hike at a park (perhaps because gas prices have gone up). For most of the county, there is a less than 10 percent likelihood of people traveling to Mission Trails. Conversely, the likelihood increases substantially for people living near Mission Trails since they are now less likely to travel to a more distant park.

As with assigning attraction values, the distance decay function and parameter you use should be based on data collected about the use of the facilities. For example, you may have information about the address or at least the ZIP Code of customers at each store. You can use this data to determine the rate at which trips to a store drop off as distance from the store increases. Similarly, you may have visitor survey data for parks showing how far people have traveled to each park. With this data, you can refine your model and use it to predict future sales or run various scenarios.

Using a cutoff distance

If there is a distance beyond which travel is highly unlikely, you can specify this distance in your model. For example, if your research shows that most hikers will travel no more than ten miles to a park to hike, you'd set the cutoff distance to ten miles. For each park, any demand points (such as census block groups) more than ten miles from the park will be assigned a probability value of 0.

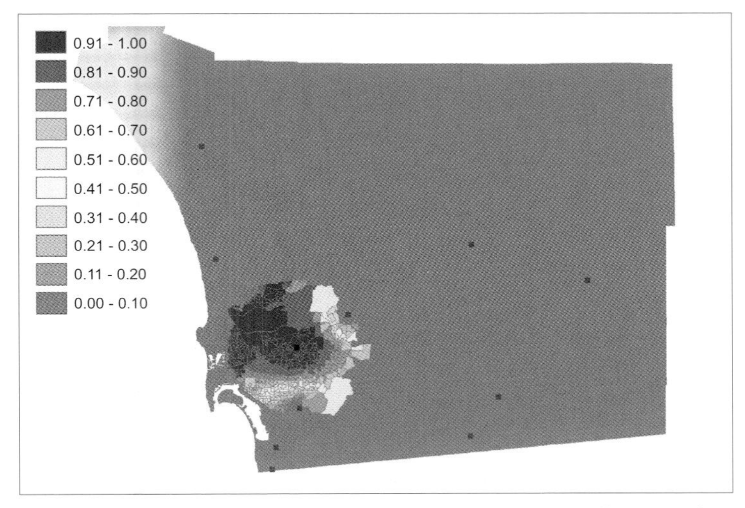

Probability values for hikers traveling to Mission Trails Park, using a cutoff distance of ten miles. Block groups farther than ten miles from the park are assigned a probability value of 0. Within the ten-mile radius surrounding the park, probability values are lower for the block groups to the northeast and south due to the presence of nearby parks. (Hikers in these block groups are more likely to travel to the closest park under the other constraints of this model.)

Specifying Euclidean versus network distance

You have the option of using a network dataset (such as streets or highways) to calculate travel distances, as opposed to using straight-line, or Euclidean, distance between the demands points and the facilities. (Indeed, the Network Analyst Market Share model requires a network dataset.)

The advantage of using a network dataset is that you can more accurately model travel between demand points and facilities by including one-way streets, prohibited turns, closed streets, and so on. Networks also take into account situations where travel to a facility is blocked by a barrier. A pet store may be a quarter mile away from a particular census block straight across a river (Euclidean distance) but might be several miles away traveling over streets and bridges. In this case, you'd want to use a network for your model to get accurate results. In addition, networks allow you to use measures of travel other than distance, such as time or monetary cost, as discussed earlier in the section "Allocating demand to facilities."

The type of application will also help determine whether you should use a network dataset. If there are many facilities that are near each other in an area where the street network is dense, you will want to use a network dataset. Using Euclidean distance may give misleading results. With fewer facilities that are far apart, the difference between using Euclidean and network distance is less.

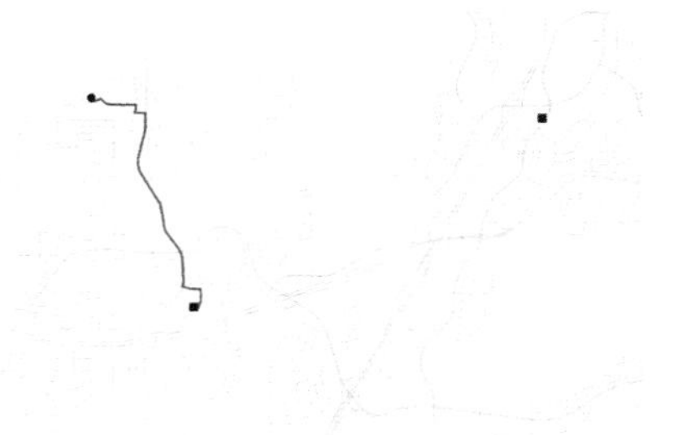

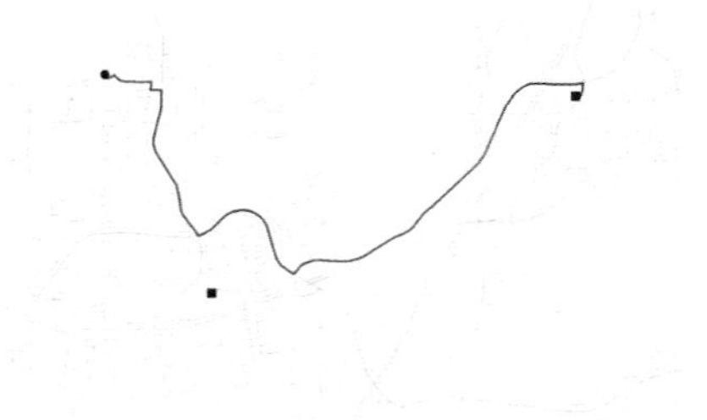

The diagrams show the distance between a demand point (dot) and two facilities (squares). The dark green lines in the diagram on the left represent the straight-line, or Euclidean, distance to both facilities. The distance to the farther facility is about one and one-half miles longer than the distance to the closer facility—about twice as far. The diagrams in the center and on the right show the route a person would actually travel over streets to each facility. When distance over the street network is used, the distance to the farther facility is about three miles longer than to the closer facility, or two and one-half times farther. In this application, you would want to calculate distances using a street network to achieve a more realistic result.

EVALUATE THE RESULTS

The results of the model are highly dependent on the attraction values of the facilities, the travel distances, and the distance parameters you specify. You'd expect, for example, facilities with a high attraction value to draw people from a wider area. If the results are not what you expect, check the attraction values to make sure they're assigned correctly. It is sometimes difficult to assign accurate attraction values, especially if you're using surrogate values (such as the size of a park to represent the total miles of trails and the variety of landscapes at the park).

In addition, the distance friction parameter is a fairly coarse estimate of what the average person would do when making a decision on how far to travel to a facility. These values can be difficult to pin down and often require extensive research. As mentioned previously, the attraction values and distance parameters should be based on survey data or other research.

Keep in mind the model results are an estimate of which facilities people are likely to use. You're attempting to model human behavior so there are a lot of factors involved. For example, while you may be attempting to model which park hikers as a whole would go to, for each individual hiker, myriad factors may go into the decision—whether it's a time of day when there's heavy traffic on some roads, which park the hiker went to last time, the season (whether it's too cold to hike at a certain park in the winter or too hot in the summer), the current price of gas, and so on. Some of these you can account for in your model: for example, you could model park visitation in summer versus winter, and if the price of gas is high, you can adjust the distance friction parameter. Other factors you can't account for, but it is still useful to model which parks, on average, hikers are more likely to use, other things being equal. The patterns of usage may give you a better sense of where visitors to the parks are coming from, how far they're traveling, and which areas of the county may be underserved. You can also model what-if scenarios—how would usage change if you added a new park or expanded an existing one.

DISPLAY AND APPLY THE RESULTS

You can use the probabilities calculated by the model to perform additional analysis, such as calculating visitation for a facility over a given time period, defining the market area for each facility, selecting locations for new facilities, or creating alternate scenarios.

Calculating visitation or sales at a facility

Summing the allocated demand for a facility (that is, the demand allocated to each facility from each demand point) will give you the total visitation for that facility. There is an implied time period for the values. If, for example, you assume each hiker will hike twice per month, on average, multiply the demand by two to get estimated monthly visitation, or by twenty-four to get annual visitation.

Similarly, if you're modeling sales at pet stores, the demand points might be ZIP Codes and the demand associated with each ZIP Code the amount spent on pets annually. By multiplying the amount spent in each ZIP Code by the probability value for that ZIP Code—for a particular store—you'd get the amount spent in the ZIP Code at that store per year.

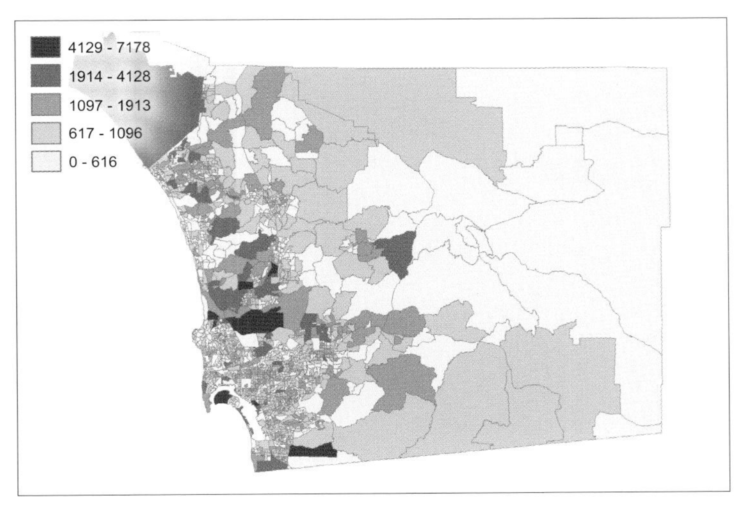

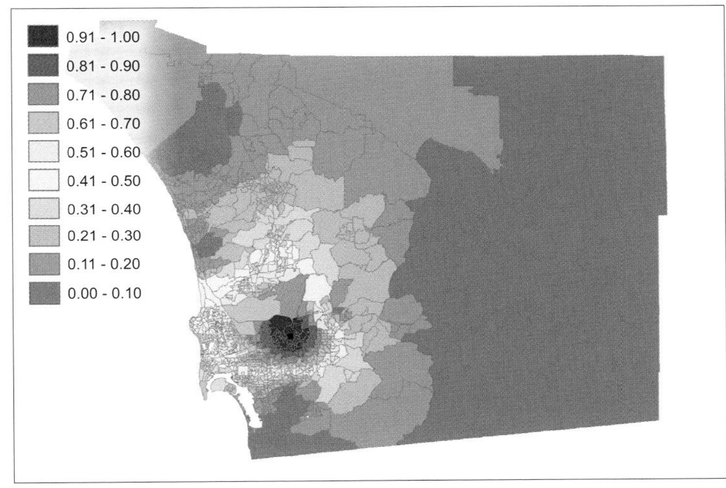

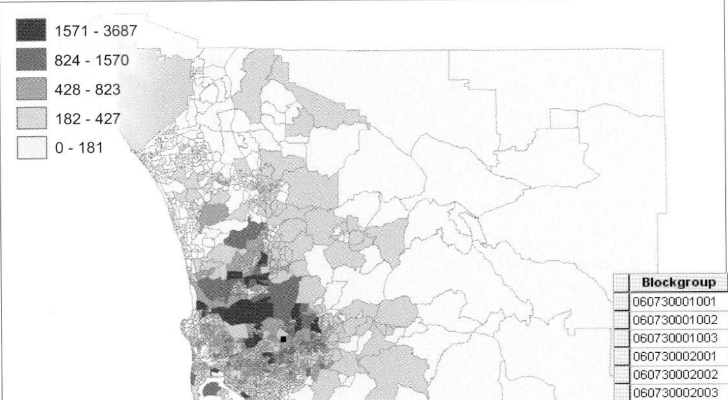

The upper left map shows the number of hikers in each block group, while the upper right map shows the likelihood of hikers traveling to Mission Trails Park from each block group. By multiplying the number of hikers in each block group by the probability value for that block group, you get the number of hikers likely to travel from each block group to Mission Trails Park (map to the left). As you might expect, block groups that are near Mission Trails and that have large numbers of hikers also provide more hikers to the park. The map essentially shows you where trips to the park originate.

Blockgroup	Hikers	Mission Trails prob	Mission Trails visits
060730001001	274	0.48	130.39
060730001002	234	0.47	109.14
060730001003	624	0.46	286.31
060730002001	456	0.48	218.29
060730002002	1324	0.46	615.58
060730002003	743	0.45	336.07
060730002004	779	0.47	364.45
060730003001	341	0.48	163.58

Another way of displaying the demand is to create a probability surface for the facility from the demand point probabilities. The demand points are used to create the surface.

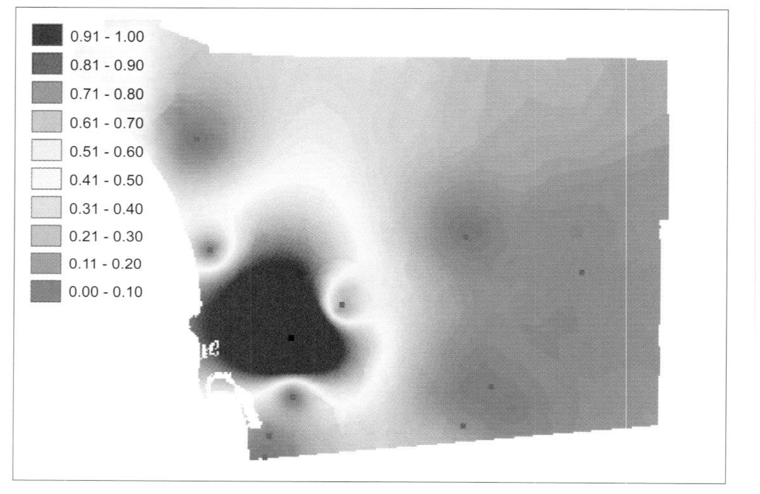

Probability values for hikers traveling to Mission Trails Park, displayed as a continuous surface. In this example, the demand points are block group centroids. Park locations (gray squares) are also shown. While the overall pattern is the same as with the map of block groups (on the previous page), slight differences exist. The pattern of the surface will vary depending on the interpolation method you use.

Calculating service/market areas

The probabilities calculated by the model can also be used to map each demand point color-coded by the facility for which it has the highest probability value. This essentially defines service or market areas. While similar to the service areas created using a location-allocation model, the areas in this case are based on the preferences of the potential clients (as modeled by the attraction-distance interaction) and not simply on which facility they're closest to.

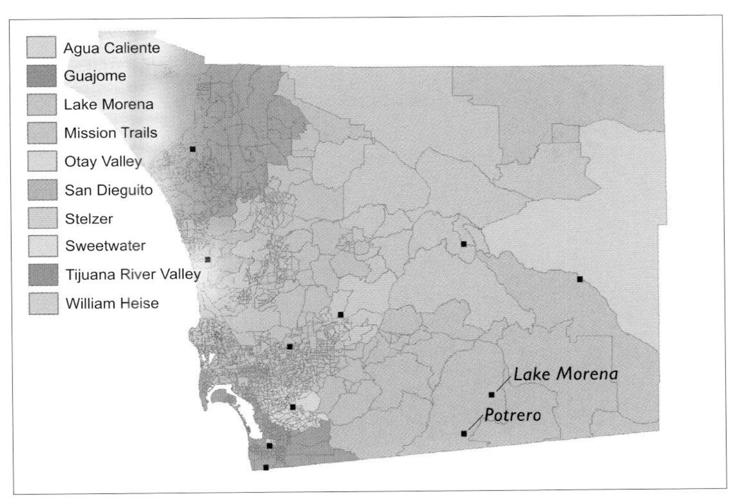

Market areas for hikers traveling to major regional parks. One smaller park, Potrero, has no market area since it is close to Lake Morena, which has a much larger attraction value. (In fact, about 5 percent of hikers in the block group are likely to go to Potrero, but 85 percent are likely to go to Lake Morena.) The situation is similar for the other block groups surrounding the two parks. Market areas are not necessarily contiguous—several block groups in the upper right and upper center of the map are part of the Lake Morena market area, even though there are closer parks. For the block group in the upper right, for example, while 28 percent of hikers are likely to go to Lake Morena, William Heise Park is a close second, with 25 percent of hikers likely to go there. However, since more hikers are likely to go to Lake Morena, the block group is assigned to that park's market area. Factors such as the road network in rural areas and the location of demand points (block group centroids) come into play in the results.

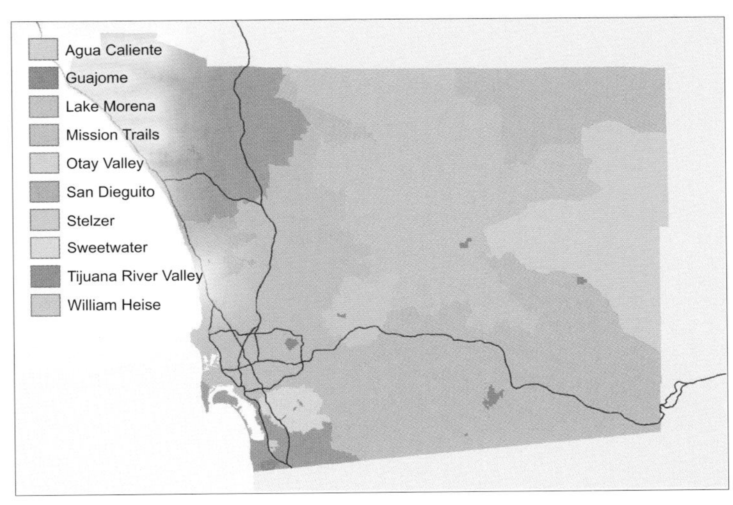

This map was created by dissolving the boundaries between individual census block groups categorized by market area. The market areas are shown here with park areas (in dark green) and interstate highways, for context.

Choosing locations

You can use the model to choose the best location for a new facility from a set of candidates, just as you do with a location-allocation model. In this case, though, the chosen location will be the one with the highest probability of attracting potential clients, given the locations of other existing and competing facilities.

Once the location is chosen, you can see how visitation or sales at the new facility compare to existing facilities and competitors. In fact, you can see how opening the new facility would affect visitation or sales at existing and competing facilities by comparing the visitation values before and after the new facility is added. Similarly, you could estimate how sales would be affected at your stores if a competing store were to open nearby.

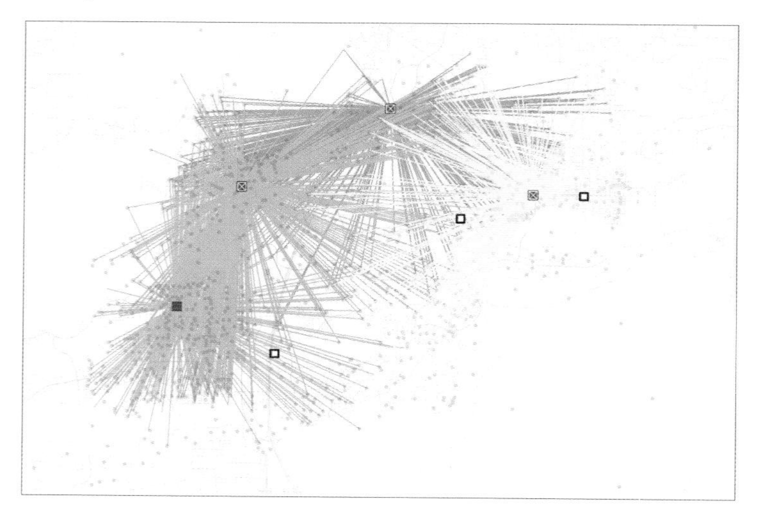

In this model, the location for a new pet store (solid square) has been chosen from a set of potential locations. The locations of competing stores is shown (boxes with Xs) as well as candidate locations that weren't chosen (hollow boxes). The lines show at which store people in the various census blocks (orange dots) are likely to shop. The square footage of each store was used as the attraction value (the larger the store, the more products they carry).

Creating alternate scenarios

You can use the model to create alternate scenarios. For example, you might create scenarios to see how visitation at existing youth camps changes when you add a new camp at each of several parks and how well each scenario serves the population.

Similarly, you can change the attraction value for an existing facility to see what the impacts would be. For example, you might see how visitation would change at existing parks if a new hiking trail is built at one of the parks, thus increasing its attraction value. This would help you decide whether it is worth the cost of building the trail.

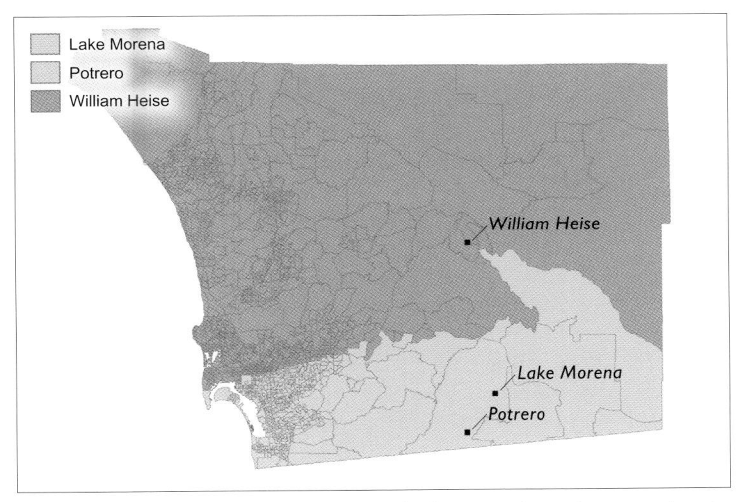

Market areas for three youth group campgrounds. Campground capacity was used as the attraction value (48 campers at Lake Morena, 45 at Potrero, and 60 at William Heise).

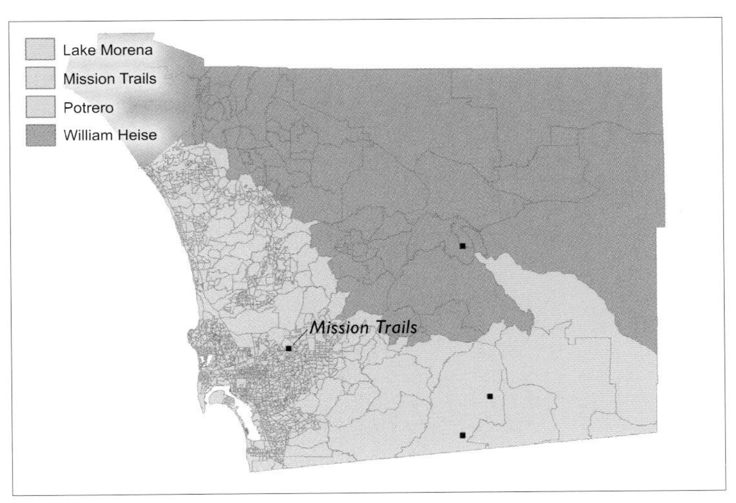

Predicted market areas with a new youth group campground at Mission Trails Park (with a proposed capacity of 40 campers).

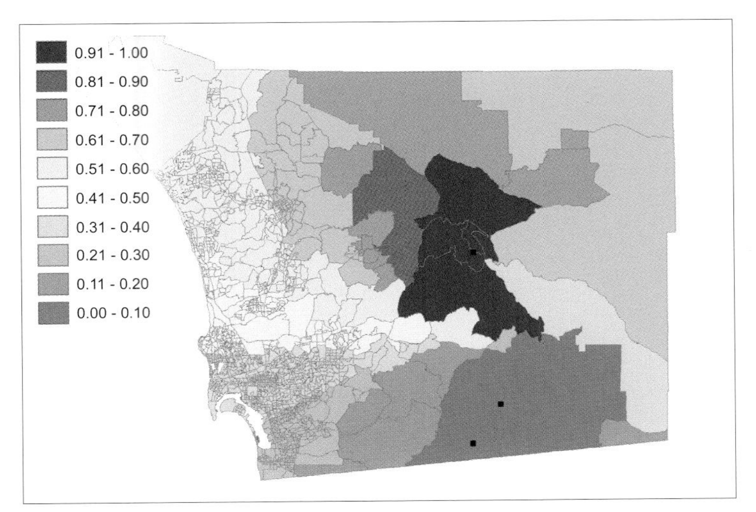

Probability values for the youth group campground at William Heise Park, given the three existing campgrounds.

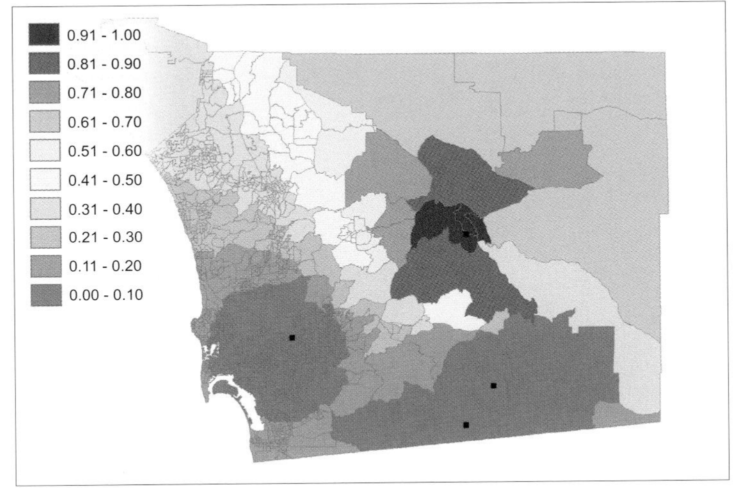

Probability values for the youth group campground at William Heise Park, given the addition of a new campground at Mission Trails Park.

Fotheringham, A. S. and M. E. O'Kelly. 1989. *Spatial Interaction Models: Formulas and Applications.* Boston, MA: Kulwer Academic Publishers. Discusses the design of spatial interaction models and describes their application in demographic and retail analysis.

Haynes, Kingsley E. and A. Stewart Fotheringham. 1984. *Gravity and Spatial Interaction Models.* Beverly Hills, CA: Sage Publications. Presents an overview of spatial interaction models, including applications for retail analysis and facility location analysis.

Huff, David. 1963. "A Probabilistic Analysis of Shopping Center Trade Areas." Land Economics, Vol. 39. Presents the concepts underlying the Huff Model.

Huff, David and Bradley M. McCallum. 2008. "Calibrating the Huff Model Using ArcGIS Business Analyst—An Esri Whitepaper." Esri. Describes the process for estimating attraction values and distance friction parameters from customer survey data.

Wang, Fahui. 2006. *Quantitative methods and applications in GIS.* Boca Raton, FL: CRC/ Taylor & Francis. Includes discussions of estimating distance and travel time, as well as measures of accessibility.

Zeiler, Michael. 2010. *Modeling Our World: The Esri Guide to Geodatabase Concepts.* Redlands, CA: Esri Press. Describes in depth the various data models used by ArcGIS. Includes a discussion on the concepts behind the network databases used for transportation applications.

Appendix

Many of the concepts in this book apply to GIS in general. However, the examples were created using ArcGIS, Esri's multifaceted geographic information system. ArcGIS includes a wide range of tools for evaluating locations and analyzing movement. Some tools are available as part of the basic ArcGIS Desktop product while others are included with Desktop extension products that require purchase of an additional license.

Following is a list of the ArcGIS analysis tools referenced in this book, by chapter.

Finding suitable locations using selection

Add Join	Joins a table to another table based on a common field
Add Field	Adds a new field to a table
Calculate Field	Calculates the values of a field in a table
Select Layer by Attribute	Selects features based on an attribute query
Select Layer by Location	Selects features based on a spatial relationship to features in another layer

Finding suitable locations using overlay

Feature overlay method

Buffer	Creates buffer polygons around input features to a specified distance
Raster to Polygon	Converts a raster dataset to polygon features
Union	Creates new features where input features overlap; output extent is equal to the full extent of input layers
Intersect	Creates new features where input features overlap; output extent is equal to the overlap area
Dissolve	Aggregates features based on specified attributes
Eliminate	Eliminates polygons by merging them with neighboring polygons
Select Layer by Attribute	Selects features based on an attribute query

Raster overlay method

Slope	Identifies the slope (gradient or rate of maximum change in z-value) from each cell of a raster surface
Euclidean Distance	Calculates the Euclidean distance from each cell to the closest source
Polygon to Raster	Converts polygon features to a raster dataset
Reclassify	Reclassifies (or changes) the values in a raster
Con	Performs a conditional if/else evaluation on each of the input cells of an input raster
Combine	Combines multiple rasters so that a unique value is assigned to each unique combination of input values

CHAPTER 3 RATING SUITABLE LOCATIONS

Rating locations using weighted overlay

Polygon to Raster	Converts polygon features to a raster dataset
Reclassify	Reclassifies (or changes) the values in a raster
Con	Performs a conditional if/else evaluation on each of the input cells of an input raster
Slope	Identifies the slope (gradient, or rate of maximum change in z-value) from each cell of a raster surface
Aspect	Derives the slope direction (north, northeast, east, and so on) from a raster surface
Area Solar Radiation	Derives incoming solar radiation from a raster surface
Viewshed	Determines the raster surface locations visible to a set of observer features
Euclidean Distance	Calculates the Euclidean distance from each cell to the closest source
Path Distance	Calculates, for each cell, the least accumulative cost distance to the nearest source, using advanced factors
Weighted Overlay	Overlays several rasters using a common measurement scale and weights each by importance

Rating locations using fuzzy overlay

Reclassify	Reclassifies (or changes) the values in a raster
Fuzzy Membership	Transforms an input raster into a 0 to 1 scale indicating the strength of a membership in a set
Fuzzy Overlay	Combines fuzzy membership rasters together, based on specified overlay type

Modeling a path over a network

Modeling visits to locations

Calculate Field	Calculates the values of a field in a table
Make Route Layer	Makes a route network analysis layer and sets its analysis properties
Add Locations	Adds network analysis objects to a network analysis layer
Solve	Solves the network analysis layer problem based on its network locations and properties

Modeling deliveries

Calculate Field	Calculates the values of a field in a table
Make VRP Layer	Makes a Vehicle Routing Problem network analysis layer and sets its analysis properties
Add Locations	Adds network analysis objects to a network analysis layer
Solve	Solves the network analysis layer problem based on its network locations and properties

Modeling an overland path

Reclassify	Reclassifies (or changes) the values in a raster
Weighted Overlay	Overlays several rasters using a common measurement scale and weights each by importance
Cost Distance	Calculates, for each cell, the least accumulative cost distance to the nearest source
Path Distance	Calculates, for each cell, the least accumulative cost distance to the nearest source, using advanced factors
Cost Path	Calculates the least-cost path from a source to a destination
Corridor	Calculates the sum of accumulative costs for two input accumulative cost rasters
Extract by Attributes	Extracts the cells of a raster based on a logical query

CHAPTER 5 MODELING FLOW

Modeling accumulation over a surface

Fill	Fills sinks in an elevation surface raster to remove small imperfections in the data
Set Null	Sets identified cell locations to NoData based on specified criteria
Flow Direction	Creates a raster of flow direction from each cell to its steepest downslope neighbor
Flow Accumulation	Creates a raster of accumulated flow into each cell; a weight factor can optionally be applied
Reclassify	Reclassifies (or changes) the values in a raster
Stream to Feature	Converts a raster representing a linear network to features representing the linear network
Stream Order	Assigns a numeric order to segments of a raster representing branches of a linear network
Stream Link	Assigns unique values to sections of a raster linear network between intersections
Snap Pour Point	Snaps pour points to the cell of highest flow accumulation within a specified distance
Watershed	Determines the contributing area above a set of cells in a raster
Zonal Geometry as Table	Calculates geometry measures for each zone in a dataset and reports the results as a table
Zonal Statistics as Table	Summarizes the values of a raster within the zones of another dataset and reports the results as a table
Tabulate Area	Calculates cross-tabulated areas between two datasets and reports the results as a table

Tracing flow over a network

Add Junction Flag	Identifies a network junction as a trace origin
Add Edge Flag	Identifies a network edge as a trace origin
Add Junction Barrier	Identifies a network junction as a barrier for the trace
Add Edge Barrier	Identifies a network edge as a barrier for the trace
Trace Upstream	Specifies that the trace will be upstream from the flagged location(s)
Trace Downstream	Specifies that the trace will be downstream from the flagged location(s)
Find Upstream Accumulation	Calculates the accumulation of a quantity associated with network features during an upstream trace
Solve	Runs the specified trace function and displays the selected network features

(Note: tools for modeling flow over a network are included in the Utility Network Analyst toolbar)

Allocating demand to facilities

Calculate Field	Calculates the values of a field in a table
Make Location-Allocation Layer	Makes a location-allocation network analysis layer and sets its analysis properties
Add Locations	Adds network analysis objects to a network analysis layer
Solve	Solves the network analysis layer problem based on its network locations and properties

Modeling travel to facilities

Make Location-Allocation Layer	Makes a location-allocation network analysis layer and sets its analysis properties
Add Locations	Adds network analysis objects to a network analysis layer
Solve	Solves the network analysis layer problem based on its network locations and properties
Huff Model	Calculates interaction-based probabilities of consumers at each origin location patronizing each facility

(Note: The Huff Model tool can be downloaded from the ArcGIS Resource Center)

Data Credits

The following organizations and individuals provided GIS datasets used to create the examples throughout the book.

Oregon data is used with permission of the Oregon Geospatial Data Clearinghouse.

Data from the Portland RLIS dataset is used with permission of Metro Regional Services, Data Resource Center, Portland, Oregon.

Tillamook, OR region soils data - Soil Survey Staff, Natural Resources Conservation Service, United States Department of Agriculture. Soil Survey Geographic (SSURGO) Database for Oregon. Available online at http://soildatamart.nrcs.usda.gov . Accessed 5/31/2011.

Northwest Oregon Roads data is from Data and Maps for ArcGIS 2010, courtesy of Tele Atlas.

Highways for western Oregon data is from Data and Maps for ArcGIS 2008, courtesy of Esri.

Highways data is from Data and Maps for ArcGIS 2010, courtesy of Esri.

Mountain summit point features is from Data and Maps for ArcGIS 2010, courtesy of USGS – GNIS.

Poway, California streets data is from Data and Maps for ArcGIS 2002, courtesy of GDT and Esri.

Poway, California schools data is from Data and Maps for ArcGIS 2010, courtesy of USGS-GNIS.

Poway, California business sites, fire station locations, library locations, census blocks, municipal, adult day care locations, regional parks - GIS data derived from SanGIS/SANDAG downloadable data - www.sangis.org Copyright SanGIS 2009.

San Diego County, California census block groups data is from Data and Maps for ArcGIS 2010, courtesy of Tele Atlas, U.S. Census, Esri (Pop2010 field).

San Diego County, California streets data is from Data and Maps for ArcGIS 2010, courtesy of Tele Atlas.

San Diego, California area highways data is from Data and Maps for ArcGIS 2010, courtesy of U.S. Bureau Transportation Statistics.

Sewer data used with permission of the City of Riverside, California.

Fulton County, Georgia seniors data is from Data and Maps for ArcGIS 2010, courtesy of Tele Atlas, U.S. Census, Esri (Pop2010 field).

Ft. Pierce, Florida streets data is from Data and Maps for ArcGIS 2010, courtesy of Tele Atlas.

Ft. Pierce, Florida Storm water data from NorthStar Geomatics, Inc., accessed October 2006.

Deer sightings data provided with permission from the Wyoming Game and Fish Department.

Wyoming data used with permission from the Wyoming Geographic Information Science Center.

DEM is courtesy of U.S. Geological Survey.

U.S. Counties data is from Data and Maps for ArcGIS 2010, courtesy of Esri, derived from Tele Atlas, U.S. Census, Esri (Pop2010 field).

Ocean data is from Data and Maps for ArcGIS 2010, courtesy of Esri.

Paris street data Copyright © 2009 NAVTEQ. All rights reserved.

Paris metro entrances, stations, and lines Copyright © 2006 RATP. All rights reserved.

Index